[illegible]TY OF ROCK SLOPES

PROCEEDINGS

Thirteenth Symposium on Rock Mechanics

held at

The University of Illinois

Urbana, Illinois

August 30-September 1, 1971

Edited by

EDWARD J. CORDING

Department of Civil Engineering

University of Illinois

Urbana, Illinois

SYMPOSIUM SPONSORED BY

The U.S. National Committee for Rock Mechanics

of the

National Academy of Sciences—National Academy of Engineering

and

The University of Illinois

Published by

AMERICAN SOCIETY OF CIVIL ENGINEERS

345 East 47th Street

New York, N.Y. 10017

1972

FOREWORD

The Thirteenth Symposium on Rock Mechanics was the first conducted under the auspices of the United States National Committee for Rock Mechanics, sponsored by the National Research Council of the National Academy of Sciences, National Academy of Engineering. The USNC/RM was formed on November 15, 1971 by merging the Intersociety Committee on Rock Mechanics with the former USNC/RM. The topic selected for the Symposium was "Stability of Rocks Slopes," and four of the five sessions centered on various aspects of rock slope stability.

The stability of natural and cut slopes is one of the major problems facing engineers on many mining and civil works projects. Specific examples are the evaluation of landslide hazards on natural slopes, the design of slopes for open pit mines, and the design of spillway cuts on hydroelectric projects. The evaluation of a rock slope stability problem may be categorized according to the following four essential steps:

(1) Identification of potentially unstable slopes and critical geologic features.

(2) Determination of shear strength parameters in terms of effective stresses on potential failure surfaces.

(3) Determination or estimation of pore pressures acting on the potential failure surfaces.

(4) Analysis of the static stability by three-dimensional methods which account for the actual geometry of the geologic discontinuities in relation to the slope surfaces.

The first two sessions of the 13th Symposium contained papers which touched upon three essential steps (1, 2, and 4) necessary for the evaluation of rock slopes. Although the estimation of pore pressures in jointed rock slopes is essential for the evaluation of rock slope stability, the lack of field data on this subject was reflected by an absence of papers on this topic. Thus it appears that a fruitful area of research would be the field measurement of pore pressures in real rock slopes.

Papers in Session I described methods both for identifying potential failure surfaces and for evaluating the shear strength on these surfaces. Results of laboratory and in-situ direct shear tests were presented. Residual shear strength parameters determined from laboratory direct shear tests were summarized for a variety of rock types. In Session II, several methods for analyzing slopes were presented. Discussions and papers in both Sessions II and III described the limitations as well as the applicability of present analytical techniques in approximating the actual conditions of a rock slope containing a three-dimensional pattern of discontinuities and having complex shear strength-displacement parameters.

In Sessions III and IV interesting and valuable case histories of slope failures were presented which illustrate several key aspects of rock slope behavior. Discussions in Session IV centered on the need for complete case histories in which the following items are documented:

(1) The complete geometry of the sliding surface and the slope surface.

(2) The effective shear strength parameters along the failure surfaces.

(3) The pore pressure distribution on the sliding surface during failure.

(4) Deformation records indicating the pattern and sequence of slope movements.

The development of "key" case histories which will include all the above requirements presents a strong challenge to the rock mechanics profession. The difficulty of the task is apparent in the fact that at present, no such complete case histories (to the writer's knowledge) are documented. The inclusion of all these essential elements of information in case histories should provide the basis for major advances in the state of the art of rock slope design.

A. J. Hendron, Jr.
Professor of Civil Engineering

E. J. Cording
Associate Professor of Civil Engineering

University of Illinois
at Urbana-Champaign

CONTENTS

PUBLISHED PROCEEDINGS OF PREVIOUS SYMPOSIA ON ROCK MECHANICS IN THIS SERIES

"Third Symposium on Rock Mechanics," *Quarterly of the Colorado School of Mines*, Vol. 54, No. 3, July 1959.* $3.35

"Fourth Symposium on Rock Mechanics," *Bulletin 76, Mineral Industries Experiment Station, Pennsylvania State University* November 1961† $5.00

"Sixth Symposium on Rock Mechanics" (Oct. 1964), Preprints only, University of Missouri at Rolla, Mo. 65401. $15.00

"Seventh Symposium on Rock Mechanics," Pennsylvania State University (June 1965), Preprints out of print. Most papers published in *Transactions, Society of Mining Engineers of AIME* (1965-1966).

"Status of Practical Rock Mechanics—Ninth Symposium on Rock Mechanics," Colorado School of Mines, April 1967. Published by the Society of Mining Engineers of AIME, 1968.
ASCE & AIME Members $16.50
Nonmembers $19.50

"Basic and Applied Rock Mechanics—Tenth Symposium on Rock Mechanics," University of Texas at Austin, May 1968. Published by the Society of Mining Engineers of AIME 1972.
ASCE & AIME Members $17.50
Nonmembers $23.00

"Rock Mechanics—Theory and Practice—Eleventh Symposium on Rock Mechanics," University of California, June 1969. Published by the Society of Mining Engineers of AIME, 1970.
ASCE & AIME Members $18.00
Nonmembers $24.00

"Dynamic Rock Mechanics—Twelfth Symposium on Rock Mechanics," University of Missouri-Rolla, November 1970. Published by the Society of Mining Engineers of AIME, 1971.
ASCE & AIME Members $16.00
Nonmembers $21.00

"Stability of Rock Slopes—Thirteenth Symposium on Rock Mechanics," University of Illinois, Urbana, Illinois, Aug.-Sept. 1971. Published by the American Society of Civil Engineers, 1972.
ASCE & AIME Members $ 5.00
Nonmembers $10.00

"New Horizons in Rock Mechanics—Fourteenth Symposium on Rock Mechanics, Pennsylvania State University, College Park, Pa., June, 1972. Published by the American Society of Civil Engineers, 1973.
ASCE & AIME Members $10.00
Nonmembers $15.00

* Available from:
Colorado School of Mines
Department of Publications
Golden, Colorado 80401

† Available from:
Pennsylvania State University
118 Mineral Sciences Building
University Park, Pennsylvania 16802

NOTE:
Persons desiring to obtain out-of-print Proceedings of earlier symposia, which are not listed herein, are advised to refer to the Engineering Societies Library, 345 East 47 Street, New York, N.Y. 10017, regarding microfilm or photocopies of such publications.

CUMULATIVE SUMS TECHNIQUE: A NEW APPROACH TO ANALYZING JOINTS IN ROCK

By Douglas R. Piteau* and Lindsay Russell**

SYNOPSIS

The cumulative sums technique for analyzing joints in rock was developed as part of an extensive slope stability study of Nchanga pit. It was used successfully to determine the joint orientation trends, the pattern of their behavior and whether the joint information could be extrapolated to other areas in which slopes are proposed. This technique is illustrated with reference to the Nchanga study.

INTRODUCTION

The cumulative sums technique for analyzing joints in rock was developed as part of an extensive slope stability analysis of the hanging-wall of the Nchanga open pit in Zambia. This technique was used successfully to define the characteristic features of the joints, most particularly to determine the pattern of their behavior from one part of the Nchanga syncline, where the pit is situated, to the other. A description of cumulative sums technique is

---

* Formerly, Consultant, Nchanga Consolidated Copper Mines Ltd., Chingola Division, Zambia; Presently, Consulting Engineering Geologist, Principal, Piteau Gadsby Macleod Limited, Geotechnical Consultants, Vancouver, B. C.

** Senior Operations Research Officer, Nchanga Consolidated Copper Mines Ltd., Chingola, Zambia.

given with particular reference to the joint analysis of the Nchanga pit.

Basically, the cumulative sums technique was developed to analyze the joint trends in a more definitive manner. This technique led to a better understanding of the genetic relationship of the joints occurring within the overall synclinal structure. Ultimately, predictions were made as to whether the joint data acquired from the existing hanging-wall pit face could be extrapolated (a) to an area some 300 ft behind the existing hanging-wall face, where the final slope is to be located and (b) to areas east of the existing face, where the pit is to be extended another mile.

The Nchanga syncline is approximately one and one quarter miles wide and seven miles long. The Nchanga open pit is located in the southern half of the western limits of the syncline, which consists of a clearly defined sequence of mainly sedimentary rocks (i.e. argillite, siltstones, shale, sandstone, etc). The sediments strike roughly east-west. The south limb dips between 20° N to 35° N, and the north limb dips steeply to the south, forming an asymmetrical synclinal structure with an axial plane dipping steeply north. The syncline plunges between 5° and 15° to the west.

The overall approach to the structural analysis of the hanging-wall slopes was basically straightforward /Piteau (6)/. Discontinuities in the rock were systematically measured along over three miles of benches, using the continuous detail line survey method as described by Piteau (5). The joint data were statistically analyzed, initially using rectangular, histogram, cumulative sums and other analysis methods to determine their nature and distribution. For purposes of this discussion "joint" is meant to include any naturally-occurring structural discontinuity in the rock mass.

CONSIDERATION OF THE ROCK MASS JOINT MODEL

An objective of the joint analyses is to obtain a schematized concept or model of the joints in the rock mass and to establish certain criteria which indicate where this model changes. Also, one seeks to establish confidence limits in areas where the model is considered to apply, regardless of whether it is in areas of extensive, limited or no sampling.

When designing open pit slopes on a rational basis, an important, if not the most important, consideration in most geological environments is the determination of the attitude, geometry and spatial distribution of the joints in the boundary of the proposed excavation. Thus, for purposes of rationally analyzing a rock slope, such a study must be dependent upon assessing three main factors, namely (1) the nature and structural arrangement of the joints; (2) the strength parameters of the joints; and (3) their relationships to possible failure surfaces. Based on this approach, the geological factors and certain geological premises are given by Piteau (4) and (5), methods of structural interpretation by Robertson (8) and mathematical theories for stability calculations by Jennings (3). This discussion deals exclusively with assessment of factor (1).

Of the three factors listed above, the first is the most important, as the two others are of little consequence if the structural interpretation, and hence the jointing model of the rock mass, does not represent the actual situation in a statistical sense. The first requirement of the model, whether it is of a physical, graphical or mathematical nature, is that it be true, and that a statistical sampling of any property will give a representative picture of the whole situation. The second is that any calculations made for a representative portion or section of the model apply to the model as a whole.

Thus, on the basis of the jointing model, and with due consideration of the strength parameters and kinematically possible failure modes for that particular structural situation, the stability of the slope can be theoretically determined /Jennings (3)/.

CONSIDERATIONS AT NCHANGA LEADING TO THE DEVELOPMENT OF THE CUMULATIVE SUMS TECHNIQUE

The present dimensions of the pit are 9,600 ft along strike, but will extend, eventually, along strike for three miles, after the extension of the pit eastwards is completed. It is presently approximately 2,500 ft wide at its present depth of 750 ft. However, it is planned to go to 1,000 ft, and possibly even to 1,200 ft depth, the result being a final width of about 3,000 ft.

The structural mapping was conducted on the hanging-wall face of the pit. The problem involved trying to determine whether the same or a different structural situation can be expected to exist in the hanging-wall slope when the final depth of 1,000 ft is achieved. The final hanging-wall slope will be at least 300 ft farther in from the existing face as the pit is advanced northwards. The existing and approximate final locations of the hanging-wall, along with some salient geological features, are shown in Fig. 1.

A print-out of the raw joint data representing greater than 3,000 joints from the hanging-wall is shown in a rectangular plot in Fig. 2. Horizontal rows indicate similar angle of dip, whereas vertical rows on the upper half and lower half of the plot indicate joints with similar direc-

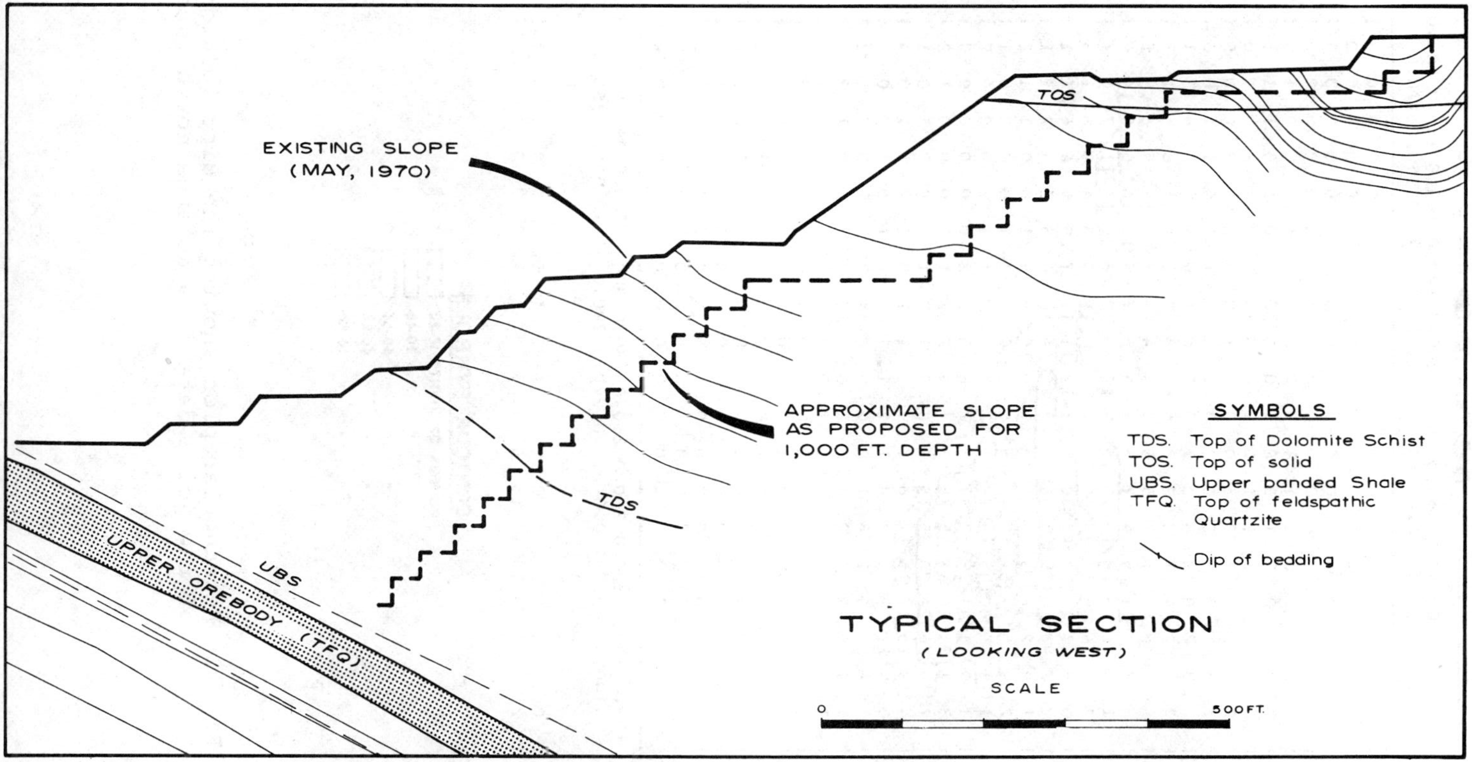

Fig. 1 Typical geological section (looking west), showing the existing and approximate final location of the Nchanga pit hanging-wall.

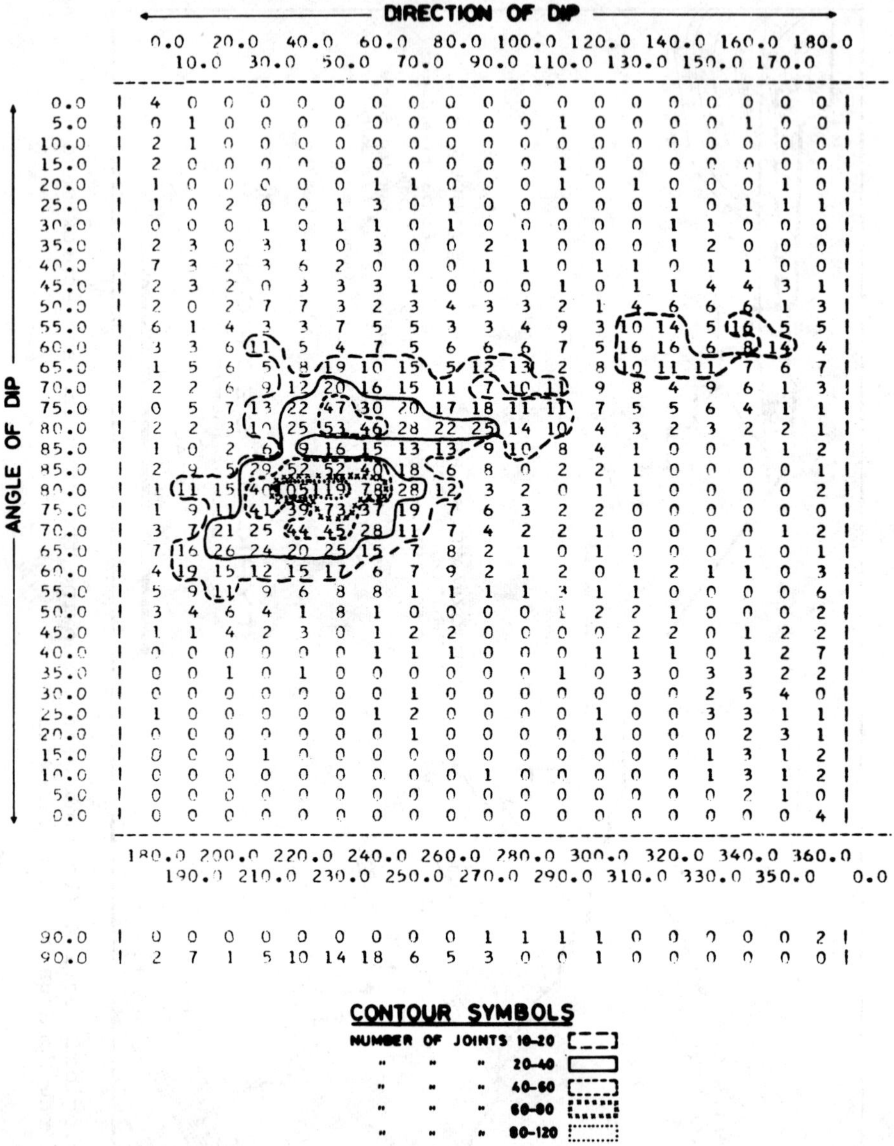

Fig. 2 Rectangular plot showing the distribution of raw joint data from Nchanga north face

tion of dip* /Robertson (8)/. It can be seen that about 70 percent of the joints are highly concentrated, occurring within a direction of dip interval of 50° (i.e. between 20° and 70°, and 200° and 250° in the upper and lower halves, respectively), and within an angle of dip interval of 55° (i.e. between 70° and 90°, and 65° and 90° in the upper and lower halves, respectively). That is, the peak concentration of these joints is centred about a strike of approximately 310°, and they dip steeply both to the northeast and southwest.

Detailed examinations of drag, monoclinal and major recumbent folds, both locally and at other points around the syncline, revealed that their axial planes were in fact striking about 300° to 310°, and not east-west, as might be expected from the approximately east-west orientation of the Nchanga syncline proper. From the results in Fig. 2 it can be seen that the peak concentration of the joints approximately parallel this tectonic fold axis.

Further study indicated that the topography of underlying basement granite dome structures to a large extent controlled the overall synclinal shape and did not control the tectonic fold process proper. For purposes of extrapolation and, ultimately, for assessing the significance of the joints with respect to slopes developed at different locations in the syncline, a more definitive knowledge of the joint behavior in regards to both dip and strike trends was required.

---

* Direction of dip of a joint is the strike plus or minus 90°, depending upon whether the joint dips in a clockwise or counter-clockwise direction.

THE CUMULATIVE SUMS TECHNIQUE

GENERAL FEATURES

Cumulative sums, or "cusums" as they are also called, have been used extensively in industrial quality control /Woodward and Goldsmith (9)/. They have also been used for studying long-term trends in natural phenomena, such as river volume flows and silt deposition /Hurst et al (2)/. As far as is known, these techniques have been applied only to series of events equally spaced in time. In the analysis of joints, however, we have used these methods to study events occurring, not in time, but in an irregular sequence in space. This analysis is sequential in that the dip or strike values of the joints are considered in the order in which they are derived along the survey line.

The cumulative sums technique provides a rapid and a precise method of determining major trends above or below a particular reference value which is selected, and for ascertaining both the magnitude and location of these variations. The main uses of such an analysis method can be summarized briefly as follows:

(a) To detect general changes in joint orientation above and below the mean level of the joint orientation data;

(b) To determine where changes in joint orientation take place in the rock mass;

(c) To determine a reliable estimate of the mean orientation of the joints at any point along the surveyed pit face;

(d) To predict the average orientation of a particular joint set, or group of joints, in other parts of the mass where information is not available.

METHOD OF COMPUTATION

Basically, the approach is simple, consisting merely of subtracting a constant quantity, which at Nchanga was taken to be the mean value of either the strike or dip, from each value of strike or dip in the series, and accumulating the differences as each additional value is introduced. Successive accumulated differences are designated the "cumulative sums" of the original sequence of joint orientation values. The resulting graph of these sums is designated the "cumulative sum joint orientation plot".

When large numbers of joints are to be analysed, it is convenient to create cusum plots by computer methods, methods to which the analysis is ideally suited. Plots on the line printer, using a width of 100 characters, have proven to be an excellent medium for this method of analysis. In order to make the plots comparable, however, it is necessary to use the same cusum range and mean for all plots.

Let us suppose that we have a series of joint strike values acquired from a continuous detail joint survey. We will denote these values by $X_1, X_2, \ldots X_r$, recorded in that order along the pit face. From each X we subtract a reference value K, the mean strike of the joints. We then add these deviations to form a series of partial sums:

$$S_1 = X_1 - K$$

$$S_2 = (X_1 - K) + (X_2 - K) = S_1 + (X_2 - K)$$

$$S_3 = S_2 + (X_3 - K)$$

The general equation for the cumulative sums can thus be written as follows:

$$S_r = S_{r-1} + (X_r - K) = X_1 + X_2 + \ldots\ldots X_r - rK$$

S_1, S_2, S_3, ... S_r is the cumulative sum series (or cusum) of the joint strike series. The plot of S against position in the sequence (S_r vs r) is the cumulative sum joint orientation plot.

The random spacing of the joints presents no problem, so long as the position, not the distance, in the sequence is used. It does not matter if the interval between observations changes. In this computation, strike or direction of dip data must be converted so that only values from either the 0° to 180° or the 180° to 360° intervals are calculated in the same analysis.

METHOD OF INTERPRETATION

If there is no trend in the strike of the joints, some of the difference terms (X_r - K) will be positive and others negative, with the result being that the cusum will be basically constant. But, if the current or local mean strike value is slightly greater than K (the overall mean), more of the differences will be positive, and the cusum will then be a straight line or curve sloping upwards. The reverse will occur if the current mean is less than K.

The actual distance of the plotted cusum curve from the horizontal is irrelevant; the interpretation is based exclusively on the average slope of the curve. The steeper the curve, the further the mean strike of the joints within any particular location is from the mean value K. The slope of the line (and hence the amount of deviation of the current mean strike from the overall mean value) can be easily calculated. The slope of the plotted line joining, let us say, the mth point and an nth point further along in the series indicates the average difference from the reference value of all the results from X_{m+1} to X_n inclusive. The mean strike ($\overline{X}$) over any interval of the cumulative sum joint orientation plot is given by

$$\overline{X} = K + \frac{\text{change in cumulative sum}}{\text{change in n}}$$

When conducting this type of analysis considerable care should be taken in selecting a suitable reference value (K). One important feature of this analysis method is that relatively small changes, say in the current mean value of the joint strike, will appear as clearly different slopes. However, changes from one positive value to another in the slope of the cusum plot are not nearly so discernable as a reversal of the sign of the slope, i.e. a change from a situation in which the mean strike of the joints is above the reference value, to one in which it is below. The reference value K should be chosen as a reasonable target from which the results are expected to vary. Also, erratic variations or "noise" in the data are smoothed out. This is a significant factor when looking for trends and patterns, particularly when analyzing data from natural phenomena such as joints.

This technique is best used to determine long-term trends. Interpretation becomes difficult if attempts are made to include short-duration effects.

COMPARISON WITH TIME TREND ANALYSES

Several techniques, adapted from time series analysis, have been used extensively to analyze sets of geological data which are arranged as a series in space [Harbough and Merriam (1)]. They include moving average methods, harmonic analysis, spectral analysis and auto-correlation. All but the first of these are concerned with acquiring information from rapid fluctuations present in all data.

The moving average techniques (including polynomial

trend analysis) tackle a problem similar to that discussed here. However, they assume that the underlying variations sought are continuous functions and will smooth out any sudden breaks. The analyst is presented with a plethora of results which are difficult to interpret.

In contrast, cusums are best used to highlight step changes in the underlying function, and are excellent for displaying slow cyclic variations. A comparison of cusums with other techniques used to detect slow variations is given by Hurst et al (2). It is interesting to note that the cusum of a series of equally spaced events is a convenient aid in calculating the simple moving average, particularly when a number of base lengths are to be examined.

APPLICATION OF THE CUMULATIVE SUMS ANALYSIS AT NCHANGA

METHOD OF APPROACH

All joints occurring within 30° of either side of the tectonic fold axis (which for analysis purposes was taken to be 300°) were considered in the analysis (i.e. joints with a direction of dip of 0° to 60° and 180° to 240°). In order that the joint data be representative of different parts of the pit slope, the hanging-wall was sub-divided into 14 arbitrary areas of approximately similar size going from west to east.

Analysis Using One Mean

One cusums technique consisted of analyzing the direction of dip data of all joints within the limits defined. Those joints with direction of dip of 0° to 60° were converted to 180° to 240° by adding 180° to their respective values. Thus, all joints could be analyzed together in the 180° to 240° range.

An example of the method of interpretation of the cumulative sums is shown in Fig. 3. The actual direction of dip orientations, as calculated from strike measurements in the field, are shown in Fig. 3(a). The resulting cumulative sums joint orientation plot of the raw data in Fig. 3(a) is shown in Fig. 3(b). Fig. 3(c) shows a Manhattan diagram, depicting the degree of deviation of the current mean strike above or below the overall mean strike K, according to the curves plotted in Fig. 3(b). See Fig. 4 for details of Fig. 3.

In Fig. 5, Manhattan diagrams of the cumulative sums of the analyses of the entire hanging-wall area that was surveyed, are shown. The various bench levels and subdivided areas of the hanging-wall (i.e. 1 to 14) are denoted accordingly. The bottom Manhattan diagram in Fig. 5 gives, for each of the 14 areas, the current mean deviation of the strike* of the joints about the mean. This is determined by calculating the mean deviation for all the benches occurring in a particular area.

Analysis Using Four Means

This analysis technique consisted of determining the cusums of both the direction of dip and angle of dip of the joints dipping to the northeast and of those dipping to the southwest. The respective mean direction of dip and mean angle of dip for each group were used. Hence, four K values are required, giving four cusum plots. The four K values applying to the Nchanga data are as follows:

---

* The cumulative sums plots in Fig. 3, of direction of dip data, is converted to actual strike values in Fig. 5 for purposes of clarity.

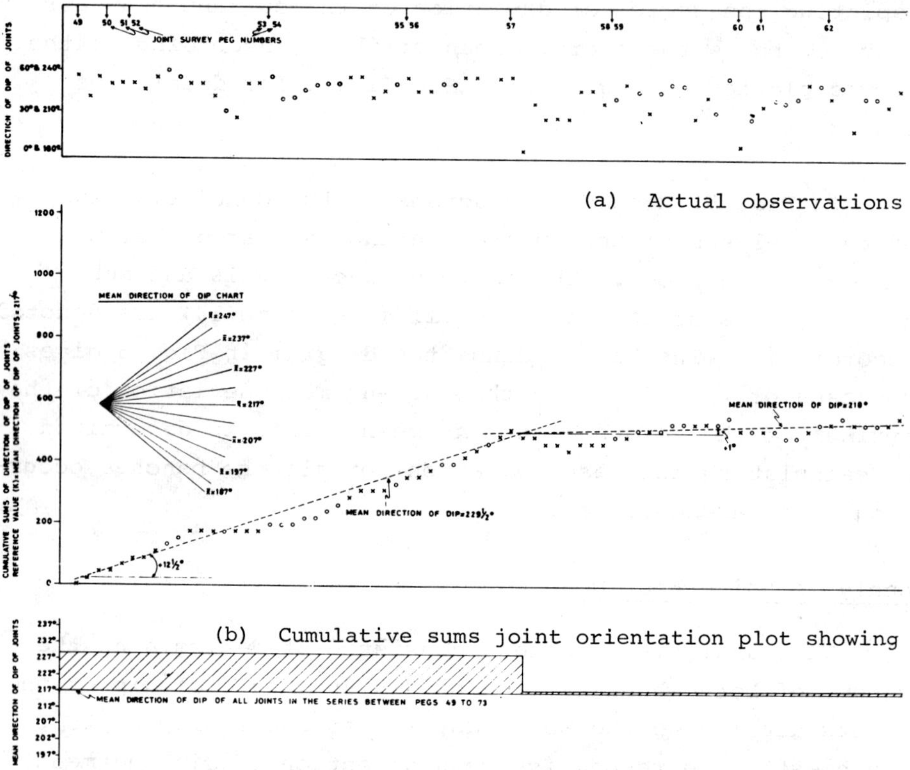

(a) Actual observations

(b) Cumulative sums joint orientation plot showing

(c) Manhattan diagram showing the current mean direction of

Fig 3 Illustration of cumulative sums technique

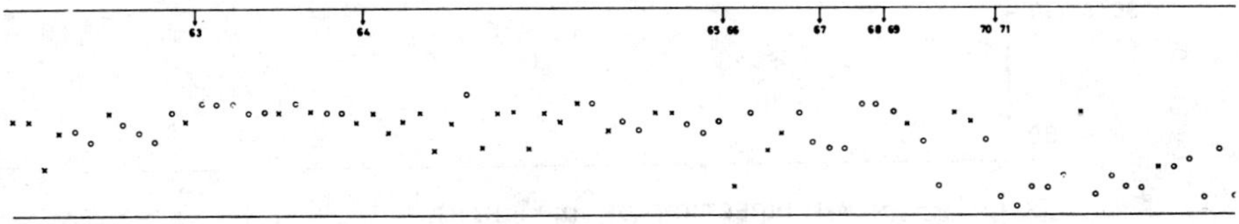

recorded in the joint survey

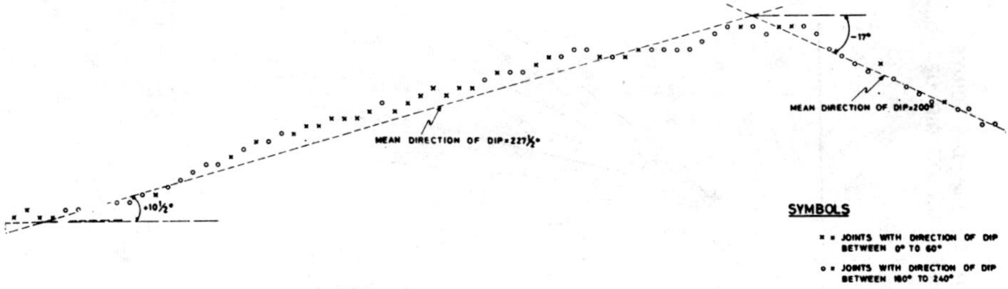

deviations of the direction of dips of the joints

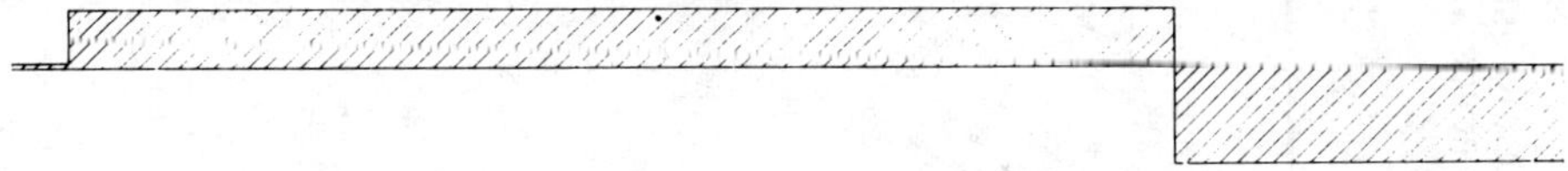

dip of the joints at progressive positions along the survey line

for analysing joint direction of dip trends

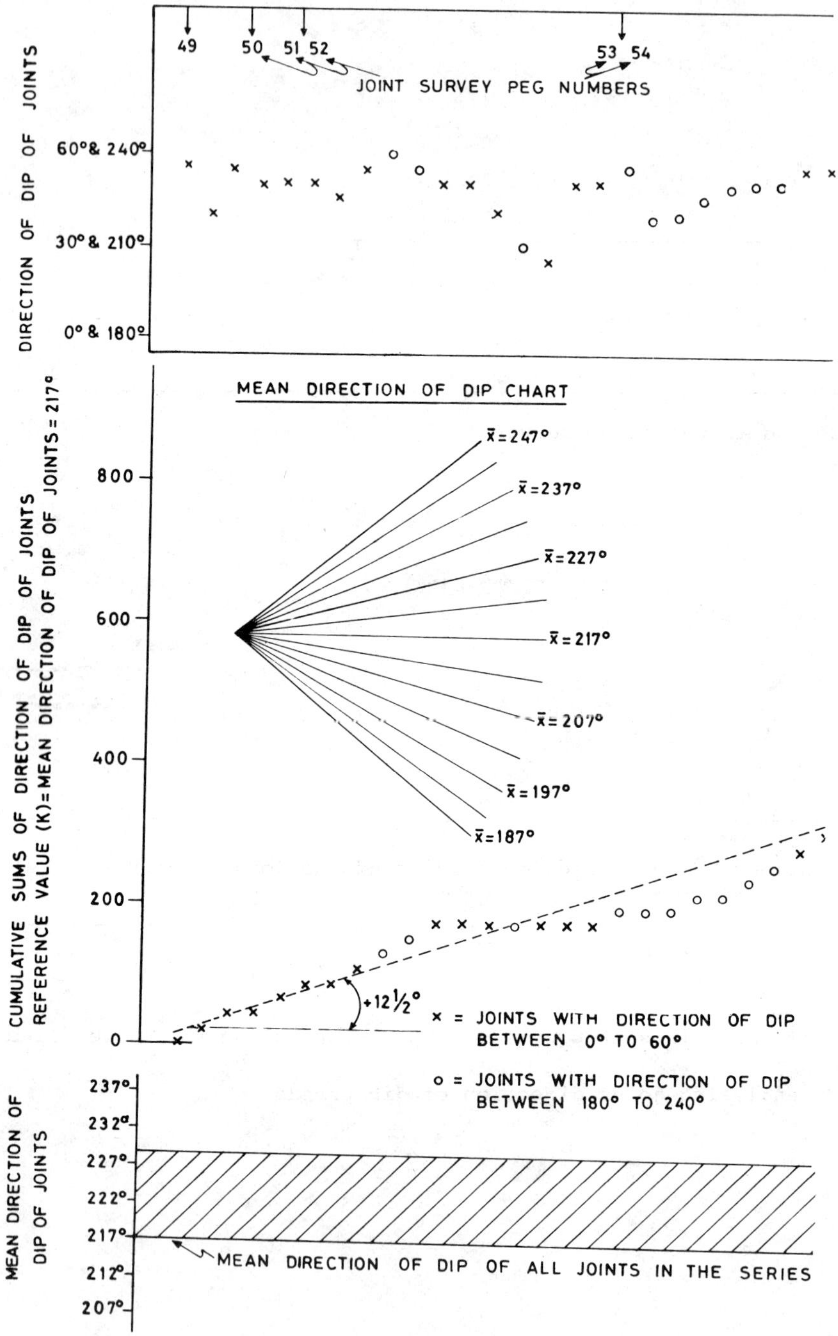

Fig. 4 Enlarged portion of the left side of Fig. 3 showing detail clearer

Joints Considered	Orientation Data	K Value
Direction of Dip $0^{\circ} - 60^{\circ}$	Strike	310°
" " " $0^{\circ} - 60^{\circ}$	Angle of Dip	74°
" " " $180^{\circ} - 240^{\circ}$	Strike	306°
" " " $180^{\circ} - 240^{\circ}$	Angle of Dip	78°

The general results are similar to those illustrated in Fig. 3, except that four cusum joint orientation plots are produced. Two plots apply to strike and two to angle of dip, although only one strike chart and one angle of dip chart are required. Four Manhattan diagrams must be considered in the same manner. The Manhattan diagrams of this analysis are given in Fig. 6. See Fig. 7 for some details of Fig. 6.

Applying this general form of cumulative sums analysis, boundaries to structural regions (i.e. areas of similar jointing characteristics in a statistical sense) were also determined. Cusum techniques used for this purpose will be published elsewhere.

DISCUSSION OF THE RESULTS

Counter-clockwise Rotation

In Fig. 5 it can be seen that the mean strike of all the joints is 307°. There is, however, a counter-clockwise rotation in the current mean strike, going from west to the east side of the hanging-wall. It rotates from about 317° in areas 1 and 2 to about 297° in areas 9 to 14. Around areas 5 and 6 the current mean strike is about the same as K.

Effects of Major Fault

As shown in Fig. 5, the rate of change of this rotation is greatest in areas 3 to 7. In Fig. 6, where the northeast and southwest dipping joints are analyzed separately, it can be seen that this phenomenon is due largely to the rotation

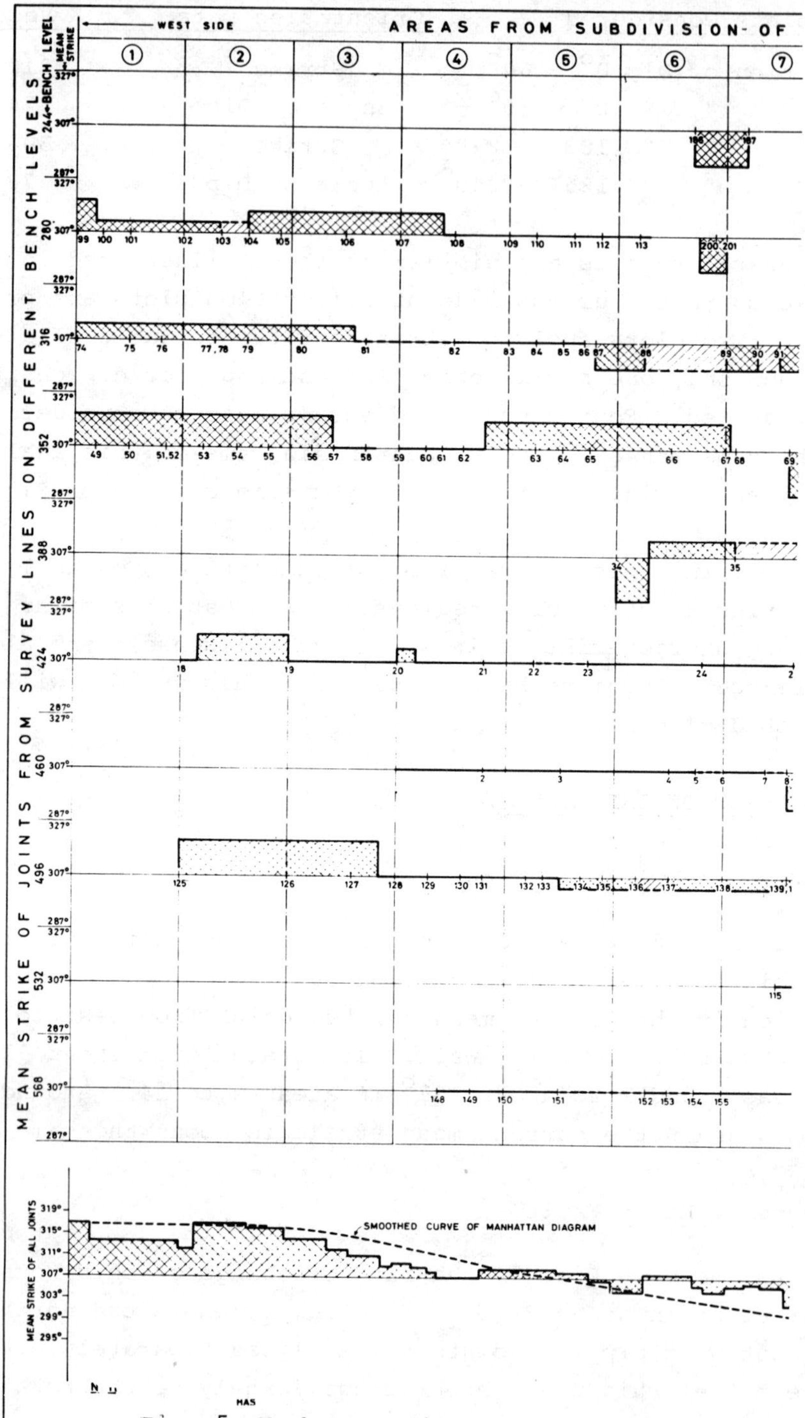

Fig. 5 Manhattan diagrams of strike trends

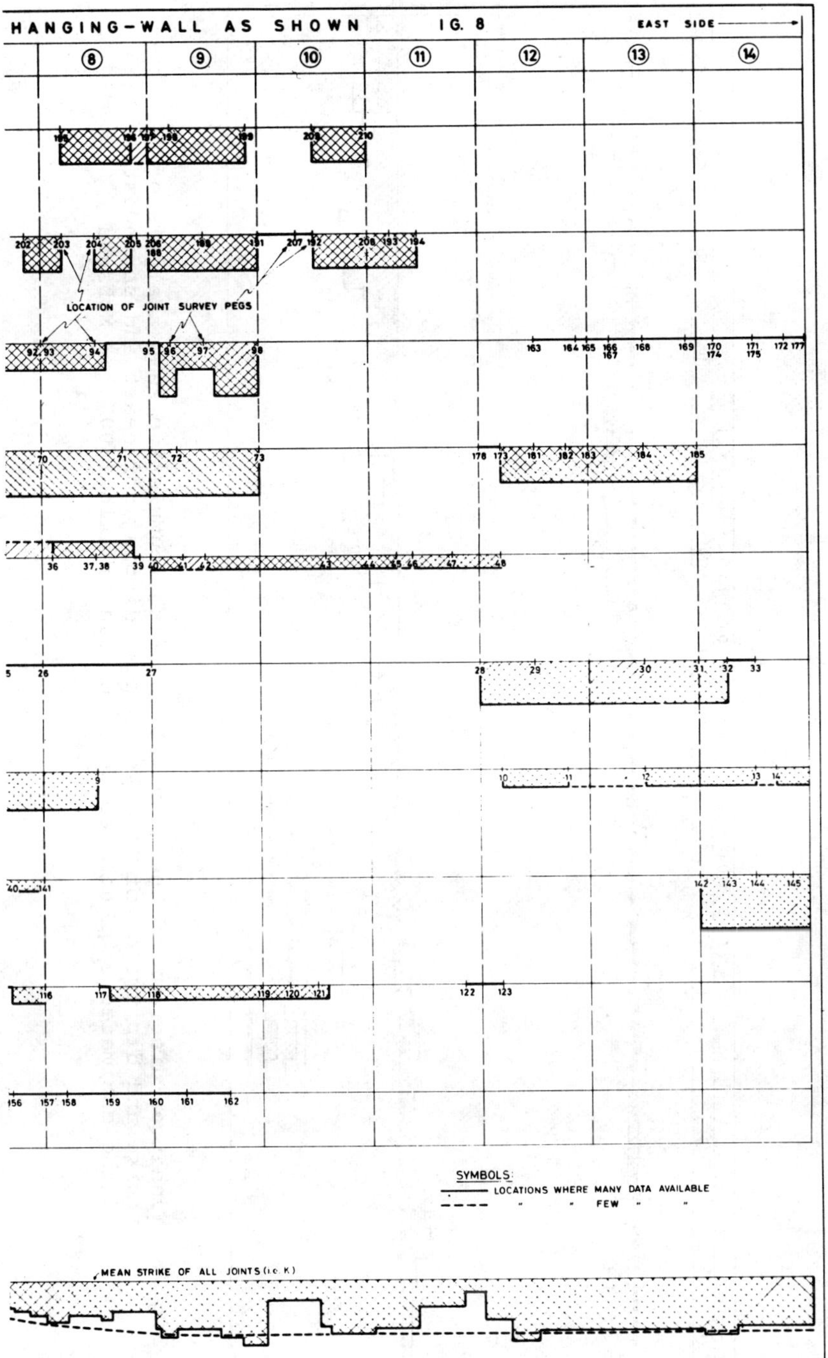

of joints occurring within ± 30° of 300°

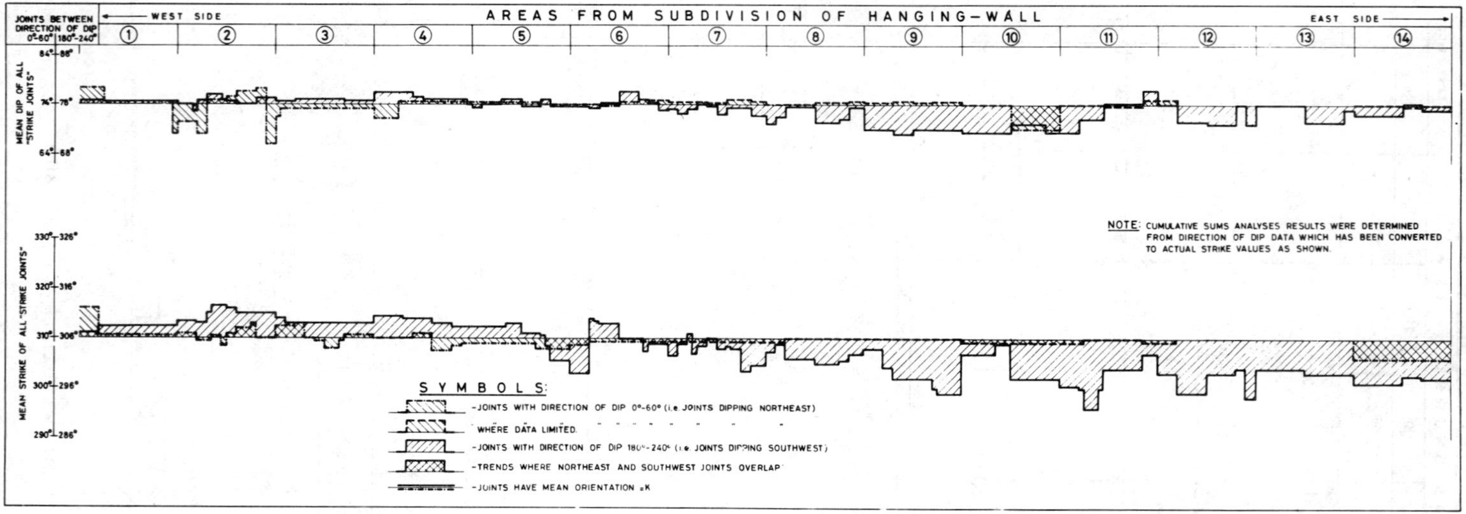

Fig. 6 Manhattan diagrams of the mean values of the cumulative sums analysis of strike and dip trends of "strike joints" dipping northeast and southwest of the estimated regional tectonic structural axis (300°) at Nchanga

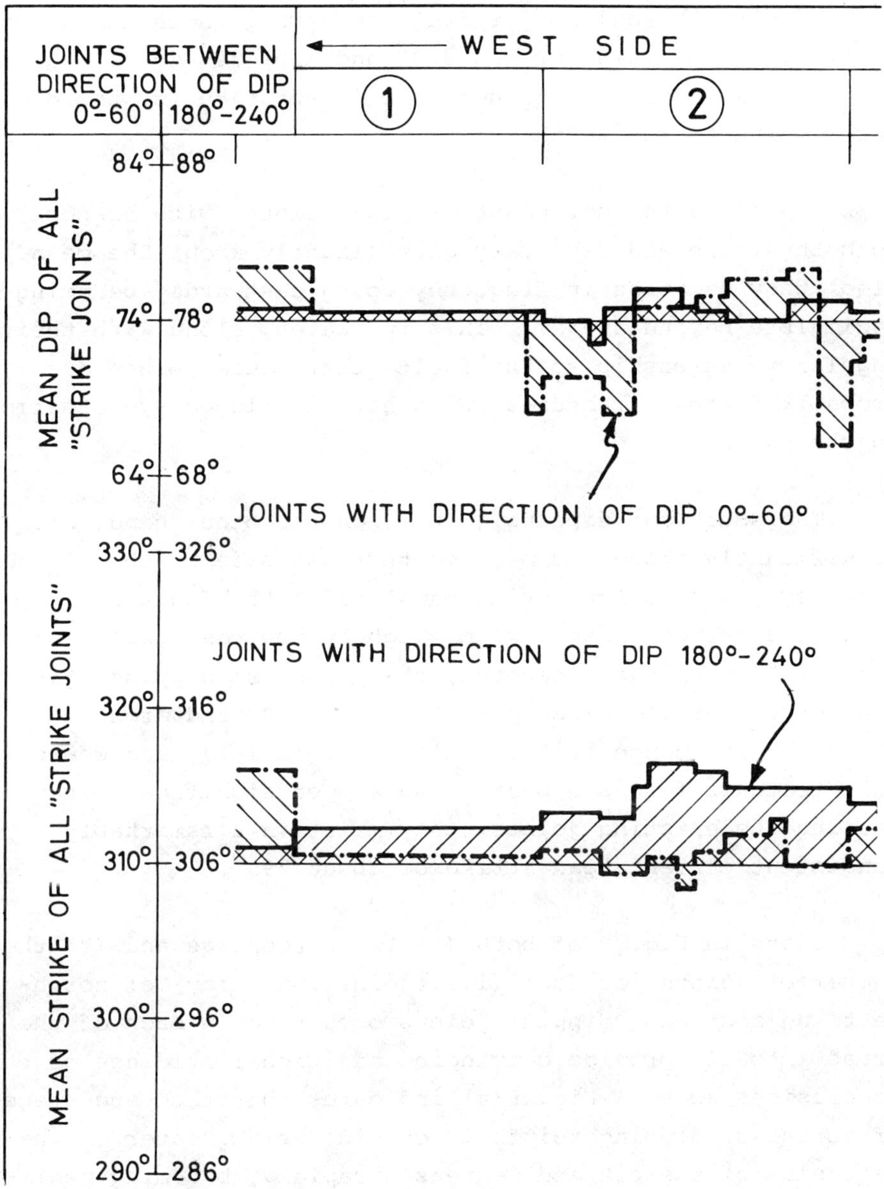

Fig. 7 Enlarged portion of the left side of Fig. 6 showing detail clearer

of the southwest dipping joints. This rotation is due to a major fault, the only major fault occurring in the area considered. This fault strikes 320° and dips 80° to 85° SW. The vertical component of net slip is about 80 ft, the downthrow being to the southwest.

In Fig. 6 the northeast dipping joints, with respect to both strike and dip, vary only slightly about the mean. Also, they decrease in frequency going eastwards, becoming negligible beyond area 7. This indicates, along with their angular relationship to the fault, that these joints are probably feather fractures which have developed sympathetic to the fault.

The southwest dipping joints, on the other hand, are significantly above the mean on the west side of the pit. Here, sympathetic fracturing parallel to the fault has swung the current mean strike slightly towards that of the fault. Further east, however, the southwest dipping joints rotate counter-clockwise past the mean. Beyond area 9, where the influence of the fault is negligible, and where only tectonic forces appear to have been significant in causing the existing joints, they maintain a remarkably consistent current mean strike of about 295°.

Plots in Fig. 8 of both (1) the percentage and (2) the number of joints per foot (i,e, joint intensity) of northeast and southwest dipping joints occurring in each of the areas 1 to 14, provide convincing additional evidence of the conclusions above. Fig. 8(a) indicates that the percentage of northeast dipping joints is considerably greater on the west side of the pit and decreases rapidly, becoming negligible east of area 7. The opposite is true for the southwest dipping joints. Area 4 is the changing point where one or the other dominates. In Fig. 8(b) the influence of the fault can be seen clearly. The intensity of the northeast dipping joints is excessively high in areas 1 to 4,

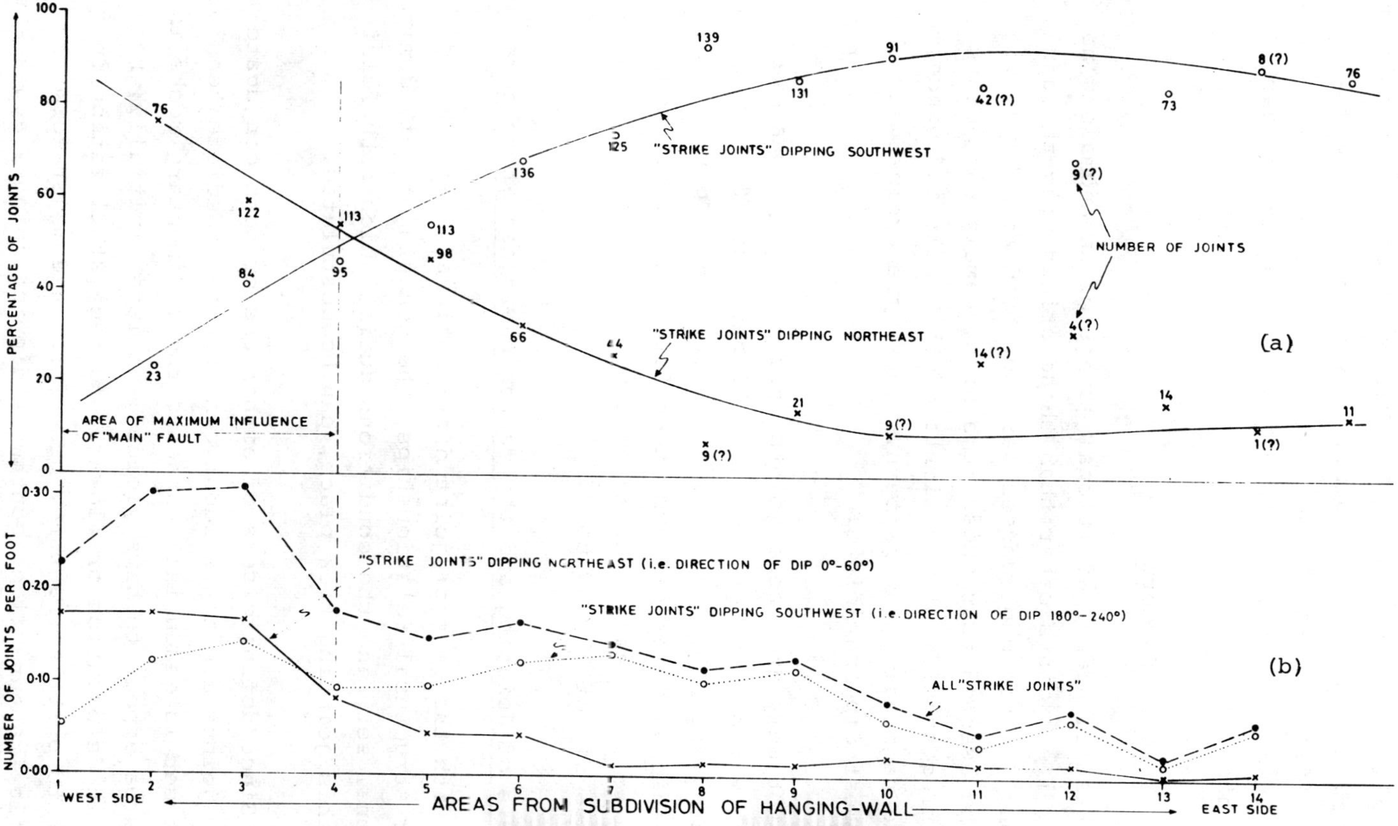

Fig. 8 Plot of the percentage (a) and frequency (b) of "strike joints" dipping southwest and northeast of the synclinal axis for different areas across the hanging-wall

but negligible beyond area 7. The southwest dipping joints, however, increase in frequency only slightly across the fault-affected area and maintain a fairly constant, though slightly decreasing, frequency going from west to east beyond this area.

Regional Joint Pattern

East of area 7 additional joints of an anomalous nature, (i.e. joints other than those originating through regional tectonic processes) are not evident. Thus, it must be assumed that those remaining, namely the joints occurring outside of the limits of the fault influence, are of the regional joint pattern. These are exclusively the southwest dipping joints.

Not only are these joints part of the regional pattern, but they represent greater than 80 per cent of the regional pattern (see Fig. 2 and 8). Greater than 80 per cent of the regional joint pattern, therefore, can be defined, approximately, as having an average strike of 295° and an average dip of 72° SW to 76° SW.

Genesis of Jointing and the Tectonic Process at Nchanga

With the general joint distributions in Fig. 2 and other structural relationships, the genesis of this dominant regional set, and the conditions during which both folding and this jointing took place, can be postulated.

Since definite sets of either one or both conjugate shear joints are not evident, and the intermediate tension joint set is essentially absent, the sedimentary rocks in the area appear to have yielded, at least initially, by plastic deformation or flowage and recrystallization in contrast to brittle fracture. The first and major form of brittle fracture (i.e. southwest dipping joints) appears to have developed as a result of elastic rebound of the originally highly compressed materials after both the temperature and pressure subsided. The southwest dipping joints, therefore, appear to be tension joints, having developed due to elastic rebound after the principal tectonic force had terminated. The principal form of deformation was that of crustal shortening. The type of folding was related primarily to those of horizontal tectonics, i.e. to processes of deformation wherein the maximum principal stress (tectonic stress) acted horizontally.

EXTRAPOLATION OF JOINT DATA

For purposes of making slope stability evaluations for the pit faces advancing both northwards and eastwards at Nchanga, the question of the reliability of applying information acquired from the existing hanging-wall slope to other parts of the mass where information is not available and where the advancing and final pit faces are to be located, is an important consideration. If any degree of confidence is to be achieved in proposing slope designs based to a large extent on these results, it must be shown whether the joint characteristics can be expected to be the same or to differ, and in what way to differ, in other parts of the mass where information is not available. The question of the extrapolation of joint properties when designing engineering structures in rock, and basic considerations relating to this problem, are discussed by Piteau (7).

Results in Fig. 6 show that both the current mean strike and current mean angle of dip are, statistically speaking, remarkably consistent east of area 8. This is particularly so with respect to the current mean strike. In either case, the deviation about the overall mean strike and mean dip orientation in this area is plus or minus three degrees. The history of folding in the majority of the syncline, and at least within the confines of the proposed final pit limits, is expected to be reasonably similar. Since the joints in question are genetically related to this folding process (in that they are rebound features which developed normal to the principal tectonic stress), based on the results of the cumulative sums analysis described above (see Fig. 6), there is good reason to believe that southwest dipping joints with similar orientations will exist in the proposed eastern extension areas of the pit.

For comparative purposes it is fortunate that at Nchanga an extensive joint survey had been conducted on the hanging-wall of the pit in 1966. The pit face at the time was 250 ft to the south of its present location, but the joint survey was conducted at approximately the same elevation and same relative location as that of the present survey. Hence, an ideal situation exists for determining whether the joint patterns are similar between the two survey lines and, accordingly, whether extrapolation of such structures is reasonable over this same distance in the opposite direction.

The 1966 survey results were available on stereographic projections, hence the peak concentration of the southwest dipping joints was easily measured. This information was compared directly to the cusum results for the respective areas across the pit. Except for minor variations, in general a remarkable similarity was found between the two separate survey results. Since the history of deformation is expected to be similar within the final pit limits, this indicated that the results from the present analysis would probably apply also behind the existing face in areas where the advancing pit faces are to be eventually located. These results also confirmed the conviction that the joint trends will be maintained in areas further east of the pit in which the extension is proposed.

CONCLUSIONS

The cumulative sums technique, illustrated with particular reference to an extensive joint analysis of the Nchanga open pit hanging-wall, provides an efficient and definitive method of examining joint dip and/or joint strike data in the order in which the joints are derived along the survey line. Unlike most joint analysis methods, this technique smooths out "noise" effectively. Also, both step changes in the underlying function and slow cyclic variations are readily displayed.

Basically, it is used to determine:

(a) the current deviation of either the joint dip or strike above or below some level of the orientation data or reference value (K) (i.e. in the Nchanga analysis K was taken to be the mean of the orientation data used);

(b) where these particular changes take place along the pit face; and

(c) the current mean orientation or simple moving average at any point in the consecutive sequence of the joints.

The behavior of a particular group of joints can be ascertained with respect to such characteristics as imperceptible rotation, both in the horizontal (i.e. strike) and vertical (i.e. dip) planes. In that the plots depicting this behavior are statistically significant, they can assist in predicting whether the information from the exposed pit face can be extrapolated with confidence to other parts of the mass where information is limited, but where pit slopes are to be eventually located. In this respect a knowledge of the geological history and the genetic relationship of the joints to the regional structure is important.

ACKNOWLEDGEMENTS

The authors extend their thanks to Nchanga Consolidated Copper Mines Limited - Chingola Division, for the co-operation which was given throughout the course of this work. The authors particularly wish to express their appreciation to C. Kovacevic, cartographer, for drawing not only these figures but numerous others in the Nchanga report; to A. Carbray, for helping to process the data; to F. Chisela, for supervising the survey; and to J. Hellings, for his general co-operation. Finally, the authors acknowledge with grateful thanks J. E. Jennings, O. K. H. Steffen and A. MacG. Robertson, with whom the senior author worked closely in recent years and who, although not directly involved with the development of this new technique, contributed significantly to important earlier development work, of which the results in this paper are an outgrowth.

REFERENCES

1. Harbough, J. W. and Merriam, D. F., "Computer Applications in Stratigraphic Analysis," John Wiley and Sons, New York, 1968.

2. Hurst, H. E., Black, R. P. and Simaika, Y. M., "Long Term Storage," Constable, London, 1965.

3. Jennings, J. E., "A Mathematical Theory for the Calculation of the Stability of Slopes in Open Cast Mines," Open Pit Mining Symposium, S. Afr. Inst. Min. Metal, Sept., 1970.

4. Piteau, D. R., "Engineering Geology Contribution to the Study of Stability of Slopes in Rock with Particular Reference to De Beers Mine," Ph.D. Thesis, University of the Witwatersrand, July, 1970.

5. Piteau, D. R., "Geological Factors Significant to the Stability of the Slopes Cut in Rock," Open Pit Mining Symposium, S. Afr. Inst. Min. Metal, Sept., 1970.

6. Piteau, D. R., "Analysis of the Genesis and Characteristics of Jointing in the Nchanga Open Pit for Purposes of Ultimately Assessing the Slope Stability," Report for Nchanga Consolidated Copper Mines Limited - Chingola Division, Dec., 1970.

7. Piteau, D. R., "The Question of Characterizing and Extrapolating Joint Properties in Engineering Practice," 20th Colloquium on Geomechanics, Salzburg Soc. Geomech., Salzburg, Sept., 1971 (in print).

8. Robertson, A. MacG., "The Interpretation of Geological Factors for Use in Slope Theory," Open Pit Mining Symposium, S. Afr. Inst. Min. Metal, Sept., 1970.

9. Woodward, R. H. and Goldsmith, P. L., "Cumulative Sum Techniques," Oliver and Boyd, Edinburgh, 1964.

FRACTURE EVALUATION BY THE BOREHOLE LOGGING METHOD

by

John I. Myung[1] and R. W. Baltosser, P. E.[2]

INTRODUCTION

The successful extraction of petroleum at depth in the ground requires a knowledge of the rock characteristics related to porosity, permeability and fractures. The borehole logging tools that have been developed to obtain this information are adaptable to other problems that require a knowledge of the rock conditions. Among these are the problems associated with dam sites, foundations, mines, etc. The importance of obtaining acceptable solutions to these problems is demonstrated by the number of papers published on these subjects in recent years.

In most of the problems involved in the design of a slope or, for that matter, all problems involved with subsurface engineering, a mechanical basis for anticipating the behavior of the rocks is required. It is necessary to know the location, aperture width, dip and the orientation of the fractures including bedding planes, joint systems and other weak zones in the rocks, such as vugs, etc.

---

[1]Senior Log Analyst, Birdwell Division of Seismograph Service Corporation, Tulsa, Oklahoma.

[2]Senior Mining Engineer, Birdwell Division of Seismograph Service Corporation, Tulsa, Oklahoma.

One approach is to make core analyses to investigate the spacing and attitude of fractures. Unfortunately, core samples do not necessarily always represent the true fractures due to crumbling or splitting of the rock during the process of coring. To investigate the actual fractures "in situ," two acoustical logging tools have been used to locate and evaluate the fracture systems penetrated by the drill. The use of these logging systems enables the rock mechanics engineer to project the dominant fracture and joint systems at depth within the rock and from this information be better able to evaluate his projected design requirements.

One of these logging tools is called the "Seisviewer." It provides a high resolution acoustical picture of the borehole wall. The dip angle and orientation of fractures or bedding planes cut by the hole can be determined from the resulting acoustical picture.

The other logging tool is the 3-Dimensional Velocity log, which displays photographically the amplitude of elastic waves as a function of time and of the depth of the tool in a borehole. A full complement of elastic waves including compressional, shear, and boundary waves are recorded. The fractures can be recognized on the 3-D Velocity log by investigating the changes in the compressional and shear wave arrivals.

Dzeban,[2] Morris,[8] Pickett,[9] and Walker,[10] have investigated the effect on the amplitude of the compressional and shear waves due to fractures. These investigations have shown that the angles at which a fracture plane cuts a borehole affects the amplitudes of acoustical signals. Morlier[7] derived a mathematical equation showing the relationship of the fracture porosity and the pressure.

The combination of these two logging tools plus a detailed surface map of the joint and fracture systems and, where possible, the co-ordination of this data with core analysis enables the rock mechanics engineer to develop a three-dimensional concept of the joints and fractures in the area under study.

THEORETICAL INVESTIGATION

The presence of fractures, joints or other discontinuities in the rocks will interfere with the transmission of any type of wave energy through the rock. It is generally accepted that fractures can be recognized by the reduction of the amplitude of the compressional and shear waves. Experimental work by Morris[8] has shown that the angle at which a fracture plane crosses a borehole affects the attenuation of acoustical signals. Theoretically, horizontal fractures should cause little attenuation of the compressional waves; on the other hand, the shear waves are significantly attenuated by these same fractures. Knopoff, et al,[4] has calculated curves of transmission coefficients for plane compressional and shear waves across an infinitely thin lubricated crack in an infinite medium. Figure 1 is taken from Knopoff's curves for a hard carbonate formation with a Poisson's ratio of 0.3. This figure indicates the amplitude of the shear wave should be more diagnostic in identifying fractures when the fractures lie at a very low or very high dip angle. Conversely, the amplitude of the compressional waves should be more sensitive to fracture dips between 33 degrees and 78 degrees. A vertical fracture should not affect either the compressional or the shear wave.

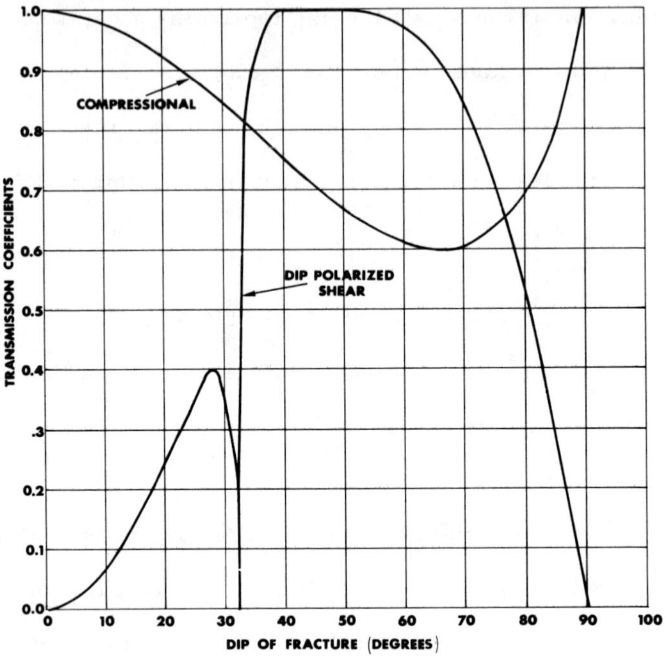

Figure 1

Another investigator of the problem is I. P. Dzeban[2] of the USSR Ministry of Geology. In the series of tests outlined in this report entitled "Elastic Wave Propagation in Fractured and Vuggy Media," the amplitude of both the pressure and shear waves is noted as well as the effect of the angle of incidence on the passage of a wave across a fracture. In general, the results of these tests agree with those of Knopoff[4] as reported by Morris[8] from Figure 1. Figure 2 from Dzeban's paper is the result of model studies and is a plot of the ratio of the elastic wave amplitudes, Afr/Ao, versus an angle of incidence at the fracture. Afr is the amplitude recorded on a fractured model while Ao is that of a nonfractured model. In this figure, the ratio of Afr/Ao is plotted on the vertical axis and the angle of incidence on the horizontal axis.

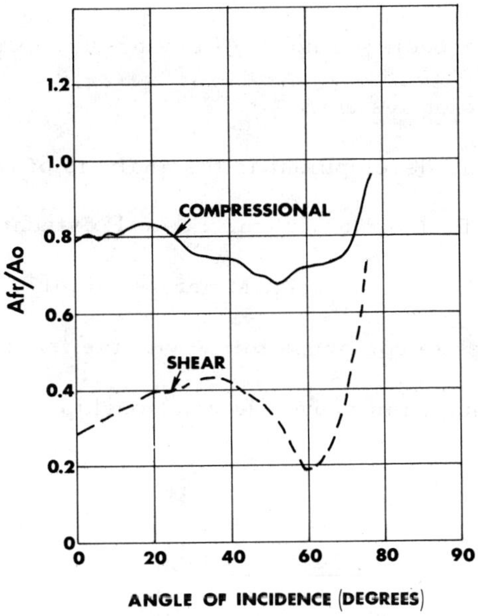

Figure 2

According to Dzeban's experimental data, the ratio of the shear wave amplitudes, Afr/Ao, decreases from 0.4 at 30° to 0.2 at a 60° angle of incidence and the maximum variation of the ratio of the compressional wave amplitudes (Afr/Ao) decreases 1.3 times, in Dzeban's calculation, between the same angles of incidence. The sharp rise of the curves at incidence angles of more than 60° is probably due to the presence of distributed wave conversions; most of the energy does not cross the fractures but is propagated along the fracture bedding plane of the model.

Pickett,[9] et al discusses the application of amplitude logs in locating fractures. This technique has proved to be useful in fracture detection when properly applied. Experience has shown that it is necessary to know the lithology and rock types, the general geology of the area and the condition of the borehole - amount of caving, for example - in evaluating the fracture and joint system.

A relatively recent development in the analysis of fractures in rocks is the work by P. Morlier[7] of the Ecole Polytechnique reported in the July 15, 1969 Revue of Industrial Minerals. In this investigation, an attempt is made to relate compressibility with the fracturing of the rock. An equation was developed to show this relationship:

$$e = \frac{3(1-2\nu)}{E} P - \eta(P) \qquad (1)$$

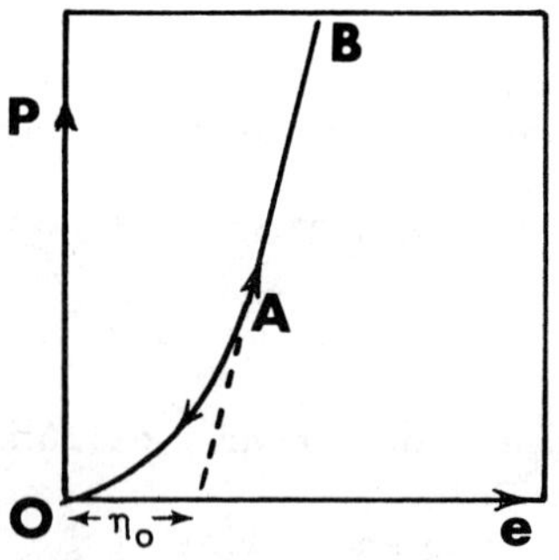

Figure 3

Where: e=Fracture porosity
ν=Poisson's ratio
E=Young's modulus
P=Pressure

In this equation the intersection of the line A-B with the e axis gives the volume porosity, η_o. The straight line A-B of Figure 3 represents the deformation of the unfractured matrix rock, while the curve from

O to A represents the reduction of fracture volume η_o as the pressure increases.

Since both Young's modulus and Poisson's ratio are readily obtained from the 3-D Velocity Log, we are able to compute the fracture porosity using Equation 1.

An alternate method of calculating underground stresses is to assume that the stress field is nearly hydrostatic with the three principal stresses approximately equal, although the extent of fracturing due to faulting or folding suggests that there is a high degree of mobility in the rocks and hence that unequal principal stresses exist.

In the theories of hydraulic fracturing treatment of oil wells, the assumption is made that the rock will fracture in the plane perpendicular to the least principal stress. This provides a mechanical basis for anticipating the mechanical behavior of rocks. However, it is necessary to know something concerning the stress states which may be expected in the rocks.

SEISVIEWER SYSTEM

The Seisviewer is an acoustical device that produces a continuous photographic display of the borehole wall. It consists of a downhole scanning instrument, or logging probe, with its multi-conductor cable and surface panel. The downhole scanning instrument, comprising a motor-driven acoustical transducer and a fluxgate magnetometer, rotates at 3 revolutions per second to scan the borehole. The resulting log is oriented to a north marker signal from the fluxgate magnetometer.

The acoustical transducer is pulsed at a rate of 2,000 times a second, with the acoustical beam focused on the wall. The amplitude of the signal reflected from the wall depends upon the acoustical impedance of the wall rock and associated physical properties of the wall. Fractured or highly disturbed zones are recognized by a poor or no reflected signal. The advantage of this system is that this tool can be operated in any type of fluid, and in holes from two inches to twelve inches in diameter.

The detected acoustical signals are amplified along with the magnetic north marker pulse and transmitted to the surface through a multiconductor cable.

The surface panel combines this downhole information to produce an acoustical picture of the borehole wall. The actual operation might be compared to a somewhat simplified television presentation. An oscillograph sweep is triggered by the north marker signal with the sweep intensity modulated by the detected return signal from the wall. The horizontal line spacing is controlled by the logging speed. The completed picture is a compilation of a multitude of transducer scans which when photographed combine to make a log picture of the borehole wall. The acoustical transducer rotates clockwise with the left edge of the picture oriented at magnetic north. In effect, the picture of the borehole is a cylinder of the borehole split at magnetic north.

Figure 5 shows the complete tool with the acoustical transducer near the bottom, the fluxgate magnetometer in the center and the electronic cartridge at the top.

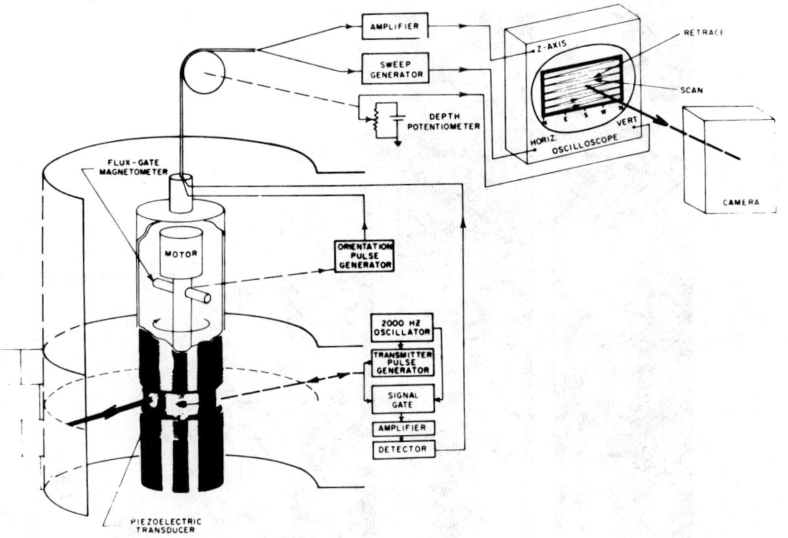

Figure 5. Block Diagram of the Seisviewer Logging System (After Zemanek)

Figure 6 is a portion of a log recorded in a mud-filled rotary-drilled hole. The section is highly fractured, the fractures showing as the dark sinusoidal features. The angle of dip varies greatly, as shown by the sharply dipping fractures in both the right and left strips of the log display while relatively low angle dips are shown in the middle strip. On the left section at a depth of about 4, 960 feet, the section is highly fractured and would be interpreted as a faulted section. The black areas indicate gouge and the light areas indicate the brecciated fragments in the gouge. The middle strip shows some gouge near 4, 980 feet. Immediately below this is either a series of low angle fractures, or bedding planes. Obviously, when the logs show this degree of complexity, positive

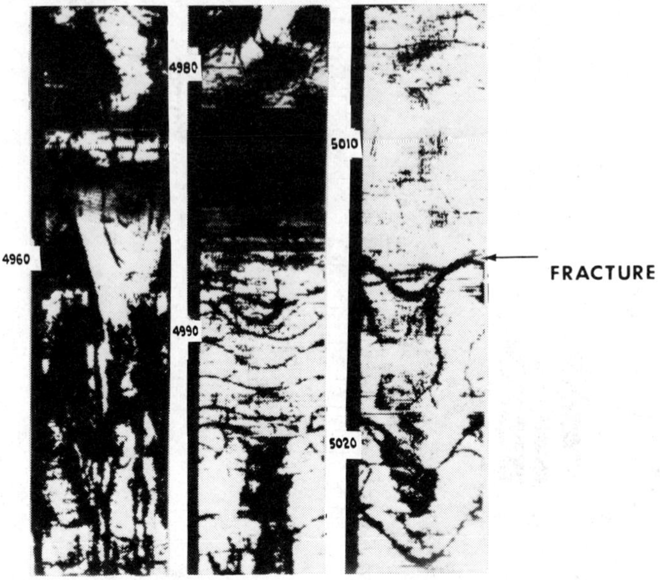

Figure 6. Highly Fractured Section
(After Zemanek)

identification of the breaks should be based on correlation with cores from the section under study. Some of the darker lines normal to and connecting the sinusoidal lines may be jointing in the rock. It should also be noted that the fractures are dipping to the south. In this area the rock appears to be competent with a minimum of joints and fracture filling. These fractures would be considered hazardous to a slope design if the stresses on the slope were parallel with the southerly dipping fracture system. Ideally, the stresses imposed by a load should be normal to any condition which weakens the rock. In this example, it is the fracture system. Assuming that any load imposed can meet this specification, the secondary weak zones in the rock would be the joint system.

These joints appear to be discontinuous and would not be considered a design problem provided the inherent physical strength of the rock was able to resist crushing or granulation under the imposed load.

The dip and strike of fractures, joints or bedding intersecting the borehole can be easily determined as shown in Figure 7.

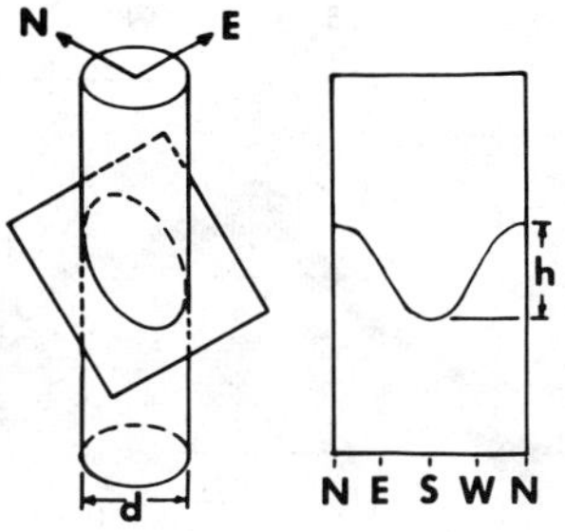

Figure 7. Isometric Drawing of Fracture Plane Intersecting the Borehole

The angle of dip can be computed from the following equation:

$$\text{Angle} = \tan^{-1} \frac{h}{d} \qquad (6)$$

In this equation, d is the diameter of the hole and h is the distance from the trough to the crest of the sinusoidal waveform measured on the log. The trough of the wave form points in the direction of the dip which can be read from the log orientation.

Figure 8 provides an example of the determination of dip angle and bearing from a log. There are two high-angle fractures which nearly intersect within the borehole. The traces of the fractures indicate that the two planes cross each other a short distance beyond the borehole. The two fractures illustrate the orientation and angle of dip of two sets of fractures that extend over 40' of hole. The calculated angle of dip of the fracture between 5561 and 5562 feet is 58° and, from the low point of the trough, the direction of dip is N 70 E. The other fracture which occurs between 5562 and 5564 feet has a calculated dip of 74° and a dip direction of N 45 W.

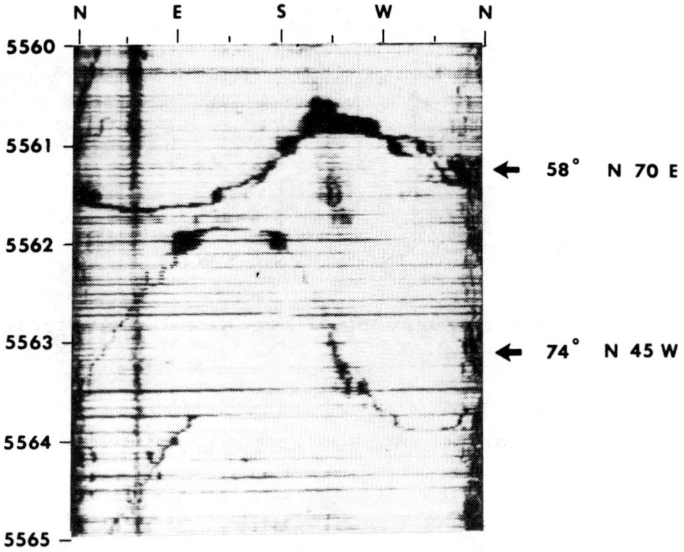

Figure 8. High Angle Fractures Intersecting Near Borehole

3-D VELOCITY SYSTEM

The 3-D Velocity tool has a magnetostrictive transmitter separated from the piezoelectric receiver by an acoustical isolator. The receiver detects changes in pressure resulting from the interaction of the compressional wave at the interface boundary between the fluid and the borehole wall and transforms the changes to an electrical signal for the surface recorder. The 3-D Velocity system provides a continuous display, on film, of the complete wave trains. Previous papers by Lawrence and Baltosser,[5] Christensen,[1] Geyer and Myung[3] describe the principle of operation, the logging equipment, and its applications to subsurface engineering. A schematic diagram of the downhole 3-D Velocity tool is shown in Figure 9.

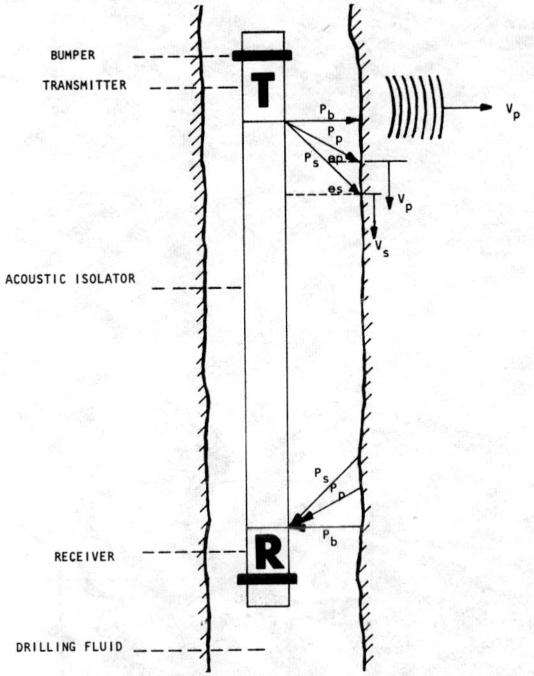

Figure 9. Diagram of 3-D Velocity Tool

Figure 10 shows an intensity-modulated 3-D Velocity log run in a series of coal beds to analyze carbon, ash, and moisture content, and the deformation modulus. The pressure and shear waves and the three coal seams are indicated on the log.

The 3-D Velocity log is usually recorded in an intensity-modulated form as shown in Figure 10. An alternate presentation would be a

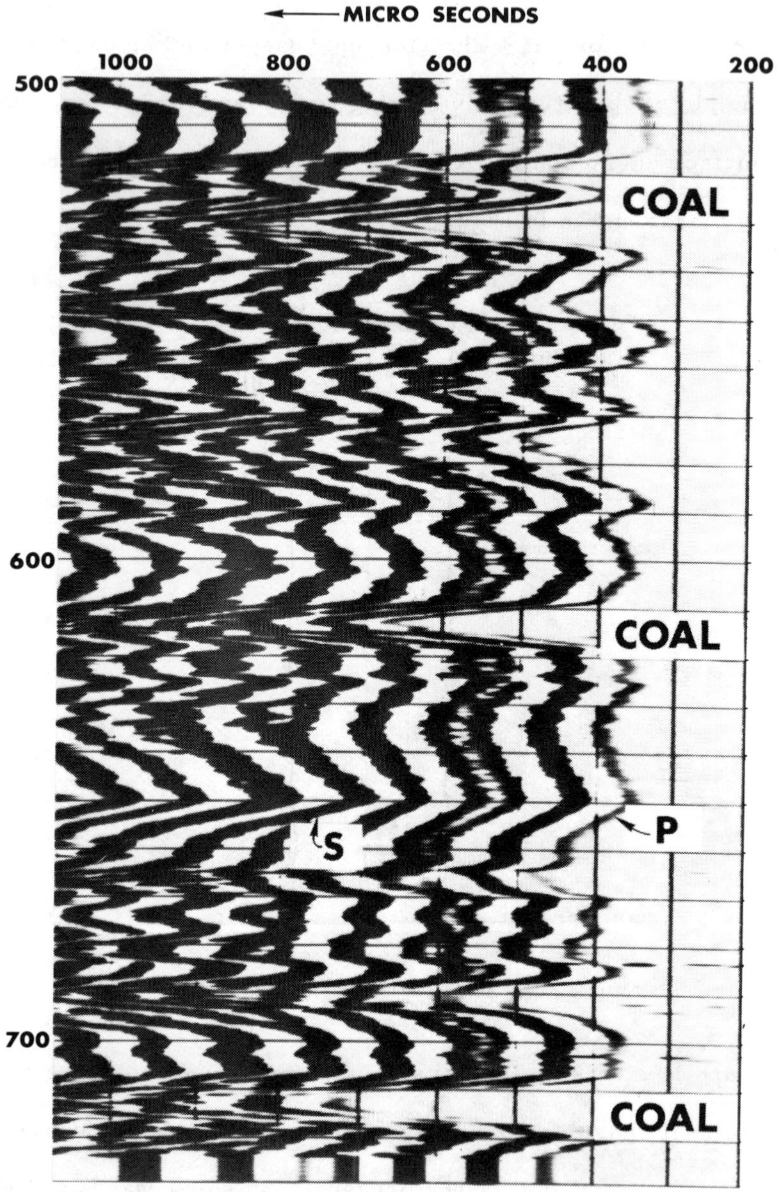

Figure 10. Intensity Modulated 3-D Velocity Log

signature log which actually records the wave form, as shown in Figure 11.

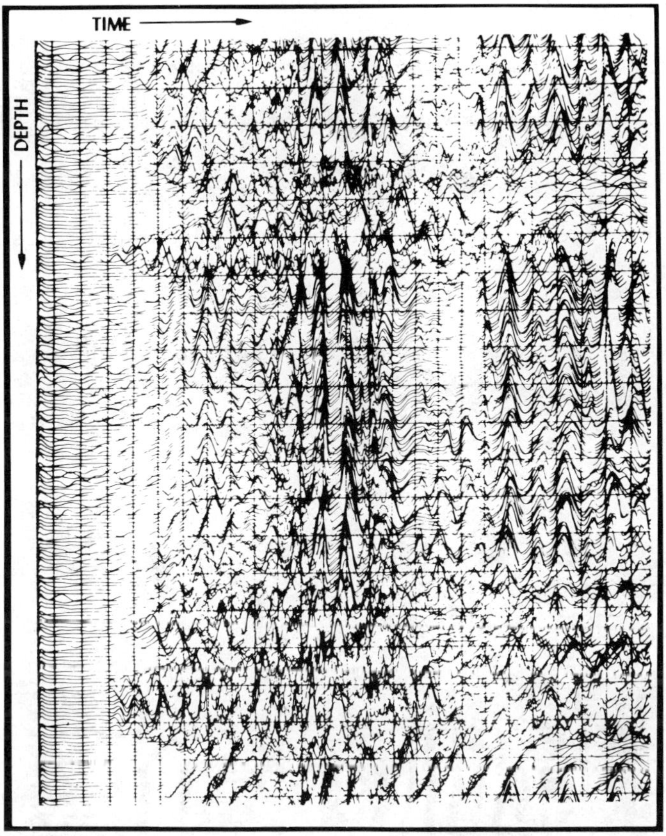

Figure 11. 3-D Signature Log

For specific purposes, there are advantages in each of these alternate presentations. The intensity-modulated log indicates amplitude by increased density while the signature log records the relative amplitude of each wave train. Where both logs are desirable, they can be recorded simultaneously, overlapping each other.

FIELD LOGS

Figure 12 is a log from granite in the state of New Hampshire. In Zone C of this log, the compressional wave is not attenuated while the shear wave amplitude is severely reduced due to a low angle or horizontal fracture, which agrees with Knopoff's calculated curves shown in Figure 1.

The strong energy arrivals for both compressional and shear waves in Zone B indicate that no fracture is present. In Zone A, both the compressional and shear waves are attenuated, probably due to an oblique fracture.

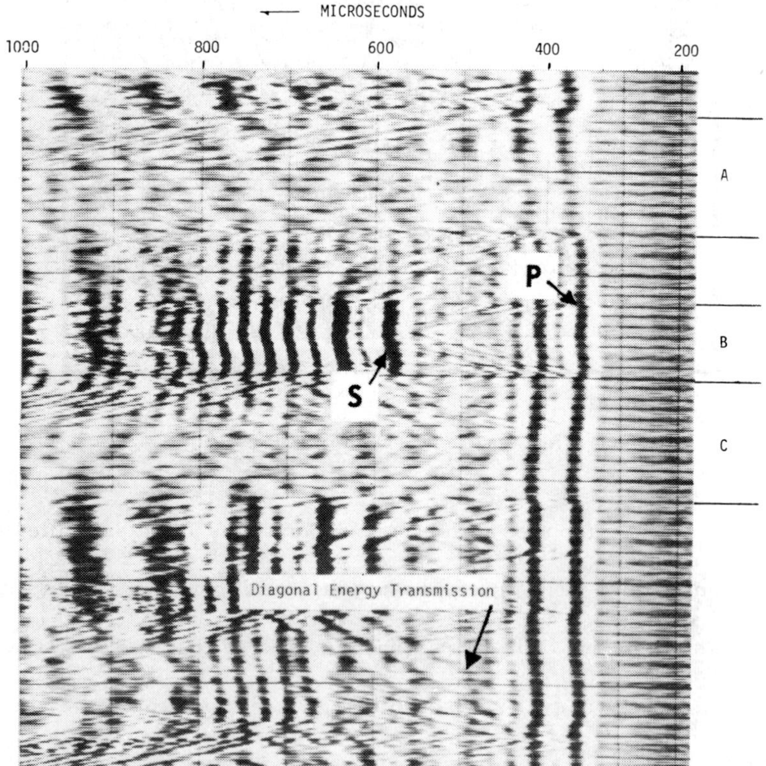

Figure 12. 3-D Velocity Log in Granite

A further analysis of the geometry of a fractured zone would suggest that fractures act as thin beds separated by other rock materials with a strong acoustical impedance mismatch to the surrounding rock. Figure 12 shows a distinctive diagonal energy transmission below Zone C as the tool moves up to and by a reflector surface adjacent to the borehole. Lawrence [6] discusses the geometry of these wave paths in detail.

As stated earlier, a vertical fracture does not affect either the compressional or shear wave energy. Figure 13 compares a Seisviewer log and a 3-D Velocity log run in Clinton Sand in Ohio. In the interval between 4000 and 4030, the Seisviewer log shows a vertical fracture that strikes almost due east-west. On the 3-D Velocity log there appears to be little effect on either the compressional or shear wave energy in this same interval. It is interesting to note that the 4010 to 4016 feet interval is affected by minor caving due to clays in the zone. This is observed on both the 3-D Velocity and the Seisviewer logs.

Figure 14 shows a typical group of logs from an NX hole. These are the logs required in the evaluation of rock properties and fracture systems in the science of rock mechanics. This project is involved in a study of the Chattanooga Limestone in Alabama. The combination consists of a 3-D Velocity log, made with a 3 foot isolator, a Seisviewer log, a Gamma-Gamma Density log and a Caliper log together with the computed elastic properties. Normally, the isolator dimension between the transmitter and receiver is of the order of 0.5, 1, 3, or 6 feet. The shorter the isolator the better the thin fracture zones are defined. The following symbols are used to identify the elastic properties listed on the

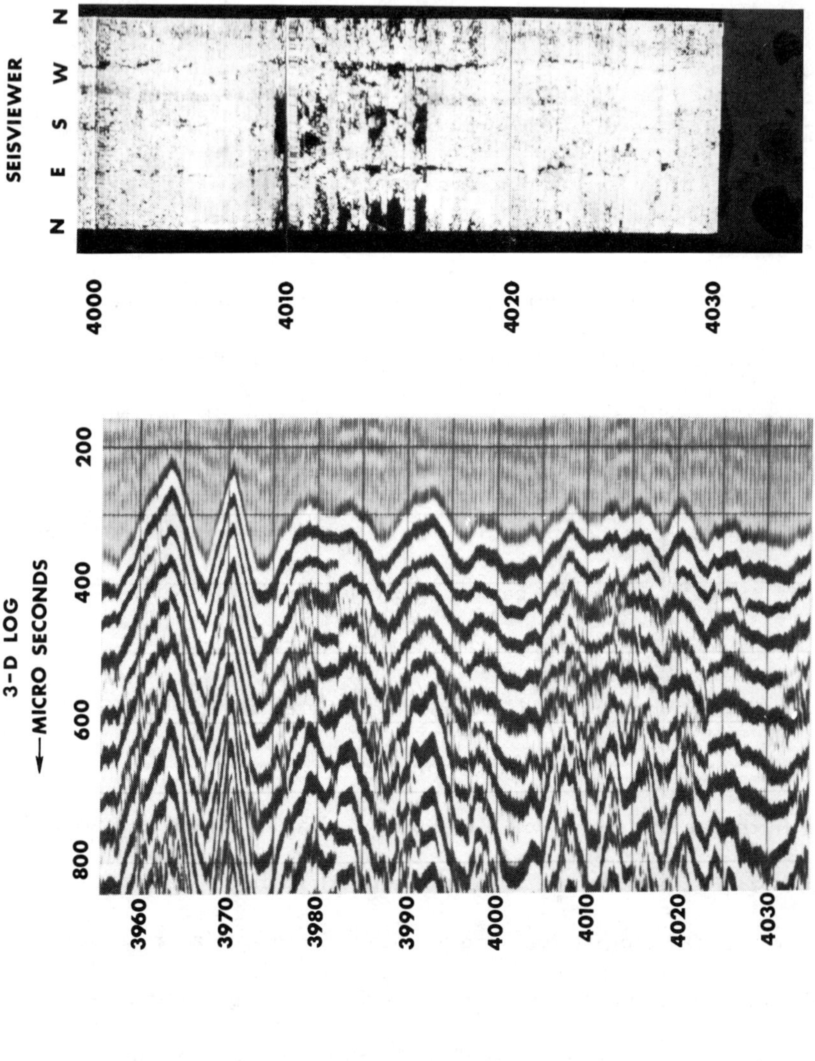

Figure 13. Comparison of 3-D Velocity Log and Seisviewer

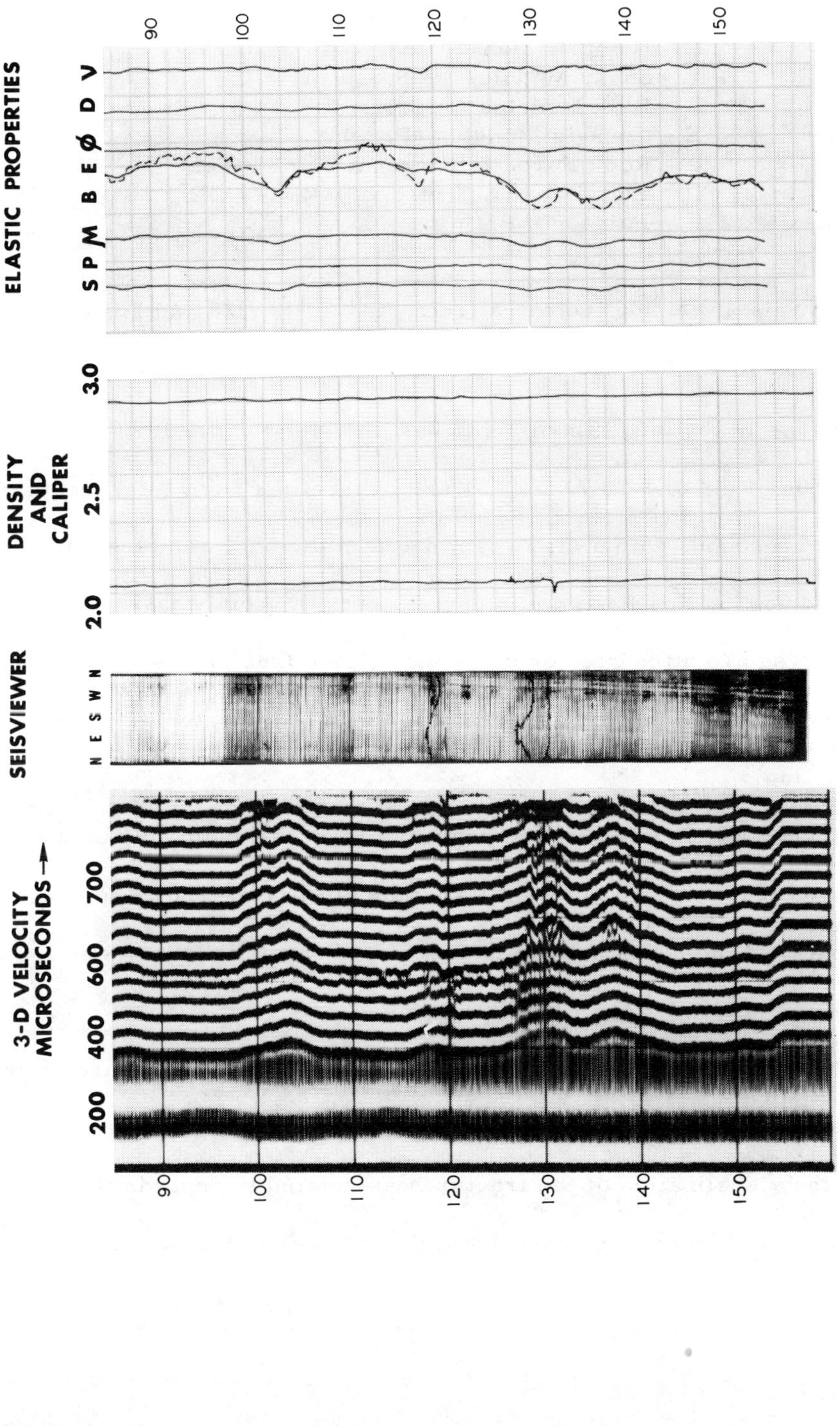

Figure 14. Composite Logs in Chatanooga Limestone

Log heading:

S - Shear Wave Velocity (ft/sec)
P - Compressional Wave Velocity (ft/sec)
μ - Shear Modulus (P.S.I.)
B - Bulk Modulus (P.S.I.)
E - Young's Modulus (P.S.I.)
Φ - Rock Porosity (Percent)
D - Bulk Density (gm/c.c.)
ν - Poisson's Ratio

The Seisviewer log indicates fracturing at 119, 128 and 131 feet. The fracture at 119 feet has a calculated dip of 78° and direction of dip of S 70W while that at 128 foot depth has a calculated dip of 81° and orientation of dip of S 70W. At 131 feet, there is a horizontal fracture which corresponds with a clay seam noted in the core samples. On the 3-D Velocity log for the same intervals, both the compressional and shear waves are attenuated or distorted. Those fractures shown by the Seisviewer log are in good agreement with those indicated by the 3-D Velocity log.

Figure 15 shows the cores recovered from the same hole at the intervals discussed in Figure 14. A study of the core shows good agreement between the cores and the data obtained from the Seisviewer log and the 3-D Velocity log.

The polar coordinate presentation shown at the top of Figure 15 is an alternate means of presenting the fracture pattern. It is designed to assist the visualization of the fractures as related to depth in the hole. The circular pattern is a logarithmic plot of the depth in the hole.

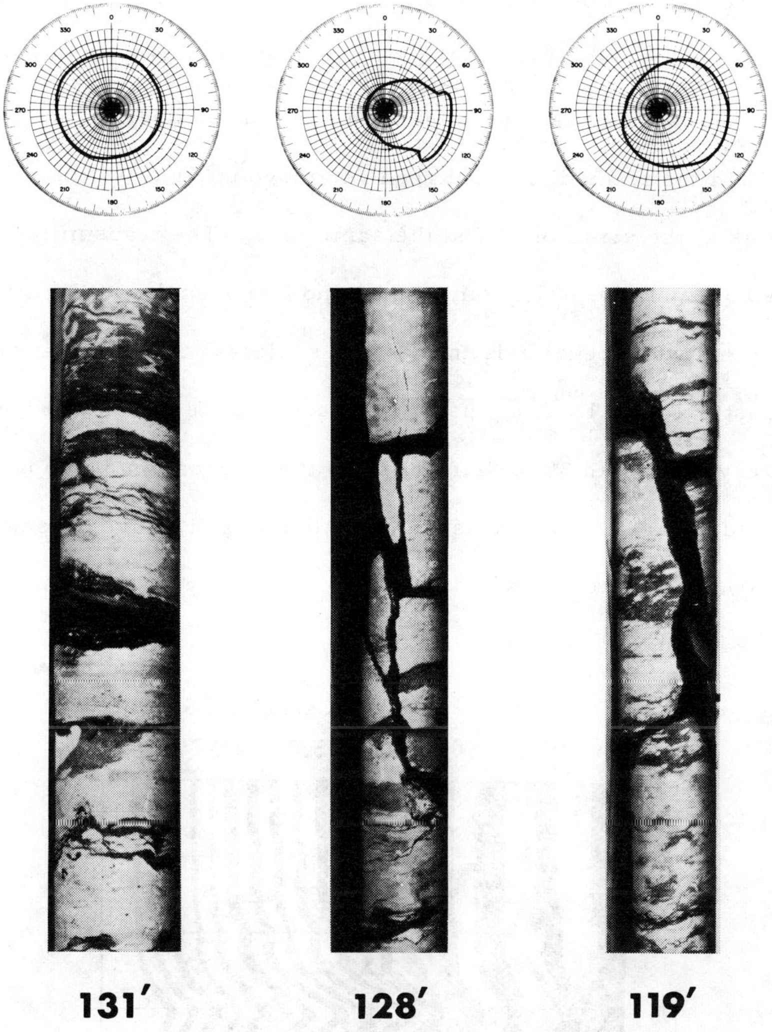

Figure 15. Core Samples and Polar Coordinate Plots

HOLE-TO-HOLE TECHNIQUE

There is a further application of the 3-D Velocity logging procedure which is of use in fracture studies. This is the hole-to-hole method. In this, the transmitter is run in one hole and the receiver run in an adjacent hole. Both the transmitter and receiver move up or down their respective holes at the same level and the same rate. The transmitted signal travels horizontally. The competency and homogeneity of the material between the two boreholes is in this way subjected to investigation by the acoustical signal. A log using this technique is shown in Figure 16. The interval between 85 to 92 feet shows the attenuation of the compressional and shear wave arrivals probably due to a fracture present in the area between the two holes.

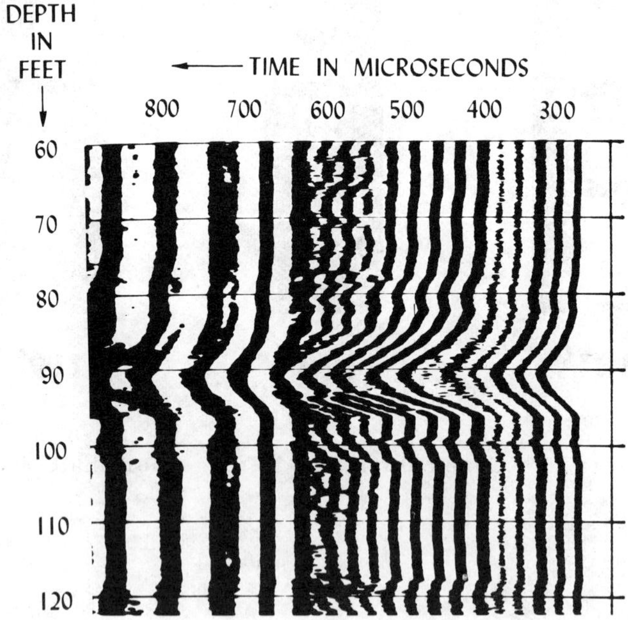

Figure 16. Hole-to-Hole 3-D Velocity Log.

Figure 17 is another method of recording the 3-D Velocity log by the hole-to-hole technique. In this figure, the receiver remained at 65 feet, as the transmitter moved up in the other hole. The distortion showing on the P-wave at 37 feet indicates the fractured zone.

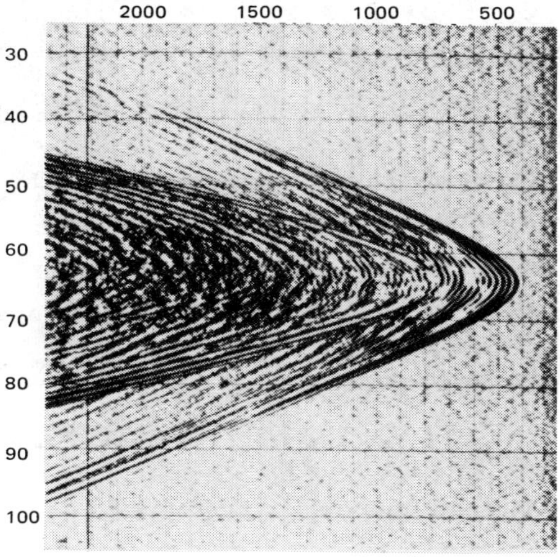

Figure 17. Hole-to-Hole 3-D Velocity Log.

The distance over which a signal can be transmitted hole-to-hole is dependent on the intervening material and the size of the transmitter used. The transmitter specifications are as follows:

Transmitting Transducer O. D.	Minimum Hole Size	Expected Range In Competent Rock
2-1/4"	3"	30 feet
3-5/8"	4-1/2"	75 feet
5-3/8"	6-1/2"	200 feet

This technique is recommended for evaluating fractures existing between two holes as well as those cut by the borehole.

SUMMARY AND CONCLUSIONS

Two acoustical logging devices, the Seisviewer and 3-D Velocity systems can be successfully used in fracture evaluation.

The dip and strike of fractures as determined from the Seisviewer log can be satisfactorily correlated with the core samples from the same hole while the 3-D Velocity log will assist in identifying the fracture as well as providing reliable information on the elastic properties of the rocks

The experimental data of Dzeban and Knopoff regarding the attenuation of amplitude of both the compressional and shear waves across fractures seems to compare satisfactorily with the field results obtained by 3-D Velocity log and Seisviewer systems.

The hole-to-hole technique with the 3-D Velocity system can be used to investigate fracture planes which are not penetrated by the borehole.

Seisviewer and 3-D Velocity logs will provide "in situ" information regarding the fractures and the elastic properties of rocks in slope stability studies as well as other design problems involved with subsurface engineering.

BIBLIOGRAPHY

1. Christensen, Dean M., "A Theoretical Analysis of Wave Propagation in Fluid Filled Drill Holes for the Interpretation of the 3-Dimensional Velocity Log," Fifth Annual Logging Symposium, May, 1964.

2. Dzeban, I. P. "Elastic Wave Propagation in Fractured and Vuggy Media," IZV Earth Physics, No. 10, 1970 (translated by D. G. Fry).

3. Geyer, Robert L., and Myung, John I., "The 3-D Velocity Log, a Tool for In Situ Determination of the Elastic Moduli of Rocks," Twelfth Annual Symposium on Rock Mechanics, University of Missouri, Rolla, November, 1970.

4. Knopoff, L., and MacDonald, G. H. F., "Attenuation of Small Amplitude Stress Waves in Solids," Geophysics, Vol. 34, No. 4, October, 1958.

5. Lawrence, H. W., and Baltosser, R. W., "Engineering Problems and Downhole Geophysical Solution," Circular 101 Exploration for Mineral Resources, June, 1969, New Mexico Bureau of Mines and Mineral Resources.

6. Lawrence, H. W., "Reflection, Refraction and Energy Mode Conversion as Seen on 3-D Velocity Logs," 35th Annual Meeting of the Society of Exploration Geophysicists, Dallas, Texas.

7. Morlier, P., "Spectre de Fissuration et Celerite des Ondes," Revue de L'Industrie Minerale, 15 Juillet, 1969.

BIBLIOGRAPHY (Continued)

8. Morris, R. L., Grine, D. R., and Arkfeld, T. E., "Using Compressional and Shear Acoustic Amplitudes for the Location of Fractures," Journal of Petroleum Technology, June, 1961, pp 623-632.

9. Pickett, G. R., and Reynolds, E. B., "Evaluation of Fractured Reservoirs," Society of Petroleum Engineers Journal, Vol. 9, No. 1, March, 1969.

10. Walker, Terry, and Riddle, George, "Field Investigation of Full Wave Acoustic Wave Recording," presented at Fourth Annual Logging Symposium, SPWLA, Oklahoma City, Oklahoma, 1963.

11. Zemanek, Joe, "The Borehole Televiewer - A New Logging Concept for Fracture Location and Other Types of Borehole Inspection," Mobile Research and Development Corporation, Field Research Laboratory, Dallas, Texas.

ENGINEERING PROPERTIES OF A PENNSYLVANIAN SHALE

By

GHOLAMREZA MESRI[1] and ROCCO GIBALA[2]

SUMMARY

Intact specimens of a Pennsylvanian shale were tested in one-dimensional compression, triaxial compression and direct shear. Drained and undrained triaxial compression tests with pore water pressure measurements were performed under constant rate of strain and constant load conditions. The anisotropic characteristics of the shale were investigated by the proper orientation of the consolidation and triaxial specimens with respect to the plane of laminations. Drained direct shear tests were performed on intact, pre-cut intact and remolded specimens. The shear specimens were subjected to large shear displacements by forward and backward motion of the shear box.

Contrary to expectations, the drained specimens experienced volume decrease and undrained specimens developed positive pore water pressures during shear and drained compressive strength was twice the undrained strength at all water contents. Drained and undrained triaxial tests indicate that compressive strength can be correlated with initial water content. Consolidation and triaxial tests show that the initial modulus in the direction of laminations is 2 to 3

---

[1] Asst. Prof. of Civ. Engrg., University of Illinois at Urbana, Urbana, Ill.

[2] Res. Asst., University of Illinois at Urbana, Urbana, Ill.

times the initial modulus normal to laminations. A comparison of initial moduli determined from various laboratory and field tests indicates that constant load drained triaxial test and one-dimensional consolidation test in which maximum drainage and creep is allowed give the lowest modulus values whereas pressure meter test which is undrained, short term, and measures modulus in the direction of laminations gives the highest modulus. Direct shear tests show that the residual shear strength which is measured immediately after peak strength does not necessarily correspond to the minimum residual shear resistance which is reached after large shear displacements. In case of the shale tested, the minimum residual shear resistance was not influenced by the method of remolding and remolded and pre-cut intact specimens gave the same residual friction angle.

GEOLOGICAL DESCRIPTION AND INDEX PROPERTIES OF THE SHALE

The shale specimens used in this investigation were cored at the Mississippi River crossing of Interstate 280 near Rock Island, Illinois, in connection with the design of piers for a continuous, multiple-span highway bridge. At this location alluvial deposits from the Mississippi River ranging from 4 to 40 ft in depth overlie Pennsylvanian period sedimentary rocks with thicknesses of 40 to 60 ft, which in turn overlie Middle Devonian period limestone. The Pennsylvanian rocks (Caseyville Formation) consist primarily of gray silty shale and dark-gray to black fissile shale, with lesser amounts of coal, underclay, sandy shale, siltstone, sandstone, and conglomerate.

The gray shales which were used in this investigation tend to be somewhat massive except for abundant plant material found along many bedding planes. Grain size analysis show compositions of 3 to 6 percent sand, 58 to 73 percent silt, and 23 to 39 percent clay (< 0.002 mm). Clay mineralogy analysis reveal the clays to be mainly illite and chlorite with lesser amounts of kaolinite and mixed-layer clay minerals. Typical Atterberg limits are liquid limit, 37 percent, and plastic limit, 22 percent. The shale has a tendency to crack upon exposure to air and to disaggregate in water or slake.

DESCRIPTION OF TESTING PROCEDURE AND TEST RESULTS

One-Dimensional Consolidation Tests

The one-dimensional consolidation tests were performed on undisturbed specimens of shale trimmed into 2.50-inch diameter by 0.750-inch high Wykeham Farrance consolidation rings. A pressure ratio of two was used. Stress-strain curves from one-dimensional consolidation tests are shown in Fig. 1. The consolidation specimens were trimmed such that in one specimen the plane of laminations was perpendicular to the axis of the ring, so that the consolidation pressure was applied normal to the plane of laminations, whereas in the second specimen the consolidation pressure was applied parallel to the plane of laminations. It is apparent that the initial modulus of the shale in the direction of laminations is three times its modulus normal to the plane of laminations. The stress-strain curve of the specimen loaded in the direction of laminations has a large hysteresis between the loading and unloading curves with almost two percent unrecoverable strain at zero stress. On the contrary the stress-strain curve in the direction of normal to the plan of laminations shows nearly elastic behavior, indicating that the shale had been preloaded in the

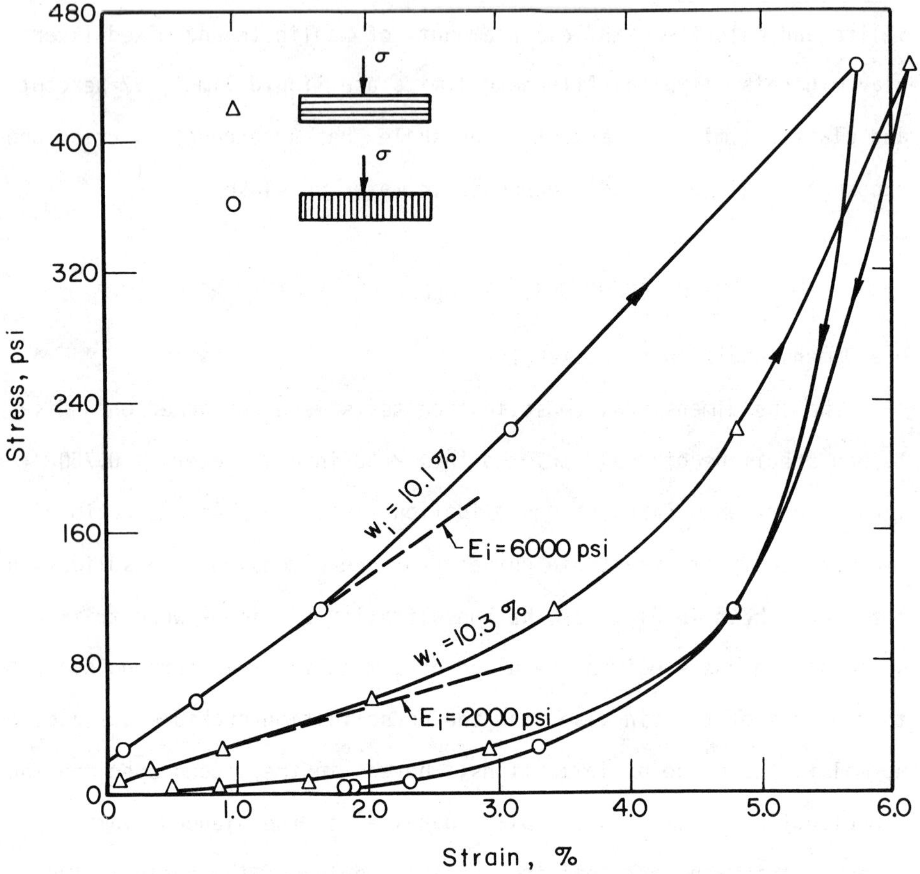

FIG. I STRESS-STRAIN CURVES FROM ONE-DIMENSIONAL CONSOLIDATION TESTS field to much higher loads than it was subjected to in this test. Also as expected, there is more swelling in the later case.

Triaxial Compression Tests

Three series of constant rate of strain triaxial compression tests were performed on intact shale specimens. Also several constant load triaxial compression tests were performed. All triaxial tests were performed in modified Wykeham Farrance triaxial cells (Mesri, 1969). The rate of strain was estimated by Blight's equations (Blight, 1963). All triaxial specimens failed by formation of a single or multiple shear planes.

The first two series of triaxial tests consisted of one series of consolidated-drained tests and one series of consolidated-undrained tests with pore water pressure measurement. The shale specimens 1.5 inches in diameter and 3.0 inches high were allowed to come to equilibrium under an all around confining pressure (40 or 80 psi) before they were subjected to shear stress. The undrained tests were back pressured by 40 psi before shear and in all cases a B-coefficient (Skempton, 1954) close to one was obtained under the last increment of back pressure. The rate of strain for drained and undrained tests were 1×10^{-5} in./min and 2×10^{-4} in./min, respectively.

Typical stress-strain curves from triaxial compression tests are presented in Fig. 2. Contrary to expectations, the drained shale specimens experienced volume decrease during shear and undrained specimens developed positive pore water pressures. The drained stress-strain curves had a sharp peak and a corresponding minimum on volumetric strain curve at the same axial strain whereas the undrained

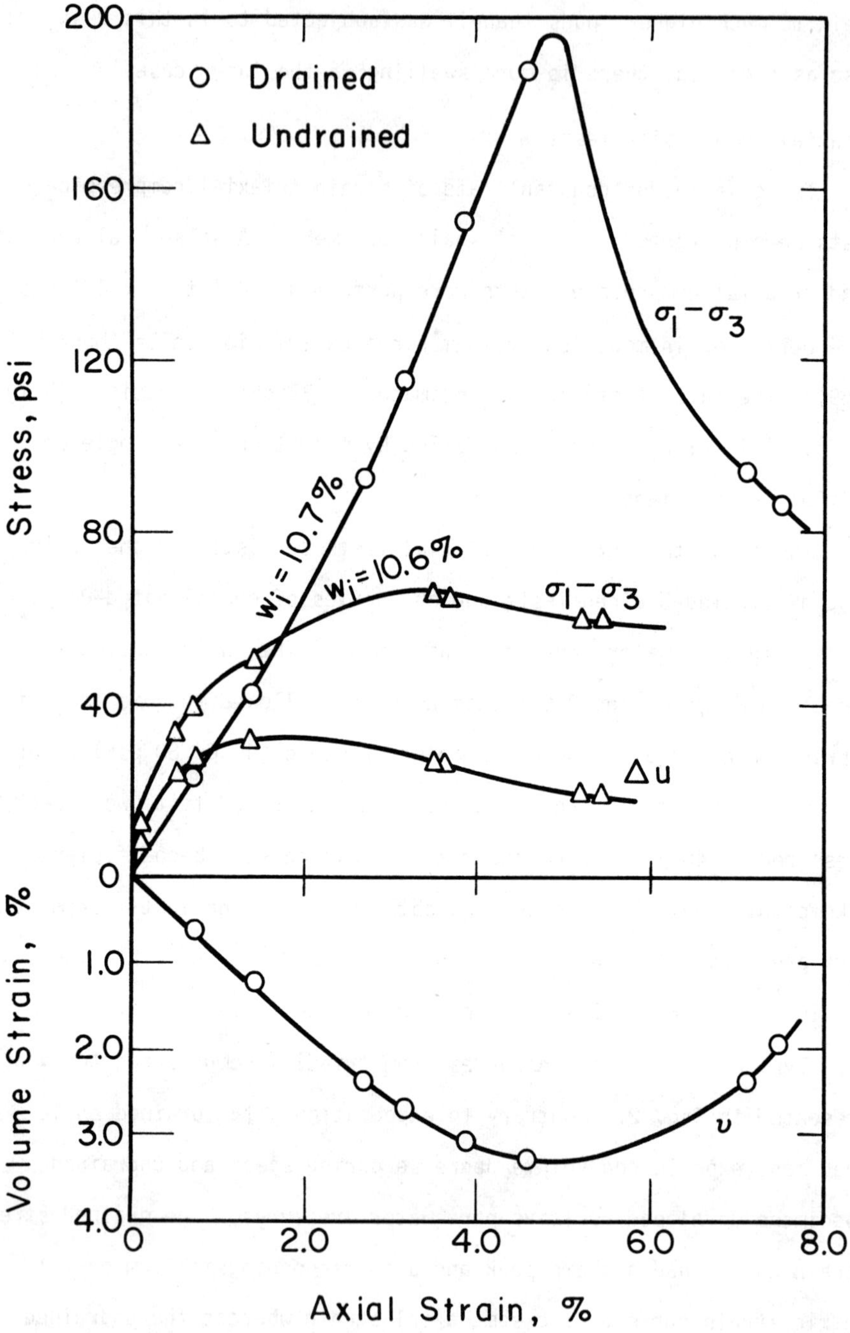

FIG. 2 STRESS–STRAIN CURVES FROM TRIAXIAL COMPRESSION TESTS

specimens did not show a peak stress and were much weaker than drained specimens.

One consolidated-drained test was performed on a specimen which was trimmed such that the plane of laminations was parallel to the axis of the specimen. The stress-strain curve of this specimen is shown in Fig. 3 and is compared with stress-strain curve of a similar sample which had been trimmed with the plane of laminations normal to the axis of the specimen. It is apparent that as in the case of one-dimensional consolidation, the shale is less compressible in the direction of laminations.

One series of unconsolidated-undrained triaxial compression tests were performed on 5.5 inch diameter by 8.0 inch high specimens under 10 psi all around pressure. The rate of strain was 4×10^{-2} in./min. The behavior of these specimens were more or less similar to the behavior of consolidated-undrained specimens.

Direct Shear Tests

Two series of drained direct shear tests were performed on shale specimens prepared in various ways. One series of tests were performed in a special direct shear machine (Coulson, 1970) using intact specimens 6 inches long and 2 inches wide. The rate of shear displacement was 4×10^{-2} in./min and these tests are believed to be drained. The peak and residual shear strength from these tests are summarized in Fig. 4. As is expected peak strength data exhibit appreciable scatter. But the friction angle of 29 degrees which is indicated from these data compares reasonably with the range of 29 to 32 degrees which is estimated from triaxial tests. In the first series of direct shear tests the residual strength was defined in the conventional way as the

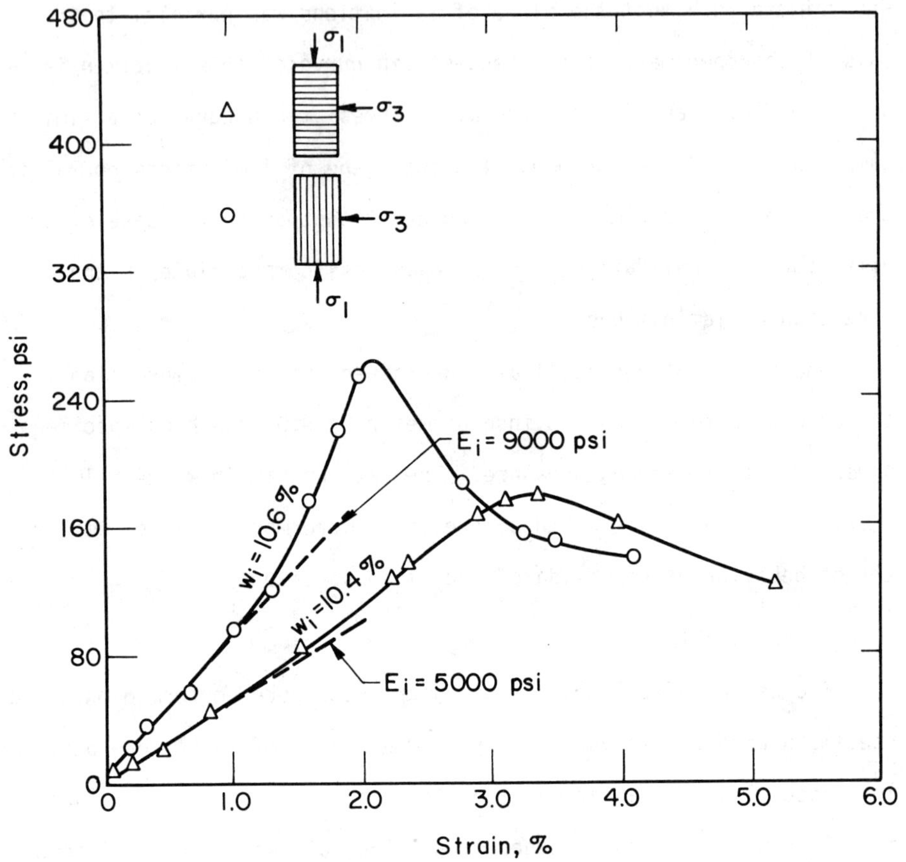

FIG. 3 STRESS-STRAIN CURVES FROM DRAINED TRIAXIAL COMPRESSION TESTS

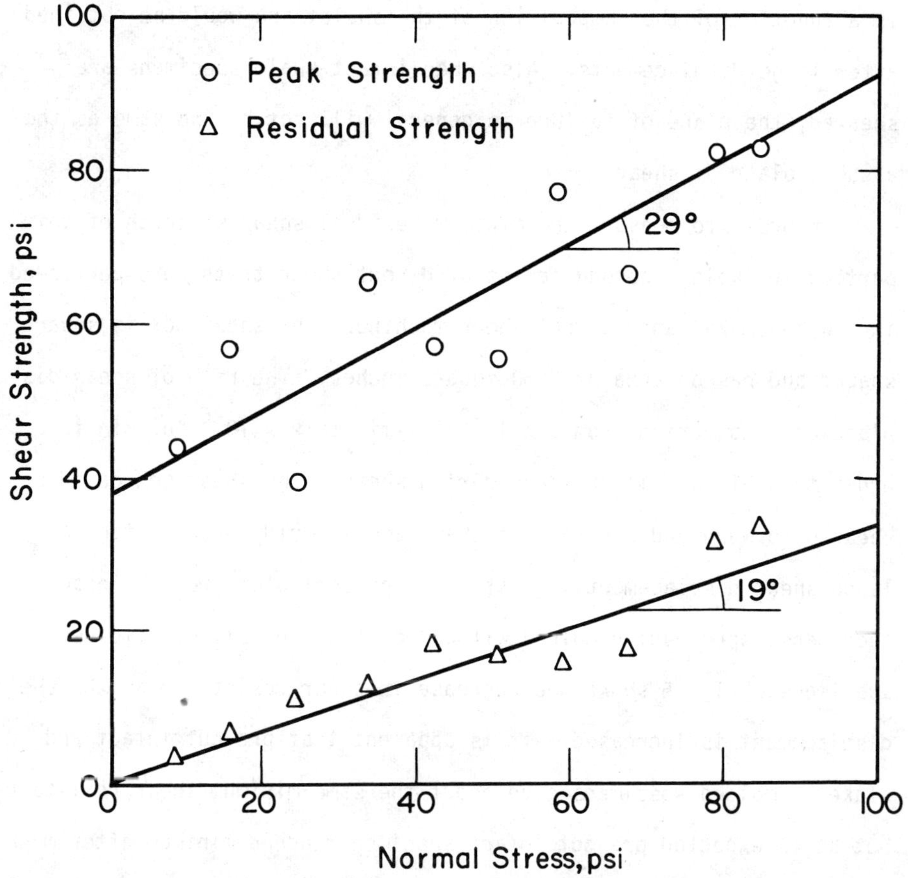

FIG. 4 DIRECT SHEAR TESTS ON INTACT SPECIMENS

constant minimum shear stress mobilized after peak stress. The shear strength defined in this manner which is reached at shear deformations less than 1/4 inch does not represent the minimum shear strength which is a function of the composition of the shale, and would be obtained after large displacements. Also when intact shale specimens are sheared, the plane of failure in general will not be the same as the assumed plane of shear.

In order to measure the minimum residual shear strength of this particular shale a second series of direct shear tests were performed in a Wykeham Farrance direct shear machine. The shear box is square shaped and has an area of 5.50 square inches. The rate of shear displacement was varied from 2×10^{-3} in./min to 2×10^{-5} in./min in order to achieve full drainage during shear. The shear box could be sheared forward and backward so that sample could be subjected to large shear displacements. Tests were performed on pre-cut intact specimens, specimens remolded without drying, and slaked-remolded specimens. Fig. 5 shows the decrease in shear resistance as the shear displacement is increased. It is apparent that pre-cut intact and slaked remolded specimens give about the same minimum shear resistance, but as is expected pre-cut intact specimen reached minimum after much smaller shear displacements. The residual shear strength from three different specimens gave a friction angle of 11 degrees, Fig. 6.

DISCUSSION OF TEST RESULTS

Clay shale is a predominant sedimentary rock in the earth's crust and because of its unique characteristics could play important role in the design and construction of many engineering structures, e.g.

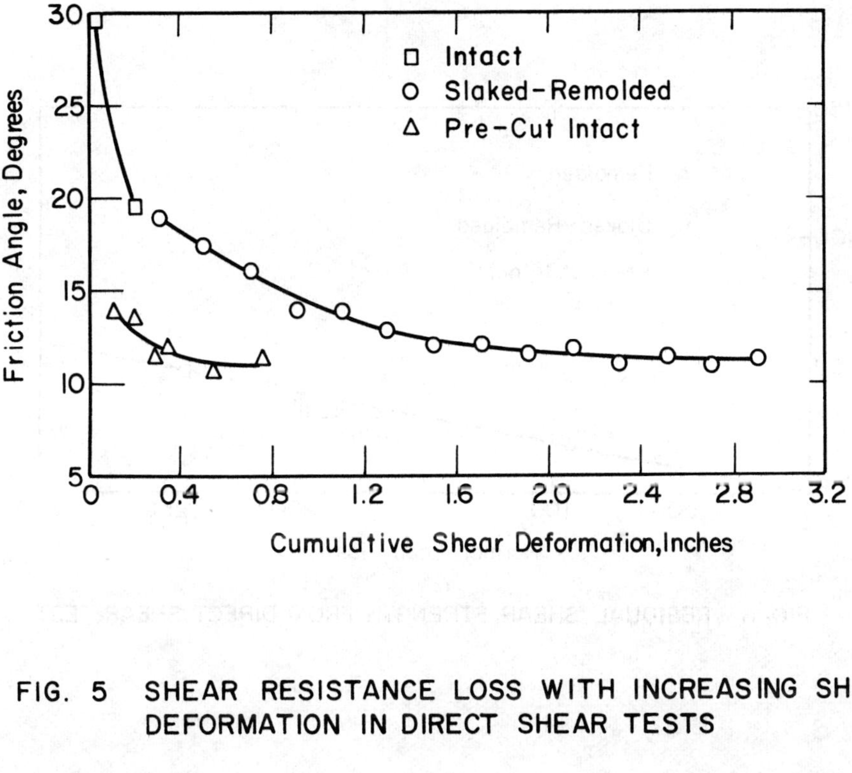

FIG. 5 SHEAR RESISTANCE LOSS WITH INCREASING SHEAR DEFORMATION IN DIRECT SHEAR TESTS

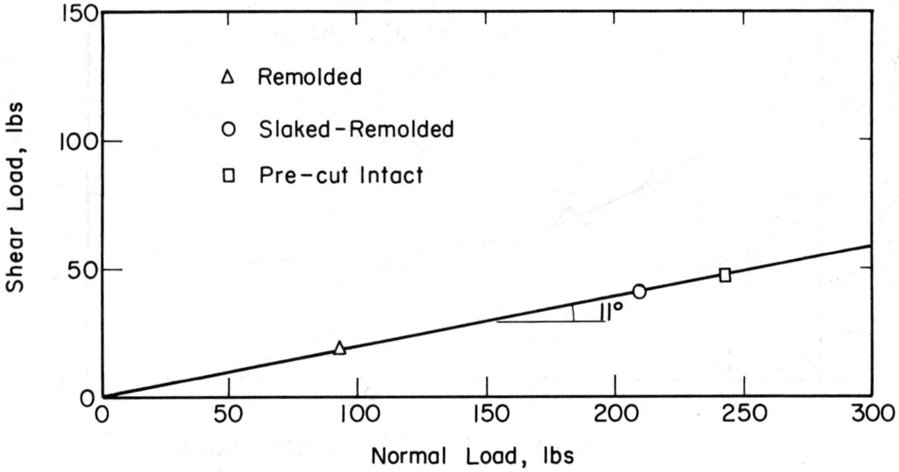

FIG. 6 RESIDUAL SHEAR STRENGTH FROM DIRECT SHEAR TEST

Panama Canal (Binger and Thompson, 1949), Fort Peck Dam (Middlebrooks, 1942), Oahe Dam (Knight, 1963) and numerous others. The engineering behavior of shales can be explained in terms of their composition, particularly mineralogy of clay content, and in terms of the chemical and geophysical conditions under which they were formed and to which they were subjected since their formation. Two important features of clay minerals which distinguish them from other rock forming minerals are their platy shape which give anisotropic properties to shales and the manner in which they interact with water which makes the shales susceptible to weathering and drastic changes in their engineering properties upon interaction with water. Because of these factors most shales are more compressible in the direction normal to the plane of laminations than in the direction parallel to it. Also along the laminations or along any plane of shear on which large shear displacements have resulted in clay particle orientation, the shear resistance is rather low. Also shales will expand upon unloading and exposure to moisture, particularly in the direction normal to laminations.

This investigation was not intended to be a basic study of the mechanisms controlling load-deformation properties of shales, rather routine laboratory tests were used to determine the engineering properties of a particular shale.

An attempt was made to correlate compressive strength with initial water content, Fig. 7. As is expected from natural shale specimens, there is a substantial scatter, but at a given water content the drained compressive strength is much higher than undrained compressive strength. It is possible that shale specimens were badly disturbed upon sampling.

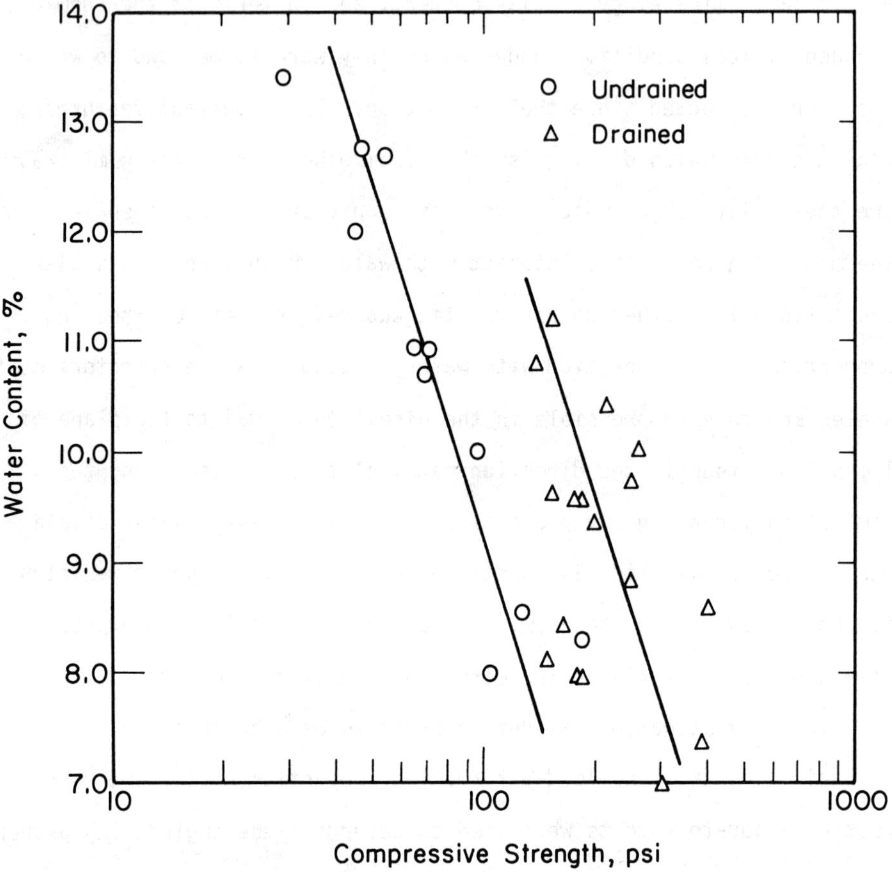

FIG. 7 RELATIONSHIP BETWEEN COMPRESSIVE STRENGTH AND WATER CONTENT FROM TRIAXIAL TESTS

One important purpose of this investigation was to compare the deformation modulus of the shale as it is determined by various testing procedures. The initial deformation modulus versus initial water content are shown in Fig. 8. Since field pressure meter tests are being increasingly used to determine deformation modulus (Hendron et al., 1969), pressure meter tests are included for comparison. In spite of the expected large scatter it is apparent that constant load triaxial test and one-dimensional consolidation test result in the lowest moduli and the field pressure meter tests give the highest values. This is not surprising since pressure meter is undrained, short term, and measures modulus in the direction of laminations. Regardless of drainage conditions the duration of the loading plays an important role since a major portion of the deformation is of creep nature. The importance of the direction of loading is demonstrated by the consolidation and triaxial tests performed in such a way that compression loads were applied parallel to the plane of laminations and are indicated by stars on Fig. 8.

Attempts have been made to correlate deformation modulus with strength of soils and rocks (Wilson and Dietrich, 1960; Deere and Miller, 1965). A plot of drained and undrained moduli versus corresponding compressive strength is shown in Fig. 9. The relationship shown indicates that the ratio of undrained modulus to the undrained compressive strength is 70:1 and the corresponding ratio for drained values is 25:1.

It is well known (Skempton, 1964) that long term stability of highly overconsolidated clays and clay shales are controlled by the residual shear strength. But there appears to be no widely accepted

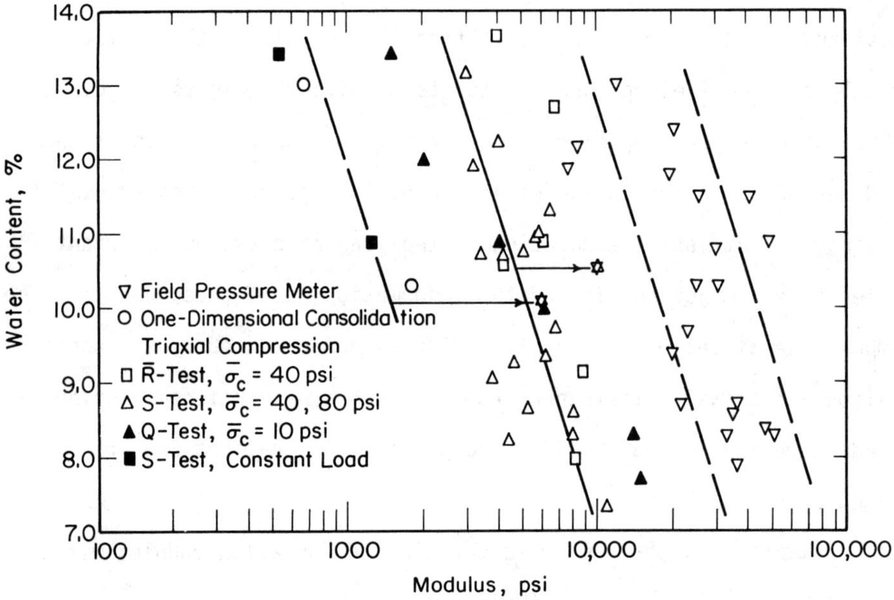

FIG. 8 INITIAL MODULUS versus WATER CONTENT FROM VARIES TESTS

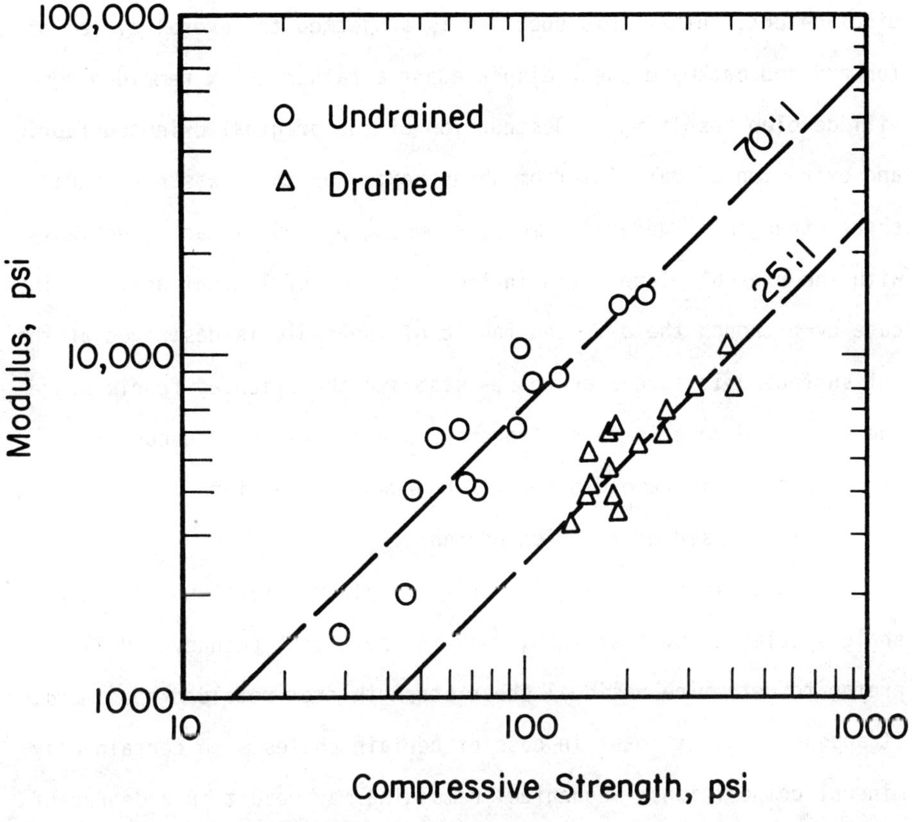

FIG. 9 DEFORMATION MODULUS versus COMPRESSIVE STRENGTH

procedure for the measurement of the residual shear. When an intact specimen is used to measure residual shear strength by subjecting it to large shear displacements, generally the first plane of shear which develops at peak stress does not coincide with the plane of shear box displacement. After this specimen is subjected to several cycles of forward and backward shear displacement a rather thick remolded zone will develop resulting in destruction of the original oriented fabric and extrusion of material from shear box. For this reason residual shear strength is generally measured using pre-cut intact specimens with the pre-cut plane being in the direction of laminations. In this case even though the oriented fabric of the shale is destroyed at the cut surfaces, it is easier to re-establish the oriented fabric since the disturbed zone is thin, therefore shear stresses are concentrated and also the thin remolded zone is bounded by flat intact shale which is itself composed of an oriented fabric.

Since it is generally difficult to obtain intact and undisturbed shale specimens and test them, many investigators (Kenney, 1967) prefer to determine residual shear strength from remolded specimens. It appears that at least in case of certain shales with certain clay mineral compositions, mechanical remolding may result in a degree of particle and fabric disaggregation which does not correspond to the remolding action of shear stresses in the field. Therefore the residual shear strength measured in this way may be a function of the method of remolding. It turns out that in case of this particular shale the measured shear resistance was not influenced by the method of remolding and remolded and pre-cut intact specimens gave the same minimum residual friction angle, Fig. 6.

REFERENCES

Binger, W. V. and T. F. Thompson (1949), "Excavation Slopes, Panama Canal," Trans. ASCE, Vol. 114, pp. 734-754.

Blight, G. E. (1963), "The Effect of Non-Uniform Pore Pressure on Laboratory Measurements of Shear Strength of Soils," ASTM Special Tech. Pub. No. 361, pp. 173-191.

Coulson, J. H. (1970), "The Effects of Surface Roughness on the Shear Strength of Joints in Rock," Ph.D. Thesis, University of Illinois.

Deere, D. U. and R. P. Miller (1965), "Classification and Index Properties of Intact Rock," AFWL-TR-65-116, Air Force Special Weapons Center, Kirtland AFB, Albuquerque, N.M.

Hendron, A. J., G. Mesri, J. C. Gamble, and G. Way (196), "Compressibility Characteristics of Shales Measured by Laboratory and In Situ Tests," ASTM Special Tech. Publ. No. 477, pp. 137-153.

Kenney, T. C. (1967), "The Influence of Mineral Composition on the Residual Strength of Natural Soils," Proc. of Geotechnical Conf., Oslo, pp. 123-129.

Knight, D. K. (1963), "Oahe Dam: Geology, Embankment, and Cut Slopes," Proc. ASCE, Vol. 89, SM2, pp. 99-125.

Mesri, G. (1969), "Engineering Properties of Montmorillonite," Ph.D. Thesis, University of Illinois.

Middlebrooks, T. A. (1942), "Fort Peck Slide," Trans. ASCE, Vol. 107, pp. 723-764.

Skempton, A. W. (1954), "The Pore Pressure Coefficients A and B," Geotechnique, December, pp. 143-152.

Skempton, A. W. (1964), "Long-term Stability of Clay Slopes," Geotechnique, Vol. 14, pp. 77-101.

Wilson, S. D. and R. J. Dietrich (1960), "Effect of Consolidation Pressure on Elastic and Strength Properties of Clay," ASCE Conf. on Shear Strength of Cohesive Soils, Boulder, Col., pp. 419-435.

SHEAR STRENGTH OF FLAT SURFACES IN ROCK

By James H. Coulson,[1]

INTRODUCTION

Each discontinuity, or joint, in a rock mass derives its shear strength from two sources: (1) the frictional resistance offered by two smooth surfaces sliding relative to one another, and (2) the resistance to sliding imposed by the geometry of irregularities on the surface of sliding. On natural rock surfaces sliding can be initiated by riding over surface asperities, or by shearing them off, depending upon the geometry of the opposing surfaces, the normal force, and the strength of the intact rock. Either method of initiating movement requires more energy than that needed for sliding on smooth flat surfaces.

The relationships for the geometrical component of shear strength have been discussed by Newland and Allely (21), Ripley and Lee (26), Withers (35), and Patton (23). The objective of this investigation is to define the basic shear strength values for smooth, flat surfaces in rock, to which the geometrical component of shear strength, as defined by field mapping or core studies, can be added to obtain the shear strength along natural joints.

Tests were performed on flat rock surfaces prepared in the laboratory. Because of the random nature of small asperities on a finely ground surface, the artificial "micro-roughness" of such a

[1]Civil Engineer, Civil Design Branch, Tennessee Valley Authority, Knoxville, Tennessee 37902

surface lacks the intimate interlocking required to develop the geometrical component of shear strength, and the basic frictional behavior of the rock can be studied independently. To determine the effects of the micro-roughness itself, flat surfaces on 10 rock types with roughnesses ranging from nearly polished to sandblasted were sheared wet and dry at normal pressures from 10 to 1000 psi.

DESCRIPTION OF ROCK TYPES AND TESTING APPARATUS

Rock Types.--Index properties for each of the 10 rock types used in this testing program are given in Table 1.

Table 1
Index Properties of Rock Types Tested

Rock Type	γ, pcf	q_u, psi	E_t, psi x 10^6	R	V_p, fps	A, %
Lower Granite Basalt	170.2	33,000	7.28	57	17,300	1.92
Oneota Dolomite	153.0	12,600	6.38	43	16,300	5.39
Bedford Limestone	137.7	7,400	4.14	33	12,800	8.14
Solenhofen Limestone	163.6	35,500	9.26	54	18,900	1.34
Fine-Grained Grand Coulee Granite	160.5	28,200	7.96	53	15,200	0.39
Coarse-Grained Grand Coulee Granite	164.0	23,800	7.60	52	13,400	0.31
Berea Sandstone	136.2	10,700	2.80	42	8,700	8.27
Navajo Sandstone	125.8	6,300	2.22	30	8,400	11.63
Schistose Gneiss	174.4	13,450	2.90	35	9,000	0.17
Hackensack Siltstone	162.0	17,800	3.81	47	13,100	0.65

Notes: γ = Unit weight, pcf
q_u = Unconfined compressive strength, psi
E_t = Tangent modulus of elasticity at 50% of ultimate strength, psi
R = Schmidt rebound hardness
V_p = Sonic velocity at 100 psi axial stress, fps
A = Absorption, % of dry weight

Specimens were manufactured from 6-inch lengths of NX core by sawing the core along its axis. After surface preparation the two half cylinders for each specimen were cast into individual shear boxes with quick setting grout.

Testing Apparatus.--The direct shear machine designed for this study is depicted schematically in Fig. 1. The machine was capable of applying normal forces from 60 to 15,000 lb., and it allowed shear displacements of up to 3 in. During a test the lower half of the specimen was pushed from under the upper half, while the upper half was restrained from shear displacement by two tie bars which transmitted the shear force to load cells. With this system the load cells measured only the shear strength of the surface of sliding, and no correction was necessary for machine friction. The shear strength and displacement were monitored continuously during each test with an X-Y plotter.

Each specimen was sheared 2 in. under the specified normal pressure. If residual shear strength was not reached during the first shear run, successive runs were made until the residual condition was reached. All specimens were sheared either oven dried or submerged in water.

SURFACE ROUGHNESS

Measurement.--Surface roughness as a function of rock type and surface preparation was obtained with a Brush Surfanalyzer 1200 System. True profiles and arithmetic average roughness values were plotted for surfaces lapped with #600 silicon carbide grit and then polished with tin oxide, surfaces lapped with #600 silicon carbide grit, surfaces lapped with #400 silicon carbide grit, surfaces lapped with #80

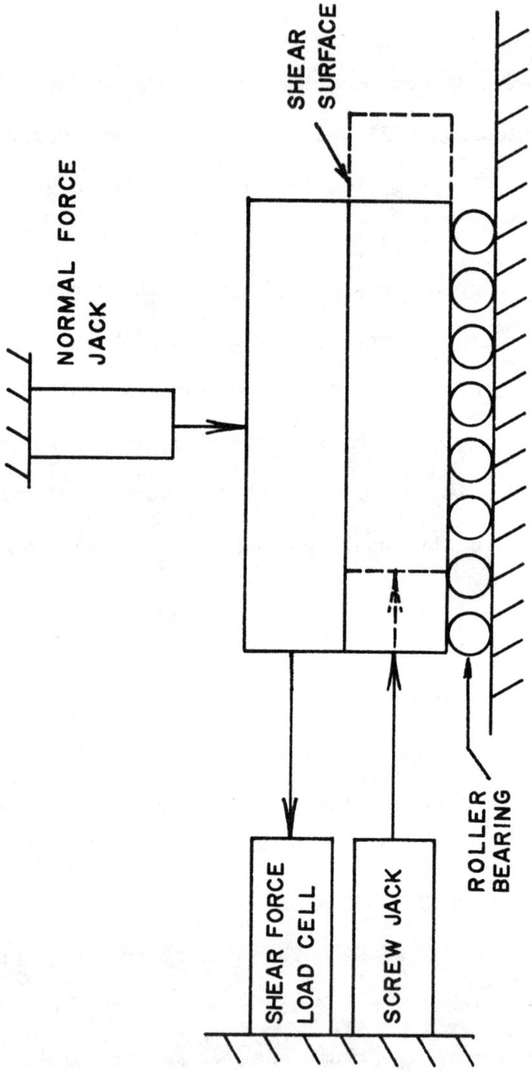

NOTE: In actuality shear force load cell and screw jack act in same plane as shear surface.

FIG. 1.--SCHEMATIC OF DIRECT SHEAR MACHINE

silicon carbide grit, and surfaces lapped with #80 silicon carbide grit and then sandblasted with Ottawa sand.

Profiles.--In general, surface profiles fall into two groups according to rock type. Group A rocks - Lower Granite basalt, Solenhofen limestone, coarse- and fine-grained Grand Coulee granite, schistose gneiss, and Hackensack siltstone - are nonporous, have unconfined compressive strengths greater than 13,400 psi, and have absorption values less than 2%. Group B rocks - Oneota dolomite, Bedford limestone, Berea sandstone, and Navajo sandstone - exhibit open surface pores, have unconfined compressive strengths less than 12,600 psi, and have absorptions greater than 5.0%.

Representative profiles from each group are illustrated in Fig. 2. Lapped surfaces on Lower Granite basalt, from Group A, were characterized by a uniform texture with concordant asperities spaced at regular intervals. The texture of the sandblasted surface was similar to that of the #80 surface except that asperities were larger and no longer concordant. Polished surfaces tended to be rougher than #600 surfaces because of the ability of the flexible corduroy polishing pad to round asperities and "wipeout" slightly less resistant areas. Grain boundaries and mineralogy had no influence on profiles on Group A rocks.

Profiles for Bedford limestone, representative of Group B rocks, show the effects of porosity. The #600, #400, and polished surfaces had the appearance of a series of concordant truncated peaks between major valleys created by surface pores. The #80 grit was large enough to erode pore walls and increase the distance between concordant peaks. Individual grains were visible only on the Berea and Navajo sandstones. Grains were lapped flat with #400 and #600 grits on Berea sandstone, but the #80 grit was large enough to dislodge grains at topographic

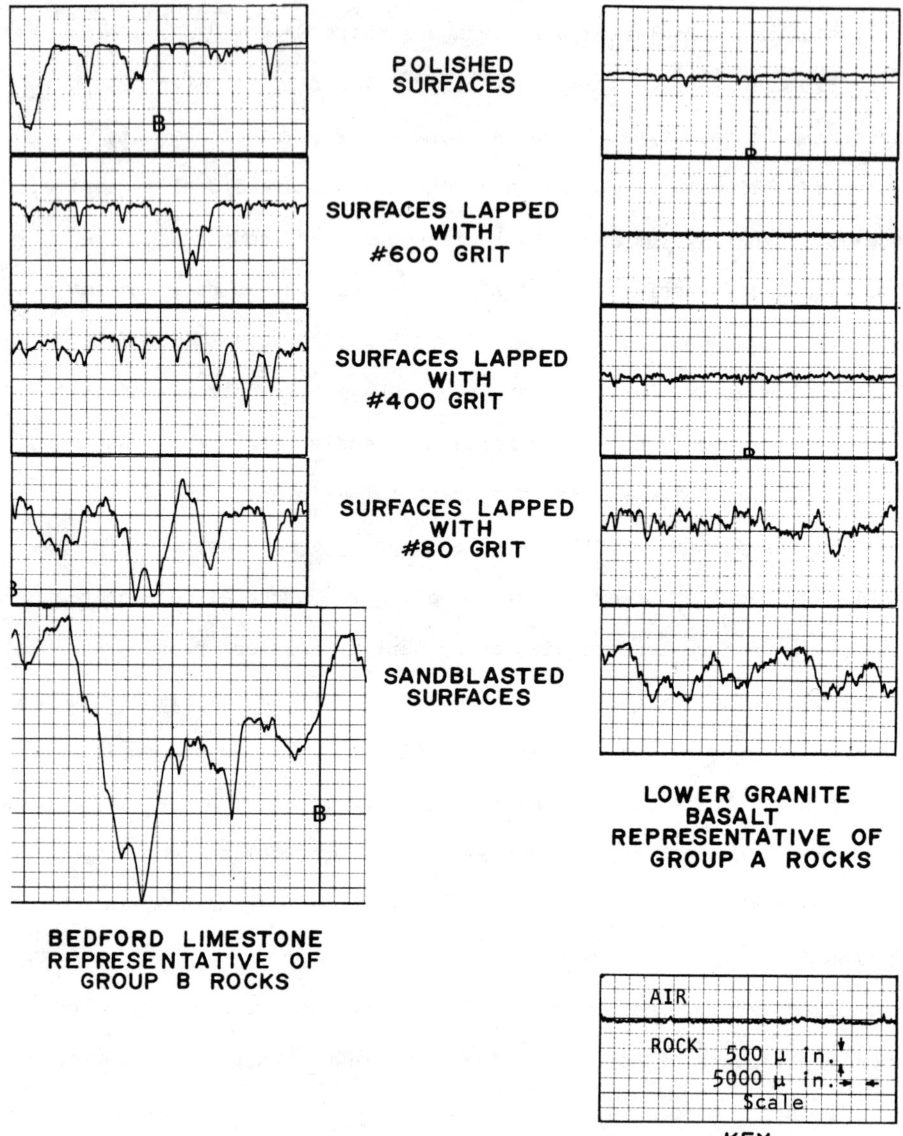

FIG. 2.--SURFACE ROUGHNESSES COMPARED FOR GROUP A AND GROUP B ROCKS

highs rather than lap them. The friable Navajo sandstone disintegrated during sandblasting and could not be lapped with grits smaller then #80.

On the basis of the roughness profiles the #600, #80, and sandblasted surfaces were chosen as the standards for this testing program. The #600 surfaces were chosen rather than polished surfaces because they were the flattest, most uniform, and often the smoothest surfaces. The #400 surface preparation was eliminated as being essentially the same as the #600. The #80 surfaces provided a coarse but flat surface, and sandblasting provided a very coarse surface texture superimposed upon random waviness.

SURFACE DAMAGE

The following surface damage features were observed after shearing: (1) rock flour, either loose or compacted, (2) gouge zones, (3) polished areas, and (4) indurated crusts. All surfaces suffered some degree of damage during shearing. In general damage increased with normal pressure, and often two types of damage were evident.

Rock Flour.--In all cases rock flour consisted of small discrete particles. Each particle could be separated from its neighbors under a microscope, and no fusion between particles was evident even when they were highly compacted in gouge zones. Byerlee (7, 8) observed compacted rock flour on Westerly granite sheared in a triaxial cell at normal pressures up to 9 kilobars. Even at these high pressures he found that debris particles were distinct, optically continuous, angular, and appeared to be the result of tensile failures. Particles ranged in size from 0.1 mm to 1 micron. Rock flour for all surface preparations on all 10 rock types consisted of particles similar to those described by Byerlee.

Rock flour was generated on all rock types at all normal pressures and for all surface preparations, and occurred alone as well as in association with gouging and polishing. The only surfaces which had no rock flour were those sheared wet on Oneota dolomite and Bedford limestone where an indurated crust developed.

Gouge Zones.--For Hackensack siltstone, Navajo sandstone, Berea sandstone, Solenhofen limestone, Bedford limestone, and Oneota dolomite gouging began at normal pressures from 80 to 250 psi and became more intense as normal pressure was increased. For the remainder of the rock types, except the fine-grained Grand Coulee granite which exhibited no gouging, normal pressures in excess of 500 psi were required to develop significant gouging. For schistose gneiss, Navajo sandstone, and Berea sandstone gouging was equally well developed on wet and dry surfaces. For the remainder of the rock types gouging was more advanced on dry surfaces.

Figure 3 shows a gouge trough developed on a sandblasted Solenhofen limestone surface sheared dry under a normal pressure of 1177 psi. Although gouge troughs varied in size, all exhibited the characteristics of the one shown in Fig. 3. The surface shown in the photos moved in the direction of the arrow while its mate above remained stationary.

Part A of the figure shows the gouge trough filled with fine, highly compacted and slickensided rock flour as it appeared after the test. Part B shows the same gouge trough after removing the compacted rock flour with a jet of water. The gouge trough ranged in depth from about 0.0045 in. to 0.0225 in., and steps in the bottom of the trough dipped from 17-1/2° to 21°, averaging about 20°. Part C of Fig. 3 is a schematic cross section of the features shown in the photos. It shows relative displacement and orientation of the damage features.

a) Gouge zone filled with compacted debris. Note chatter marks.

b) Base of gouge zone after blasting with a jet of water to remove compacted debris.

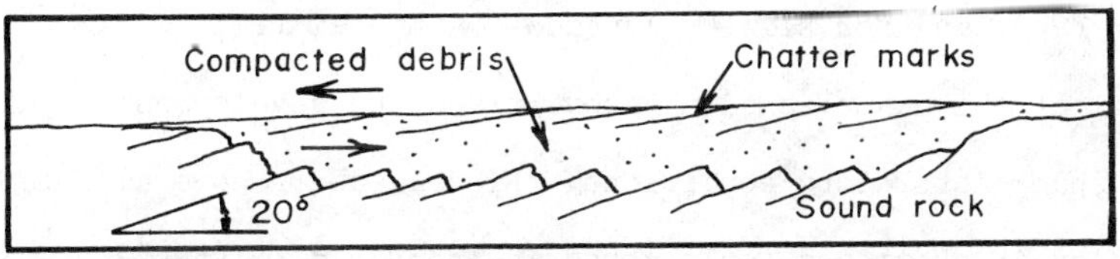

c) Schematic cross section of gouge zone.

FIG. 3.--CHARACTERISTICS OF A TYPICAL GOUGE ZONE. SOLENHOFEN LIMESTONE SHEARED DRY UNDER 1177 PSI NORMAL PRESSURE

Both the chatter marks in the rock flour and the step-like features in the bottom of the gouge trough were oriented so that shear across them was in the rough direction. In other words, if one were to run his finger over the surface in the direction of shear, he would be moving in the rough direction on the slickensides. This is contrary to previous references in structural geology which state that the direction of movement along slickensides is in the smooth direction in order not to grind off the step-like features (Billings, 1; Lahee, 16).

Failure planes of the same orientation can be produced by the Closs-Riedel experiment where a layer of clay is placed on two boards and sheared by sliding one board relative to the other. These experiments were first performed by H. Cloos (10) and Riedel (25), and the resulting en-echelon "Riedel shears" were oriented as shown for the gouge trough in Fig. 3. Similar experiments have been reported by Lee, Chen, and Lee (18), Blyth (2), E. Cloos (9), Skempton (30), and Morganstern and Tchalenko (20).

Step-like features oriented so as to oppose the direction of shear were reported for rock surfaces by Paterson (22), Jaeger (14), and Riecker (24), and for plaster surfaces by Lajtai (17).

All of the examples cited above trace the development of Riedel shears under laboratory conditions. Riedel shears have also been identified in shear and fault zones in the field, Blyth (2), Tjia (33), Skempton (30), and Tchalenko (31).

There is a distinct difference in the formation of Riedel shears in the Cloos-Riedel experiment and the development of inclined shear planes in gouge zones generated during the course of this testing program. In the Cloos-Riedel experiment, Riedel shears form first, followed by shears parallel to the direction of slip which coalesce to form the principal slip surface. The Riedel-like features formed in gouge zones are secondary features, the prepared surface of sliding being the primary surface. However, because of the close relationship of the gouge zone features and the shears generated in the Cloos-Riedel experiment, the shears in the gouge zones can be positively identified as Riedel shears, and probably form by the same mechanism. It would appear that at least on a local basis the prepared rock surfaces are pressed together with such intimacy that the rock behaves as an intact mass, and it becomes easier for Riedel shears to form than for slip to continue on the pre-existing surface. If this is indeed the case, the Riedel shears once again can be considered primary features.

It is generally agreed that the dip of the Riedel shears with respect to the surface of sliding is $\phi/2$, where ϕ is the angle of internal friction of the rock. On the dry Bedford limestone surfaces sheared under a normal pressure of 1000 psi, the average dip of the Riedel shears was approximately 10°. The apparent ϕ for the limestone was then 20°. Triaxial experiments by Schwartz (28) show the angle of internal friction for Bedford limestone to decrease from 28° to 14° as confining pressure increases from 1000 to 3000 psi, so ϕ predicted from the orientation of Riedel shears is in general agreement with ϕ from the triaxial test data.

A similar situation is evident on Solenhofen limestone with an angle of internal friction of approximately 40° (Robertson, 27; Handin, Heard, and Magouirk, 12). Riedel shears had an average dip of 20°.

Indurated Crust.--Surface damage in the form of an indurated crust was recognizable only on Oneota dolomite and Bedford limestone surfaces sheared wet. It appeared as a continuous, thin, amorphous crust which cracked upon drying after the test. The indurated crust replaced the deep gouging formed during dry shear on these two rock types.

Boozer, Hiller, and Serdengecti (3) postulated that the adsorption of water by calcite crystals in Bedford limestone decreases the surface energy and strength of the individual crystals. They observed twinning and intracrystalline gliding extending to a depth of several grain diameters below shear surfaces generated in a triaxial cell. For the Bedford limestone and Oneota dolomite sheared wet during this testing program, internal deformation of crystals near the surface of sliding, and recrystallization of debris particles, lead to the development of the indurated crust at the expense of deep gouging and the formation of Riedel shears.

Surface Polish.--Except for the Oneota dolomite, for which only dry surfaces were polished, all polishing took place on rocks containing a significant percentage of minerals with Mohs' hardness of 6-1/2 or 7. The calcite rocks, Bedford limestone and Solenhofen limestone, and the Hackensack siltstone with its clay mineral content did not polish. Polishing occurred on wet and dry #600 surfaces at all normal pressures.

In summarizing the data for polished metallic surfaces, Bowden and Tabor (5) show that the chief mechanism involved is surface flow produced by frictional heating. Hardness apparently plays a minor role in the polishing of metals. When two metals of different hardnesses are rubbed together, the one with the lowest melting point will polish.

Many of the rock surfaces which polished during shear had an extremely fine powder on them, indicating brittle behavior and mechanical abrasion on an extremely fine scale. If polishing is indeed an abrasion process, mineral hardness should be of some importance.

This may explain why Solenhofen limestone, Mohs' hardness 3, did not polish while quartz-rich rocks approaching a hardness of 7 did. If the flow mechanism was paramount in polishing rock, one would expect limestone with its melting point of 800° C to polish before quartz which melts at about 1500° C.

SHEAR STRENGTH VERSUS DISPLACEMENT

Curve Type.--Each of the X-Y recorder plots of shear strength versus displacement can be characterized by one of the curves in Fig. 4. The Type 1 curve, which shows no change in shear strength with

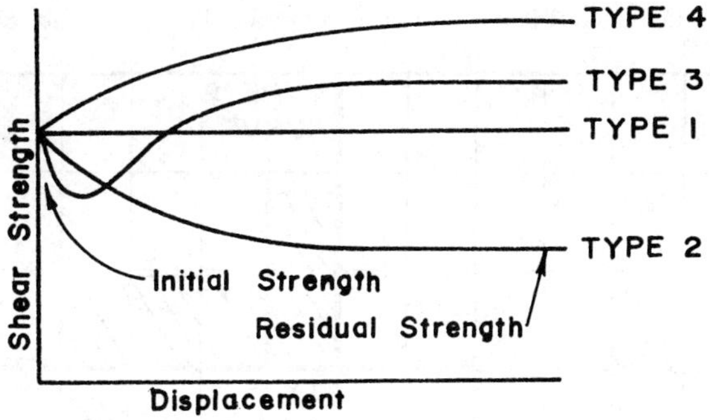

FIG. 4.--TYPICAL SHEAR STRENGTH VERSUS DISPLACEMENT CURVES

displacement, occurred for only 3.9% of the tests, always at normal pressure less than 60 psi. The Type 2 curve, for which shear strength decreases with displacement, occurred at low normal pressures on dry surfaces, and was common over the whole range of normal pressures from 10 to 1000 psi for tests on wet surfaces. For Type 3 and Type 4 curves which encompass over 60% of the tests, residual shear strength was greater than initial shear strength. These two curve types appeared at intermediate to high normal pressures for both wet and dry surfaces.

In general residual strengths were reached after less than 5 cm displacement for tests at normal pressures greater than 400 psi. However, at lower normal pressures Bedford limestone, Solenhofen limestone, fine-grained Grand Coulee granite, coarse-grained Grand Coulee granite, and Berea sandstone required up to 20 cm displacement before a residual condition was reached. The type of shear strength - displacement curve and displacements required for residual strength are given for individual tests in Appendix I.

Correlation of Curve Type and Surface Damage.--Correlation of shear strength - displacement curves with surface damage provides insight into the effectiveness of the various surface damage processes in either increasing or decreasing shear strength. The bar diagrams in Fig. 5 show the distribution of curve types with surface damage.

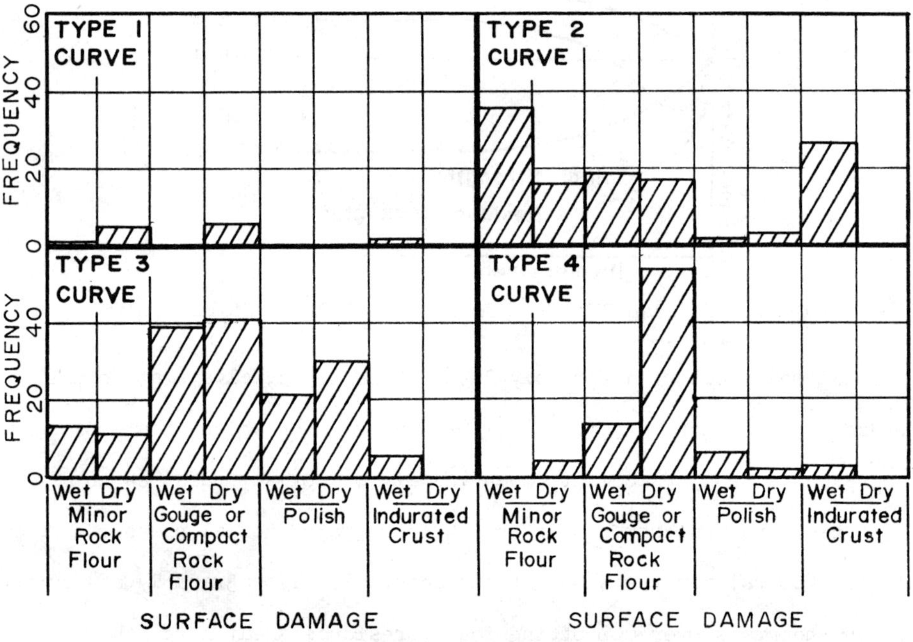

FIG. 5.--CORRELATION OF SURFACE DAMAGE WITH THE SHAPE OF THE SHEAR STRENGTH VERSUS DISPLACEMENT CURVE

The generation of small amounts of rock flour was most often associated with Type 2 curves. Since only a limited amount of debris was generated, particle interference was low, and to some extent rolling friction acted to decrease strength. As more rock flour was generated particle interaction was augmented and the Type 3 curve appeared as an extension of the Type 2 curve. Residual strength was obtained when particle interference reached its maximum and the rock surfaces were completely separated by a highly compacted layer of rock flour.

Gouging and the formation of Riedel shears gave rise chiefly to Type 3 and Type 4 curves. Thus, shear strength generally increased with the development of advanced surface interaction and gouging.

Polishing on both wet and dry surfaces produced Type 3 curves. The initial dip in the Type 3 curve probably occurred with the rapid development of powder as tiny surface asperities were sheared off. Shear strength then increased with further displacement as polishing progressed and the real area of contact increased. Residual strength which was also the maximum strength, was reached when polishing had brought the surfaces into contact over the maximum possible area.

The development of the indurated crust was almost entirely associated with decreasing shear strength with displacement, Type 2 curves.

SHEAR STRENGTH

Initial and residual coefficients of friction ranged from 0.5 to 0.9 (equivalent angles of friction 26° to 42°) for nearly all tests on sandblasted and #80 surfaces. Values ranged from approximately 0.4 to 0.8 (22° to 38°) for #600 surfaces. The calcite rocks, Bedford limestone and Solenhofen limestone, had the highest coefficients of friction, while the schistose gneiss with its 40% mica generally had the lowest values. Representative initial and residual coefficients of friction for each of the 10 rock types are given in Appendix I.

The influence of surface roughness and normal pressure on initial and residual coefficients of friction follow the same trends for most rock types. The plots of coefficient of friction as a function of normal pressure for Lower Granite basalt are typical, Fig. 6. This type of plot was chosen in preference to the Mohr diagram to more adequately represent minor changes in shear strength for normal pressures less than 200 psi.

Initial Coefficient of Friction.--Several well developed trends are established in Fig. 6. Surface roughness effectively increases initial friction for all rock types. The smooth #600 surfaces consistently had the lowest coefficients of friction while #80 and sandblasted surfaces had successively higher values. However, there was generally little difference between the initial coefficients of friction for the #80 and sandblasted surfaces.

For normal pressures less than 50 psi the initial coefficient of friction for #600 surfaces was normally about 0.20 less than that for #80 and sandblasted surfaces. As normal pressure was increased to 1000 psi the effect of surface roughness was reduced, and in most cases the initial coefficient of friction for #600 surfaces moved to within 0.10 of the value for the other two surface preparations.

A second result of increasing normal pressure was the reduction of the initial coefficient of friction for all surface roughnesses from 5 to 20%. Notable exceptions to this generalization were the #80 and #600 surfaces on Bedford limestone and the #80 surfaces on Navajo sandstone for which there was an increase in initial friction with increasing normal pressure. The porosity and low strength of these two rock types allowed relatively deep penetration of asperities into the opposing surface upon application of N, and severe surface damage began at zero displacement. Asperity penetration and surface

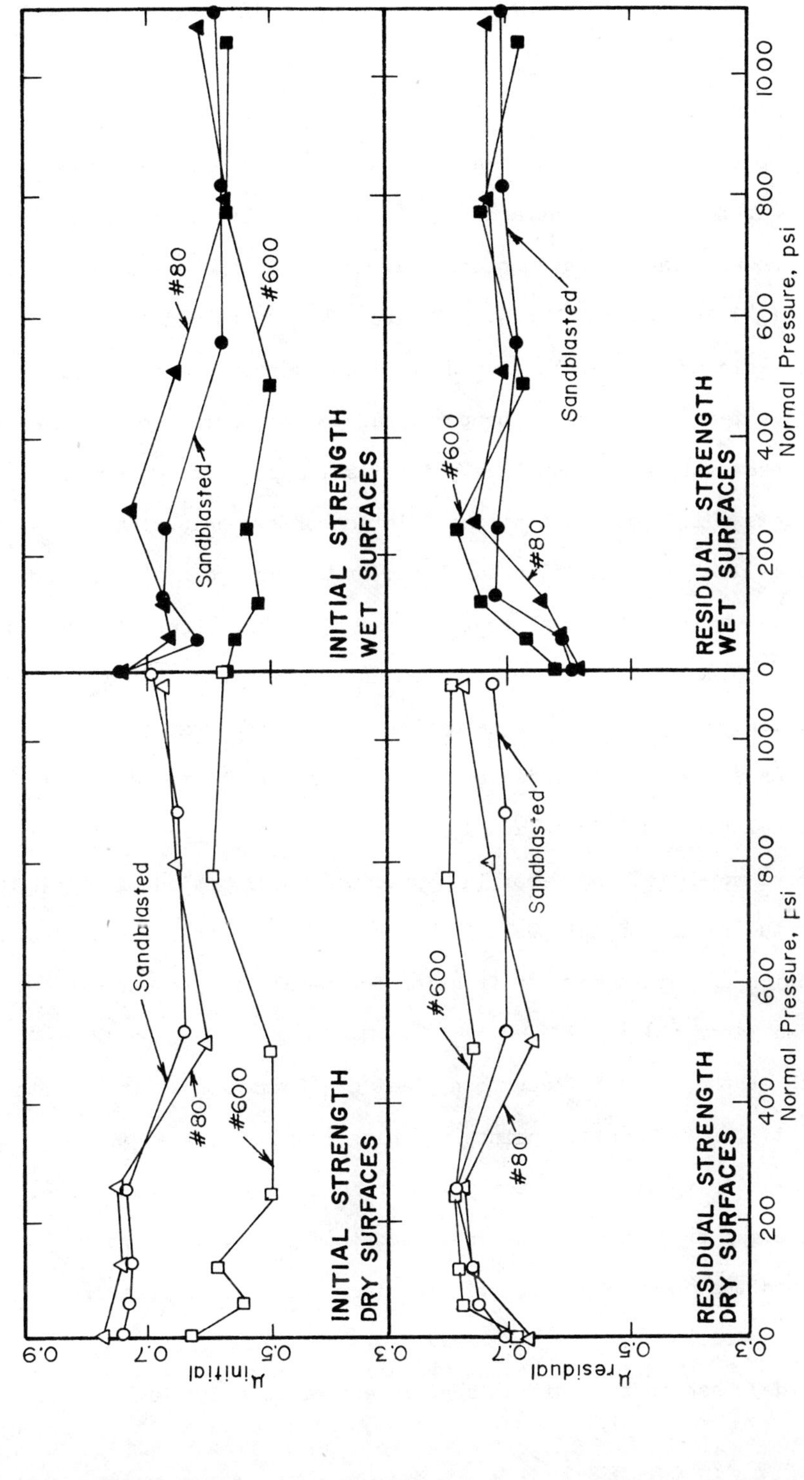

FIG. 6.--INITIAL AND RESIDUAL COEFFICIENTS OF FRICTION FOR LOWER GRANITE BASALT

interaction increased with normal pressure, thus increasing initial friction.

Amontons' law of friction states that the coefficient of friction between two sliding surfaces remains constant as normal force increases and is independent of the apparent contact area. The most widely accepted models explaining Amontons' law have been presented by Terzaghi (32) and Bowden, Moore, and Tabor (4). Both models rely upon two basic assumptions: (1) the real contact area across two opposing surfaces is directly proportional to the normal force, and (2) the shear strength of contact points is independent of normal force. Terzaghi theorized that the friction of two unlubricated surfaces is derived from molecular bonds at contact points, and that friction is equal to the real contact area times the shear strength of the bonds. Since the real contact area is only a minor fraction of the apparent contact area, yielding and plastic flow occur at the contact points. As a result, the real contact area and friction are directly proportional to normal force and independent of the apparent contact area. Bowden, Moore, and Tabor (4) explained the friction of metals as being the result of cold welding of the contact points. Experimental verification of this theory of friction for metals is presented by Bowden and Tabor (5).

Both of these models rely upon plastic deformation at the points of contact. If the contact points are elastic and brittle, total contact area is no longer directly proportional to the normal force, N, but becomes proportional to $N^{2/3}$. Amontons' law is no longer obeyed, and the coefficient of friction decreases with increasing normal pressure.

The decrease in the initial coefficient of friction with

increasing normal pressure for all rock types except the Bedford limestone and Navajo sandstone supports the idea that a large part of the deformation upon application of the normal and shear forces is elastic, and that brittle fracture rather than ductile flow is the chief mode of asperity failure.

Mauer (19) also found the coefficient of friction to decrease with increasing N for rock, and attributed the effect to brittle and elastic behavior. Seal (29) found the shear strength of polished diamond to be a function of $N^{0.8}$. The coefficient of friction decreased from 0.15 to 0.05 with increasing normal pressure. Like Mauer, Seal attributed this decrease in friction to elastic behavior.

Residual Coefficient of Friction.--Significant variations in the residual coefficient of friction as a function of roughness occurred most frequently at normal pressures less than 100 psi where surface damage did not develop to the extent necessary to mask the initial surface roughness. At higher normal pressures residual coefficients of friction for the three surface roughnesses generally agreed to within about 0.05.

For Lower Granite basalt, schistose gneiss, fine-grained Grand Coulee granite, and Berea sandstone polishing occurred on #600 surfaces, and they often produced the highest residual coefficients of friction.

In most instances residual friction increased with normal pressure as shown in Fig. 6. This is a direct result of more advanced surface damage at higher pressures. The effect is analagous to the development of Type 3 and Type 4 shear strength - displacement curves where shear strength increases with surface damage.

Effects of Water.--Water potentially has several effects on the sliding friction of rock surfaces. It can appreciably change the coefficient of friction of smooth mineral surfaces; it can lower the surface energy of individual crystals and decrease crystal strength; pore pressures can be developed in its presence; and in specific cases water can significantly alter the mode of surface damage, as evidenced by the formation of the indurated crust on Bedford limestone and Oneota dolomite.

Horn and Deere (13), Tschebotarioff and Welch (34), Bromwell (6), and King and Tabor (15) found from tests on cleavage planes and polished surfaces that the frictional characteristics of smooth surfaces are affected by the presence of water, and that the effect is a function of the crystal structure of the mineral. On smooth surfaces water generally behaves as an antilubricant on massively structured minerals, and as a lubricant on minerals with a layer-lattice crystal structure. Bromwell (6) found the effectiveness of water in increasing friction on quartz to be a function of surface roughness. For surfaces with an average roughness greater than 60 μ in., water no longer increased strength, and both wet and dry surfaces had the same coefficients of friction.

The average roughness for #600 surfaces on Group A rocks was 70 μ in., the approximate upper limit for the water effect. Although the average roughness for #600 surfaces on Group B rocks was considerably greater, the individual grains which came into contact across the surface of sliding approached the smoothness of Group A rocks (Fig. 2).

Coefficients of friction for wet and dry surfaces are compared in Fig. 7. Initial coefficients of friction for wet surfaces were generally within 10% of those for dry surfaces for all three surface preparations. For several tests on #600 surfaces on Berea sandstone, fine-grained Grand Coulee granite, and Bedford limestone, the presence of water on

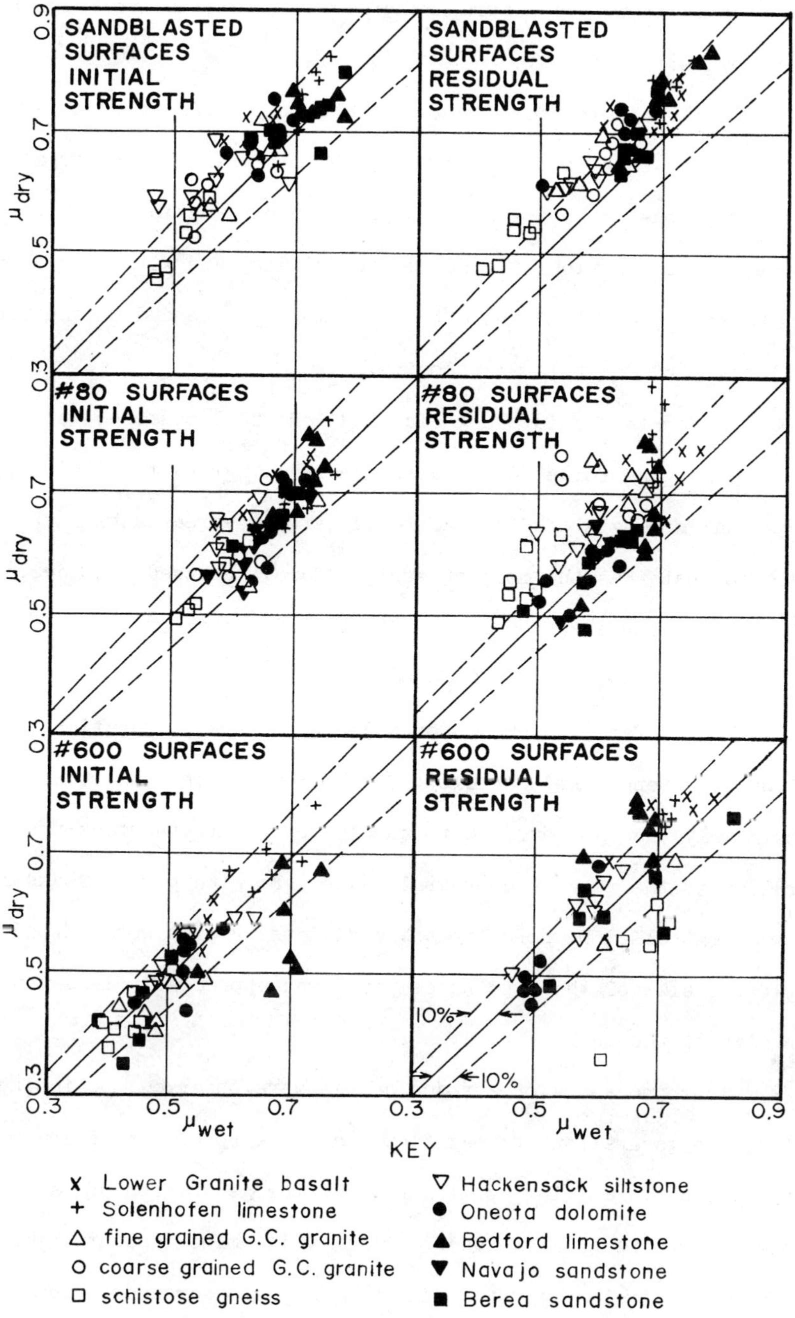

FIG. 7.--CORRELATION OF COEFFICIENTS OF FRICTION FOR WET AND DRY SURFACES

the massively structured minerals increased the initial coefficient of friction by more than 10%.

Residual strengths for #80 and sandblasted surfaces were generally greater for dry than for wet surfaces, the difference being greatest for the nonporous Group A rocks. A possible explanation is that pore pressures developed along the surface of sliding and within the surface damage material, and that the effect was most pronounced on the Group A rocks because of their lower porosity.

Except for the schistose gneiss and Bedford limestone, wet and dry residual strengths for #600 surfaces generally agreed to within 10%. The residual strength of the Bedford limestone decreased with the addition of water as the mode of surface damage changed from deep gouging to the formation of a superficial indurated crust.

CONCLUSIONS

1. The surface roughness for a given surface preparation is a function of strength and porosity. The strong, nonporous rocks had approximately the same surface roughness for a given surface preparation regardless of the rock type or grain size. For the porous rocks roughness was controlled by porosity and individual grains were lapped flat only if the size of the lapping compound was appreciably less than the grain size of the rock.

2. Four processes of surface damage were observed; polishing, gouging, the generation of rock flour, and the formation of an indurated crust. Gouging, polishing, and the generation of large amounts of rock flour increased friction while minor amounts of rock flour and the development of an indurated crust decreased friction. For over 60% of the tests residual friction exceeded initial friction.

All processes of surface damage except the development of the indurated crust indicated brittle behavior.

3. The initial coefficient of friction for flat rock surfaces generally increased as surface roughness increased from that of nearly polished to that of a sandblasted surface. The increase was a function of normal pressure, and ranged from approximately 0.2 at 10 psi to 0.1 at 1000 psi. The initial coefficient of friction tended to decrease with increasing normal pressure, as expected when surface damage is by brittle rather than ductile processes.

4. For normal pressures less than 100 psi surface roughness affected the residual coefficient of friction in much the same manner that it affected initial friction. At greater normal pressures surface damage tended to neutralize the effects of surface roughness. Residual strength was reached after from 0.5 to 5.0 cm displacement for most tests, but up to 20 cm displacement was occasionally required.

5. The presence of water was of only minor practical significance, with coefficients of friction for wet and dry surfaces normally differing by 10% or less for the same rock type and surface preparation. The most dramatic effect of water was for Bedford limestone and Oneota dolomite where it changed surface damage from brittle to ductile.

RECOMMENDATIONS

When facilities are not available for shear strength testing, the often used assumption that $\phi = 30°$ ($\mu = 0.53$) along flat joint surfaces is generally conservative for design unless the field surface is slickensided or polished, or unless the rock contains an appreciable percentage of mica, or other platey mineral. For limestone ϕ can generally be increased to 35°. The geometrical component of shear strength as determined by methods described by Patton (23) and others can be added directly to these valves.

Special consideration should be given to polished or slickensided surfaces, or rocks containing mica. In these instances Fig. 7 indicates that values of ϕ from 22° to 25° (μ = 0.40 to 0.47) are generally conservative.

Shear strength tests on rock surfaces should be performed with direct shear apparatus in order that the displacements required to obtain residual shear strength can be accommodated. Flat surfaces should be lapped with #80 silicon carbide grit unless the field situation warrants a polished or nearly polished surface. Surfaces should be tested in the same range of normal pressures as imposed upon field joints. Since the effects of water vary with rock type and surface roughness, tests should be performed under moisture conditions approximating those of the field situation. Generally the minimum coefficient of friction should be used, regardless of whether it is an initial or residual valve.

ACKNOWLEDGMENTS

The testing program described herein was carried out in the rock mechanics laboratory of the University of Illionis at Urbana under the supervision of Professor D. U. Deere. Funds were provided by the U. S. Army Corps of Engineers, Missouri River Division Laboratory, under grant No. DACA-45-67-C-0009. The writer gratefully acknowledges J. C. Gamble and P. Arellano, graduate research assistances, for their help in performing many of the tests. Special thanks goes to Mr. G. Wilson, Clevite Corporation, El Monte, California, for the use of a Brush Surfanalyzer 1200 System for making surface profiles.

APPENDIX I.--DATA

The data presented in the following tables have been chosen to define the behavior of each rock type. A more complete listing of data can be found in Coulson (11).

TABLE 2 COEFFICIENTS OF FRICTION FOR SURFACES LAPPED WITH #600 GRIT

	Oven Dry Surfaces					Saturated Surfaces				
Rock type	N, psi	i	res	res, cm	Curve type	N, psi	i	res	res, cm	Curve type
Lower Granite Basalt	12	0.62	0.69	2.1	3	12	0.57	0.62	2.3	3
	122	0.57	0.77	2.3	3	122	0.51	0.75	3.8	3
	488	0.49	0.76	5.0	3	489	0.49	0.67	3.0	3
	1163	0.57	0.79	4.1	3	1057	0.57	0.69	1.7	3
Oneota Dolomite	12	0.57	0.49	0.9	2	12	0.59	0.49	2.1	2
	116	0.53	0.48	2.7	3	117	0.53	0.50	2.9	3
	466	0.44	0.55	3.3	3	467	0.45	-	-	3
	1011	0.51	0.68	4.4	3	1009	0.52	0.60	2.0	3
Bedford Limestone	12	0.50	0.69	9.5	3	12	0.55	0.57	0.2	1
	120	0.50	0.76	2.7	3	122	0.71	0.69	3.0	2
	403	0.60	0.77	3.6	4	401	0.69	0.67	3.5	2
	1010	0.69	0.78	2.7	4	1010	0.75	0.67	3.1	2
Solenhofen Limestone	12	0.69	0.75	4.0	3	12	0.72	0.70	3.5	3
	82	0.63	0.73	1.5	3	82	0.64	0.69	3.8	3
	420	0.70	0.77	1.9	4	401	0.66	0.70	5.5	3
	1048	0.67	0.75	0.6	4	1090	0.60	0.69	3.5	4
Fine-Grained Grand Coulee Granite	12	0.41	0.55	2.0	4	13	0.49	0.61	3.6	4
	124	0.49	0.68	4.0	3	125	0.57	0.73	3.6	3
	486	0.43	-	-	3	484	0.47	0.71	4.0	3
	1081	0.40	-	-	3	1065	0.50	0.71	3.2	3
Berea Sandstone	11	0.53	0.75	7.9	3	11	0.51	0.81	12.9	2
	81	0.46	0.56	10.7	3	81	0.46	0.71	9.1	4
	386	0.42	0.48	3.1	3	385	0.39	0.53	6.6	4
	985	0.42	0.60	3.2	4	988	0.48	0.61	3.0	4
Schistose Gneiss	12	0.41	0.35	3.6	2	12	0.42	0.60	7.8	3
	118	0.46	0.58	1.9	3	118	0.44	0.72	1.8	3
	472	0.42	0.55	2.3	3	469	0.46	0.68	3.3	3
	1054	0.41	-	-	3	1059	0.39	-	-	3
Hackensack Siltstone	12	0.51	0.50	3.1	1	12	0.50	0.46	4.2	1
	119	0.57	0.56	3.5	3	117	0.55	0.57	1.0	3
	477	0.48	0.61	3.4	3	477	0.48	0.59	4.5	3
	1031	0.49	0.63	2.5	4	1017	0.49	0.55	1.7	3

TABLE 3 COEFFICIENTS OF FRICTION FOR SURFACES LAPPED WITH #80 GRIT

Rock type	Oven Dry Surfaces N, psi	i	res	res, cm	Curve type	Submerged Surfaces N, psi	i	res	res, cm	Curve type
Lower	13	0.77	0.67	3.5	2	13	0.73	0.59	3.3	2
Granite	127	0.73	0.76	4.6	3	125	0.67	0.65	3.1	3
Basalt	500	0.59	0.66	4.5	3	507	0.65	0.71	4.5	3
	1098	0.66	0.77	3.2	3	1083	0.62	0.74	5.1	3
Oneota	12	0.72	0.56	5.0	2	12	0.73	0.52	3.7	2
Dolomite	116	0.65	0.62	3.6	2	117	0.71	0.59	3.6	2
	465	0.55	0.50	3.2	2	467	0.63	0.55	4.1	2
	1011	0.63	0.61	3.2	4	1008	0.65	0.62	1.2	2
Bedford	12	0.67	0.51	2.0	2	12	0.67	0.57	6.5	2
Limestone	120	0.72	0.67	2.0	3	119	0.74	0.70	3.9	2
	404	0.74	0.75	2.0	3	406	0.75	0.70	4.0	2
	1008	0.80	0.79	5.0	3	1010	0.73	0.67	3.6	2
Solenhofen	12	0.73	0.89	8.8	3	12	0.77	0.68	8.4	2
Limestone	82	0.68	0.80	9.8	4	82	0.70	0.69	4.5	2
	394	0.68	0.73	2.7	4	402	0.73	0.68	2.4	2
	1295	0.64	0.71	2.2	4	1184	0.69	0.68	2.8	2
Fine-	12	0.69	0.75	5.0	4	12	0.74	0.59	3.1	2
Grained	81	0.58	0.68	14.6	3	81	0.64	0.65	10.0	3
Grand Coulee	402	0.54	0.74	5.0	4	396	0.63	0.66	10.0	3
Granite	1042	0.57	0.70	7.8	3	1042	0.60	0.68	5.0	3
Coarse-	13	0.73	0.77	4.5	3	12	0.73	0.54	7.0	2
Grained	86	0.59	0.68	16.0	4	86	0.65	0.60	8.0	2
Grand Coulee	430	0.56	0.69	5.6	4	422	0.60	0.68	20.3	3
Granite	1162	0.59	0.67	5.2	3	1162	0.61	0.65	3.8	3
Berea	11	0.67	0.47	8.8	2	11	0.69	0.59	2.1	2
Sandstone	81	0.67	0.56	7.2	2	81	0.68	0.55	7.5	2
	383	0.70	0.63	2.5	2	390	0.72	0.64	1.1	2
	987	0.70	0.65	2.3	2	987	0.69	0.64	4.8	2
Navajo	15	0.53	0.49	2.2	4	15	0.63	0.54	0.50	2
Sandstone	74	0.56	0.60	2.4	4	73	0.56	0.59	1.5	4
	577	0.61	0.63	3.7	4	577	0.64	0.62	3.3	2
	1214	0.58	0.61	2.0	4	1042	0.62	0.60	1.1	2
Schistose	12	0.62	0.47	3.7	2	12	0.63	0.44	2.5	2
Gneiss	117	0.61	0.61	3.7	3	118	0.58	0.50	2.3	2
	467	0.49	0.57	2.5	3	471	0.51	0.53	2.6	3
	1163	0.51	0.60	4.5	3	1024	0.54	0.59	2.9	3
Hackensack	12	0.65	0.64	3.4	1	12	0.57	0.49	0.7	2
Siltstone	117	0.58	0.58	3.3	3	122	0.58	0.53	1.2	2
	477	0.59	0.63	4.4	3	480	0.60	0.62	3.9	3
	1061	0.60	0.61	0.6	4	1038	0.57	0.59	1.6	3

TABLE 4 COEFFICIENTS OF FRICTION FOR SANDBLASTED SURFACES

Rock type	Dry Surfaces N, psi	i	res	res, cm	Curve type	Wet Surfaces N, psi	i	res	res, cm	Curve type
Lower	14	0.73	0.71	3.2	2	13	0.74	0.60	4.1	2
Granite	129	0.72	0.77	4.7	3	139	0.67	0.72	3.5	3
Basalt	516	0.63	0.71	4.4	4	558	0.57	0.68	3.0	4
	1208	0.69	0.73	2.9	4	1113	0.59	0.71	3.7	4
Oneota	12	0.75	0.61	4.2	2	12	0.67	0.51	4.1	2
Dolomite	117	0.72	0.75	1.3	1	118	0.70	0.69	0.6	2
	471	0.63	0.71	3.9	4	469	0.64	0.65	4.0	4
	1016	0.66	0.73	1.9	4	1020	0.59	0.66	3.1	4
Bedford	12	0.72	0.84	0.2	4	12	0.79	0.77	2.1	2
Limestone	82	0.74	0.77	1.9	4	81	0.71	0.71	3.1	2
	405	0.73	0.75	1.0	4	405	0.73	0.71	2.7	2
	1008	0.73	0.76	1.2	4	1009	0.72	0.70	1.8	2
Solenhofen	14	0.83	0.78	5.2	1	13	0.75	0.71	2.8	2
Limestone	67	0.79	0.80	0.9	1	67	0.74	0.69	7.7	2
	437	0.76	0.75	1.1	4	441	0.70	0.69	1.3	2
	1178	0.65	0.75	0.7	4	1173	0.68	0.69	1.5	4
Fine-	12	0.72	0.60	2.7	2	12	0.64	0.57	3.5	2
Grained	121	0.67	0.70	4.0	3	122	0.66	0.60	1.1	2
Grand Coulee	485	0.57	0.62	4.1	3	490	0.54	0.56	4.1	3
Granite	1062	0.57	0.71	2.5	4	1051	0.57	0.66	5.0	3
Coarse-	12	0.63	0.60	4.3	2	12	0.67	0.59	3.1	2
Grained	120	0.67	0.69	4.1	1	122	0.63	0.62	4.6	3
Grand Coulee	480	0.62	0.67	4.2	4	489	0.53	0.60	2.7	4
Granite	1040	0.58	0.64	3.4	4	1059	0.54	0.66	4.3	3
Berea	12	0.80	0.63	13.4	2	11	0.79	0.59	14.4	2
Sandstone	81	0.67	0.67	2.2	2	83	0.75	0.68	2.1	2
	387	0.70	0.66	6.3	2	395	0.66	0.66	2.6	2
	998	0.68	0.66	3.9	2	1031	0.63	0.63	3.6	2
Schistose	12	0.47	0.48	0.5	1	12	0.49	0.43	2.1	2
Gneiss	118	0.53	0.56	1.8	3	118	0.51	0.45	0.9	2
	476	0.49	0.54	1.1	3	476	0.49	0.49	4.3	3
	1153	0.46	0.53	3.1	3	1125	0.48	0.48	4.2	3
Hackensack	12	0.61	0.63	3.6	1	12	0.69	0.59	1.1	2
Siltstone	115	0.62	0.62	2.0	3	117	0.57	0.56	3.1	3
	490	0.59	0.65	2.3	4	481	0.53	0.58	1.4	4
	1057	0.57	0.61	0.7	4	1019	0.47	0.52	0.7	4

APPENDIX II.--REFERENCES

1. Billings, M. P., Structural Geology, Prentice-Hall, Englewood Cliffs, New Jersey, 1954.
2. Blyth, F. G., "The Sheared Prophyrite Dikes of South Galloway," Quarterly Journal of the Geological Society of London, Vol. 105, 1949, pp. 393-423.
3. Boozer, G. D., Hiller, K. H., and Serdengecti, S., "The Effects of Pore Fluids on the Deformation Behavior of Rocks Subjected to Triaxial Compression," Proceedings 5th Sumposium on Rock Mechanics, Edited by C. Fairhurst, Pergamon Press, 1963, pp. 579-625.
4. Bowden, F. P., Moore, A. J. W., and Tabor, D., "The Ploughing and Adhesion of Sliding Metals," Journal of Applied Physics, Vol. 14, 1943, p. 80.
5. Bowden, F. P., and Tabor, D., The Friction and Lubrication of Solids, The Clarendon Press, Oxford, 1950.
6. Bromwell, L. G., "The Friction of Quartz in High Vacuum," thesis presented to the Massachusetts Institute of Technology, at Cambridge, Massachusetts, in 1966, in partial fulfillment of the requirements for the degree of Doctor of Science.
7. Byerlee, J. D., "The Frictional Characteristics of Westerly Granite," thesis presented to the Massachusetts Institute of Technology, at Cambridge, Massachusetts, in 1966, in partial fulfillment of the requirements for the degree of Doctor of Philosophy.
8. Byerlee, J. D., "Frictional Characteristics of Granite Under High Confining Pressure," Journal of Geophysical Research, Vol. 72, No. 14, 1967, p. 3639.
9. Cloos, E., "Experimental Analysis of Fracture Patterns," Geological Society of America Bulletin, Vol. 66, 1965, pp. 241-256.
10. Cloos, H., "Experiments zur inneren Tektonik," Centralblatt fur Mineralogic Geologic und Palaontologie, 1928, pp. 609-621.
11. Coulson, J. H., "The Effects of Surface Roughness on the Shear Strength of Joints in Rock," thesis presented to the University of Illinois, at Urbana, Illionis, in 1970, in partial fulfillment of the requirements for the degree of Doctor of Philosophy.
12. Handin, J., Heard, H. C., and Magouirk, J. N., "Effects of the Intermediate Principal Stress on the Failure of Limestone Dolomite, and Glass at Different Temperatures and Strain Rates," Journal of Geophysical Research, Vol. 72, 1967, pp. 611-640.
13. Horn, H. M., and Deere, D. U., "Frictional Characteristics of Minerals," Geotechnique, Vol. 12, 1962, pp. 319-335.
14. Jaeger, J. C., "The Frictional Properties of Joints in Rock," Geofisica Pure E Applicata-Milano, Vol. 43, 1959, pp. 148-158.
15. King, R. F., and Tabor, D., "The Strength Properties and Frictional Behavior of Brittle Solids," Proceedings Royal Society of London, A223, 1954, p. 225.
16. Lahee, F. H., Field Geology, McGraw-Hill, New York, 1961.
17. Lajtai, E. Z., "Mechanics of Second Order Faults and Tension Gashes," Geological Society of America Bulletin, Vol. 80, 1969, pp. 2253-2279.
18. Lee, J. S., Chen, C. H., and Lee, M. T., "Experiments with Clay On Shear Fractures," Bulletin of the Geological Society of China, Vol. 28, 1948, p. 25.
19. Mauer, W. C., "Shear Failure of Rock Under Compression," Journal Society of Petroleum Engineers, Vol. 5, No. 2, 1965, p. 167.

20. Morgenstern, N. R., and Tchalen,o, J. S., "Microscopic Structures in Kaolin Subjected to Direct Shear," Geotechnique, Vol. 17, 1967, pp. 309-328.
21. Newland, P. L., and Allely, B. H., "Volume Changes in Drained Triaxial Tests on Granular Materials," Geotechnique, Vol. 7, 1957, pp. 17-34.
22. Paterson, N. S., "Experimental Deformation and Faulting in Wombeyan Marble," Geological Society of America Bulletin, Vol. 69, 1958, p. 465.
23. Patton, F. D., "Multiple Modes of Shear Failure in Rock and Related Materials," thesis presented to the University of Illinois, at Urbana, Illinois, in 1966, in partial fulfillment of the requirements for the degree of Doctor of Philosophy.
24. Riecker, R. E., "Fault Plane Features: An Alternative Explanation," Journal of Sedimentary Petrology, Vol. 35, 1965, pp. 746-748.
25. Riedel, W., "Zur Mechanik Geologischer Brucherscheinungen," Centralblatt fur Mineralogic Geologic und Palaontologie, 1929, pp. 354-368.
26. Ripley, C. F., and Lee, K. L., "Sliding Friction Tests on Sedimentary Rock Specimens," Communication 8, 7th Congress of Large Dams, Vol. 4, 1961, pp. 657-671.
27. Robertson, E. C., "Experimental Study of the Strength of Rocks," Geological Society of America Bulletin, Vol. 66, 1955, pp. 1275-1314.
28. Schwartz, A. E., "Failure of Rock in the Triaxial Shear Test," Proceedings 6th Symposium on Rock Mechanics, University of Missouri at Rolla, 1964, pp. 109-151.
29. Seal, J., "The Abrasion of Diamond," Proceedings of Royal Society of London, 248A, 1958, pp. 379-393.
30. Skempton, A. W., "Some Observations on Tectonic Shear Zones," Proceedings of the First Congress of the International Society for Rock Mechanics, Lisbon, Portugal, Vol. 1, 1966, p. 329.
31. Tchalenko, J. S., "Similarities Between Shear Zones of Different Magnitudes," Geological Society of America Bulletin, Vol. 81, 1970, pp. 1625-1639.
32. Terzaghi, K., Erbaumechanik, Wien, Franz Deuticke, 1925.
33. Tjia, H. D., "Slickensides and Fault Movements," Geological Society of America Bulletin, Vol. 75, 1964, pp. 683-685.
34. Tschebotarioff, G. P., and Welch, J. D., "Lateral Earth Pressures and Friction Between Soil Minerals," Proceedings 2nd International Conference on Soil Mechanics and Foundation Engineering, Vol. 7, 1948, pp. 135-138.
35. Withers, J. H., "Sliding Resistance Along Discontinuities in Rock Masses," thesis presented to the University of Illinois, at Urbana, Illinois, in 1964, in partial fulfillment of the requirements for the degree of Doctor of Philosophy.

METHODS OF PERFORMING AND INTERPRETING IN SITU SHEAR TESTS

By Michael L. Haverland[1] and Edward J. Slebir[2]

INTRODUCTION

For years, bedrock has represented the ultimate of stability; however, increased loadings of larger structures and several major foreign disasters have directed more and more attention to conditions which produce instabilities in rock foundations, rock masses, and slopes. The optimum ultilization of rock foundations for dams and other heavy structures and of jointed rock masses for underground powerplants and tunnels requires the best possible data concerning rock characteristics and the best possible methods of analysis.

Rock is the direct foundation for many of our structures. Yet, the physical properties and structural behavior of rock masses, which usually include joints and discontinuities, are imperfectly known. Rock mechanics problems fall into two classes: those which depend upon deformation characteristics of the rock mass and those which depend upon the strength characteristics of the rock itself. Foundations for concrete dams, pressure tunnels, and similar projects are in the first class. The stability of underground openings, stability of rock slopes, and similar problems are in the second class. Studies are being made to

---

[1] Structural Engineer, Division of General Research, U.S. Bureau of Reclamation, Denver, Colorado.
[2] Supervisory Structural Engineer, Division of General Research, U.S. Bureau of Reclamation, Denver, Colorado.

define which rock properties are needed to predict the structural action of the rock mass, how the properties can be obtained, and how these data should be included in the design of safe, economical structures.

At present the Bureau of Reclamation utilizes several methods of shear testing. Triaxial shear tests are performed on NX or 6-inch-diameter cores to provide data for developing Mohr's envelope. Direct shear tests are performed on NX or 6-inch-diameter cores to which desired normal loads are applied to healed discontinuities or rock (if no discontinuities exist) and shear loads are applied to determine shear strength parameters. Sliding friction tests are done on cores from the aforementioned direct shear tests or from core specimens in which the discontinuities were open (not healed). Sliding loads are applied while the core is under normal load. The final method is the in situ shear test. This test is done at the project site. It is conducted on larger samples and in areas where the discontinuity is difficult to sample and test in the laboratory or where the feature will be disturbed by removal. The in situ shear test will be the topic of discussion for the remainder of this paper.

Several successful in situ field tests were recently completed by the Bureau of Reclamation. The authors' usage of the term, "successful tests," relates not only to the data obtained but also to all phases of the tests, such as excavation of the test block, instrumentation of the test setup, block integrity (block remained intact during the entire test), etc. Only when all phases of an in situ test are successful can the results be considered as reliable, even though some test values still may not agree with accepted theory.

TEST METHODS

To obtain shear parameters for undisturbed rock, in situ shear tests on 15- by 15-inch-square by 10-inch-deep test blocks were conducted at two major construction sites. These were Auburn damsite, Auburn, California, and the Third Powerplant at Grand Coulee, Washington. For purposes of this presentation, six tests will be discussed.

Four of the tests were at the site of Grand Coulee Third Powerplant Forebay Dam. These were performed on a rock outcrop just downstream from the dam on the right abutment. A 90° segment of the top of this outcrop was removed, as shown in Figure 1. This provided access to many geologic discontinuities that were considered representative of those in the construction area as shown in Figures 2, 3, and 4.

Two of the tests were at the Auburn damsite. These were at selected locations in the right abutment exploratory tunnels.

Two distinct, yet similar, test procedures were used. The primary difference between the two procedures was the method of application of the normal load to the discontinuity. For tests conducted at the surface such as those at Grand Coulee, the normal load was applied by utilizing a cable grouted in the rock mass beneath the discontinuity being tested. A center pull hydraulic ram applied the normal load to the rock by applying tension to the cable, as shown in Figure 5.

At Auburn, where the tests were conducted beneath ground in exploratory tunnels, the hydraulic ram used to apply the normal load reacted against an opposing wall of the tunnel. Figure 6 illustrates the tunnel setup. Both methods use a frame to contain the rock block and provide an application surface for the shearing (or sliding) load. The shear load was applied above the plane of the discontinuity and was angled downward 15° to intersect the center of the test surface. This loading arrangement was used to reduce the overturning moment.

After the discontinuity to be tested was located by NX drill holes, all but 10 inches of rock above the discontinuity was removed. Subsequent preparation of the test block was virtually the same for both sites. At Auburn, a concrete pad thick enough to provide an even surface, was placed to provide bearing for the normal load plate. The load plate, the hydraulic ram, and wood cribbing were installed at this time and a small normal load was applied. Next the test block was cut free from surrounding rock by sawing on each of the four sides of the block. The surrounding rock was removed leaving the test block remaining. The block frame was then fitted and grouted in place.

At Grand Coulee, an NX hole, perpendicular to the joint and in the center of the proposed test block, was drilled after all the rock 10 inches above the joint was removed. This hole was about 20 feet deep. A 1-inch-diameter cable, 25 feet long, was grouted into the bottom 15 feet, leaving approximately 5 feet of cable protruding from the hole. A small amount of aluminum powder was used in the grout to make an expansive mixture to insure good bond. A thin grout pad was then installed on top of the test block. The normal load plate and ram were set in place and cable clamps were attached to the cable. The ram was activated to apply a slight tension to the cable before sawing to isolate the block. The sawing operation is shown in Figure 7. Material around the block was then removed and the block frame grouted in place. The equipment at Auburn, fabricated from steel, was very heavy and required a large crew to install it. New equipment for use above ground at Grand Coulee was fabricated from aluminum to reduce weight and was easily handled by two persons. Figures 8 and 9 show each field setup.

Both systems work satisfactorily and provide good results, if proper site preparation and instrumentation techniques are maintained.

At each site, a series of tests was conducted on each block. The first load applications to the block, or actual shear test, was performed at a normal load approximating the actual load which would be applied by the completed structure. At Auburn, this was 300 psi and at Grand Coulee, 100 psi. The shearing load was applied in the same direction that important shear loads from the structure are expected to act. The general direction at Auburn was downstream, and at Grand Coulee it was toward the river. After the initial shear test was completed, a series of sliding tests was performed at normal loads of 50, 100, 200, and 400 psi at Grand Coulee and 300, 500, 700, and 1,000 psi at Auburn. As mentioned previously, the shear load was inclined downward 15° which produces a component of force normal to the shear surface. At both sites, as inclined sliding load was increased, the normal load was reduced to compensate for the additional normal component of this inclined load. This procedure allowed the normal load to be maintained as near as possible to the design load. The test procedure provides for the sliding load to be applied slowly, a test generally taking about 30 minutes to run. From these repeated runs, values of sliding and normal stresses are plotted and a curve of sliding resistance is obtained by a least squares fit in the form $S = C + N \tan \phi$, the familiar Coulomb-Navier equation:

where S = Shearing stress in direction of movement (shear load/surface area)

C = Apparent cohesion

N = Stress normal to direction of movement (normal load/surface area)

ϕ = Angle of friction.

For all tests, loads and displacements were measured by electronic transducers and recorded on XY plotters to provide a continuous record of events during the tests. From these records, initial and maximum shear and sliding values were obtained.

INTERPRETATION

Various methods of interpretation of shear test data have been proposed in the past. Frequently, the results of a series of tests on a single specimen are used to define a line intended to represent the shear strength of in situ material. The authors believe that the values which have greatest significance, when conducting a shear (as opposed to sliding) test on rock, are derived from the initial test. After this, all data are based on a disturbed or otherwise nonrepresentative surface.

Using the linear regression (Coulomb-Navier) method of analysis, which provides the formula for a straight line representation of shear strength, may produce misleading results. To obtain the results shown in Table 1, data from the initial test for a given block were generally omitted from the calculations when determining the slope of the line, tan ϕ. Exceptions were tests where data indicated that initial shear strength was virtually nonexistent. From observations, specimens with wide, nonangular, and/or filled joints usually fall in this group. For a tight, angular, healed, and/or otherwise intact joint, the initial values will almost always be greater than subsequent results and therefore should not be included in the linear regression equation. After the slope of the line defining the sliding resistance is determined, this line is translated upward until it passes through the point from the initial test. The intercept of this line with the ordinate axis determines a value defined as the primary cohesion intercept. Two equations are shown: the initial sliding equation represents load conditions when movement first occurred and maximum values represent the greatest shear force recorded. At Grand Coulee, some of the maximum normal values are numerically less than the initial values. This occurred because some of the normal load was relieved to compensate for the normal component of the inclined load as the shearing or sliding load increased.

At this time, the authors would like to present a new and presumably different concept for analyzing and reporting shear and sliding data for use in design of structures. It will be referred to as the Angle-Envelope Method. In this method the following assumptions are made:

1. After the initial shear test (initial run) cohesion is virtually zero and all subsequent runs should indicate this.
2. Sliding resistance is independent of the surface area of the block shear and cohesion is dependent on surface area.
3. Shear contributes very little to the sliding resistance. Friction, not cohesion, predominantly affects those values.
4. The surface condition is changed after the initial run; therefore, the contact area changes and should not be applied to the cohesion intercept for later runs.

Results tabulated in Table 2 are presented in units of stress as determined from applied loads divided by initial contact area. For subsequent runs, however, data were analyzed in terms of applied forces. This simplification was possible because of the above assumption that the cohesion is virtually zero for subsequent runs. The relationship between the normal and sliding loads becomes a dimensionless quantity independent of the loaded area. The plots of the Angle-Envelope Method results show two triangular-shaped envelopes radiating from the origin. The lower triangular envelope represents initial sliding values; the upper triangular envelope represents the maximum values. The envelopes are generated by drawing lines from the origin (cohesion equal to zero) to points producing the largest and the smallest angles for both initial and maximum values of test results. When a tight competent joint is tested, the initial run has always fallen outside the envelopes. An initial point falling within its triangular envelope indicates a joint lacking in cohesive strength. Flat envelopes (low slopes) show low values of sliding resistance. The actual mechanics of the Angle-Envelope Method are developed in Figure 10 utilizing the data for Test Site Location IS-2B-311, with the steps itemized in Figure 10A. Designers are able to select a friction angle based on what they are designing and apply this angle to the initial test (first run) values to obtain a primary cohesion intercept.

DATA DISCUSSION

Figures 11 through 16 illustrate the familiar graphic method of presenting shear test results utilizing the Coulomb-Navier theory. Figures 17 through 22 illustrate the same test results utilizing the Angle-Envelope Method.

Establishing a correlation between these two methods of data interpretation is not the objective of introducing the Angle-Envelope Method. Rather, this new method is considered superior for most practical problems and design applications. A designer is provided with a range of friction angles and cohesion intercepts to consider. The structure under consideration can be analyzed to determine which of the strength parameters are most critical. Then, these strength parameters may be selected from the possible range of values to reflect the required factor of safety and design philosophy. Naturally, it is desirable when introducing and utilizing a new concept to have an accepted theory, such as the Coulomb-Navier, with which to compare. The range of results for the angle of friction and cohesion intercept, obtained from both methods, compare relatively closely to each other. The authors share the opinion that a range of shear parameters from an in situ test will be more desirable and usable than just one set of parameters.

CONCLUSION

No new or varied approach to solving rock mechanics problems is foolproof or perfect. Space-age technology and instrumentation must sometimes be utilized in a caveman's environment. The preparation of an undisturbed rock specimen is in itself a monumental task. The authors have utilized limited in situ test data in presenting the Angle-Envelope Method. However, it is hoped that other investigators will apply the Angle-Envelope Method to their test data and compare results with those previously obtained. This method can also be applied to laboratory direct shear tests on core specimens. The application does not require any test procedural changes, only a new interpretation of the test results.

Figure 1. In situ test site showing block locations at Grand Coulee. Photo PX-D-64247

Figure 2. Open Joint No. 1 at Grand Coulee. Photo PX-D-64248

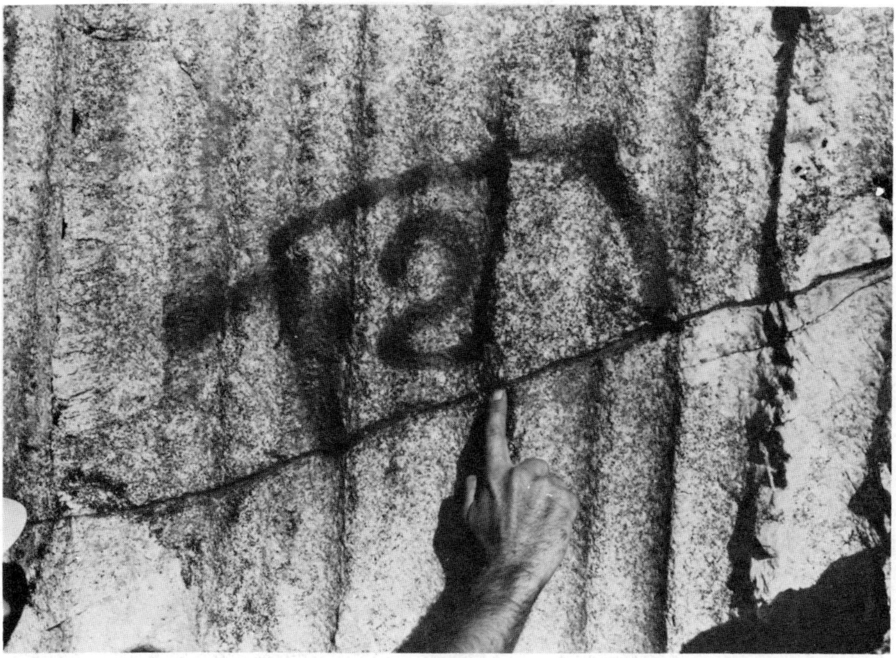

Figure 3. Open Joint No. 2 at Grand Coulee. Photo PX-D-64249

Figure 4. Tight Joint No. 5 at Grand Coulee. Photo PX-D-64251

Figure 5. Normal load applied through tensioning cable by hydraulic Ram (plastic covered) at Grand Coulee. Photo PX-D-64257

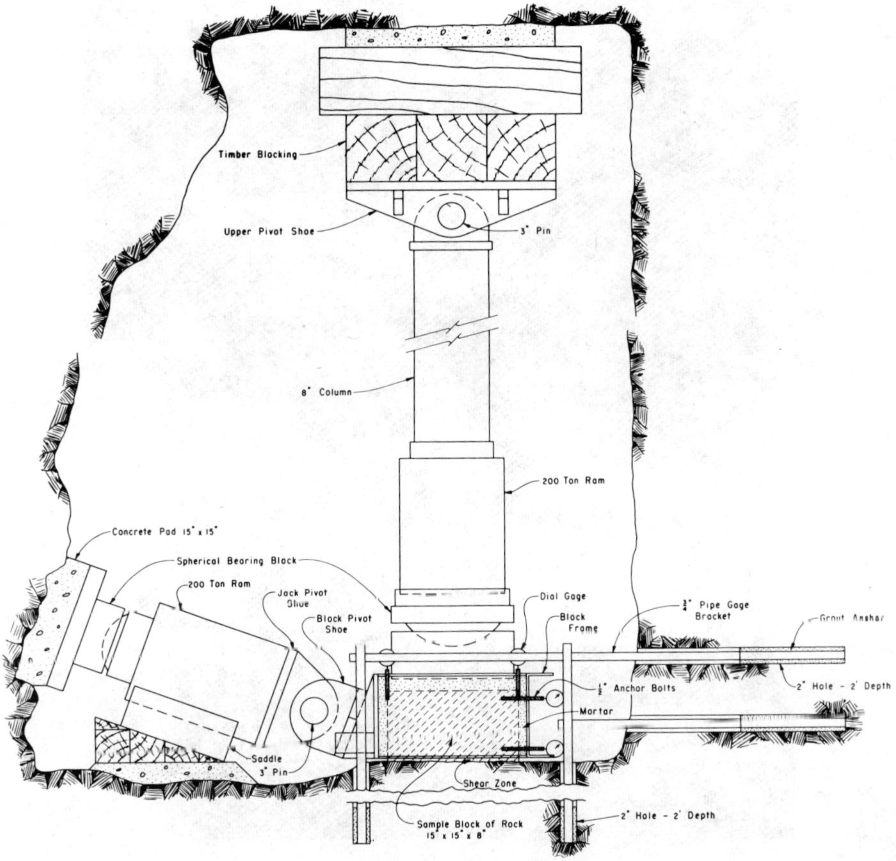

FIGURE 6 - SHEAR AND SLIDING FRICTION TESTING EQUIPMENT
AUBURN DAMSITE

Figure 7. Sawing to isolate test block at Grand Coulee. Photo PX-D-64258

Figure 8. Grand Coulee shear test setup ready for testing. Photo PX-D-64260

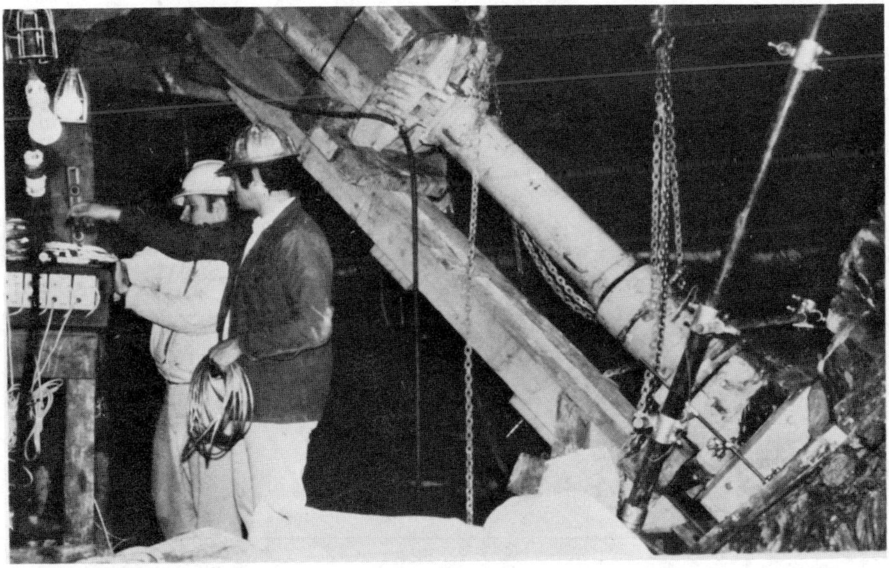

Figure 9. Auburn shear test setup ready for testing. Photo C-859-245-1487

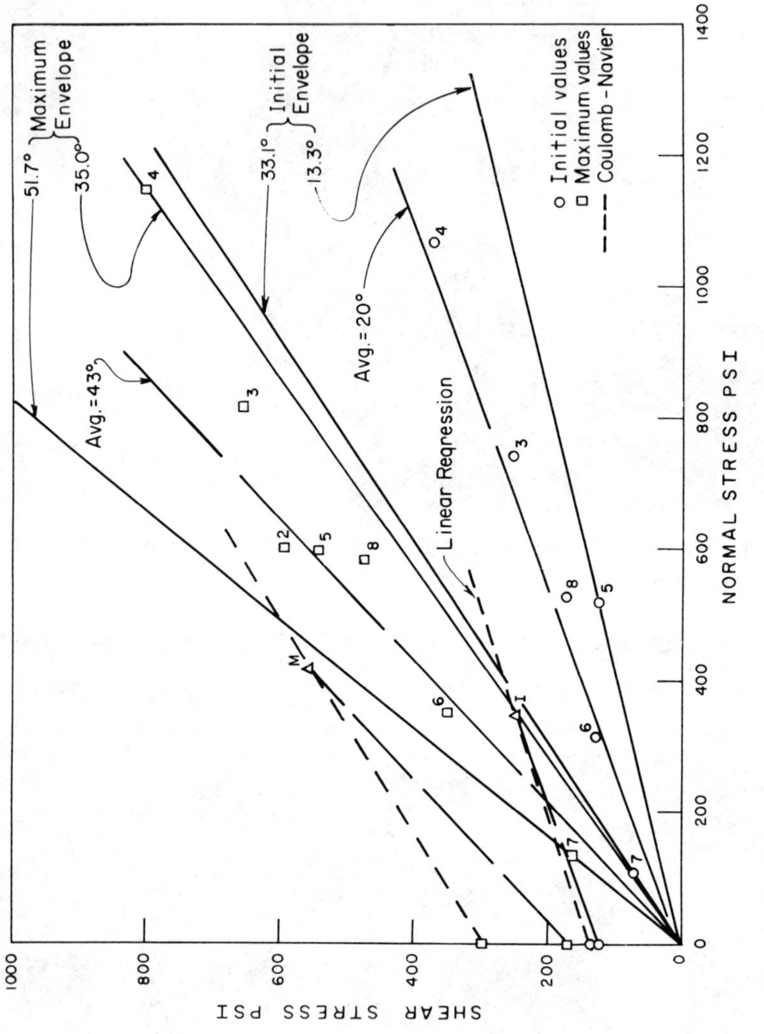

FIGURE 10-AUBURN SLIDING FRICTION TEST SITE IS-2B-311

1. Plot test results of all runs, Table 2, Site Location IS-2B-311.

2. For the "initial sliding" set of values for each run draw a triangular envelope from the origin (0,0) passing through the minimum and maximum, initial sliding points, excluding the initial run noted Δ I.

3. For the "maximum values" for each run draw a triangular envelope from the origin (0,0) passing through the minimum and maximum, maximum value points, excluding the initial run noted Δ M.

4. At this point we have the initial and maximum envelopes. The designer must now determine whether he wishes to utilize high or low angles of friction with corresponding low or high values of cohesion respectively.

5. After selecting a high, low, or average friction angle from the initial or maximum envelopes, this slope is passed through the corresponding Δ I or Δ M to obtain the primary cohesion as shown in Figure 10.

NOTE: Figure 10 also shows the Linear Regression Curve obtained from Coulomb-Navier passing through Δ I and Δ M, for relative comparison. It can graphically be seen that Coulomb-Navier has one solution, the Angle-Envelope a range of solutions.

Figure 10A - Development of Angle-Envelope Method

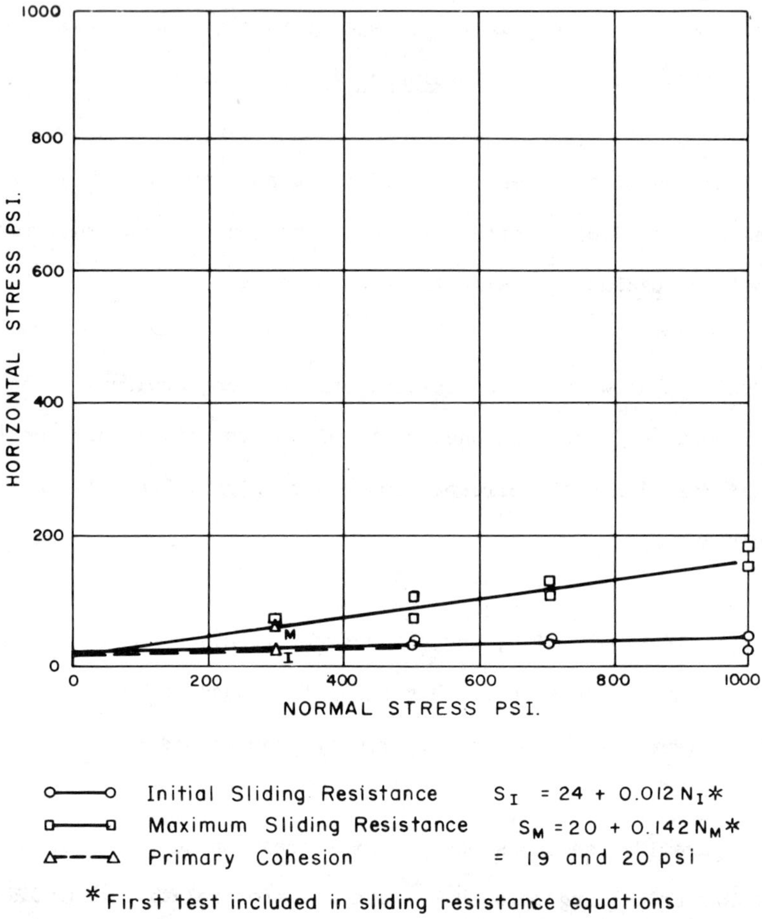

FIGURE 11 – GRAPHIC RESULTS OF SHEAR TEST AT SITE IS-1-730 AUBURN DAM (COULOMB-NAVIER)

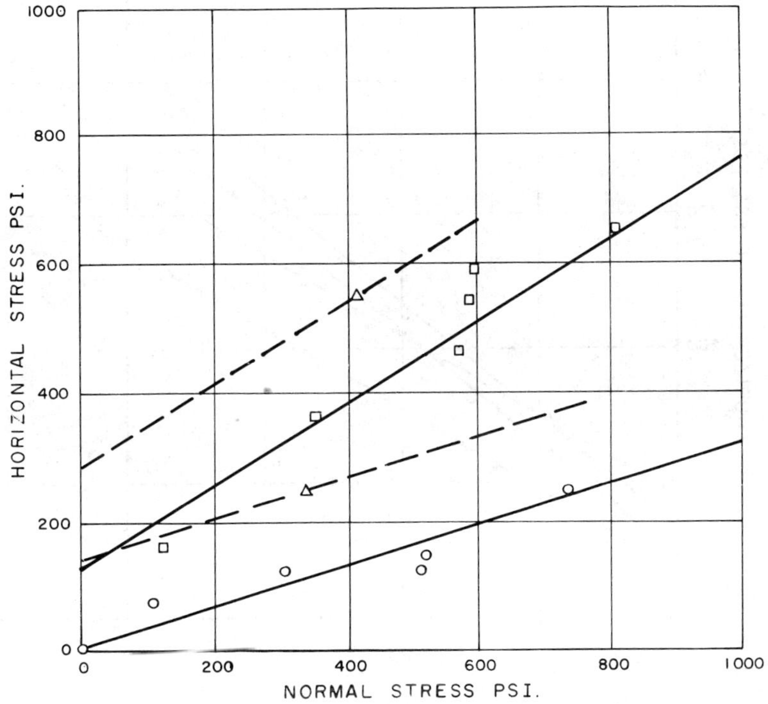

o——o Initial Sliding Resistance $S_I = 9 + .313\, N_I$*

□——□ Maximum Sliding Resistance $S_M = 124 + 0.633\, N_M$*

△– – –△ Primary Cohesion 135 and 285 psi

* First test not included in the sliding resistance equations

FIGURE 12 — GRAPHIC RESULTS OF SHEAR TEST AT SITE IS-2B-311 AUBURN DAM (COULOMB-NAVIER)

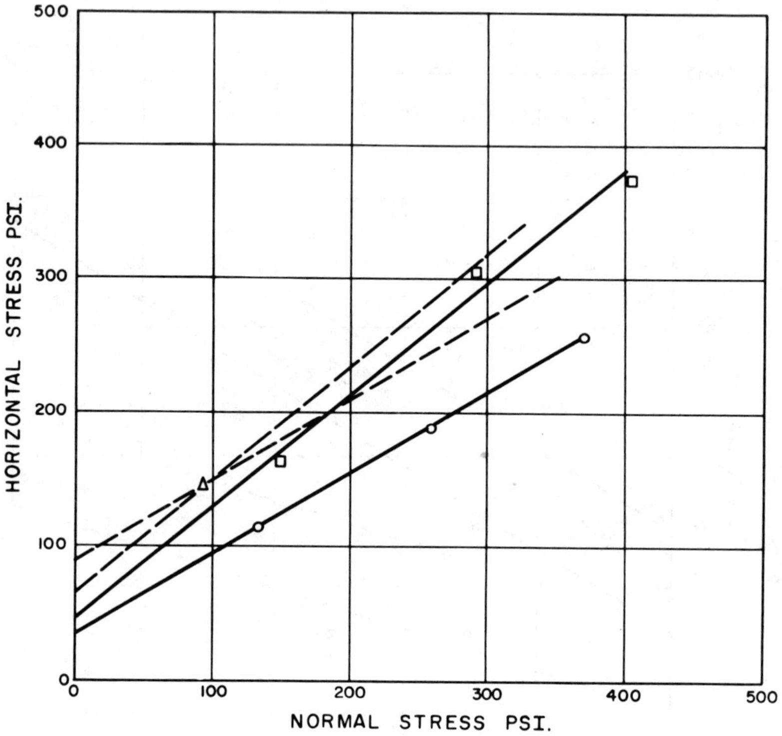

o——o Initial Sliding Resistance $S_I = 33 + 0.603N_I$*

□——□ Maximum Sliding Resistance $S_M = 44 + 0.839N_M$*

Δ- - -Δ Primary Cohesion = 90 and 69psi

* First test not included in the sliding resistance equations.

FIGURE 13— GRAPHIC RESULTS OF SHEAR TEST AT SITE BLOCK 4-A GRAND COULEE DAM (COULOMB-NAVIER)

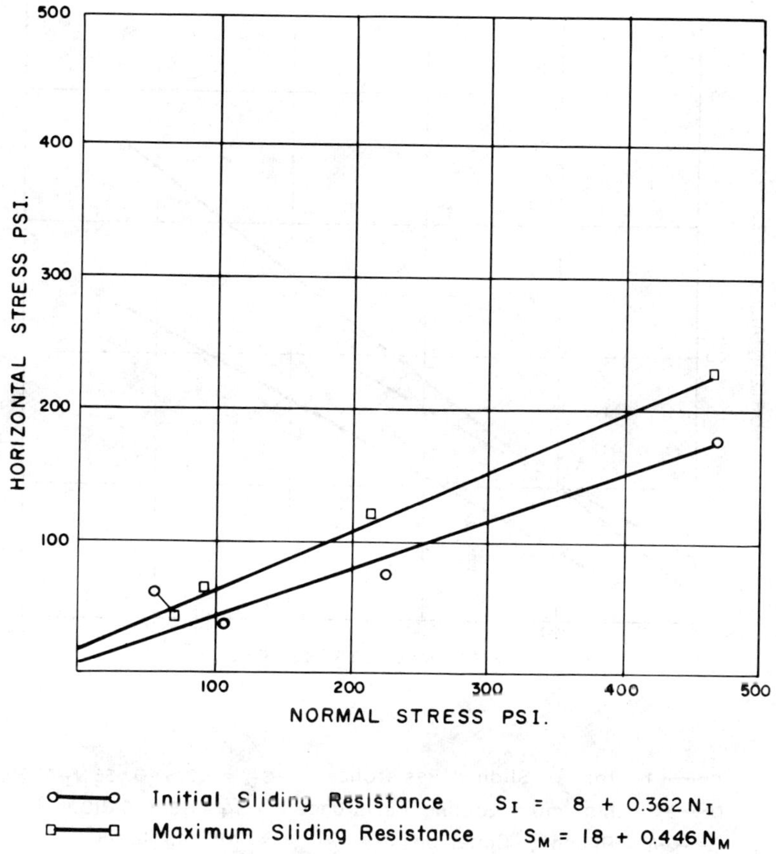

FIGURE 14 — GRAPHIC RESULTS OF SHEAR TEST AT SITE BLOCK 2-B GRAND COULEE DAM (COULOMB-NAVIER)

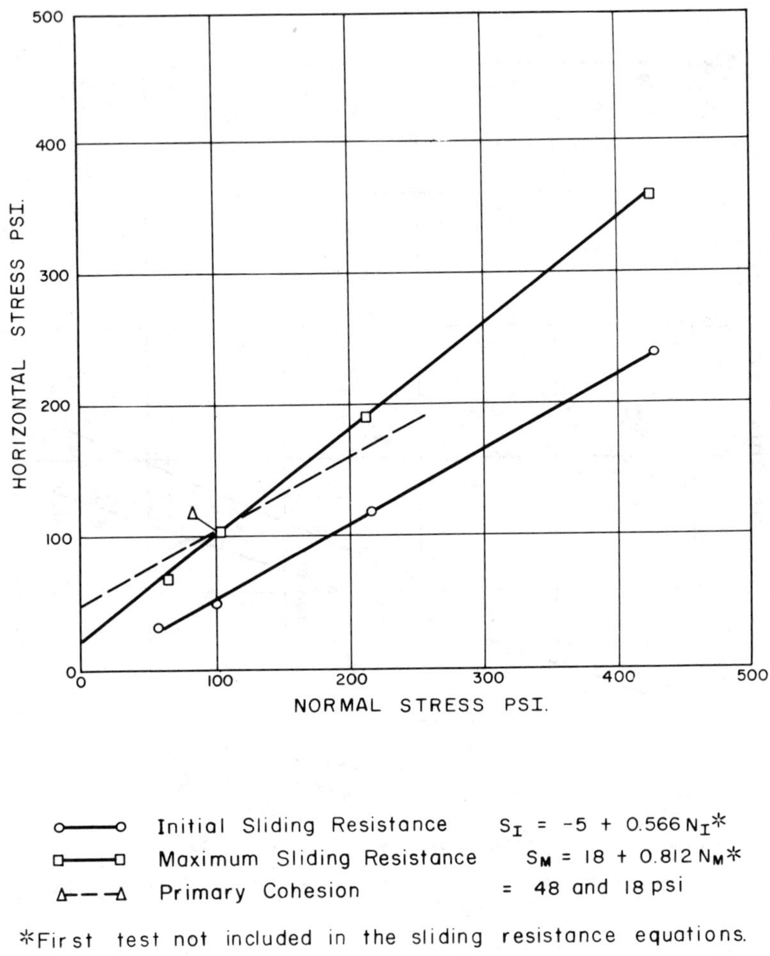

FIGURE 15 — GRAPHIC RESULTS OF SHEAR TEST AT SITE BLOCK I-C GRAND COULEE DAM (COULOMB-NAVIER)

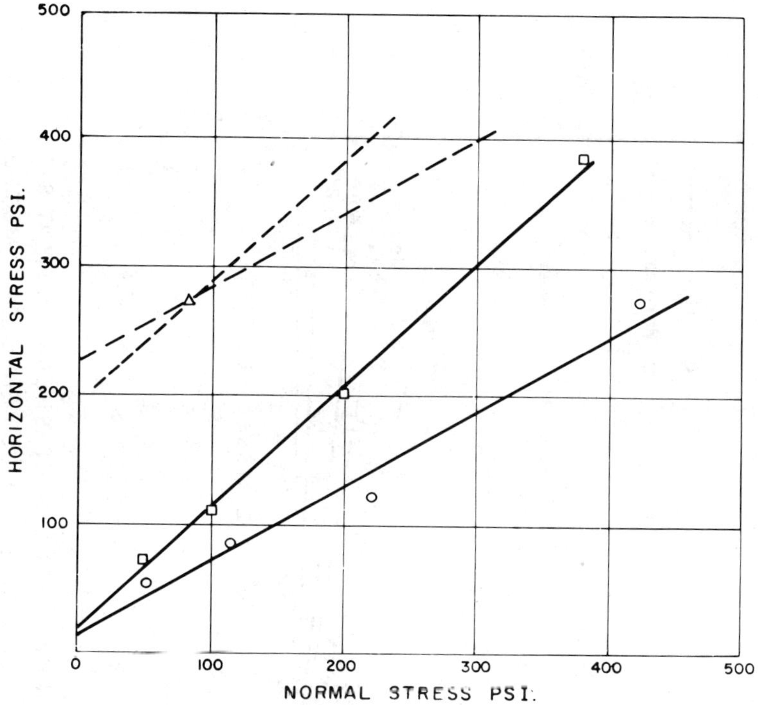

o——o Initial Sliding Resistance $S_I = 14 + 0.592 N_I$*

□——□ Maximum Sliding Resistance $S_M = 19 + 0.954 N_M$*

△- - -△ Primary Cohesion = 227 and 195 psi

* First test not included in the sliding resistance equations

FIGURE 16 — GRAPHIC RESULTS OF SHEAR TEST AT SITE BLOCK 5-D GRAND COULEE DAM (COULOMB-NAVIER)

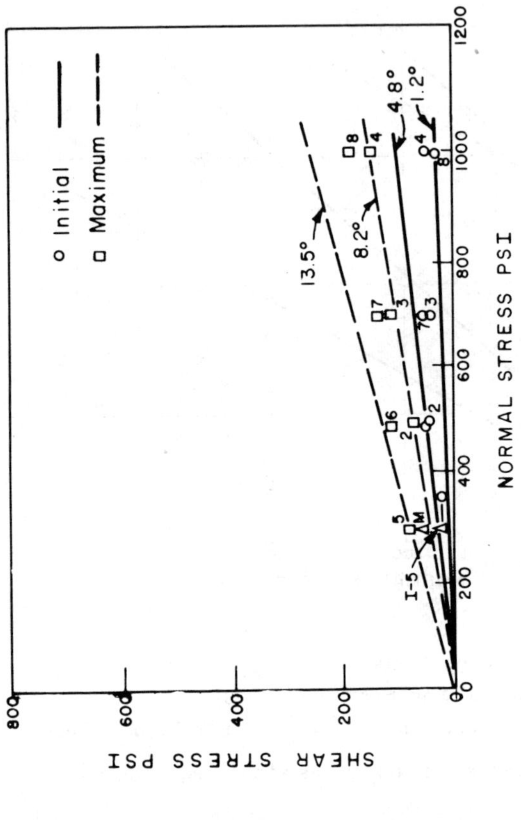

FIGURE 17 – GRAPHIC RESULTS OF SHEAR TEST AT SITE IS-1-730 AUBURN DAM (ANGLE-ENVELOPE METHOD)

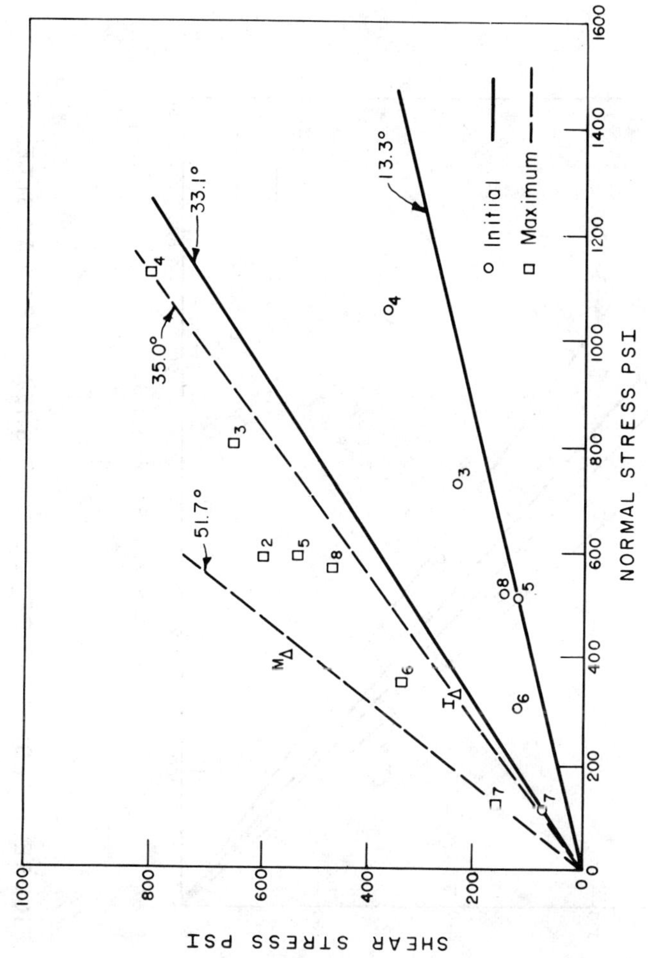

FIGURE 18 - GRAPHIC RESULTS OF SHEAR TEST AT SITE IS-2B-311 AUBURN DAM (ANGLE-ENVELOPE METHOD)

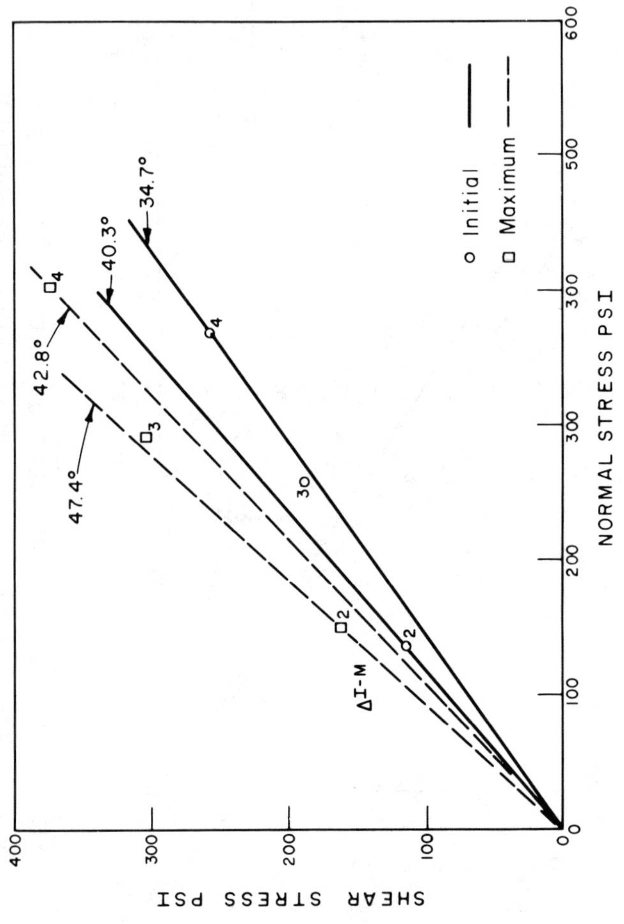

FIGURE 19 — GRAPHIC RESULTS OF SHEAR TEST AT SITE 4-A GRAND COULEE DAM (ANGLE-ENVELOPE METHOD)

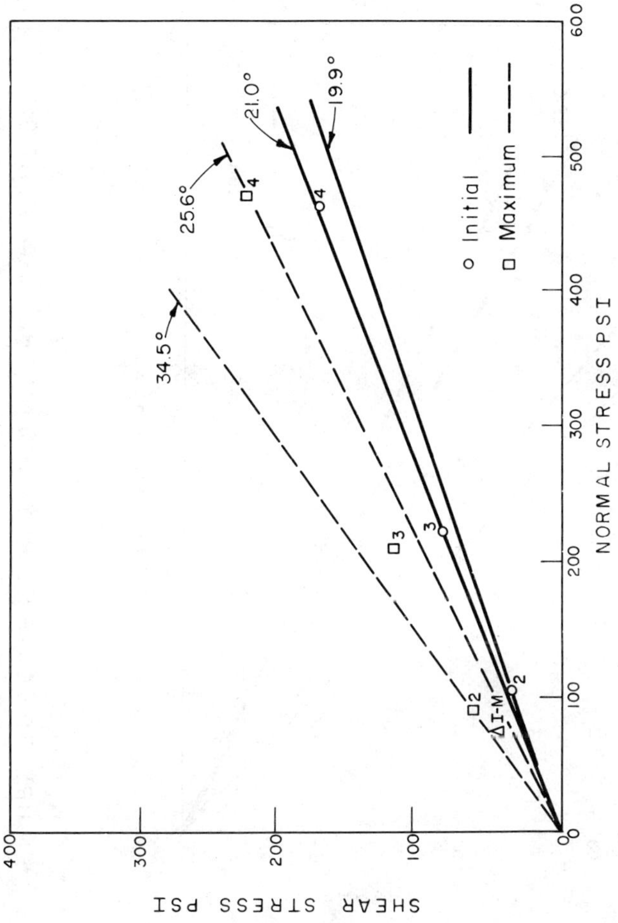

FIGURE 20 – GRAPHIC RESULTS OF SHEAR TEST AT SITE 2-B GRAND COULEE DAM (ANGLE – ENVELOPE METHOD)

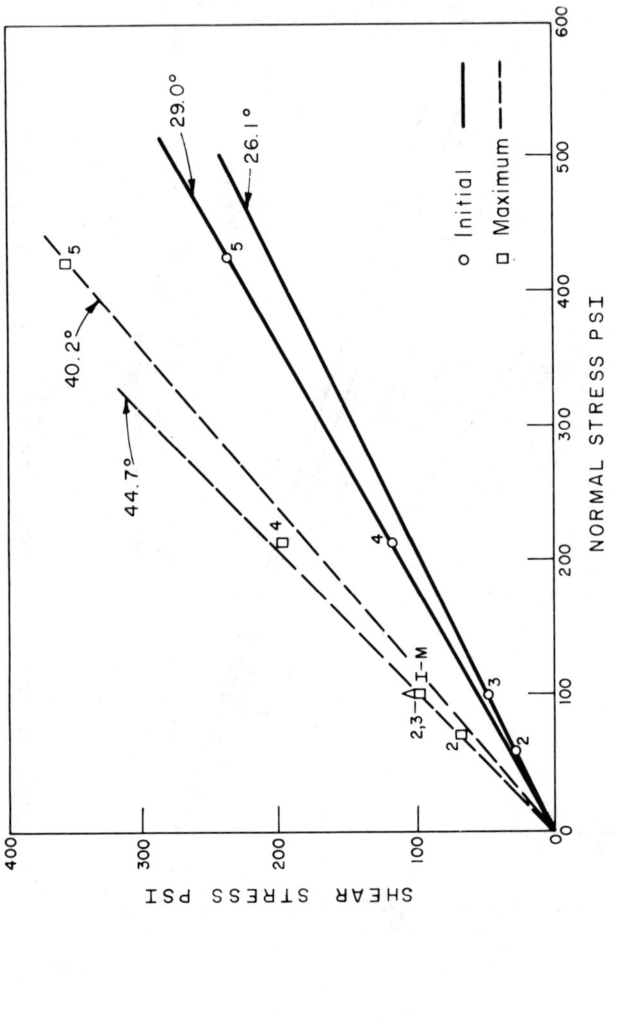

FIGURE 21 — GRAPHIC RESULTS OF SHEAR TEST AT SITE I-C GRAND COULEE DAM (ANGLE-ENVELOPE METHOD)

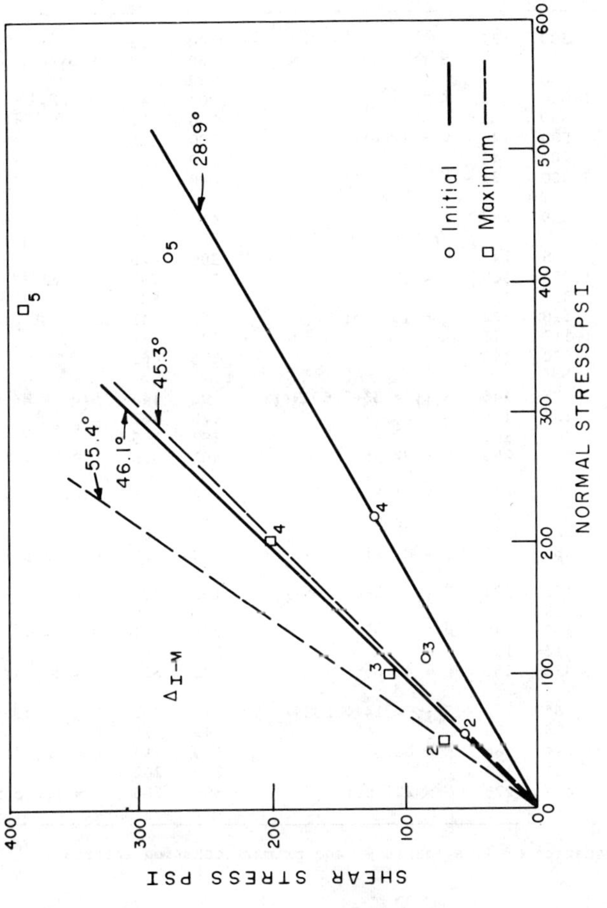

FIGURE 22— GRAPHIC RESULTS OF SHEAR TEST AT SITE 5-D GRAND COULEE DAM (ANGLE-ENVELOPE METHOD)

Table 1

TABULATED RESULTS BY COULOMB-NAVIER METHOD
(S = C+N tan ∅)

Site Location	Run Number	Initial Sliding N(I) (psi)	Initial Sliding S(I) (psi)	Initial Sliding Results*	Maximum Values N(M) (psi)	Maximum Values S(M) (psi)	Maximum Values Results*
	Initial	300	21		300	59	
	2	500	32	$S_{(I)} = 24+0.012N_{(I)}$	500	72	$S_{(M)} = 20+0.142N_{(M)}$
	3	700	32		700	106	
IS-1-730	4	1,000	42	∅ = .7°	1,000	149	∅ = 8.1°
(Auburn)	5	300	21		300	72	
	6	500	42	C = 19 psi	500	106	C = 20 psi
	7	700	42		700	127	
	8	1,000	21		1,000	180	
	Initial	339	243		412	545	
	2	-	-	$S_{(I)} = 9+0.313N_{(I)}$	597	583	$S_{(M)} = 124+0.633N_{(M)}$
	3	735	243		806	645	
IS-2B-311	4	1,053	365	∅ = 17.4°	1,128	791	∅ = 32.3°
(Auburn)	5	516	122		590	535	
	6	310	122	C = 135 psi	357	341	C = 285 psi
	7	112	73		127	161	
	8	520	146		576	462	
	Initial	93	146	$S_{(I)} = 33+0.603N_{(I)}$	93	146	$S_{(M)} = 44+0.839N_{(M)}$
4-A	2	137	116	∅ = 31.1°	149	162	∅ = 40.0°
(Coulee)	3	261	189		292	305	
	4	374	259	C = 90 psi	405	375	C = 69 psi
	Initial	74	43	$S_{(I)} = 8+0.362N_{(I)}$	74	43	$S_{(M)} = 18+0.446N_{(M)}$
2-B	2	105	38	∅ = 19.9°	93	64	∅ = 24.0°
(Coulee)	3	222	85		216	122	
	4	466	177	C = 8 psi	478	229	C = 18 psi
	Initial	103	104	$S_{(I)} =-5+0.566N_{(I)}$	103	104	$S_{(M)} = 18+0.812N_{(M)}$
	2	59	30		70	69	
1-C	3	100	49	∅ = 29.5°	102	101	∅ = 39.1°
(Coulee)	4	216	119		212	198	
	5	428	237	C = 48 psi	422	356	C = 18 psi
	Initial	84	273	$S_{(I)} = 14+0.592N_{(I)}$	84	273	$S_{(M)} = 19+0.954N_{(M)}$
	2	53	55		49	71	
5-D	3	114	86	∅ = 30.6°	100	113	∅ = 43.7°
(Coulee)	4	223	123		200	202	
	5	425	276	C = 227 psi	381	386	C = 195 psi

* Results include equation of line, angle ∅, and primary cohesion intercept.

Table 2

TABULATED VALUES FOR IN SITU TESTS

		Initial Sliding*			Maximum Values*		
Site Location	Run Number	N(I) (psi)	S(I) (psi)	Angle (I) (degrees)	N(M) (psi)	S(M) (psi)	Angle (M) (degrees)
	Initial	300	21	-	300	59	-
	2	500	32	3.7	500	72	8.2
	3	700	32	2.6	700	106	8.6
IS-1-730	4	1,000	42	2.4	1,000	149	8.5
(Auburn)	5	300	21	4.0	300	72	13.5
	6	500	42	4.8	500	106	12.0
	7	700	42	3.4	700	127	10.3
	8	1,000	21	1.2	1,000	180	10.2
	Initial	339	243	-	412	545	-
	2	-	-	-	597	583	44.3
	3	735	243	18.3	806	645	38.7
IS-2B-311	4	1,053	365	19.1	1,128	791	35.0
(Auburn)	5	516	122	13.3	590	535	42.2
	6	310	122	21.5	357	341	43.7
	7	112	73	33.1	127	161	51.7
	8	520	146	15.7	576	462	38.7
	Initial	93	146	-	93	146	-
4-A	2	137	116	40.3	149	162	47.4
(Coulee)	3	261	189	35.9	292	305	46.2
	4	374	259	34.7	405	375	42.8
	Initial	74	43	-	74	43	-
2-B	2	105	38	19.9	93	64	34.5
(Coulee)	3	222	85	21.0	216	122	29.5
	4	466	177	20.8	478	229	25.6
	Initial	103	104	-	103	104	
	2	59	30	27.0	70	69	44.6
1-C	3	100	49	26.1	102	101	44.7
(Coulee)	4	216	119	28.9	212	198	43.0
	5	428	237	29.0	422	356	40.2
	Initial	84	273	-	84	273	-
	2	53	55	46.1	49	71	55.4
5-D	3	114	86	37.0	100	113	48.5
(Coulee)	4	223	123	28.9	200	202	45.3
	5	425	276	33.0	381	386	45.4

* Values tabulated above have been corrected for the inclined angle of S and divided by the area of the sliding surface.

PROGRESSIVE FAILURE OF EXCAVATED ROCK SLOPES.

by Nicholas Barton *

SUMMARY

The stability of a rock slope is largely controlled by the presence of discontinuities in the rock. Their presence means that failure is generally of a translational type, and is therefore amenable to simple methods of analysis. The most unstable situation is chosen; one of the joint sets dipping into the slope with a strike direction parallel to the slope face. This situation is amenable to a two dimensional approach.

A limit equilibrium method is used to analyse a simple plane failure. Three refinements are then incorporated; the division of the unstable rock mass into slices (representing an additional set of vertically dipping joints), the assumption of zero tensile strength across these slices, and analysis of the effect of excavation on the assumed self weight stress distribution acting on the joints exposed by the excavation. The stability or instability of different parts of the slope is characterised by forces acting parallel to the failure surface. The depth of failure can be calculated without recourse to computing methods.

The concept of pre-failure shear displacements and increased weathering of overstressed joints is introduced. This progressive failure mechanism leads to a possible stepped portion of the failure surface. The predicted multi-linear slide scar is characterised by a vertical scarp passing through the crest of the slope, a stepped portion on which the vertical joints open, with sliding on the inclined joints, and shear failure on the inclined joint passing through the toe.

The stepped portion is caused by progressive failure, and residual shear strength parameters are adopted in this region for design purposes. This is considered to be a more realistic solution than a global assumption of residual strength. The method is illustrated by worked examples, in which the progressive failure mode is shown to reduce the failure depth considerably. A further reduction in stability is caused by transient water pressures. The pessimistic assumption of a full tension crack, and steady seepage reducing to zero exit pressure at the toe is used as an illustration.

---

* Rock Mechanics Engineer, Norwegian Geotechnical Institute, Oslo

INTRODUCTION

The majority of rock slopes warranting stability investigations will be unstable due to one of two inherent structural features. This could be the chance interaction of the slope with an unfavourably dipping fault surface or with clay filled joints, or alternatively, the eventual overstressing of one or more sets of adversely orientated joints dipping into the slope. The overstressing mechanism may be one of joint water pressure increases caused by poor drainage, or simply due to excavation of the slope to steeper angles or increased cut heights. The failure mechanism may also be initiated by dynamic loading from earthquakes or very large blasts, but this type of loading will not be considered here.

The type of failure to be investigated will be limited to that occurring on adversely dipping joints, associated with at least one further joint set. A two dimensional approach, and consideration of progressive failure lead to the development of a multi-linear failure surface, which has the appearance of a circular slide scar. However, this belies the fact that the mode of failure is purely translational. (Skempton and Hutchinson)[1] There is no rotational component.

The stability of rock slopes has been the subject of several recent papers, in which the three dimensional graphical method of analysis has been utilized (Wittke,[2] Londe, Vigier and Vormeringer[3,4], and John[5]). There is no doubt that in any open pit excavation there are more potentially adverse rock tetrahedra defined by the intersection of two joint sets, than there are unstable wedges defined by single joint sets. However, the frequency of occurrence is somewhat overshadowed by the fact that potential failures delineated by two shear planes are inherently more stable than those delineated by one shear plane given similar joint dips. At present the graphical methods do not incorporate any stress or limit analysis for determining the position of tension cracks in the unstable tetrahedra, should vertical joints be present. Consequently, the results obtained are dangerously under-conservative.

The two dimensional approach which will be adopted here, incorporates sliding on combinations of a single joint set and opening of tension cracks in a secondary joint set. This is the most pessimistic failure mode and consequently the one demanding most attention. Figure 1 shows

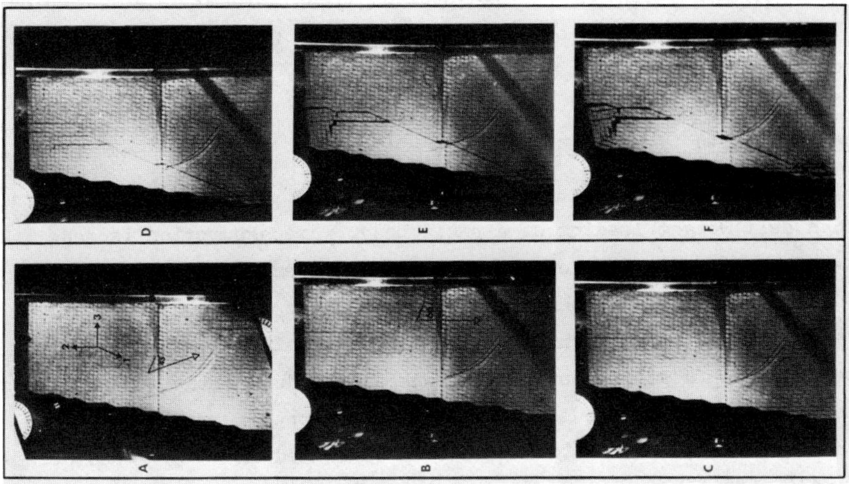

(a)

Figure 1. Translational shear and tension crack opening in jointed model slopes that were slowly rotated to induce failure.

(b)

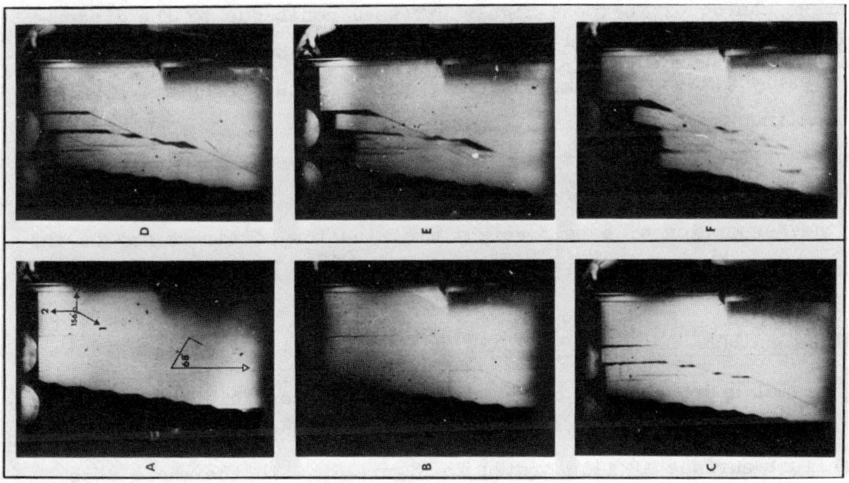

the sequential development of this type of translational failure, which was illustrated by models of jointed slopes that were slowly rotated to induce failure.

Stresses induced by self weight

It is assumed in the present analysis that all stresses are initiated through self weight loading of a rigid body. This assumption is made in spite of the current fashion for computer produced elastic stress distributions. Mueller and John[6] once suggested that the concept of overburden depth and self weight stresses used in shallow soil slopes was not applicable to steep unsupported rock sloes, due to the severe stress concentrations around the toe of these steep slopes.

These stress concentrations have been indicated in numerousphotoelastic and finite element studies of elastic materials. They have also been demonstrated by rock bursts near the base of steep fjord slopes in Norway. However, fjords generally lie at the foot of mountain ridges which are frequently in excess of 5000 feet high. It is well known that such areas are tectonically active. In addition and more importantly, most steep fjord slopes have extremely sparse jointing limited almost entirely to relief or 'sheet' type jointing (Bjerrum and Jorstad[7]).

Recent studies have indicated that elastic stress distributions can be very misleading when applied to normally jointed rock, as encountered in most open pit locations. St. John[8] has suggested that a redistribution of stress can occur in these highly stressed zones due to the discontinuous nature of the jointed rock. This redistribution of stress has been clearly demonstrated by Cundall[9], from finite difference studies of discontinuous block models of rock slopes.

Large physical models of jointed, excavated rock slopes which were loaded horizontally (triangular distribution) and by gravity, indicated that the assumption of a self weight distribution of stress may be the most valid one, even for steep slopes (Barton[10]).

1. SIMPLE FAILURE ON A SINGLE PLANE

A trivial approach will be followed initially as this helps to clarify the essential elements of the problem. Failure is assumed to occur on the failure surface AB illustrated in Figure 2, with the whole mass sliding coherently on this plane. The only stresses acting on this failure plane are assumed to be self weight stresses caused by gravity. No joint water pressures are considered for the present.

The shear strength characteristics of the failure plane are assumed (for the purposes of this presentation) to be adequately described by the parameters (c) cohesion intercept, and (ϕ) angle of friction. A linear shear strength envelope will be assumed, taken as the best piecewise linear approximation to the actual curved envelope.

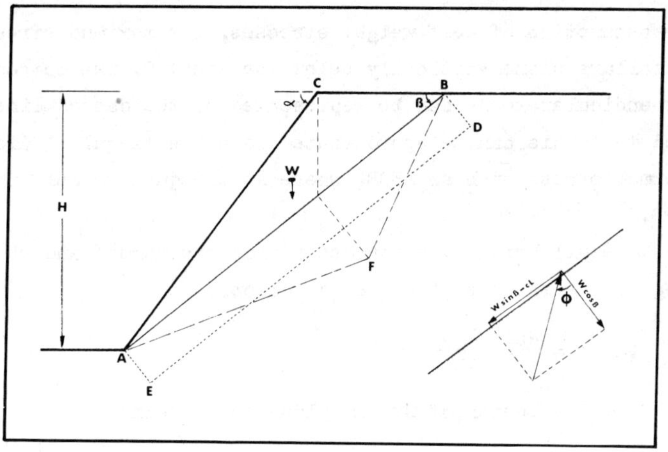

Figure 2. The equilibrium of forces on a planar joint induced by a self-weight stress distribution.

The problem will be further simplified by selecting one pair of (c) and (ϕ) values as being most relevant to the range of stresses involved. It is frankly admitted that this approach is seldom justified in practice, since an important aspect of slope stability studies is that for any given slope the range of normal stress extends to zero. It is in this region that the shear strength of rock joints is most uncertain, and the traditional (c), (ϕ) Coulomb assumptions are most in question. However, the errors involved can be significantly reduced by using the relevant curved strength envelope and a computer, and basing the programme on the simple principles which are to be outlined.

From the geometry of Figure 2 the weight of material W above the assumed failure plane AB can be calculated as follows:

$$CB = H(\cot\beta - \cot\alpha)$$
$$W = \frac{\gamma H^2}{2}(\cot\beta - \cot\alpha) \qquad (1)$$

where: H = the height of the slope
γ = the average density of the rock mass
β^o = the inclination of the assumed failure plane
α^o = the inclination of the overall slope

Based on the assumption of self weight stresses, the maximum stress occurs on the failure plane vertically below the crest C. The component of stress perpendicular to AB can be represented by the dashed lines AFB. It is assumed that this can be approximated to a 'rectangular' distribution of normal stress such as AEDB, where AE is equal to half the height of AFB.

The limiting equilibrium equation describing the normal and shear forces acting on the failure plane is as follows:

$$\tan\phi = \frac{W\sin\beta - cL}{W\cos\beta} \qquad (2)$$

where L = the length of failure plane AB = $H/\sin\beta$

(The third dimension is unity in this two dimensional approach)

Substituting for W and L in equation 2 gives the following relationship:

$$\tan\phi = \tan\beta\left[1 - 2c/\gamma H \cdot \frac{1}{(\cot\beta - \cot\alpha)\sin^2\beta}\right] \qquad (3)$$

Simple Method of slices

Following this trivial approach the wedge of material above the plane AB is now split into parallel-sided slices. Figure 3 illustrates four slices which are chosen such that the dimensions DE and EC are equal, and similarly CF and FB.

Since GF is parallel to AB, triangle CFG is geometrically similar to triangle CBA. Therefore the weights of the four slices are related as follows:

$$W_2 = 3W_1 \qquad W_3 = 3W_4$$

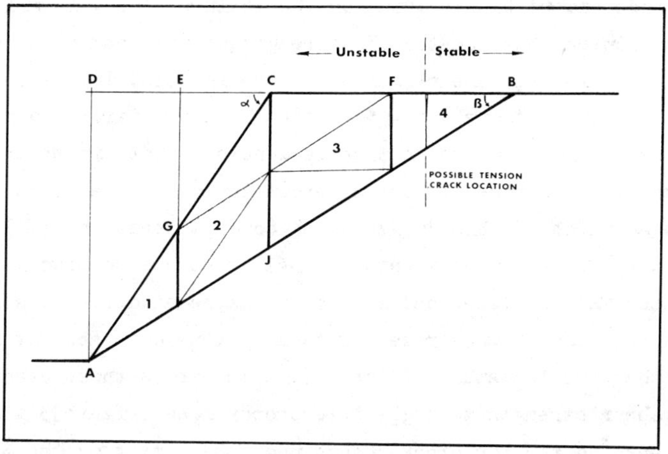

Figure 3. The geometry of four slices dividing an unstable 'wedge'.

It is assumed that the stress distribution beneath each slice can be approximated to a rectangular distribution. The validity of this assumption clearly is increased for larger numbers of slices. The limiting equilibrium equations for the four slices can be obtained by substitution in equation 2. The equations for slices 1 and 2 are as follows:

$$\tan\phi = \tan\beta\left[1 - {}^{4c}/_{\gamma H}\cdot\frac{1}{(\cot\beta - \cot\alpha)\sin^2\beta}\right] \quad \text{———} \quad (4)$$

$$\tan\phi = \tan\beta\left[1 - {}^{4c}/_{3\gamma H}\cdot\frac{1}{(\cot\beta - \cot\alpha)\sin^2\beta}\right] \quad \text{———} \quad (5)$$

Hence if treated as separate blocks lying on the failure plane, slices 1 and 2 must have different values of cohesion intercept (c) or angles of friction (ϕ), acting across the failure plane for them to both be at limiting equilibrium. Alternatively it can be said that slice 1 is inherently more stable than slice 2 given the same shear strength parameters. The same can be said of slice 4 and slice 3.

<u>No-tension concept</u>

This simple idea can be extended by dividing the two triangles ACJ and CBJ into a larger number of slices. Given certain conditions, several of the lighter weighted slices could be stable if they could be considered as independent blocks lying on the failure plane. It will be assumed from now on that the tensile strength of the rock mass in a direction perpendicular to the slice boundaries is zero. This assumption is readily justified if one considers the slice boundaries to represent a parallel set of more or less vertically dipping joints intersecting the rock mass with a strike direction perpendicular to the plane of the figure. The adoption of this "no-tension" concept leads to partial separation of the 'unstable' mass into a possible stable portion lying at the top of the failure plane, and a less stable remainder. The separation between the two parts will be referred to as a tension crack. It is assumed that the tensile nature of the movements across these cracks reduces their shear strength to negligible proportions. Figure 3 illustrates the concept. No tension crack occurs near the bottom of the failure plane since the slice boundaries in this zone will be under compression, if displacements can occur.

It is important to realise that a stability analysis based on the assumption of a coherent wedge sliding on a plane failure surface will generally produce a stability estimate that is under-conservative. That is, the stability of the slope will be overestimated. A simple 'no-tension' assumption is a first step towards improving this estimate, provided that the orientation of one of the joint sets makes the assumption reasonable.

Excavation to increased slope heights

The division of the unstable zone into slices will now be extended to include different depths of excavation. Figure 4 illustrates the critical failure planes for four depths of excavation, each with the same slope angle. The steeply dipping planes can be regarded as a persistent set of parallel joints. At each stage of excavation a different joint will be most critically stressed.

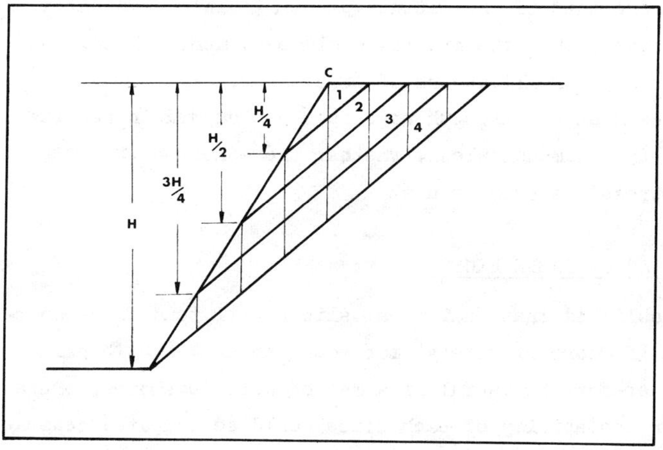

Figure 4. The critical joint planes at four depths of excavation.

It is assumed that the presence of joints does not affect the stress distribution compared to an unjointed model. This would probably be valid if no shear displacements occurred on any of the joints. The degree to which the initial stress distribution collapses during progressive failure is of course a subject in itself, and cannot be accounted for in any simple presentation of this type. However, in a global sense the total stress across such failure surfaces remains the same during progressive failure. The discontinuous nature of a rock slope should not be forgotten, when large stress concentrations are imagined. Jointed rock has a considerable potential for redistributing concentrations of stress (Cundall[9]).

The four critical failure planes illustrated in Figure 4 delineate four geometrically similar triangles, each having as their third apex the crest position (C). It is a simple matter to modify equation 3 to account for the different limit equilibrium equations describing the slope instability at each depth of excavation. Suppose an unchanged angle of friction was assumed for each failure plane. Then for limiting conditions at each depth of excavation the ratios of cohesion values would need to be as follows.

$$c_4 = {}^4/_3 \cdot c_3 = 2c_2 = 4c_1$$

This is clearly a hypothetical way of viewing the relative stability at each excavation depth, but it illustrates the obvious point that stability is successively reduced as excavation proceeds. It is pertinent to point out here that if the failure planes (joints) exhibited zero cohesion intercept, then the problem would be dimensionless, and the stability identical for all depths of excavation.

However, since peak shear strength envelopes are curved in reality the problem is far from dimensionless, whether the cohesion intercept (for zero normal stress) is zero or not.

2. THE CONCEPT OF AN UNSTABLE EXCESS

It has been established that individual slices will tend to be operating at different 'factors of safety' for equal shear strength parameters. It would therefore be useful if a method were developed, whereby the stability or instability of each slice could be computed separately. Integration of all the slices below the tension-crack would then give a more realistic interpretation to the problem.

An unstable excess concept is proposed, whereby the limit equilibrium equation (equation 2) is generalised to include states of stability other than the limiting case. This method was used for estimating the stability of slices during back analyses of failed slopes (Barton[11]).

The stability of any slice is dependent on the magnitude of the forces acting perpendicular and parallel to the failure plane beneath the slice, and on the shear strength exhibited by the failure plane. If the sign of the resultant force P is positive, this implies that a force of magnitude P, acting down the plane, is needed to bring that slice to limiting equilibrium. In other words the slice is stable. If on the other hand, P is negative, this implies an unstable slice which requires a force of magnitude P acting up the plane to keep the slice from sliding down. The slice has an 'unstable excess' of magnitude P. It is assumed that any unstable excess will be transmitted to the slice immediately below, but this aspect will be dealt with shortly.

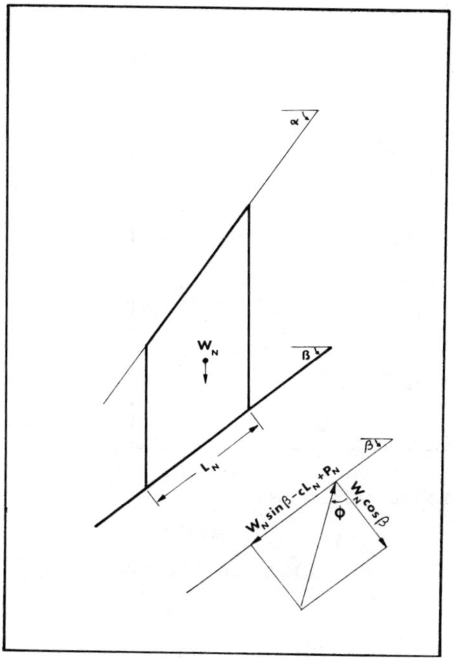

Figure 5. The limting equilibrium of forces beneath one slice.

Figure 5 illustrates the equilibrium of forces beneath an individual slice (number n) of weight W_n. The limiting equation is as follows:

$$\tan\phi = \frac{W_n \sin\beta - cL_n + P_n}{W_n \cos\beta}$$

$$\text{or } P_n = W_n \cos\beta(\tan\phi - \tan\beta) + cL_n \qquad (6)$$

where L_n = the length of sloping base beneath the slice.

Returning to the geometry of Figure 2 the new limiting equilibrium equation for the whole slope, from equation 6, will be as follow follows:

$$P = \frac{\gamma H^2}{2}(\cot\beta - \cot\alpha) \cdot \cos\beta\left[\tan\phi - \tan\beta\right] + \frac{c \cdot H}{\sin\beta} \qquad (7)$$

It has already been proposed that when splitting a part of the slope into slices, the width of the slices should be made equal to one another. Thus, in the upper triangle CBJ (see Figure 6) all the slices are of equal width. Likewise, in the lower triangle ACJ, the widths of the slices are also equal, though not necessarily equal to those of the upper triangle.

It is most convenient to treat the upper and lower triangles separately. It only remains to formulate the two limit equilibrium equations for these two triangles, for the 'unstable excess' values to be obtained for all the individual slices dividing each triangle.

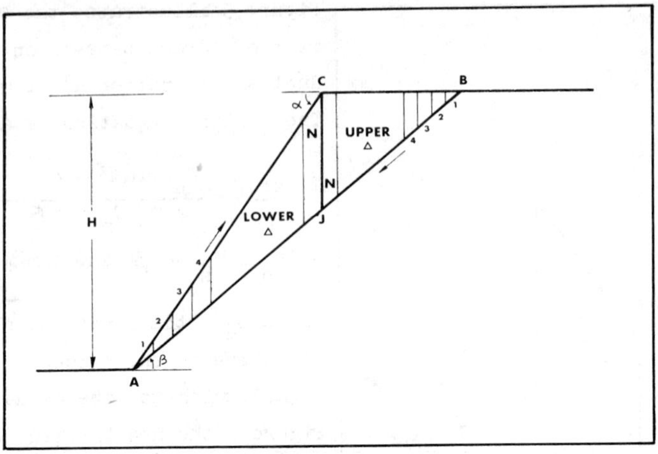

Figure 6. The nomenclature of the slices dividing the upper and lower triangles.

From the geomtry of Figure 6 the following limiting equilibrium equations can be obtained:

$$\underset{\triangle}{\text{UPPER}} : \quad P = \frac{\gamma H^2}{2} \cdot (1-\cot\alpha \tan\beta)\cdot(\cot\beta - \cot\alpha)\, \cos\beta \left[\tan\phi - \tan\beta\right] + \frac{cH\,(1-\cot\alpha \tan\beta)}{\sin\beta} \qquad (8)$$

$$\underset{\triangle}{\text{LOWER}} : \quad P = \frac{\gamma H^2}{2} \cdot (1-\cot\alpha \tan\beta)\cot\alpha \cos\beta \left[\tan\phi - \tan\beta\right] + \frac{cH\cot\alpha \tan\beta}{\sin\beta} \qquad (9)$$

If both the upper and lower triangles are divided into N slices each, then the individual lengths of the sloping bases beneath each slice are as follows:

$$\underset{\triangle}{\text{UPPER}} : \; 1/N.H\,\frac{(1-\cot\alpha \tan\beta)}{\sin\beta} \qquad \underset{\triangle}{\text{LOWER}} : \; 1/N.\frac{H(\cot\alpha \tan\beta)}{\sin\beta}$$

The other variables which are dependent on the slice widths are the weights of individual slices. From the geometrical similarity of Figure 3 it can be verified that the weights of the slices as proportions of the respective upper or lower triangles are:

$$1/N^2,\; 3/N^2,\; 5/N^2,\; \ldots\ldots\ldots\ldots\ldots\ldots \frac{2N-1}{N^2}$$

where the slices are taken in order (1 to N) as in Figure 6.

Thus, with the nomenclature of Figure 6 the limiting equilibrium equations giving the values of 'unstable excess' for each slice are:

UPPER Δ:

$$P_{1-N} = \left[\frac{1,3,5\ldots 2N-1}{N^2}\right]\frac{H^2\gamma}{2}.(1-\cot\alpha\tan\beta)\cdot(\cot\beta-\cot\alpha)\cos\beta\left[\tan\phi-\tan\beta\right] + \frac{1}{N}\cdot\frac{cH(1-\cot\alpha\tan\beta)}{\sin\beta} \qquad (10)$$

LOWER Δ:

$$P_{1-N} = \left[\frac{1,3,5\ldots 2N-1}{N^2}\right]\frac{H^2\gamma}{2}.(1-\cot\alpha\tan\beta)\cot\alpha\cos\beta\left[\tan\phi-\tan\beta\right] + \frac{1}{N}\cdot\frac{cH(\cot\alpha\tan\beta)}{\sin\beta} \qquad (11)$$

where: H = the height of the slope
γ = the average density of the rock mass
β^o = the inclination of the assumed failure plane
α^o = the inclination of the overall slope
c = the cohesion intercept exhibited by the failure plane
ϕ = the angle of friction exhibited by the failure plane
N = the number of slices dividing each triangle

Solution for P is simply a matter of substitution. Slices with P values that are positive are stable individually, and those with negative P values are unstable. The choice of N is arbitrary, since once equations 10 and 11 have been reduced to simple expressions by substitution, it takes only a few minutes longer to compute the P values for twenty slices (total of forty) than for ten slices (total of twenty).

The theoretical existence of a tension crack will be revealed by the P values obtained for the upper triangle. Taking the slices in order (from 1 to N), if a tension crack exists the P values will reduce successively from positive values, through zero to increasingly negative values - the greatest of these lying beneath the crest of the slope (the most overstressed position). The position of the tension crack is where the P value becomes zero. The rock mass above this position will theoretically be stable independently, and failure of the remainder of the slope will be characterised by a slide scar having a vertical rear face (depending on the second joint set) and a planar slide surface.

The method will be illustrated by some worked examples before moving on to considerations of progressive failure, which may be induced by excavation.

Worked example No. 1.

Firstly, an example will be chosen which illustrates the different results that are obtained when using a slices method, and when analysing the slope as a coherent 'wedge'. Figure 7 illustrates the problem to be considered. Substituting the data in equation 3 (the limiting equilibrium equation for a single coherent 'wedge' lying on a plane failure surface), it is found that failure will occur on one of the steeply dipping joints when the depth of excavation (H) exceeds 1000 feet.

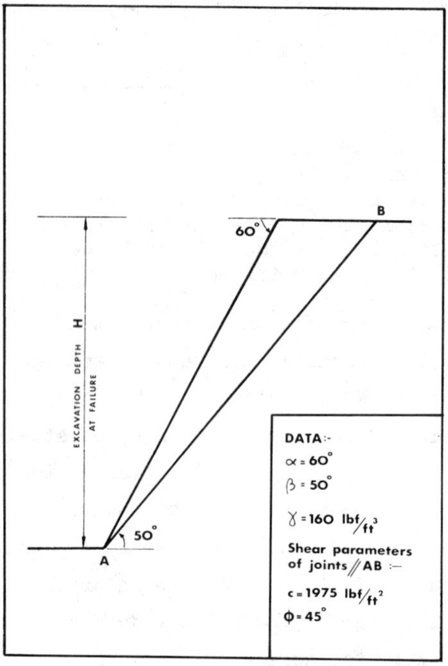

Figure 7. Data for worked example No. 1.

However, a method of slices analysis using equations 10 and 11 (assuming that the rock mass exhibits zero tensile strength in a direction perpendicular to the slice boundaries) predicts failure at a depth of excavation appreciably less than 1000 feet, as will be shown.

Substituting the data in equations 10 and 11 the following relations are obtained:

$$\text{UPPER}\,\Delta P_{1-N} = \frac{1,3,5\ldots 2N-1}{N^2}(-8.05\text{x}10^5) + \frac{1}{N}\cdot(8.05\text{x}10^5)\frac{\text{lbf}}{\text{ft}}$$

$$\text{LOWER}\,\Delta P_{1-N} = \frac{1,3,5\ldots 2N-1}{N^2}(-17.75\text{x}10^5) + \frac{1}{N}\cdot(17.75\text{x}10^5)\frac{\text{lbf}}{\text{ft}}$$

Choosing an N = 10 analysis (total of 20 slices), the P values of all the slices can be rapidly obtained and are as follows:

(Note: slice nomenclature as in Figure 6.)

P NUMBER (1 to N)	UPPER Δ	LOWER Δ
	Units: lbf/ft x 10^5	
P1	[+0.724	+1.598
P2	+0.564	+1.242
P3	+0.402	+0.887
P4	+0.242	+0.532
P5	+0.080]	+0.177
P6	-0.080	-0.177
P7	-0.242	-0.532
P8	-0.402	-0.887
P9	-0.564	-1.242
P10	-0.724	-1.598

It is important to realise that the striking symmetry of results is strictly a function of the data chosen for this example. The P values are for slices dividing a 1000 feet slope which is at limiting equilibrium, and their summation (P) is equal to zero as would be anticipated.

Since the failure surface is planar the total unstable excess is correctly interpreted as the summation of the individual P values of each slice. However, as pointed out before, the first few slices of the upper triangle may be stable (positive P values). These results have been backeted in the table above. Thus, in a 'no tension' analysis the overall unstable excess is the summation of all P values with the exception of the values in brackets.

Hence for H = 1000 ft, $\Sigma P = -2.013 \times 10^5$ lbf. per foot*
i.e. the slope has theoretically failed.

(* per foot refers to the third dimension)

Calculation of failure depth by intrapolation

A total of 20 slices (N = 10) have been chosen to divide the 1000 ft. slope. From previous arguments it will be apparent that, due to the uniform spacing of the slices, eighteen (N = 9) slices of unchanged width will cover 900 feet, sixteen (N = 8) slices will cover 800 feet, and so on. (See Figure 4.) Thus, within the results for twenty slices there exists all the information required for calculating the actual depth at which failure occurs. Hence the following table:

No. of slices included (N)	8	9	10
Equivalent depth (feet)	800	900	1000
ΣP lbf/foot x 10^5	+2.110	+.309	-2.013

The three results tabulated in the bottom row are sufficient for plotting a graph of depth of excavation versus unstable excess (ΣP). Where the curve crosses the $\Sigma P = 0$ line, the equivalent depth of excavation will be that at which limit equilibrium is reached. (See Figure 8.) The shear stress along the failure plane will just have reached the available shear strength.

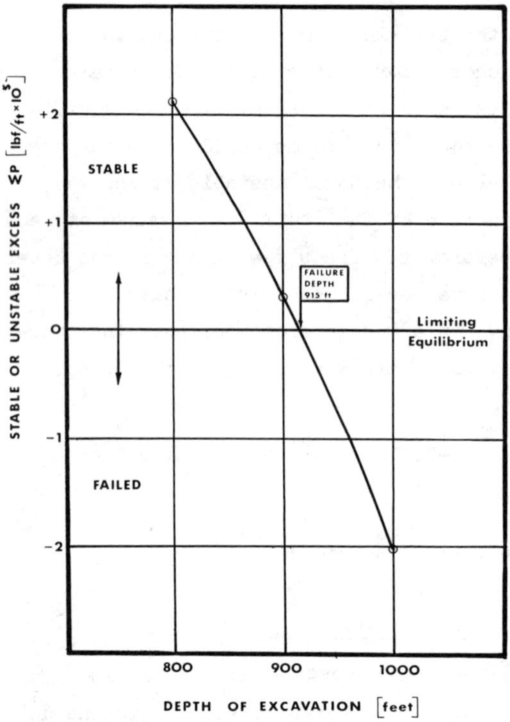

Figure 8. Graph of depth of excavation versus unstable excess

RESULTS

1) The intrapolated failure depth is 915 feet.
2) The tension crack is approximately 110 feet back from the crest of the slope. (Equation 12)*

(Markland[12] has shown that by minimising the length of failure surface lying beneath the corresponding weight of slope, the position of the tension crack can be found more directly. The distance of the tension crack behind the crest of the slope (t) is found to be purely dependent on geometry, and independent of the shear strength parameters: *

$$t = H\left[\sqrt{\cot\beta \cdot \cot\alpha} - \cot\alpha\right] \quad (12)$$

The angles (α) and (β) refer to the inclinations of the pit slope and failure surface as before. This method will be utilized shortly.)

Worked example No. 2 (using metric units)

Problem: At what depth will a pit fail, if excavated at 60° in jointed rock, where one of the joint sets dips into the pit, at an average of $52\frac{1}{2}^\circ$. The shear strength parameters of the joint set are:- $c = 2.5$ tons/m^2, $\phi = 45^\circ$. Density of the rock mass is equal to 2.5 tons/m^3.

As previously shown, the final result depends upon intrapolation of the results from a particular depth of excavation. The first step is therefore to make an informed guess of the depth at which failure will occur using equation 7. It is important that this value of H (or a guessed one) is greater than the actual depth of failure, since the result cannot be extrapolated, only intrapolated from the P values.

Try H = 100 metres. For an N = 10 analysis, substitution in equations 10 and 11 gives:

UPPER △ $P_{1-10} = \frac{(1,3,5,\ldots 19)}{100}(-109.0) + 7.8$ Tons/metre

LOWER △ $P_{1-10} = \frac{(1,3,5,\ldots 19)}{100}(-329.0) + 23.8$ Tons/metre

The P values are therefore as follows:

P NUMBER (1 to N)	UPPER △	LOWER △
	Units:	Tons/metre
P1	+6.7	+20.5
P2	+4.5	+13.9
P3	+2.3	+ 7.3
P4	+0.2	+ 0.8
P5	-2.0	- 5.8
P6	-4.2	-12.4
P7	-6.4	-19.0
P8	-8.6	-25.5
P9	-10.7	-32.2
P10	-12.9	-38.7

Slices 1,2,3 and 4 of the upper triangle are assumed to be independently stable as before. By selective summation of the P values for different depths of excavation, the second table of results are obtained, from which the failure depth can be intrapolated.

No. of slices included (N)	5	6	7	8
Equivalent depth (metres)	50	60	70	80
ΣP tons/metre	+34.7	+18.1	-7.3	-41.4

A plot of the four P values against equivalent depth of excavation indicates that a zero value of P (limiting conditions) is obtained when the depth of excavation is <u>67.5 metres</u>. Figure 9 illustrates the shape of the slide scar ABCD predicted by this method. AB is the crack which opens up at the surface, BC is the zone which is overstressed (negative P values), and CD is the remaining portion of the failure surface which fails due to the unstable excess of all the slices lying above it, between C and B.

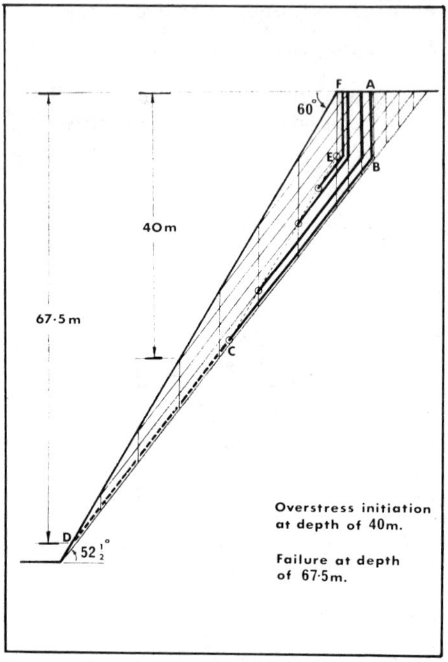

Figure 9. A diagrammatic representation of the overstressed joints predicted by worked example No. 2.

It will be apparent from the negative P values (see above table), that parts of several of the unfavourably dipping joints located above BC are also overstressed. The thick parallel lines shown in Figure 9 represent the lengths of some typical joints which are overstressed. Each of these will have a consequent tension crack opening to the surface. (This theoretical prediction appears to be frequently confirmed in practice.)

The overstressing mechanism is seen to initiate at a depth of 40 metres, and progresses with excavation until, at a depth of 67.5 metres, the total unstable excess becomes greater than the available stable excess. With the given assumptions, failure will be expected on the first joint to intersect the slope at a depth of 67.5 metres or more.

3. TWO CONCEPTS OF PROGRESSIVE FAILURE

Attention will first of all be focussed on the progression of the overstressed zone as excavation proceeds. A second, and most important factor to be considered, is the progressive failure brought about by the unstable shear strength characteristics of most joint surfaces.

A) Progression of overstressed zone

In figure 9 the lower end of each of the idealized overstressed joints is circled. It is suggested that at these points there exists the maximum unstable excess for the particular joint in question. One of these joints is illustrated in Figure 10.

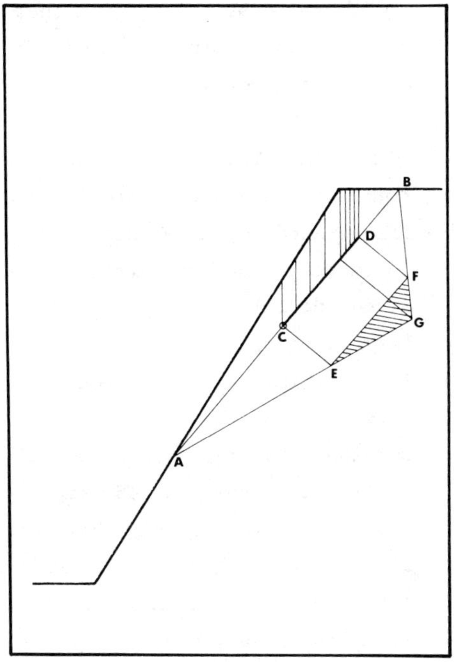

Figure 10. The distribution of shear stress assumed to act on an over-stressed joint, before displacements occur.

The overstressed portion CD, of joint AB, has negative P values. The area of the shear stress distribution triangle AGB which is in excess of the maximum shear strength level (CE = DF) is represented by the shaded area EFG. Each of the slices between C and D is overstressing the joint surface beneath it, and if shear displacements can occur with the these localized 'failures' then the unstable excess will progressively increase between D and C by simple summation. The slices between C and A will be supporting the full unstable excess of part CD, which is at a maximum at the slice interface above C. If this concept is accepted, then the slice above C can be expected to be the one most highly overstressed with respect to direct shear. For the present, the reduction in shear strength (and consequent rise in unstable excess) implied by these displacements will be ignored.

It is proposed that failure of the rock slope illustrated in Figure 9 could possibly occur on a failure surface defined by points FECD, rather than the originally proposed ABCD.

The portion EC is the assumed locus of maximally overstressed points on each critical joint that dips unfavourably into the slope. It is suggested that the rock mass might slide on a stepped surface between E and C, where each step is the secondary joint set (having zero tensile strength). In this way sliding would occur on portions of all the overstressed joints, stepping down progressively from one joint to the next immediately beneath. This concept is illustrated diagramatically in Figure 11.

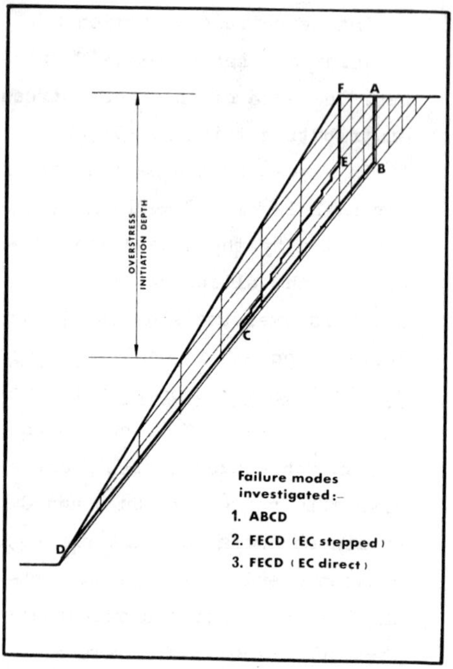

Figure 11. The multi-linear failure surfaces investigated.

It should be pointed out that the secondary joint set need not be vertical for this type of failure to be analysed by the proposed methods. The vertical slice boundaries and the vertical steps illustrated so far should just be taken as the most simple method of illustrating secondary joint opening. In reality these artificial planar boundaries between slices could be saw-toothed. (Eg. two orthogonal joint sets dipping at 45° each.) However, it is unlikely that such a system would have zero tensile strength in a horizontal direction, and in any case in practice at least one near-vertical joint set can be expected, more commonly two.

The stability of the stepped and direct failure modes illustrated in Figure 11 were checked, and compared with the original failure surface ABCD. The parameters of worked example No. 2 were used, with unchanged shear strength values:

(a) Stepped shear surface. P = +42.9 tons/metre (stable)

Note: CE is parallel to the 60° slope DF. However, the actual sliding surface is parallel to the joint dip of $52\frac{1}{2}°$. The consequent reduced length of shear surface is accounted for in the above.

(b) Direct shear surface. P = +4.6 tons/metre (just stable)

Note: It was assumed that both shear and normal components of stress would be transmitted across the vertical slice boundary above point C. The unstable excess for the upper surface (60° dip) was resolved into the dip direction of the lower surface ($52\frac{1}{2}^{o}$), by the correcting factor: cos (60° - $52\frac{1}{2}^{o}$). This method has been used previously (Barton[11]) for integrating the unstable excess of slices lying on differently inclined failure surfaces, when back-analysing failed slopes.

The latter mode of failure, although apparently predicting greater instability, is questionable because it will not be feasable unless the rock is very highly jointed. The location of the failure surface might then be independent of the structual discontinuities.

However, it is generally agreed that a 'circular stability analysis' can be usefully applied to rock slopes when the dimensions of the slope are very large compared to the joint spacing. The problem then approaches the soil mechanics end of the spectrum, and a rotational failure mode which is not structurally controlled can be anticipated, rather than failure mode No. 3 in Figure 11.

However, the more usual rock mechanics problem is one of relatively wide joint spacing. The potential failure surface is then almost totally controlled by structural discontinuities. Translational shearing then becomes the dominant mode of failure.

The progression of the overstressed zone as excavation (or erosion) proceeds has been illustrated as a possible mechanism for inducing a stepped multi-linear failure surface. (Mode No. 2 Figure 11). However, this mode of failure has been shown to be the most stable when using unchanged peak shear strength parameters.

Consequently a second mechanism of progressive failure has to be invoked to justify consideration of the stepped failure mode. This concerns the effects of displacement and weathering on the shear strength of joints in-situ.

B) <u>Unstable joint characteristics</u>

The shear strength parameters of rock joints have been idealized to those shown in Figure 12. The curved peak strength envelope is approximated to a straight line displaying (c) and (ϕ) parameters appropriate to a particular range of normal stress. Since this range is limited at low stress it may be necessary to interpret the curve in a piece-wise linear fashion.

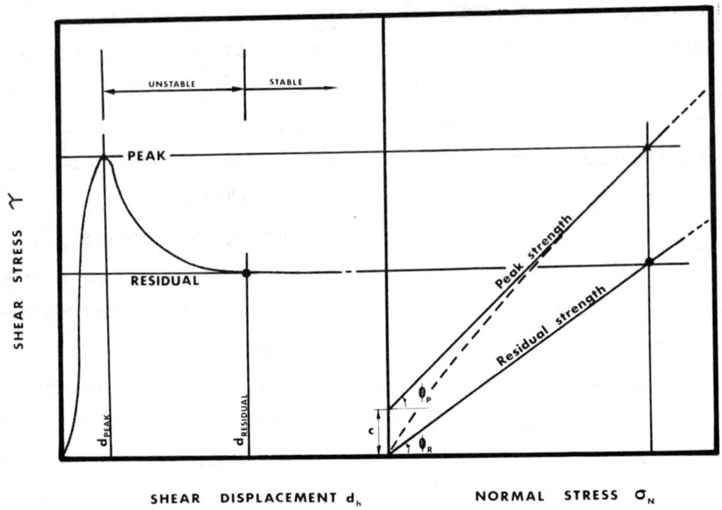

Figure 12. The simplified direct shear characteristics of an interlocking joint surface.

It will be noted that peak shear strength is reached after relatively small shear displacements. In addition the strength can drop significantly for displacements only a little greater than the peak displacements.

However, before peak or residual strength characteristics can be meaningfully applied to parts of an unstable slope, certain questions need to be considered.

1. Can relative shear displacements occur within the overstressed zone of the rock slope before final failure is approached?

2. If so, will the 'slices' at the lower (down dip) ends of the overstressed joints be those most highly overstressing the joints concerned in shear?

3. Will the fact that parts of the joints became overstressed at a relatively early stage of excavation, mean that their shear strength will have reduced by the time the failure depth is approached?

4. If displacements can occur within the rock slope, will these be large enough for a significant reduction from peak shear strength towards residual strength?

It is believed that the answer to all four questions is yes. However, it is realised that positive answers depend upon No. 1 being true. Certain observations can be cited in support.

(a) Mueller[13] has referred to slope bulging in the lower half of rock slopes of from 7 to 8% of the slope height. This exceptional figure was presumably for a slope in an advanced state of collapse.

(b) Tension cracks are regularly found at and behind the crest of slopes. Exceptionally these can be ten or more feet wide. A graben is often in evidence; the unstable portion below the crack having fallen vertically several feet. Ross Brown and Barton[14] recorded vertical and horizontal displacements of 5 and 2 metres respectively, at the crest of a 100 metre high slope that was in the process of failing. Kennedy and Niermeyer[15] recorded vertical and horizontal displacements of up to 2.5 and 5 metres respectively. This was measured on a bench roughly in the middle of the enormous Chuquicamata slide, approximately two weeks before catastropic failure occured.

(c) The excavated slope models reported by Barton[10] indicated significant down dip displacements on unfavourably dipping joints. These tended to be greatest where the overburden depth was greatest.

Down dip shearing displacements imply closure of joints in a directtion at right angles to the joints, if one assumes zero shear stresses at the free face of the excavation. Figure 10 illustrates the distribution of shear stress on a typical joint that is obtained when analysing the situation with self weight stress assumptions. The shear stress on AB builds up from zero at the free face (A) to a maximum beneath the crest of the slope. If there are joint sets intersecting AB, then closure across these can be expected, due to the incremental increase in shear stress when proceeding from A towards C. (The effect of shear stress on the pre-peak displacements occurring across tension joints and the closure of joints under normal stresses has been discussed by Barton[10].)

It has also been shown that small shear displacements induced across an interlocking joint surface can cause considerable increases in joint volume, depending upon the joint roughness. In an area of significant rainfall, this increase in effective permeability would surely accelerate the process of joint weathering. Over a period of time the shear strength of joint walls might be considerably reduced. In view of the fact that open pit mining operations are frequently in excess of 30 years duration, it is possible that rock slopes could be excavated in which overstressing of the peak strength of the joints initiated ten or more years before the final depth of the pit was reached.

It appears a practical possibility that residual strength could be approached on the critical stepped failure surfaces introduced earlier. From a design point of view such an assumption would surely be a more realistic 'worst case' than assigning a universal residual strength to the whole slope. Even in extreme cases slope failure is hardly a dimensionless problem.

An opposite point of view and one which is considered somewhat under-conservative, is that of Jennings and Robertson[16], and Jennings[17], who developed an involved 'mathematical' treatment to account for the non-continuous nature of joints. They considered a stepped failure surface involving sliding down one joint set, opening on a second joint set and shear and tensile failure of the intact rock between the two. Some of the parameters involved are almost impossible to estimate in the field.

For this reason the approach adopted by Terzaghi[18] is considered more useful. He suggested that the rock bridges between joints would in all probability be eliminated by splitting, when the shearing stresses rose significantly due to erosion or excavation of the slope concerned. It is anyway safer to design slopes on (c) and (ϕ) parameters no larger than those obtained from shear tests on continuous joints. The contribution of any failure through intact material is best ignored for design purposes.

4. ANALYSIS OF MULTI-LINEAR FAILURE SURFACE

Figure 13 represents a section through an idealized jointed rock slope. It is assumed that the strike of the joints in relation to the slope face, allows the problem to be analysed in two dimensions. As already indicated this is the worst case from the point of view of stability.

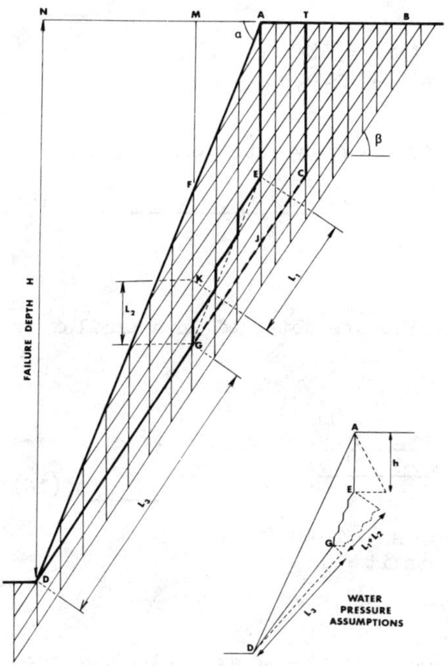

Figure 13. The detailed geometry of the multi-linear failure surface.

In the absence of computing methods, or design charts such as those developed by Hoek[19], the failure depth (H) for a fully drained block TCGD will be found by utilizing equations 10 and 11.

The position of the corresponding tension crack TC can be most easily found by using equation 12 (Markland[12]).

Thus: $AT = H\left[\sqrt{\cot\alpha \cdot \cot\beta} - \cot\alpha\right]$

The proposed failure surface AEGD conveniently divides the unstable part of the slope into two portions:

1. The block AEGF, lying on the stepped surface EG. The mean dip of EG is parallel to the slope face AF. Since residual strength is envisaged for the stepped surface (for design purposes), it will be termed the 'residual block'.

2. The 'toe triangle' FGD. Peak strength parameters are suggested for the planar surface GD.

A) Estimation of failure depth for a drained slope

(a) The dimensions of the residual block AEGF are obtained from the geometry of Figure 13.

$$AB = H\,(\cot\beta - \cot\alpha) \quad \text{and} \quad TC = TB \cdot \frac{AJ}{AB}$$

$$\therefore AE = TC = \frac{H(1-\cot\alpha \cdot \tan\beta)(\cot\beta - \sqrt{\cot\alpha \cdot \cot\beta})}{(\cot\beta - \cot\alpha)} \qquad (13)$$

$$NA = H\cot\alpha \quad \text{and} \quad MA = NA \cdot \frac{EJ}{AJ} = NA \cdot \frac{AT}{AB}$$

$$\therefore MA = \frac{H\cot\alpha(\sqrt{\cot\alpha \cdot \cot\beta} - \cot\alpha)}{(\cot\beta - \cot\alpha)} \qquad (14)$$

Since AE = FG:

1. the area of the 'residual block' AEGF = AE•MA
2. the length of shear surface beneath it = L_1
   where $L_1 = MA \cdot \sec\beta$

The unstable excess (P_1) of this block is given by equation 6, thus:

$$P_1 = W_1 \cos\beta\left[\tan\phi_r - \tan\beta\right] \qquad (15)$$

(Note residual parameters: $\phi = \phi_r$, $c = 0$.)

(b) The dimensions of the toe triangle FGD are obtained in a similar manner:

FG = AE = TC (equation 13)

$$NM = NA \cdot \frac{TB}{AB} = \frac{H\cot\alpha(\cot\beta - \sqrt{\cot\alpha \cdot \cot\beta})}{(\cot\beta - \cot\alpha)} \qquad (16)$$

1. the area of the 'toe triangle' FGD = $\frac{1}{2}$•FG•NM
2. the length of shear surface beneath it = L_3
   where $L_3 = NM \cdot \sec\beta$

If peak strength parameters are assumed for the surface DG, then the stable excess (P_3) is given by:

$$P_3 = W_3 \cos\beta\left[\tan\phi_p - \tan\beta\right] + cL_3 \qquad (17)$$

For complete failure on the multi linear surface AEGD, the unstable excess of the residual block must just exceed the stable excess for the toe triangle. At limiting conditions:

$$P_1 + P_3 = 0$$

The failure depth is obtained by evaluating equations 13 to 17 in general terms, with H the unknown. Several values of H can then be substituted (all less than for block TCGD). A graph of stable or unstable excess plotted against depth H will give an intercept $\Sigma P = 0$ at the required failure depth. (See Figure 8 for example.)

B) Estimation of failure depth for a transient water pressure assumption

It is believed that the above assumption of residual shear strength for the stepped portion of the failure surface represents a realistic 'worst case' for design purposes. However, there are many situations which also require pessimistic assumptions for possible water pressure distributions. Of the five slope failures investigated in the Rio Tinto area of southern Spain (Ross Brown and Barton[14]), three were almost certainly caused by the surface run off entering tension cracks through fractured drainage routes. The possibility of tension cracks becoming filled with water must therefore be considered in areas where flash floods and consequent transient water pressures can be anticipated.

The water pressure distribution assumed as one possible 'worst case' is shown in the inset to Figure 13. The full tension crack generates a hydrostatic pressure at its base of:

$$U_w = h\gamma_w$$

where h the depth of crack is given by equation 13, and γ_w is the density of water. It is assumed that this pressure reduces linearly in proportion to the length of drainage path, such that at D the exit seepage pressure is zero. It may be optimistic to expect constant permeability down the length of EGD. However, there are in reality an infinite variety of possible assumptions. A tension crack only partially filled with water, and a reduced permeability along EGD is an alternative which could generate equal instability to the one assumed.

Some attention may need to be paid to the equilibrium of the slide mass, since a severe assumption of water pressure can lead to zero effective pressures across the potential shear surfaces. Whether this could be expected at failure is not known.

From the geometry of Figure 13 :

1) Depth of tension crack $h = AE = TC$ (equation 13)
2) Length of inclined surface $L_1 = MA \cdot \sec\beta$ (equation 14)
3) Total length of vertical steps between E and G = L_2
   where $L_2 = EJ = AJ.\frac{AT}{AB}$

$$\therefore\ L_2 = \frac{H(1-\cot\alpha \cdot \tan\beta)\,(\sqrt{\cot\alpha \cdot \cot\beta} - \cot\alpha)}{(\cot\beta - \cot\alpha)} \qquad (18)$$

4) Length of inclined surface between G and D = L_3
   where $L_3 = NM \cdot \sec\beta$ (equation 16)

The mean water pressures assumed to act on the four surfaces above are as follows:

1) AE (length h) $\quad U_w = \frac{1}{2}h\gamma_w \qquad (19)$

   This has force components perpendicular and parallel to the dip direction (β) of:

$$\text{(a)}\quad \tfrac{1}{2}h^2\gamma_w \sin\beta$$
$$\text{(b)}\quad \tfrac{1}{2}h^2\gamma_w \cos\beta$$

2) EG (length $L_1 + L_2$)
Since the pressure at G is $\left[\dfrac{L_3}{L_1 + L_2 + L_3}\right] h\gamma_w$

the mean water pressure between E and G is:

$$\left[\frac{L_1 + L_2 + 2L_3}{L_1 + L_2 + L_3}\right]\frac{h\gamma_w}{2} \qquad \text{——— (20)}$$

(a) This is distributed on the vertical steps of length L_2 with components $\sin\beta$ and $\cos\beta$ as before.

(b) It is also distributed on the inclined surfaces of length L_1 (Perpendicular component only.)

3) GD (length L_3)
The mean water pressure is $\left[\dfrac{L_3}{L_1 + L_2 + L_3}\right]\dfrac{h\gamma_w}{2}$ ——— (21)

The failure depth for these transient water pressure assumptions is obtained by calculating the unstable excess (P_1) of the residual block AEGF, and the stable excess (P_3) of the toe triangle FGD. Once again, at failure:

$$P_1 + P_3 = 0$$

The limiting equilibrium equation for the residual block can be derived from equation 6. The tangent of the friction angle is equated to the ratio of forces acting parallel to the dip (β), and those acting perpendicular to it, thus:

$$\tan\phi = : \frac{W_1\sin\beta + \tfrac{1}{2}h^2\gamma_w\cos\beta + \left[\dfrac{L_1 + L_2 + 2L_3}{L_1 + L_2 + L_3}\right]\dfrac{h\gamma w}{2}\cdot L_2\cos\beta - cL_1 + P_1}{W_1\cos\beta - \tfrac{1}{2}h^2\gamma_w\sin\beta - \left[\dfrac{L_1 + L_2 + 2L_3}{L_1 + L_2 + L_3}\right]\dfrac{h\gamma w}{2}\cdot L_2\sin\beta - \left[\dfrac{L_1+L_2+2L_3}{L_1+L_2+L_3}\right]\dfrac{h\gamma w\cdot L_1}{2}} \qquad \text{——— (22)}$$

Hence P_1. Note: For residual parameters, $\phi = \phi_r$, $c = o$.
The limiting equilibrium equation for the toe triangle is:

$$\tan\phi_P = \frac{W_3\sin\beta - cL_3 + P_3}{W_3\cos\beta - \left[\dfrac{L_3}{L_1 + L_2 + L_3}\right]\dfrac{\gamma w h L_3}{2}} \qquad \text{——— (23)}$$

WORKED EXAMPLE

The parameters of worked example No. 1 will be used to illustrate the effects of the given failure mode on stability. (See Figure 7).

$\alpha = 60^{\circ}$ $\quad \phi_p = 45^{\circ}$ $\quad \gamma = 160$ lbf/ft^2 per ft.

$\beta = 50^{\circ}$ $\quad c = 1975$ lbf/ft. per ft.

The residual shear strength parameters will be assumed as follows:

$$\phi_r = 30^{\circ}, \quad c = 0$$

Substitution of the peak shear strength parameters resulted in the following estimates of failure depth. (See Figure 13 for lettering.)

1. Coherent wedge ABD sliding on plane DB with no tension crack. H = 1000 feet. Fully drained. Equation 7.
2. Block ATCD. Method of slices and tension crack separation. H = 915 feet. Fully drained. Equations 10 and 11.

Equations 13 to 17 were used to calculate the failure depth for the assumption of a multi-linear failure surface, with residual strength on the stepped surface between E and G. With the given geometry, and fully drained conditions:

3. AE = 0.171 H feet $\quad$ NM = 0.314 H feet

MA = 0.264 H feet $\quad$ L_3 = 0.489 H feet

$P_1 = (-)2.84\,H^2$ (lbf/foot) $\quad$ $P_3 = 965\,H - 0.524\,H^2$ (lbf/foot)

$$\therefore\ \Sigma P = P_1 + P_3 = 965\,H - 3.364\,H^2 \text{ lbf/foot}$$

A graph of ΣP versus H gives the failure depth as 290 feet

Equations 18 to 23 were used to calculate the failure depth for the transient water pressure assumptions. Once again residual parameters were taken to describe the remaining shear strength on the stepped part of the failure surface.

4. L_1 = 0.264 H feet $\quad$ $P_1 = (-)6.92H^2$ lbf/foot

L_2 = 0.143 H feet $\quad$ $P_3 = 965\,H - 1.747\,H^2$ lbf/foot

$$\therefore\ \Sigma P = P_1 + P_3 = 965\,H - 8.587\,H^2 \text{ lbf/foot}$$

A graph of ΣP versus H gives the failure depth as 115 feet

It should be noted that the above assumption of a full tension crack (for a given geometry and density) generated such high water pressures that the effective normal stress on the inclined shear surfaces below the residual block just became negative. In other words the shear strength assumptions for this surface became meaningless. However, the effective normal stresses beneath the toe triangle remained positive with the given seepage assumption.

It may be argued that this state of affairs is preferable to one in which assumptions of poor drainage cause the toe triangle to 'float'. However, this only serves to highlight the uncertainties of what form stress distributions can take at failure. An effective normal stress reducing to zero locally is certainly an ideal failure mechanism. Is it right to reject it purely from a subjective assessment of what seems possible?

CONCLUSIONS

The following observations should be made as qualifications to the multi-linear failure mode:

1. The ratio of the depth of tension crack (TC) to the depth of the shear surface beneath the crest (AJ) was found by a limit analysis. This was performed with the assumption of drained conditions, and for a particular geometry.

   The ratio determined what proportion of the total shear surface ($L_1 + L_3$) was stepped (L_1). This proportion was assumed constant for a given geometry for all states of stress. In other words, for all conditions which produced failure, a geometrically similar shear surface was assumed.

2. The concept of a multi-linear failure mode was based on the effect of excavation on the overstressing of joints. The weakening mechanisms of time and displacement were introduced to predict a failure surface with a stepped portion of reduced strength.

   The adoption of residual strength for this part of the failure surface led to considerably reduced failure depths. (i.e. 915 feet down to 290 feet). Paradoxically, if the slope was designed to the reduced height in the first instance, then in all probability the stepped/residual mode would not initiate. Consequently, the proposed methods are conservative, and serve to emphasize the severity of a global residual strength assumption.

3. The multi-linear residual failure mode has been demonstrated as being readily amenable to manual methods of analysis, when using simple assumptions of geometry and shear strength. Having established the philosophy of the method it is advantageous to computer programme the analysis so that more realistic in-situ data can be incorporated, should this be available. This might include a non-linear peak shear strength envelope based on a statistical analysis of the joint roughness (Barton[20]) and more detailed information concerning joint structures and water pressures.

REFERENCES

1. Skempton, A.W. and Hutchinson, J.N. Stability of natural slopes and embankment foundations. State of the art report. VIIth Int. Conf. Soil Mech. Found. Eng. Mexico 1969, pp. 291-341.

2. Wittke, W. A numerical method of calculating the stability of slopes in rocks with systems of plane joints. English translation: Imp. Coll. Rock Mech. Res. Rept. No. 4 Sept. 1970. (From original in German: Rock Mech. and Eng. Geol. Suppl. I. 1964)

3. Londe, P. and Vigier, G. and Vormeringer, R. Stability of rock slopes, a three dimensional study. J. Soil Mech. Found. Div. Proc. A.S.C.E. S.M.1, Jan. 1969 pp. 235-262.

4. Londe, P. and Vigier, G. and Vormeringer, R. Stability of rock slopes - graphical methods. J. Soil Mech. Found. Div. Proc. A.S.C.E. S.M.4. July 1970, pp. 1411-1434.

5. John, K.W. Graphical stability analyses of slopes in jointed rock. J. soil Mech. Found. Div. Proc. A.S.C.E. S.M.2. March 1968 pp. 497-526.

6. Mueller, L. and John K.W. Recent development of stability studies of steep rock slopes in Europe. Trans. Soc. Min. Engrs. 1963.

7. Bjerrum, L. and Jorstad, F.A. Stability of rock slopes in Norway. Norwegian Geotechnical Institute. Pub. No. 79, Oslo 1968.

8. St.John, C.M. A note on the use of stress distributions in slope stability calculations. Contribution to discussion. Symp. on Open Pit Planning. Johannesburg. 1970.

9. Cundall, P.A. The measurement and analysis of accelerations in rock slopes. Ph.D. Thesis. University of London. 1971.

10. Barton, N.R. A model study of the behaviour of steep excavated rock slopes. Ph.D. Thesis. University of London. 1971.

11. Barton, N.R. Estimation of in situ shear strength from back analysis of failed rock slopes. Symp. of Int. Soc. for Rock Mech. "Fissuration des roches" Nancy. Oct. 1971.

12. Markland, J. Private communication. Rock Mech. Dept. Imp. Coll., London 1970.

13. Mueller, L. The stability of rock bank slopes and the effect of rock water on same. Int. J. Rock Mech. Min. Sci. Vol I, 1964, pp. 475–504.

14. Ross Brown, D.M. and Barton, N.R. Rock mechanics considerations and the stability of the opencast mines at Rio Tinto, Spain. Imp. Coll. Rock Mech. Res. Rept. No. D7. Sept. 1969, pp. 1–81.

15. Kennedy, B.A. and Niermeyer, K.E. Slope monitoring systems used in the prediction of a major slope failure at the Chuquicamata Mine, Chile. Symp. on Open Pit Planning. Johannesburg 1970.

16. Jennings, J.E. and Robertson, A. MacG. The stability of slopes cut into natural rock. VIIth Int. Conf. Soil Mech. Found. Eng. Mexico 1969. pp. 585–590.

17. Jennings, J.E. A mathematical theory for the calculation of the stability of slopes in open cast mines. Symp. on Open Pit Planning. Johannesburg 1970.

18. Terzaghi, K. Stability of steep slopes on hard unweathered rock. Géotechnique. Vol. XII. No. 4. 1962. pp. 251-270.

19. Hoek, E. Estimating the stability of excavated slopes in opencast mines. Inst. Min. and Metall. Trans. Vol. 79, Oct. 1970, pp. 109A-132A.

20. Barton, N.R. A relationship between joint roughness and joint shear strength. Symp. of Int. Soc. for Rock Mech. "Fissuration des roches", Nancy, Oct. 1971.

TECTONIC STRESSES AND ROCK SLOPE STABILITY

By Jerry S. Dodd,[1] and H. Walter Anderson[2]

The finite element method of stress analysis is commonly employed in the study of the stability of rock as well as soil slopes. Generally a two-dimensional analysis is performed because this restriction allows broader applications with present computer capacities.

Two-dimensional or planar stress analyses assume either a zero stress (plane stress) or zero strain (plane strain) condition on the parallel lateral surfaces of the model. Therefore, if a two-dimensional analysis is to represent actual field conditions, the in situ stress perpendicular to the finite element model must be nil or that due only to the Poisson effect.

In situ rock stress measurements have shown that horizontal tectonic stresses are generally at least equal to and many times exceed the vertical overburden stress(1).[3] At Grand Coulee Dam the Bureau of Reclamation recently measured horizontal in situ rock stresses which are about six times the vertical overburden stress. So, if a planar finite element stress analysis is to be meaningful, excessive horizontal stress must be considered.

A technique which can be used with a two-dimensional analysis to simulate a three-dimensional condition is to apply a given lateral hydrostatic stress to the planar two-dimensional finite element model.

---

1 Head, Geotechnology Section, Division of Design, U.S. Bureau of Reclamation, Denver, Colorado 80225

2 Structural Engineer, Division of Design, U.S. Bureau of Reclamation, Denver, Colorado 80225

3 Number in parentheses refers to listed references.

The hydrostatic stress represents a known horizontal tectonic stress. For best approximations, it must be assumed that the plane of the finite element model coincides with an in situ principal stress plane - a reasonable assumption for a planar analysis oriented perpendicular to the face of a rock cut or slope.

By applying a lateral hydrostatic stress to a finite element the shear stresses at the lateral boundaries remain zero. Stress conditions for a finite element under these assumptions are shown on Fig. 1, where σ is a normal stress and τ is a shearing stress. The stress conditions are:

$$\sigma_{xx} = f(x,y)$$

$$\sigma_{yy} = f(x,y)$$

$$\sigma_{zz} = C$$

$$\tau_{xy} = f(x,y)$$

$$\tau_{yz} = 0$$

$$\tau_{zx} = 0$$

The stresses σ_{xx}, σ_{yy}, and τ_{xy} are the stresses in the plane of the finite element, and σ_{zz} is the normal hydrostatic stress on the lateral boundaries of the finite element. The stress σ_{zz} is given and remains constant for a given element, but may vary from element to element within the finite element model.

From the theory of elasticity the stress equations of equilibrium are:

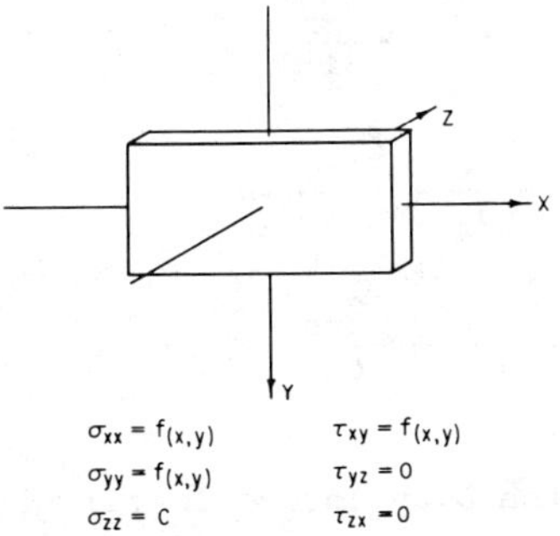

STRESSES ON A PLANAR FINITE ELEMENT

FIGURE 1

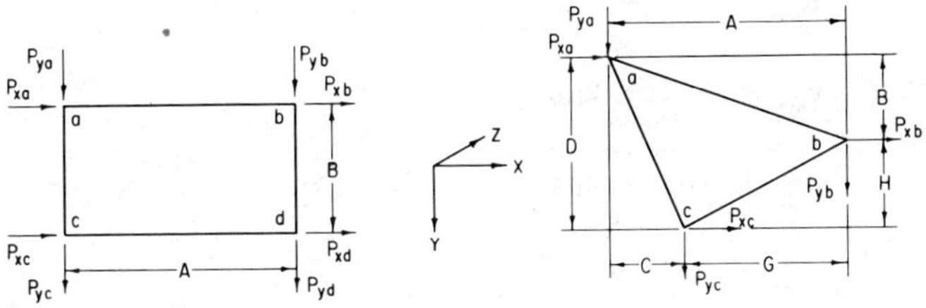

ELEMENT DIMENSIONS AND NODAL FORCES

FIGURE 2

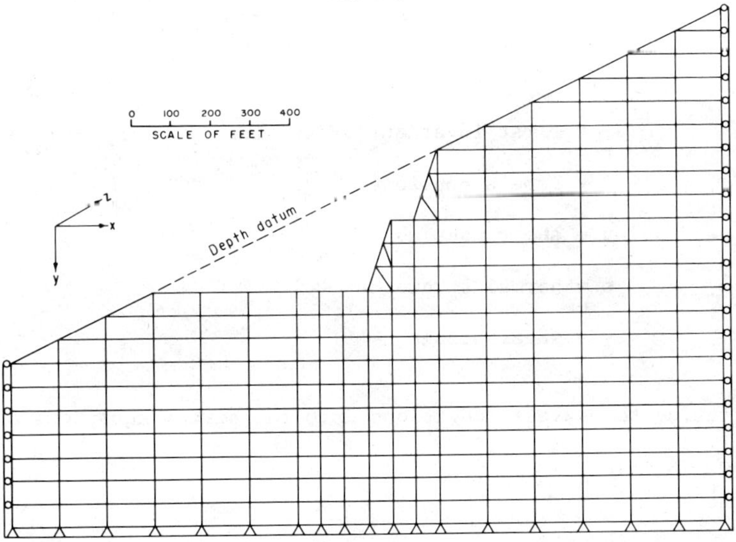

FINITE ELEMENT MESH

FIGURE 3

$$\frac{\delta \sigma_{xx}}{\delta x} + \frac{\delta \tau_{yx}}{\delta y} + \frac{\delta \tau_{zx}}{\delta z} + F_x = 0$$

$$\frac{\delta \tau_{xy}}{\delta x} + \frac{\delta \sigma_{yy}}{\delta y} + \frac{\delta \tau_{zy}}{\delta z} + F_y = 0$$

$$\frac{\delta \tau_{xz}}{\delta x} + \frac{\delta \tau_{yz}}{\delta y} + \frac{\delta \sigma_{zz}}{\delta z} + F_z = 0$$

where F_x, F_y, F_z are the body forces in the x, y, and z directions, respectively, and the stress-strain relationships are:

$$\sigma_{xx} = \lambda J_1 + 2\mu\varepsilon_{xx}$$

$$\sigma_{yy} = \lambda J_1 + 2\mu\varepsilon_{yy}$$

$$\sigma_{zz} = \lambda J_1 + 2\mu\varepsilon_{zz}$$

$$\tau_{xy} = \mu\gamma_{xy}$$

$$\tau_{yz} = \mu\gamma_{yz}$$

$$\tau_{zx} = \mu\gamma_{zx}$$

where J_1 = first invariant of strain $(\varepsilon_{xx} + \varepsilon_{yy} + \varepsilon_{zz})$

λ = Lamé's constant

μ = shear modulus

ε = normal strain

γ = shear strain

Substituting the stress conditions into the stress equations of equilibrium, we have:

$$\frac{\delta \sigma_{xx}}{\delta x} + \frac{\delta \tau_{yy}}{\delta y} + F_x = 0$$

$$\frac{\delta \tau_{xy}}{\delta x} + \frac{\delta \sigma_{yy}}{\delta y} + F_y = 0$$

and the stress-strain relations become:

$$\sigma_{xx} = \lambda J_1 + 2\mu\varepsilon_{xx}$$

$$\sigma_{yy} = \lambda J_1 + 2\mu\varepsilon_{yy}$$

$$\sigma_{zz} = \lambda J_1 + 2\mu\varepsilon_{zz} = C$$

$$\tau_{xy} = \mu\gamma_{xy}$$

$$\tau_{yz} = 0$$

$$\tau_{zx} = 0$$

From the third equation above, we can obtain the relationship:

$$\varepsilon_{zz} = \frac{C - \lambda(\varepsilon_{xx} + \varepsilon_{yy})}{\lambda + 2\mu}$$

With this value of ε_{zz} the first strain invariant J_1 becomes:

$$J_1 = (\varepsilon_{xx} + \varepsilon_{yy})\frac{2\mu}{\lambda+2\mu} + \frac{C}{\lambda+2\mu}$$

Substituting the value for J_1 in the stress-strain relations yields:

$$\sigma_{xx} = \frac{2\lambda\mu}{\lambda+2\mu}(\varepsilon_{xx} + \varepsilon_{yy}) + 2\mu\varepsilon_{xx} + \frac{\lambda C}{\lambda+2\mu}$$

$$\sigma_{yy} = \frac{2\lambda\mu}{\lambda+2\mu}(\varepsilon_{xx} + \varepsilon_{yy}) + 2\mu\varepsilon_{yy} + \frac{\lambda C}{\lambda+2\mu}$$

$$\sigma_{zz} = C$$

$$\tau_{xy} = \mu\gamma_{xy}$$

$$\tau_{yz} = 0$$

$$\tau_{zx} = 0$$

These equations are the general stress-strain relationships for plane hydrostatic stress. For the classical condition of plane stress C = 0.

In terms of the modulus of elasticity E and Poisson's ratio ν, the quantity:

$$\frac{\lambda C}{\lambda+2\mu} \text{ becomes } \frac{\nu}{1-\nu}C \text{ (the term E cancels out)}$$

These general stress-strain equations show that an analysis which considers the stress σ_{zz} is possible and is dependent only upon a two-dimensional strain (displacement) field.

To solve the above stress-strain equations by the finite element method, the fixed-end stresses within the element as well as the fixed-end nodal forces due to the lateral load on the element must be known. When the nodes, and hence the edges, of a finite element are fixed against displacement, the stress σ_{zz} uniformly distributed over the planar lateral boundaries of an element will produce nodal forces in the plane of the finite element. Under these conditions ε_{xx}, ε_{yy}, and γ_{xy} are zero, and the stress-strain equations become:

$$\sigma_{xx} = \frac{\nu}{1-\nu}C$$

$$\sigma_{yy} = \frac{\nu}{1-\nu}C$$

$$\sigma_{zz} = C$$

$$\tau_{xy} = 0$$

$$\tau_{yz} = 0$$

$$\tau_{zx} = 0$$

The nodal forces due to the above constant stresses σ_{xx} and σ_{yy} for finite elements of unit thickness are as follows, see Fig. 2.

For a rectangular element:

$$P_{xa} = P_{xc} = \alpha B$$

$$P_{xb} = P_{xd} = -\alpha B$$

$$P_{ya} = P_{yb} = \alpha A$$

$$P_{yc} = P_{yd} = -\alpha A$$

For a triangular element:

$$P_{xa} = \alpha(D-B)$$

$$P_{ya} = \alpha(A-C)$$

$$P_{xb} = -\alpha(B+H)$$

$$P_{yb} = \alpha(A-G)$$

$$P_{xc} = \alpha(D-H)$$

$$P_{yc} = -\alpha(C+G)$$

where $$\alpha = \frac{\nu\sigma_{zz}}{2(1-\nu)}$$

By loading the finite element model with these nodal forces, the nodal forces due to dead load, and any other proper nodal forces, and by considering the known fixed-end stresses within the elements, a two-dimensional plane stress finite element analysis embodies the influence of the σ_{zz} stress.

Three finite element analyses of a rock slope with varying normal stresses on the lateral boundary are shown in Fig. 4, Fig. 5, and Fig. 6 with the finite element mesh depicted in Fig. 3. In Fig 4 the

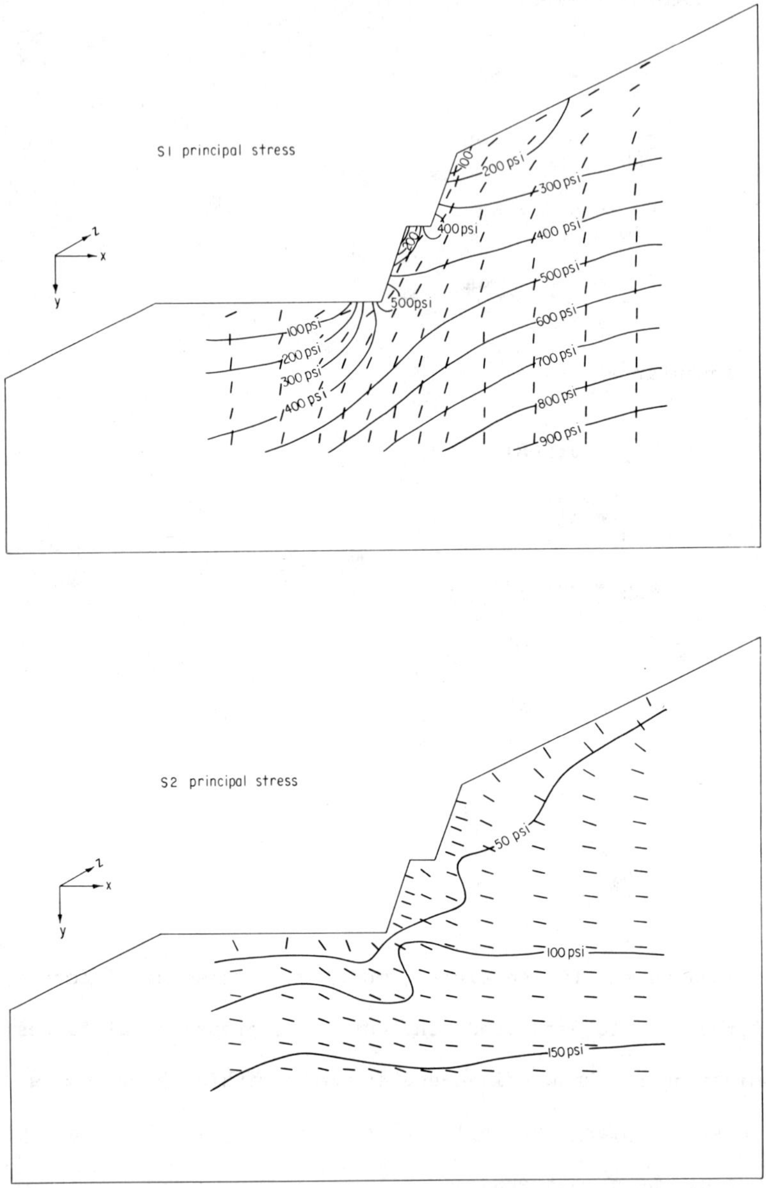

MAGNITUDE AND DIRECTION OF PRINCIPAL STRESSES
PLANE STRAIN
FIGURE 4

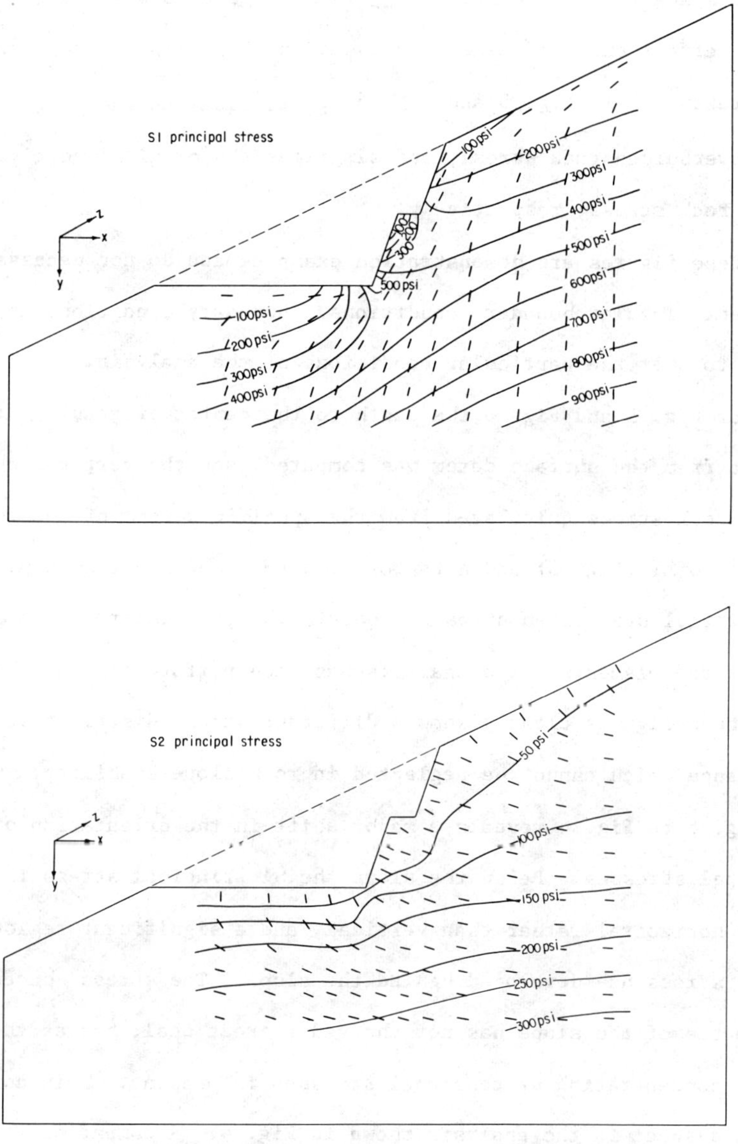

MAGNITUDE AND DIRECTION OF PRINCIPAL STRESSES
σ_{ZZ} = 1 x ORIGINAL VERTICAL OVERBURDEN ROCK STRESS
FIGURE 5

stress analysis is a plane strain analysis - the usual form of a rock mechanics analysis. The stress σ_{zz} in this case is a result of the Poisson effect only and generally is about 20 percent of the overburden rock pressure. In Fig. 5 and Fig. 6 σ_{zz} is equal to the original vertical overburden rock stress, and six times the original vertical overburden rock stress, respectively.

These figures are demonstration examples and do not necessarily represent singular boundary conditions. Boundary conditions may be varied to suit the particular conditions of the analysis.

For Fig. 5 and Fig. 6 the depth to the center of gravity of each element from the surface datum was computed, and the vertical overburden rock stress calculated from the specific weight of the rock. A factor of 1 (Fig. 5) and a factor of 6 (Fig. 6) was then applied to the vertical overburden stress to obtain the particular stress σ_{zz} for each of the elements. The analysis was then performed as described.

These figures clearly show a different stress distribution - a difference which cannot be neglected in rock slope stability. Comparing Fig. 6 to Fig. 4 reveals a major shift in the orientation of the principal stresses. Below the slope the S1 principal stress is essentially horizontal rather than vertical, and a significant region of tensional stress has developed behind the slope. The stress concentration at the toe of the slope has not changed a great deal, but at the berm a stress concentration of tensional stresses is present. This condition is not suspect in the analysis shown in Fig. 4. A comparison of Fig. 5 to Fig. 4 shows a noticeable increase in the S2 principal stress. Little change is noted in the orientation of the principal stress and the magnitude of the S1 stresses, except for a slight reduction in the stress concentrations at the toe of the slope and at the berm.

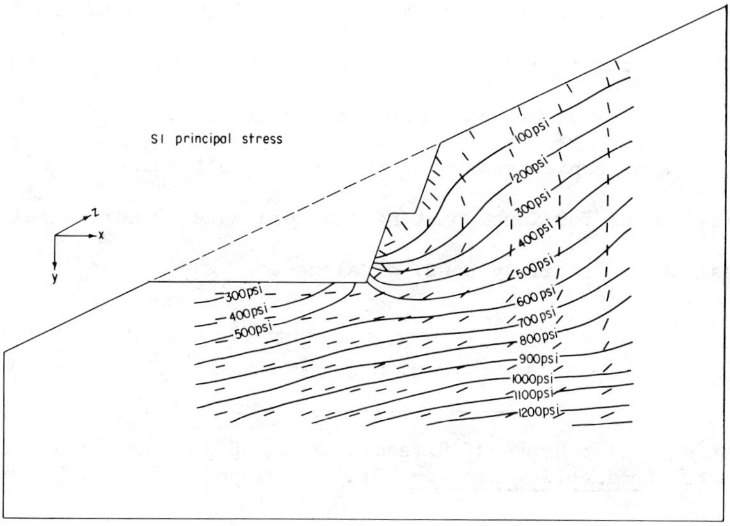

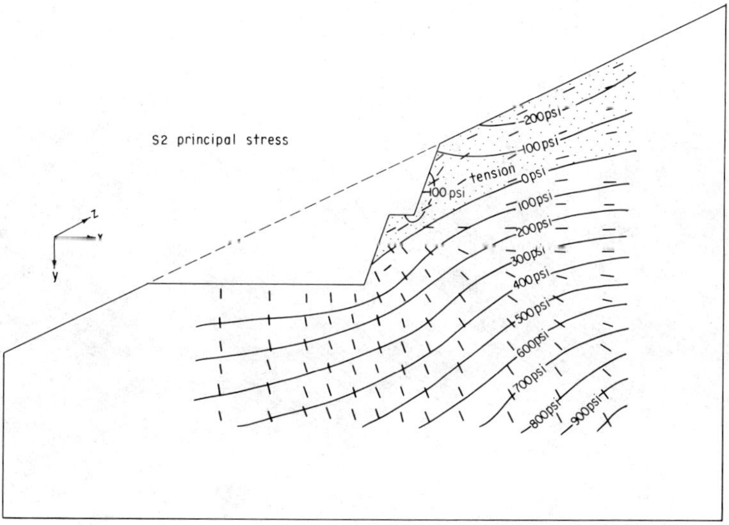

MAGNITUDE AND DIRECTION OF PRINCIPAL STRESSES
σ_{ZZ} = 6 x ORIGINAL VERTICAL OVERBURDEN ROCK STRESS
FIGURE 6

A critical evaluation will reveal that this technique does not produce exact solutions for three-dimensional stress problems and the assumptions violate known conditions. However, even with these limitations, which may not be too restrictive considering the general precision of rock mechanics analyses, the technique provides a ready and economical means for incorporating the influence of horizontal tectonic stresses into a stability study of slopes.

REFERENCE

1. Hast, N., "The State of Stresses in the Upper Part of the Earth's Crust," Engineering Geology, 2(1) 1967, pp 5-17.

DISCRETE ELEMENT TECHNIQUE FOR PIT SLOPE ANALYSIS

K. C. Ko *

ABSTRACT

The basic finite element technique is very useful in assessing the stress-strain distribution in a slope, provided that the constituent material is continuous. By introducing discontinuities in the model and modifying the solution, the technique can be used in more realistic slope models where geologic defects such as joints, bedding planes and faults control the stability. The discrete element technique, while simple in its principles, is quite effective in evaluating the influence of the discontinuities in geologic models. It is particularly useful in the stability studies on the pit slopes where a few major discontinuities control the stability. An analysis of a hypothetical slope with a planar discontinuity reveals that the unstable condition usually develops near the bottom and propagates along the discontinuity towards the top. The results give some insight into the mechanism of the initiation of the slope failure and the best mode of unloading of an unstable slope in a re-stabilization program.

Introduction

The finite element technique is gaining prominence in the analysis of stability of geologic structures. For rock slopes, the

---

* Kennecott Copper Corporation, Metal Mining Division Engineering Department, Salt Lake City, Utah

inherent geologic defects rather than the rock strength often play a greater role in the stability. Thus many attempts have been made to modify models to comply with the overall behavior of the geologic structure. A direct treatment is to have the geologic defects incorporated in the model. Additional steps are required in the procedure of finite element technique to accomodate the special treatment for the given defects.

Goodman et al (1968) introduce special elements called 'joint elements' in their models. These elements are treated with linear force-displacement relationships determined from jointed rocks. Although the element has no thickness, the overall behavior of the element is equivalent to that of a thin element with low compliance moduli. Information on normal and shear stiffnesses of the joint is required. Mathab and Goodman (1969) extended this theory to the three dimensional analysis of wedge-type slope failure.

Malina (1970) developed a technique in which an element may contain inherent directional weakness and implicit discontinuities. His program is also capable of creating new discontinuities in otherwise intact material if the stress exceeds the prescribed strength of the rock.

The 'discrete element' method (Wang and Voigt, 1969) uses dual nodes to treat discontinuities but does not require special elements. It allows local inelastic displacement (slip and separation) along the given discontinuity according to a given criterion for such displacement.

The stability of slopes in open pit mines is often affected by a few distinct discontinuous planes in the slope. The discrete element technique has been found very useful in evaluating the stability of such slopes. This paper describes the principle involved in the discrete element technique in two dimensions and its application to a hypothetical slope model. No generalization on the behavior of the slope is attempted.

Discrete Element Principle

The two dimensional structure is divided into a finite number of elements as in the usual manner. The discontinuous plane is introduced by assigning two numbers to each node along the plane (Fig. 1).

During the iteration of calculating the nodal displacements for each element in the conventional finite element method (Zienkiewicz, for example), the equilibrium at the 'dual nodes' along the plane is re-evaluated according to the given slip-or-separation criteria. In this particular study of a slope stability, the Coulomb-Navior criterion with tension cutoff was employed. Using the tension-positive rule,

$$S = c - \sigma_n \tan\phi \, , \quad T = \sigma_n \quad \ldots\ldots\ldots\ldots\ldots \quad (1)$$

where c, σ_n and ϕ are the cohesion, normal stress and friction angle of the joint (or whatever the discontinuity may be), and S and T (positive numbers) are the strength per unit length in shear and

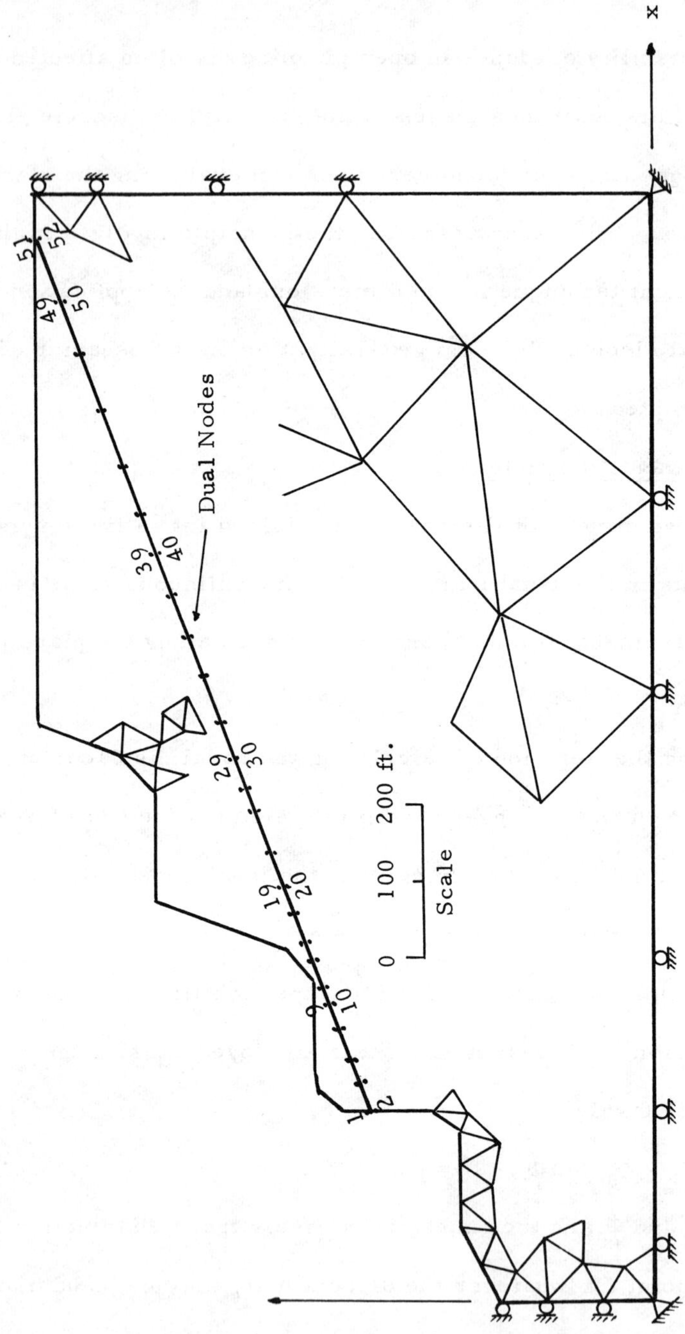

Fig. 1. Finite Element Mesh for Jointed Slope

tension mobilized at the node (Fig. 2).

Denoting a pair of initially contiguous nodes to be i and j (Fig. 3a) and the length of the contact influenced by the nodes b, the normal and tangential forces acting on the node i are (Fig. 3b):

$$\left.\begin{aligned} F_n^i &= F_x \sin\alpha + F_y \cos\alpha \\ F_t^i &= -F_x \cos\alpha + F_y \sin\alpha \end{aligned}\right\} \quad \ldots\ldots\ldots\ldots \quad (2)$$

where F is the resultant nodal force in the subscripted direction and α the dip of the discontinuity. The average stresses are calculated by averaging forces over the length b, i.e.,

$$\left.\begin{aligned} \sigma_n^i &= F_n^i/b \\ \sigma_n^i &= F_t^i/b \end{aligned}\right\} \quad \ldots\ldots\ldots\ldots\ldots\ldots \quad (3)$$

The contact stresses thus calculated are tested for slip-separation criteria.

If $\sigma_n^i \leq T$ and $|\sigma_t^i| \leq S$, the nodes remain intact acting as a single node, $F_n^i = -F_n^j$ and $F_t^i = -F_t^j$. Subsequently, $U_x^i = U_x^j$ and $U_y^i = U_y^j$.

Slip will occur between nodes i and j if $\sigma_n^i \leq T$ and $|\sigma_t^i| > S$. The forces will be readjusted to reach a new equilibrium. The relations become (Fig. 3c):

$$\left.\begin{aligned} F_n^i &= -F_n^j, \quad F_t^i = -F_t^j \\ |F_t^i| &= cb - F_n^i \tan\phi \\ \text{and} \quad U_y^i - U_y^j &= (U_x^j - U_x^i) \tan\alpha \end{aligned}\right\} \quad \ldots\ldots\ldots\ldots \quad (4)$$

The equations (4) and the force-displacement relationship are used to determine the displacements of i and j nodes. The direction of

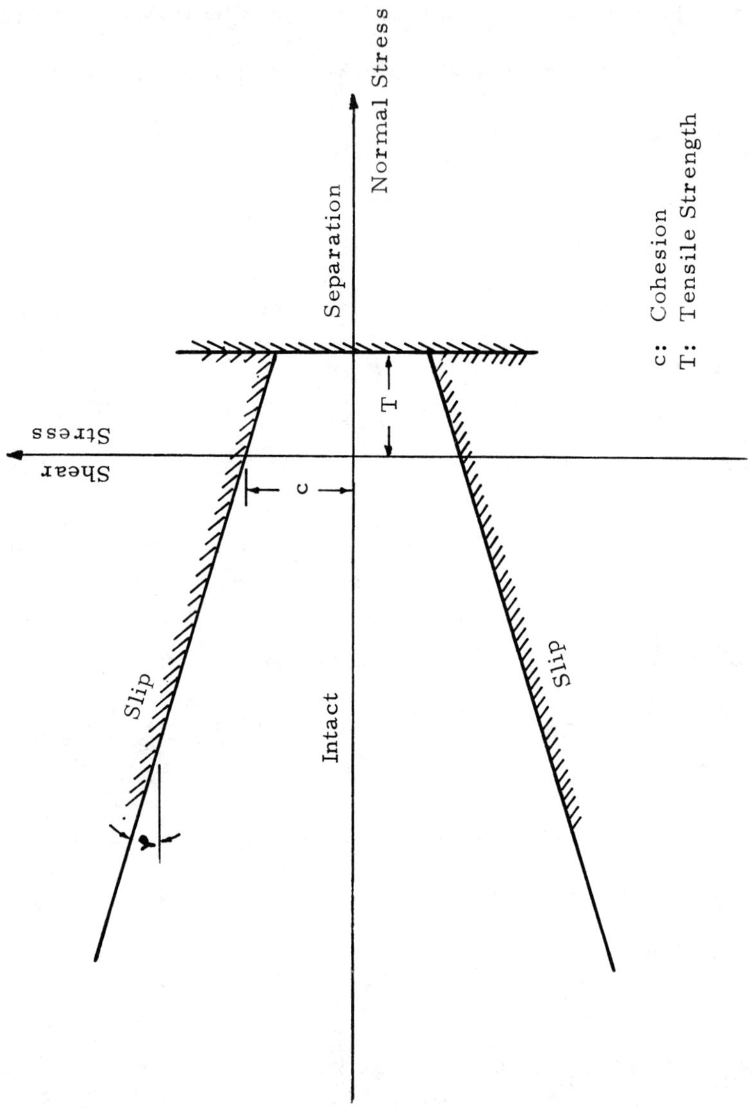

Fig. 2. Coulomb-Navier Slip Criterion with Tension Cutoff

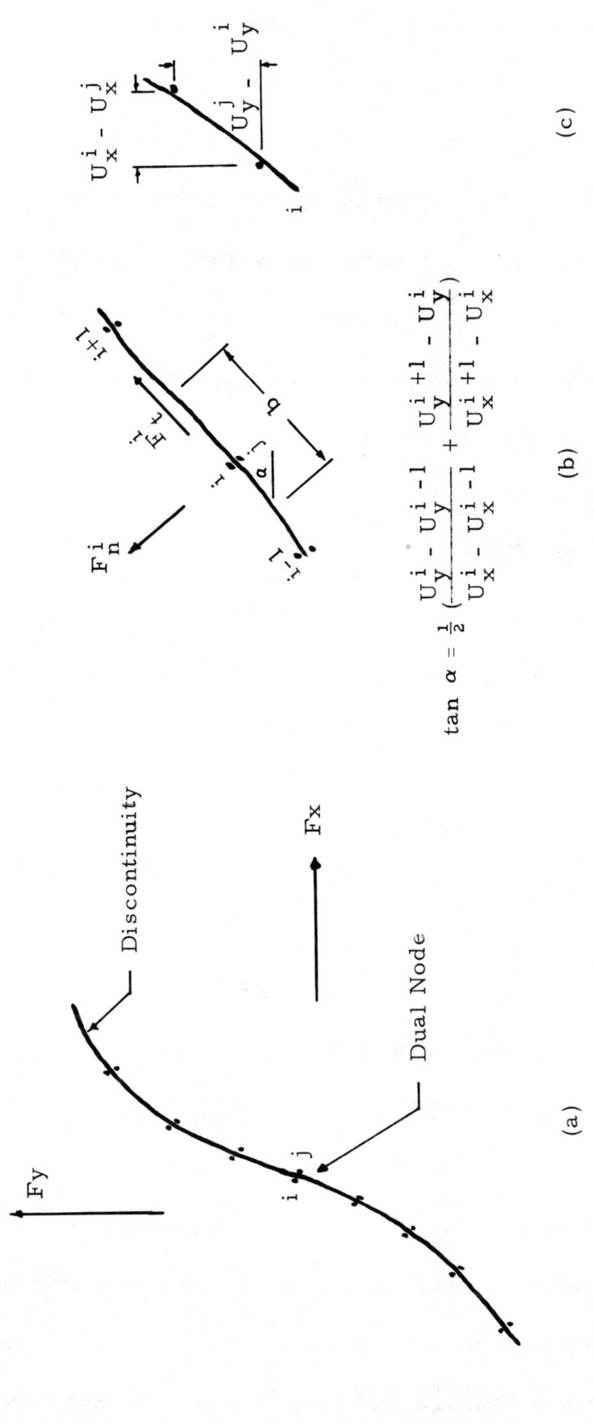

$$\tan \alpha = \frac{1}{2}\left(\frac{U_y^i - U_y^{i-1}}{U_x^i - U_x^{i-1}} + \frac{U_y^{i+1} - U_y^i}{U_x^{i+1} - U_x^i}\right)$$

Fig. 3. Notations and Forces for Dual Node System

shear force is determined by the displacement field.

When the normal tensile force is greater than the prescribed tensile strength of the joint, the dual node is separated and each node acts independently. All resultant forces on each node vanish.

This test for displacement mode is iterated each time the forces and displacements are equilibriated until the error becomes smaller than the prescribed limit. The results are the stresses, strains in the elements, and nodal displacements and forces in the final equilibrium condition.

Stability of a Jointed Slope

An example study was made on a hypothetical slope with an extending discontinuous plane (Fig. 1) at a 20° dip. The analyses were made with the following parameters:

Case	Cohesion	Friction Angle	Tensile Strength
A	20 psi	30°	20 psi
B	0	30°	0
C	0	25°	0

For all cases:

For all cases:	Specific Gravity	2.7
	Young's Modulus	1×10^6 psi
	Poisson's Ratio	.25

The first node at the lower end of the discontinuity (joint) showed separation in all cases (Fig. 4). Three nodes near the bottom showed slippage in the first case (Case A). The Case B model shows further slippage along the joint plane. For Case C,

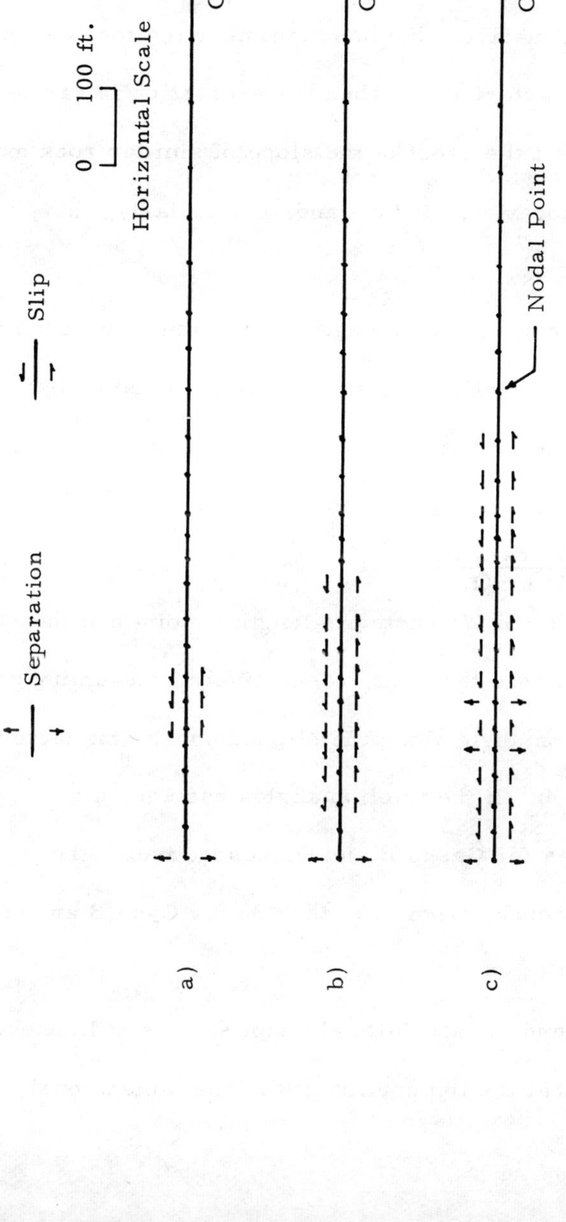

Fig. 4. Inelastic Displacements Along the Discontinuity with Various Slip Parameters

about half of the length of the joint indicated slippage and separations at two places within the slope due to buckling. Figs. 5 and 6 show the relative displacements of nodes and force distributions along the joint with respect to their position at zero-gravity condition.

The above analysis indicates that most weakness planes in a slope undergo a partial inelastic displacement near the toe, depending on the strength characteristics, when the excavation is carried out. Thus the rebound at the crest of the slope of similar rock may vary depending on the location, attitude and, particularly, the strength of the joint(s).

Once sliding occurs, the strength of the joint is reduced to a residual level. This reduction of strength can give some differences in the evaluation of the factor of safety. The factor of safety for this particular model evaluated by conventional methods is:

$$FS = \frac{cl + W \cos\alpha \tan\phi}{W \sin\alpha} = 2.04 \qquad (5)$$

for the Case A, where l and W denote the length of joint and the weight of the material above the joint, respectively. Assuming zero-cohesion and 25° friction angle along the slipped nodes, the factor of safety decreases to 1.90. If the friction angles become, after slippage, 20 and 17 degrees for Cases B and C, respectively, the factors of safety are reduced from 1.56 to 1.38 for Case B and from 1.26 to 1.05 for Case C.

It has been found, in all finite element studies at Kennecott Copper Corporation, that the top portion of the one dimensional

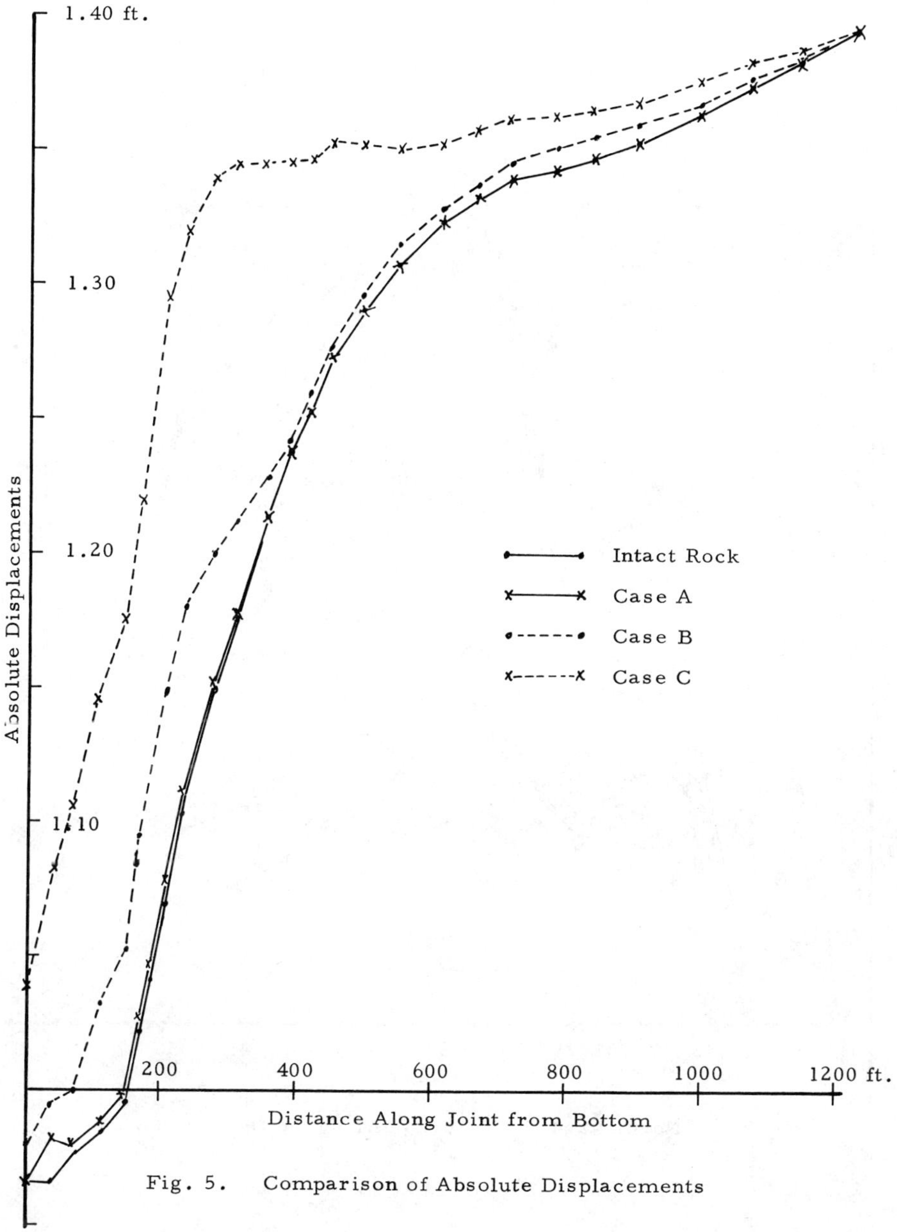

Fig. 5. Comparison of Absolute Displacements

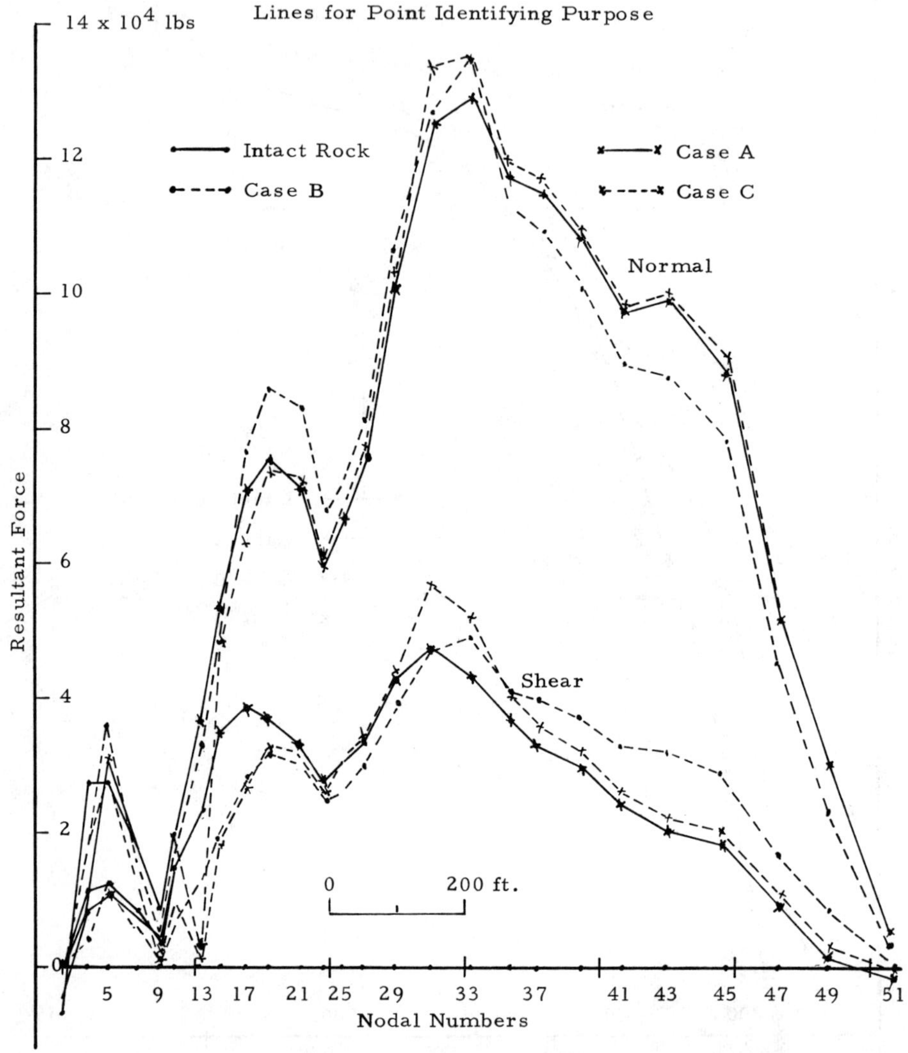

Fig. 6. Normal and Shear Forces Along the Discontinuity

discontinuity is always the most stable. If the block above the joint is continuous and elastic, the upper portion usually acts as a deterrent to sliding of an otherwise unstable block. It is conceivable that when there are near-vertical joint sets, or when the rock is very weak in tension, near-vertical tension cracks open. If the condition is such that the failure is inevitable, the lower part will slide down but the upper part usually stays intact. This has been true with the failures experienced on rock slopes of Kennecott's open pit mines. A uniform sliding of an entire block along the discontinuity, as some stability analyses suggest, is yet to be observed.

Effect of Unloading of Unstable Slopes

There are many other factors that influence the stability of slopes other than gravity forces. Their mechanisms are quite involved and not as well understood as the static force equilibrium. The unstable condition is usually first noticed by the tension cracks on the crest followed by the failure movement. Once the slope starts failing, unloading on top of the sliding block is sometimes carried out in an attempt to re-stabilize the slope. Some instances are reported as successful (Kimura and Rennie, 1968) and others unsuccessful (Kennedy et al, 1970).

Disregarding all other factors that might have caused the slide, the effect of unloading was studied on the Case C model with the shaded portion removed (Fig. 7). Considering the static force balance only, the stability depends on the mode of unloading. When

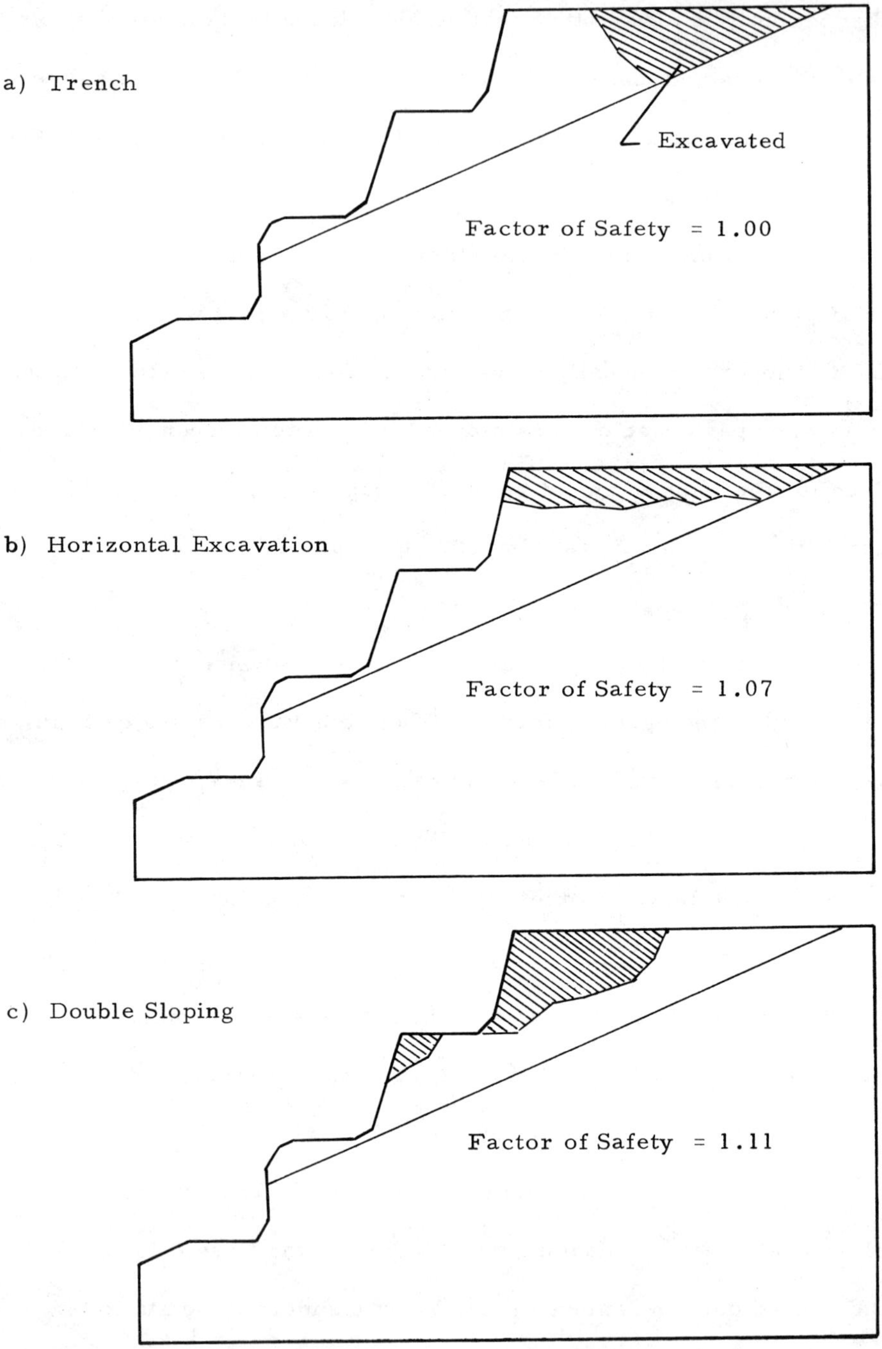

Fig. 7. Mode of Unloading and Factor of Safety

a trench was made as in Fig. 7a, the factor of safety decreased from that of the original slope with the same strength parameters. On the other hand, when the unloading was made evenly along the crest (Fig. 7b), the factor of safety remained about the same. Another case where a double slope was made (Fig. 7c) showed some increase in the factor of safety. It may be that some other factors such as blasting, played a larger role in the failure of pit slopes in unsuccessful re-stabilization programs. The only suggestion that can be made at this point is that the mode of unloading may change the probability of success.

Conclusions

The introduction of discrete nodal points in the finite element technique has some distinct advantages over the conventional method in analyzing the slope stability. The technique is simpler in usage yet gives more realistic solutions; and it can predict local inelastic (non-linear) displacements along given displacements for given strength parameters. The mechanism of failure of rock slopes may be simulated by successive modeling of the given slope with various conditions.

Although the generalization of the slope behavior is not possible, some conclusions on the stability analyses are made based on the hypothetical model:

1. Slippage (or cracking) initiates from the point near the bottom of the discontinuous plane (or

plane of weakness) in a slope.

2. The cracking may propagate upward if adverse conditions exist.

3. In case of unloading an unstable slope, making a double slope has a better probability of success in re-stabilizing the slope in terms of static force balance.

ACKNOWLEDGEMENT

The study was made with the computer program provided by Mr. Y. J. Wang of the Scientific and Engineering Computer Center, Kennecott Copper Corporation. The writer wishes to express his thanks to the Kennecott Copper Corporation for permission to publish this paper.

REFERENCES

Goodman R E, Taylor R L and Brekke T L, "A Model for the Mechanics of Jointed Rock", Journal of the Soil Mechanics and Foundation Division, ASCE. May 1968, pp 637-659.

Kennedy B A, et al, "A Case Study of Slope Stability at the Chuqucamata Mine, Chile", AIME Preprint 70-AM-81, 1970.

Kimura E T and Rennie C C, "The Craigmont Open Pit Slide." Presented at the Annual Meeting of Geo. Assoc. of Canada, Vancouver, B. C., April 1968.

Malina H, "The Numerical Determination of Stress and Deformations in Rock Taking into Account Discontinuities", Rock Mechanics, Vol. 2, 1970, pp 1-16.

Mahtab M A and Goodman R E, "Three-Dimensional Finite Element Analysis of Jointed Rock Slopes." Proceedings, 2nd Congress International Society of Rock Mechanics, Belgrade, September 1970, Vol. 3, pp 7-12.

Wang Y J and Voigt B, "A Discrete Element Stress Analysis Model for Discontinuous Materials." Proceedings, International Symposium on Large Permanent Underground Openings, Oslo, September 1969.

Zienkiewicz O C and Cheung Y K, The Finite Element Method in Structural and Continuum Mechanics, McGraw-Hill, London, 1968.

MODEL STUDY ON THE SLOPE STABILITY
AROUND AN OPEN-PIT

Y. S. Kim and R. R. Dimock, Jr.*

Department of Petroleum and Mining Engineering
New Mexico Institute of Mining and Technology
Socorro, New Mexico

Most open-pit design studies being conducted are based on completely idealized models for the necessity of developing the basic concepts from which a realistic condition in an open-pit may be induced. Both experiences and theories can be encountered for the development of such basic concepts. The present study was oriented to provide a simplified technique for investigating relative and gravity induced shear stress which may occur surrounding an operating or proposed open-pit. Based upon the recent theories of inelastic or elasticoviscous behavior of geologic substances, a scale model of an operating open pit and its surrounding topography was modeled in an elastico-viscous material. Gravity-induced displacements, horizontal and vertical, developed in the model due to the shear stress were measured with application of grid points placed

* Present address: Kenncott Copper Corporation, Nevada Mines Division, Ruth, Nevada.

on the model. Information obtained from this type of study may be used as a guide line for making decisions in slope design and determining potential areas of slope failure.

INTRODUCTION

In general, the slope angles established in most open-pits today are based upon the past practice and judgement of a design engineer. The angle of final open-pit slope is of prime importance in optimizing the economics of an open-pit. If the slope is less than the optimum angle, extra waste removal is required, thus seriously affecting the economics of the ore body. If the slope is greater than the maximum safe angle, failure will most probably occur. The failed slope would introduce additional waste into the pit, thus increasing the ore to waste ratio and consequently altering the economic considerations upon which the pit was designed. Photoelastic analysis, soil and rock mechanics, application of theory of elasticity and plasticity are a few among the many approaches which are being used for optimum slope design . However, in this study the theory of rheidity which based on both elastic and plastic theories was adopted with some basic assumptions as follows:

1. <u>Material surrounding the open pit is homogeneous</u>: The geologic materials surrounding the pit being modeled consist of sediments and intrusives typical of the forma-

tion of an orebody. These geologic units are faulted, jointed, and intersected by the water table, thus, the material surrounding the open pit is far from being homogeneous. The conditions of material surrounding an open pit are so heterogeneous that it is practically impossible to model them on a small scale using existing modeling materials. For this reason a homogeneous material was used; however, the result was then related to the various inhomogeneous conditions.

2. <u>Material surrounding the open pit is elastico-viscous</u>:

As stated by Farmer (1), rocks are subject to reversible and irreversible time dependent deformation when loaded over a finite time interval, even though the deforming stress may be well below the elastic limit of the rock. Thus a rock may be symbolized as deforming not only as an elastic solid, but also as a viscous substance, although the time-dependent effects may be due to crushing of pore space or sliding of joint/bedding plane surfaces. Nevertheless, it is possible to characterize time-dependent flow in rocks in terms of viscosity and to approach the analysis of stress in a rock in terms of fundamental rheological principles for an ideal material. This elastico-viscous property was simulated by the homogeneous modeling material. A detailed discussion of this subject will be presented in the section "Elastico-Viscous Flow Properties of Rocks".

3. Acceleration due to gravity is the same for both the original and the model:

Since both the original and the model are located on the earth's surface, the difference in gravitational forces acting upon them was neglected.

4. Acceleration in both the original and the model is so nearly zero that it is negligible:

The maximum acceleration which occurred in the model was 1.30×10^{-5} in./sec^2, and the average acceleration was much less than this value. Obviously, this value can be considered negligible. Further, Griggs' (2) work indicates even a smaller value for rocks similar to those being modeled.

5. Stresses developed in material surrounding the open pit are a result of gravity loading only:

Regional stress relief tests were not performed in the area surrounding the pit being studied, thus, it is not possible to state whether or not lateral tectonic forces exist. Studies of rock conditions in the pit walls, the condition of diamond drill core samples, and stress conditions around similar orebodies indicate that no significant tectonic forces exist. However, before applying the results of these tests to pit slope design, it would be wise to conduct regional stress relief tests to determine whether or not forces other than those due to gravity exist.

The slope design program for an open pit in a given location should consider rock types and their physical properties, structural conditions including joint set and fault orientations, topography surrounding the open pit, pit depth and geometry, normal and shear stresses developed within the pit walls, and ground water conditions, etc. The inelastic or viscous and time dependent behavior of geologic materials have been extensively studied by number of workers (2, 3, 4, 5, 6, 7, 8, 9, 10). The significant findings of these authors can be summarized as:

1. Rock can be simulated with an elastico-viscous material,
2. elastico-viscous flow or creep is a result of shear stresses,
3. in viscous flow, the greatest deformation in a given time period indicates the greatest applied force, and
4. that a viscous material will deform at a given shear stress, independent of the rate of shear strain.

Under the consideration of these theoretical concepts an open pit was modeled with a material having elastico-viscous properties such that it would flow under its own weight.

DEVELOPMENT OF THE SCALE MODEL

In a strict sense, the model used in this study was an "adequate model" which did not simulate all of the actual conditions in the open pit being modeled due to the limitations of facilities and available modeling materials. However, the model simulated the topography surrounding the open-pit, pit geometry, and normal and shear stresses developed within the pit walls.

Selection of Size

A model study which simulates a natural process depends upon degrees of similarity between the model and the true phenomena for reliability. M. K. Hubbert (3) describes these degrees of similarity as:

Geometric similarity - two bodies are said to be geometrically similar when all corresponding lengths are proportional and all corresponding angles of the two bodies are equal. Thus the model ratio or proportionality $\lambda = \frac{l_2}{l_1}$ must be constant throughout the model. l_1 is the length of the prototype and l_2 is the corresponding length of the model. This constant λ then becomes λ^2 for the area ratio and λ^3 for the volume ratio.

Kinematic similarity - if two geometrically similar bodies undergo geometrically similar changes of shape or positions, or both, the two bodies are said to be kinematically similar provided the time required for any given change in the one is proportional to that required for the corresponding change in the other. Thus, $\frac{t_2}{t_1} = T$, where t_1 is the time for the original to undergo a given transformation, t_2 the corresponding time required for the model, and T the model ratio of time.

<u>Dynamic similarity</u> - the model must have a mass distribution similar to that of the original. Thus $\frac{dm_2}{dm_1} = \mu$ where dm_1 is the mass of an element of volume dV_1, of the prototype, dm_2 that of the corresponding element of volume dV_2 of the model, and μ is the model ratio of mass. From the model ratios of mass and length, the ratio of density $\S = \mu\lambda^{-3}$ is obtained. The forces acting upon any element of mass, dm, occupying an element of volume, dV, may be divided into two separate classes, the body forces and the surface forces. The body forces are forces originating inside the body of the volume, such as those due to gravity and inertia. The surface forces are those acting only on the external surface of the element of volume considered. In the case of plastic bodies in motion, the surface forces result from or are related to, the volume change and the time rate at which the body undergoes deformation. In general, each element of mass is in a state of dynamic equilibrium, among as many of these forces as are acting simultaneously, and it is assumed that the vector sum of the body forces acting is equal and opposite to the vector sum of the surface forces. Therefore, for bodies undergoing plastic deformation we can write $F_g + F_p + F_r = -F_i$ where F_g, F_p, F_r and F_i are respectively the resultant vectors of forces due to gravity, to pressure, to viscous resistance, and to inertia. Dynamic similarity requires that on each element of mass dm_2 occupying volume dV_2 of the model,

the forces acting must be such that the motion is geometrically and kinematically similar to the corresponding motion of the element of mass dm_1 in volume dV_1 of the original. This condition is satisfied if, for each force F_1 acting on mass dm_1 in the original, there is a corresponding vector force F_2 acting upon mass dm_2 in the model, where F_2 has the same orientation as F_1 and the ratio of the magnitudes of the two forces $\frac{F_2}{F_1} = \phi$. The model ratio of force is ϕ which is constant for all corresponding points throughout the two bodies. By substituting the components of F, one can derive the basic formula

$$\phi = \mu\lambda\tau^{-2} \tag{1}$$

The necessary condition for similarity developed is that two massive bodies can be geometrically and kinematically similar only provided the masses of the one, point by point, are proportional to the corresponding masses of the other, and that the corresponding forces, point by point, have the same directions and proportional magnitudes.

When the model and the original are on the earth's surface, the acceleration due to gravity is the same for both, thus:

$$\frac{g_2}{g_1} = \gamma = 1 = \lambda\tau^{-2} \tag{2}$$

Hence, λ and τ are no longer independent and $\tau = \lambda^{\frac{1}{2}}$.

A body is said to be viscous if, when a shearing stress is applied, the body deforms continuously at a rate proportional to the applied shearing stress, approaching zero as the stress approaches zero. The combined stress under which a given material fails by fracture or flows plastically is said to be its strength.

If the prototype fails with a given mode at a certain stress, then in the model, for similarity, failure must occur with the same mode at the corresponding value of the stress. Hence the model ratio of strength must be the same as that of stress.

$$\frac{(\text{strength})2}{(\text{strength})1} = \frac{(\text{stress})2}{(\text{stress})1} = \sigma = \mu\lambda^{-1}\tau^{-2} \tag{3}$$

The rate of plastic flow in both the original and the model is so very slow that the acceleration is negligible thus producing inertial forces of only infinitesimal magnitude. In this case the forces of the plastic body are in an equilibrium condition , i.e. $F_g + F_r + F_p = 0$. Therefore, the controlling factor involved in this study is not the body force due to the inertia but is the force due to the gravity. This gravity, however, acts on the earth's surface independently of the model ratio of time employed in the present experiment. Thus, in cases where acceleration is so small, i.e. forces due to inertia are negligible, the laws of dynamic similarity can be satisfied by arbitrarily choosing the model ratios μ, λ, and τ; however, all forces must be concurrent to the model ratio $\phi = \mu \gamma_g$.

Where γg is the model ratio of acceleration due gravity and may be set to unity with a negligible error if both the prototype and model are located on the earth's surface. Therefore, $\phi = \mu$, and since the model ratio of density $\S = \mu\lambda^{-3}$, mass per unit volume, it follows that $\phi = \S\lambda^3$. By substituting these relationships into equations (1) and (3) the strength ratio which is the same as the stress ratio, can be written as:

$$\sigma = \frac{\S\lambda^3}{\lambda^2} = \S\lambda \tag{4}$$

For the present model study, the variable σ (model ratio of stress or strength), $\S$ (model ratio of density) and λ (model ratio of length) are calculated as follows:

$$\sigma = \S\lambda \quad \text{or} \quad \lambda = \sigma\S^{-1} \tag{5}$$

where,

$$\S = \left(\frac{dm_2}{dm_1}\right)\left(\frac{l_2}{l_1}\right)^{-3} \quad \text{or} \quad \frac{(dm_2)\ (l_2)^{-3}}{(dm_1)\ (l_1)^{-3}} \tag{6}$$

$$\sigma = \left(\frac{dm_2}{dm_1}\right)\left(\frac{l_2}{l_1}\right)^{-3}\left(\frac{l_2}{l_1}\right) \quad \text{or} \quad \frac{(dm_2)\ (l_2)^{-3}(l_2)}{(dm_1)\ (l_1)^{-3}(l_1)} \tag{7}$$

$$\sigma = \left(\frac{dm_2}{dm_1}\right)\left(\frac{\ell_2}{\ell_1}\right)^{-3}\left(\frac{\ell_2}{\ell_1}\right) \quad \text{or} \quad \frac{(dm_2)\ (\ell_2)^{-3}\ (\ell_2)}{(dm_1)\ (\ell_1)^{-3}\ (\ell_1)} \tag{8}$$

values for these variables are as follows:

$(dm_1)\ (l_1)^{-3} = 0.089$ lb/in^3	This density is listed in Blair (1955) for siliceous limestone which is the predominant rock type at the mine being modeled.
$(dm_2)\ (l_2)^{-3} = 0.039$ lb/in^3	Laboratory verification.
$(dm_1)\ (l_1)^{-2} \simeq 0.015$ lb/in^2	For siliceous limestone (Blair 1955).

$(dm_2)\ (l_2)^{-2} \simeq 0.015\ lb/in^2$ No strength testing equipment was available. However, one cubic inch of the material flows under its own weight, from which the strength was estimated $\approx 0.015\ lb/in^2$.

Substituting the above values, the ratios are:

$$\S = \frac{(dm_2)\ (l_2)^{-3}}{(dm_1)\ (l_1)^{-3}} = \frac{0.039}{0.089} = 0.438, \text{ and}$$

$$\sigma = \frac{(dm_2)\ (l_2)^{-2}}{(dm_1)\ (l_1)^{-2}} = \frac{0.015}{22{,}000.000} = 0.682 \times 10^{-6}$$

Thus,

$$\lambda = \frac{l_2}{l_1} = 0.682 \times 10^{-6} \div 0.438 = 1.557 \times 10^{-6}$$

or

$$l_2 = 1.557 \times 10^{-6} \times l_1$$

The size of the pit and surrounding area considered in this study is 4,000 feet by 6,000 feet, which requires a model 0.075 in. x 0.112 in. according to the model ratio. The greatest relief within the pit boundary is 2,000 feet, which is represented by 0.037 in. in the model. The maximum value of $(dm_2)\ (l_2)^{-2}$ was estimated as $0.015\ lb/in^2$, which is feasible since a lower value of $(dm_2)\ (l_2)^{-2}$ would make the required dimensions even smaller, thus no error will be resultant in the overall model study.

Obviously, building a model of this size in any detail at all is not possible. Due to the assumption of "homogeneity", the stress value calculated from a model based upon a modeling theory cannot be readily applied to an actual geologic structure which is not quite a homogeneous material. However, the results obtained from a model study can be used as a guideline for the idealized shear stresses around an open pit. Test results, however, would indicate areas of high stress concentration. Making a model of the pit out of an elastico-viscous homogeneous material to a scale which is convenient, fulfills all the model ratio requirements, except the one for computing stress ratios. Although the numeric values of displacement and stress derived from the model study cannot be used to evaluate absolute values in the pit, the relative values within the pit are valid. Grid points placed on the surface of the model, which is undergoing flow, will trace vectors, each with its characteristic length and direction. A relative study of one point to another on the model surface indicates areas of high stress concentration. Those areas where the greatest shear stress is induced by the surrounding topography will result in a greater movement, thus longer vectors. In this sense the information obtained in this study is as valuable and applicable as the information which would have been provided by an exact model ratio of strengths for the model pit.

The model was built to a scale factor of 4800 for the following reasons, (1) the horizontal extent of boundaries for the pit which would include the pertinent topography would be reasonable and workable, and (2) the 50-foot bench height in the pit could be represented by the standard 1/8 inch thick sheets of balsa wood.

Model Construction

The testing model, 400 scale, was built in three steps as follows:

1. A positive of the model of the open-pit was constructed with 1/8 in. thick balsa wood which represents the 50 ft. benches (Figure 1-a), and then poured plaster-of-paris over the positive model in order to simulate the smooth realistic ground conditions of the pit (Figure 2-b).
2. The fiberglass negative of the model was then made by the following process: (a) A very thin coat of liquid paraffin was painted over the entire surface of the model and allowed to dry. (b) A thin coat of fiberglass resin and catalyst was placed over the paraffin. (c) When the resin became tacky, several layers of fiberglass matting and resin were placed on it. (d) After the fiberglass was dry, just enough heat was applied to melt the paraffin layer, releasing the fiberglass negative from the model. The negative was then trimmed and placed on a frame with handles.
3. The final testing model was molded in the fiberglass negative by pouring the water dissolved solution gels of "Super Stuff" which solidifies by standing. This gel material, "Super Stuff" offered the most satisfactory requirements of homogeneous and elastico-viscous properties of the testing model.

(a)

(b)

Figure 1. Balsa Wood Positive of the Model

(a) Without Smoothing

(b) Smoothed with Plaster of Paris

EXPERIMENTAL

Apparatus

Motion of grid points set on the surface of the positive model of "Super Stuff" was monitored in two separate phases as follows: Phase 1: The horizontal displacement and velocity were monitored with a time-lapse 16-mm movie camera which consists of a solenoid, single-frame trip device, and a synchronized photo flood light with a black box timing device which was set to trigger the light and camera at intervals of 57 seconds (Figure 2). The velocity and direction of displacement for each grid point on the model is then recorded on movie film.

Phase 2: The vertical displacements were monitored with a micrometer fixed in a series of sliding components which traverse the model in two directions so that the micrometer can be set on any point of the model. The traversing bars were set in a frame which was fixed on 3' x 3' x 3/8" boiler plate (Figure 2). Each of the three micrometers was adapted with a brass pointer. The lengths of the pointers were chosen to allow measurement of any point around or within the pit being modeled. An attempt was made to measure vertical deformation of the model by reading the micrometer when the point touched the "Super Stuff", resulting in an accuracy of $\pm$ 0.10 of an inch. By placing a mat of rubber gasket material between the steel box containing the model and the boiler plate which supported the measuring device,

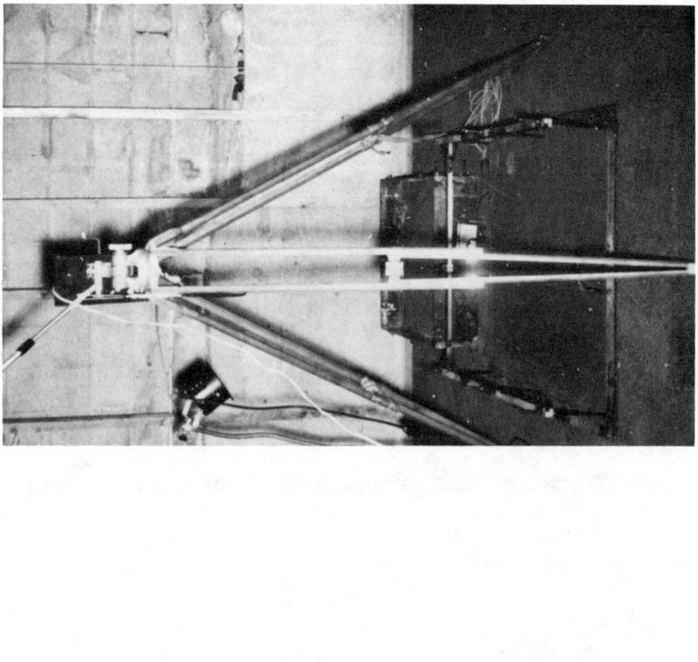

Figure 2. 16-mm Time Lapse Movie Camera and Photoflood

Figure 3. Vertical Displacement Measuring Device

and installing a galvanometer between the two pieces of metal, a completed circuit was registered when the micrometer point touched the "Super Stuff" which is 90% water. Repeatibility within 0.005 of an inch was possible with this arrangement (Figure 3).

Procedures

The fiberglass negative was placed on a wooden frame which was fixed in a square, 1/4 inch thick, steel frame in order to secure a level position for the base of the negative model. The mixed and gelled "Super Stuff" was placed over the negative and allowed to flow until it completely filled the steel frame box. A plastic cover was placed over the "Super Stuff" and a steel lid was screwed tightly to the box.

Prior to the experiment, a preliminary test film was made to determine the most visible material for grid points. Small beads did not show up on film by themselves, but when they were placed on the center of a round silver sequin, they were easily identified. The camera was placed so that its focus line was straight down to the center of the model, then the camera was raised until the required boundaries of the pit were within its screen. The position of the steel box corners was constant throughout the tests.

When all test arrangements were ready, the box containing the negative and "Super Stuff" was turned over on the rubber mat which was placed on the marked position. The bolts in the negative frame were removed and the negative was lifted straight up out of the box, leaving a positive model imprinted in the "Super Stuff" . The proposed locations of the grid points were marked on a photograph of the balsa model. After the negative was removed, the sequins with beads glued on their centers were placed at the locations indicated on the photograph . As soon as the grid points were placed, the time-lapse movie camera was turned on and the timer was set to expose one frame every 57 seconds. No further attention was required for the movie camera until the test was complete.

After the movie camera was turned on, the vertical displacement measurements were begun. A micrometer of the proper length was positioned over each grid point and the micrometer tip was advanced until a current flow was registered on the galvanometer. The micrometer was then read and recorded. A complete test of observing all grid points took approximately one hour. After waiting approximately 30 minutes, another set of vertical readings were taken. It was possible to get three sets of readings before the "Super Stuff" had completely closed the pit opening.

After a test, the bottom plate was removed from the box, the negative was bolted back in place, and the "Super Stuff" was again placed over it and allowed to flow until it filled the box. This took approximately four hours. Temperature readings were taken throughout the entire period of testing to provide information regarding the evaporation of the modeling material. A total of six tests were made, but two different procedures were used to analyze flow of the model. The first two tests consist of a complete view of the pit with the grid points arranged in rays around the pit, each one going from the pit edge to its bottom. Tests three, four, five and six were made with a telephoto lens to provide enlarged reproductions of various groups of the same rays which were photographed as a whole view in tests one and two.

RESULTS

Horizontal Displacement Data

Each frame of the 16-mm movie film was projected on tracing paper and the successive locations of each grid point were plotted on the tracing paper. It was extremely difficult to determine displacement between single successive frames, therefore, plots were made showing adequate displacement yet not missing any significant trend of the motion.

One edge of the metal box containing the model was parallel referring to North-South on the model, from which the bearings of displacement were taken while plotting the motion of grid points.

The jig which holds the micrometer was exactly one inch in width and was photographed at least once in each test; thus giving a dimensional reference to be used as a scaling factor for each test.

The grid points were placed in rays emanating from the central portion of the pit. Each ray was lettered and each grid point within a ray was numbered. Data taken from the plots of grid displacements include: (1) the frame number of each plot of a grid point, (2) the distance between successive plots of the point, (3) the cumulative displacement for the point, (4) velocity of the displacement of the point in inches per frame, and in inches per second, and (5) the total length of the straight line traced by the motion of the grid points. This total displacement was converted to the model scale and its bearing was determined and recorded in Table I. Data from the various tests were compared and analyzed using the normalized information in Table I.

Vertical Displacement Data

The vertical displacement data was obtained with micrometers which were assembled to complete an electric circuit including a galvanometer. The difference between successive readings of the micrometer provides the vertical displacement data.

Three vertical displacement monitoring devices were built, each consisting of a micrometer and point insert. Each device was capable of traversing one inch and was assembled so that an overlap occurred with the next larger micrometer. By measuring the vertical displacement of a point which fell within the overlap region with both micrometers, a conversion factor was derived for those points displaced beyond the range of a single micrometer.

An additional reading could be taken for each grid point during the telephoto tests since a lesser number of points were included in the tests.

The three sets of displacement data obtained for each ray are summarized in Table II. Agreement between vertical

TABLE L. HORIZONTAL DISPLACEMENT SUMMARY (Normalized @ 121 Frames)

		Test No. 1		Test No. 2		Telephoto Test		
Ray	Point	Length (in)	Bearing	Length (in)	Bearing	Length (in)	Bearing	Remarks
A	1	1.11	296°56'	0.95	294°00'	1.02	296°30'	Definite Trend
	2	1.16	297°06'	1.02	294°17'	1.07	294°36'	" "
	3	1.25	295°43'	1.07	295°36'	1.07	292°38'	" "
	4	1.18	296°31'	N.I.*	-	B.F.L.**	280°30'	" "
	5	1.12	296°58'	1.08	297°00'	B.F.L.	-	" "
	6	0.60	294°32'	Gone	291°24'	B.F.L.	-	Somewhat errat-ic pit bottom
	7	0.48	311°06'	N.I.	-	B.F.L.	-	Erratic-pit bottom
B	1	1.04	307°33'	0.91	303°23'	1.28	310°55'	Definite Trend
	2	1.03	306°58'	0.93	303°35'	1.27	309°56'	" "
	3	1.03	309°41'	1.03	305°49'	1.31	307°33'	" "
	4	1.00	309°31'	Gone	303°45'	1.32	304°49'	" "
	5	Gone	306°28'	N.I.	-	Gone	297°05'	Definite Trend Disappeared
	6	No Trend	-	Gone	-	B.F.L.	-	Very erratic pit bottom
C	1	0.74	310°07'	0.69	309°54'	0.90	315°46'	Definite Trend
	2	0.80	314°04'	0.73	311°30'	0.94	315°35'	" "
	3	0.83	318°13'	0.75	316°52'	0.96	316°27'	" "
	4	Gone	327°29'	0.74	316°30'	0.96	314°34'	" "
	5	0.66	332°49'	0.50	331°37'	0.42	325°28'	" "
	6	N.I.	-	No Trend	-	B.F.L.	-	Not Identifiable

* N.I. - Not Identifiable   ** Beyond Frame Limit

TABLE II. VERTICAL DISPLACEMENT DATA

		Test No. 1		Test No. 2		Telephoto Test		
Ray	Point	Mic. (in) Readings*	(in) Displacement	Mic. (in) Readings	(in) Displacement	Mic. (in) Readings	(in) Displacement	
A	1	(1).1604		(1).1826		(1).1135		
		(1).1816	+.0212	(1).1988	+.0162	(1).1175	+.0040)	+.0380
						(1).1515	+.0340)	
	4	(2).2816		(2).2656		(2).3325		
		(2).4194	+.1378	(2).4320	+.1664	(2).3443	+.0118)	+.1626
						(2).4951	+.1508)	
	6	(3).5556		(3).4475		(3).2238		
		(3).9297	+.3741	(3).7856	+.3381	(3).5020	+.2782)	+.4434
						(3).6672	+.1652)	
B	1	(1).1078		(1).1572		(1).0344		
		(1).1172	+.0094	(1).1701	+.0129	(1).0389	+.0045)	+.0046
						(1).0390	+.0001)	
	3	(2).3180		(2).2954		(2).1711		
		(2).4700	+.1520	(2).4691	+.1737	(2).3728	+.2017)	+.2984
						(2).4695	+.0967)	
	6	(3).3469		(3).2760		(3).0811		
		(3).6632	+.3163	(3).6666	+.3906	(3).2639	+.1828)	+.3320
						(3).4131	+.1492)	

* Numbers in parenthesis represent the micrometer and tip used for the measurement.

displacements in Tests No. 1 and No. 2 are quite good in most rays. Displacements in the telephoto test tend to be larger than in Tests No. 1 and No. 2. The original reading for the few points in each telephoto test was completed within approximately 20 minutes of the negative removal, whereas the original reading in Tests No. 1 and No. 2 took approximately one hour to complete. Those points which were read late in the sequence for Test No. 1 and No. 2 had been deforming for a relatively long time prior to the first reading.

For a more accurate determination of the total vertical displacements the telephoto test data should be used, whereas for the directional information all three test data are valid.

SUMMARY AND CONCLUSIONS

An attempt has been made in this study to provide a method for determining relative magnitudes of gravity induced shear stresses in the walls of a proposed open pit. This objective is accomplished by monitoring displacements of grid points placed on a three-dimensional scale model of the proposed pit and surrounding area which is molded in a flowing elastico-viscous material. Displacements in the model are a result of gravity induced shear stresses in the pit walls. Displacement of grid points in the model during a given time period is directly proportional to the shear stress causing displacement. Comparison of displacements of the various grid points the relative shear stress of these points were estimated.

The model study was conducted twice with a wide angle lens for the entire model and once with a telephoto lens for each quarter of the model. Displacement magnitudes and directions were in good agreement between the various tests. Reproducibility of results in the various tests indicates that the procedure is reliable, and application of this method for the analysis of other pit configurations should also provide reliable results.

The relatively large horizontal displacements for those grid points which were backed by mountains proves that the relief surrounding an open pit exerts a significant influence on the magnitude of forces which affect the pit walls. Results of the model study indicate that the presence of mountains behind the south pit wall resulted in shear stresses which were greater than the shear stresses in the north pit wall which has no significant relief behind it. Displacement directions indicate the direction of maximum shear stress and prove that the presence of mountains around an open pit exerts a greater influence on maximum shear stress direction than the irregular shape of pit walls.

The vast number of parameters which must be considered for complete analysis of the stresses surrounding an open pit makes it virtually impossible to model all of them. This "adequate" model study of some of the parameters indicates that the location and size of mountains surrounding an open pit excavation have a significant effect on the shear stresses and maximum shear stress directions in the walls of an open pit. This type of model study may give a guide line to locate the gravity induced shear stresses and their directions for the operating or proposed open pit configuration if gravity is the only significant body force.

However, information concerning shear stresses in the pit walls must be considered in conjunction with structural conditions existing in the open pit. The relationship between shear stresses and structural planes of weakness in an open pit is the key to stability or instability of the pit walls.

ACKNOWLEDGEMENTS

This study was partially funded by the United States Bureau of Mines at New Mexico Institute of Mining and Technology. The authors wish to thank the Kennecott Copper Corporation for providing equipment and funds used in conducting the model study.

REFERENCES

1. Farmer, I. W., Engineering Properties of Rocks, E. F. Spon Limited, London, 1968.

2. Griggs, D. T., "Creep of Rock", Journal of Geology, Vol. 47, 1939.

3. Hubbert, M. K., "Mechanical Basis for Certain Familiar Geologic Structures", Bulletin of the Geological Society of America, Vol. 62, April 1951.

4. McClain, W. C., "The Effect of Non-Elastic Behavior of Rocks", Eighth Symposium on Rock Mechanics, University of Minnesota, 1966.

5. Hills, E. S., Elements of Structural Geology, John Wiley & Sons, Inc., New York, 1963.

6. Carey, S. W., "The Rheid Concept in Geotectonics", Geological Society of Australia Journal, Vol. 1, 1953.

7. Coates, D. F., Rock Mechanics Principles, Queen's Printer and Controller of Stationery, Ottawa, Canada, 1967.

8. Obert, L. & Duvall, W. I., Rock Mechanics and the Design of Structures in Rock, John Wiley & Sons, Inc., New York, 1967.

9. Bridgeman, P. W., "Shearing Phenomena at High Pressure of Possible Importance to Geology", Journal of Geology, Vol. 44, 1936.

MONTE CARLO APPROACH TO SEEPAGE
IN JOINTED ROCK

By William E. Kelly,[1] A.M. ASCE

INTRODUCTION

The purpose of this paper is to present the continuous Monte Carlo method and to demonstrate its application to determining potential values in a jointed medium. It will be shown that the continuous Monte Carlo technique has certain characteristics which suggest it may be particularly applicable to the study of seepage in a jointed medium.

Before presenting the continuous Monte Carlo method itself, it is apropos to discuss the different definitions of permeability (6). Then the method itself will be presented along with a discussion of the accuracy which can be expected. Finally the applications will attempt to demonstrate the solution technique and some of its characteristics.

BACKGROUND

Theory and experience at this time indicate that for engineering purposes seepage through a jointed rock can, in many cases, be treated as seepage through an equivalent anisotropic medium (18, 16, 13).

---

1 Research Assistant, Department of Civil Engineering, University of Notre Dame, Notre Dame, Indiana 46556

Permeability. Starting with the expression for laminar flow between parallel plates, the permeability of a system of sets of parallel plates can be computed. The parallel plates here represent the joint sets. If there is structural symmetry at the point and the properties of the joints are the same in each direction, a symmetric second-order tensor is obtained. Thus, principal axes and permeabilities can be computed (13). With the principal values of permeability obtained there is still the problem of which permeability distribution to use (6, 9, 10, 13).

A flow net solution for an anisotropic seepage problem employs the scale transformation

$$X_T = \sqrt{\frac{Kmin}{Kmax}}\, x \qquad (1)$$

where x is the distance measured in the direction of maximum permeability,
Kmin and Kmax are the minimum and maximum permeability respectively,
and X_T is the transformed x distance.

to transform an anisotropic medium into an equivalent isotropic medium which allows a flow net to be drawn. Thus, the expression for the permeability in the direction of the resultant velocity is the expression commonly used. Using the scale transformation given by equation 1, a flow net can be drawn which is in reasonable agreement with the results of model experiments (7). However, there is contradictory evidence on the validity of using the equivalent permeability (7, 11)

$$Ke = \sqrt{Kmax\ Kmin} \qquad (2)$$

where Ke is the equivalent permeability.

Regardless of which definition of permeability is used the value of permeability that is finally computed will be some average value. In computing potentials the important thing is to determine the relative magnitudes of the permeability, or for an anisotropic medium the ratio of the principal permeabilities. With values of Kmax and Kmin determined, any of the numerical techniques could be used to solve a seepage problem. If, on the other hand, the permeability cannot be represented by its principal directions and magnitudes (2, 6), the continuous Monte Carlo method could still be used.

Monte Carlo Method. The basic techniques of Monte Carlo as applied to ground water have been presented by Scott (15) and more completely developed by LaRusso (8). Because of computer time requirements, the Monte Carlo method is recommended where only potentials at a few critical points are required (5, 15). It has been demonstrated, however, that with carefully written programs and by solving for all required points simultaneously, immense savings in computer time may be realized (8).

One distinct advantage of Monte Carlo techniques is that with a digital computer there is a minimal storage requirement. Special computers have been built to solve seepage problems using Monte Carlo techniques and employing a minimal storage capability (14).

CONTINUOUS MONTE CARLO METHOD

Previous applications of the Monte Carlo method have involved solving the finite difference approximation to the appropriate differential equation. What appears to offer distinct possibilities in savings of time and problem-formulation is a suggestion by Brown (1). Brown suggested that rather than have a particle walk on a lattice, the particle should walk to the perimeter of the largest circle that can be drawn around the point in question. The walk continues in this manner until it reaches a boundary where a success is recorded. The calculation of the potential is as it is when solving the finite difference equations.

$$POT = \frac{W_U + W_L}{W_T} \times H. \qquad (3)$$

where POT is the potential at a point W_U and,
W_L are the walks reaching the upstream and downstream boundary respectively.
W_T is the total number of successful walks and H is the difference in potential between the upstream and downstream boundaries.

A comparison with known solutions will demonstrate that this method does approximate the correct value of potential. In explaining the solution of the Laplace equation by the Monte Carlo method, the analogy with Brownian motion is often noted. This analogy is formally recognized in the growing field of probabilistic potential theory (4). Muller (12) studied this approach for the Dirichlet problem and concluded that the allowable truncation error was the key to the successful application of the approach. By truncation error is meant that a point travelling towards a boundary is not required to arrive precisely on the boundary to be termed a success. When the point comes within a specified distance of the boundary, termed the truncation area, it is called a success and the point is restarted.

The accuracy of any Monte Carlo solution is dependent on the number of trials performed. However, convergence is slow and consequently time for solution increases rapidly for a small increase in accuracy. Equations are available for computing probable error (17). LaRusso (8) has demonstrated, however, that in general the error is less than the maximum error which could occur. There is no theoretical way at present to assess the additional error that is introduced by truncating the walk. The easiest approach is to compute some potentials with various values of truncation and compare the results with a known solution. In more general cases where known solutions aren't available, the convergence of the solution may be studied.

APPLYING THE CONTINUOUS MONTE CARLO METHOD

The continuous Monte Carlo method will be applied to a series of illustrative confined flow problems. Briefly, the continuous walk technique is applied as follows:

1. The particle is located, i.e., either underneath, upstream or downstream of the dam.
2. The radius of movement is computed:
   A. For a point under the dam the radius is the minimum of the distances to the heel and toe of the dam;
   B. For a point beneath the upstream or downstream boundary the radius is equal to the current y coordinate of the particle.
   C. When wells and interior boundaries are involved they must also be considered in computing the radius.

3. The direction of movement is then chosen by generating a pseudo-random number and applying it to the appropriate permeability distribution. There are many ways this can be done and an efficient scheme is important.
4. The point is moved in the direction selected, a distance equal to the radius previously computed.
5. If the point encounters a reflecting boundary, e.g., the base of the dam, or an impermeable cutoff, it is reflected back into the medium.
6. If the particle arrives within a band of width equal to the truncation along a known potential boundary a success is recorded.
7. Potentials are computed as in the lattice solution by equation 3.

APPLICATION TO ILLUSTRATIVE PROBLEMS

Homogeneous Isotropic Medium With a Lower Impervious Boundary. The problem as solved is shown in Figure 1a. The D/b ratio is 4/3, where D is depth of permeable medium and b is one-half the dam width. Only the values for potentials along the base of the dam are shown.

The truncation was approximately 15 percent of the dam width and a minimum of 500 successes were obtained for each point. In Figure 2a the solution presented for comparison is from Harr (3).

Homogeneous Anisotropic Media With a Lower Impervious Boundary. The statement of the problem is as shown in Figure 1a. The D/b ratio is 4/3. The ratio Kmax/Kmin is 9 and Kmax is horizontal.

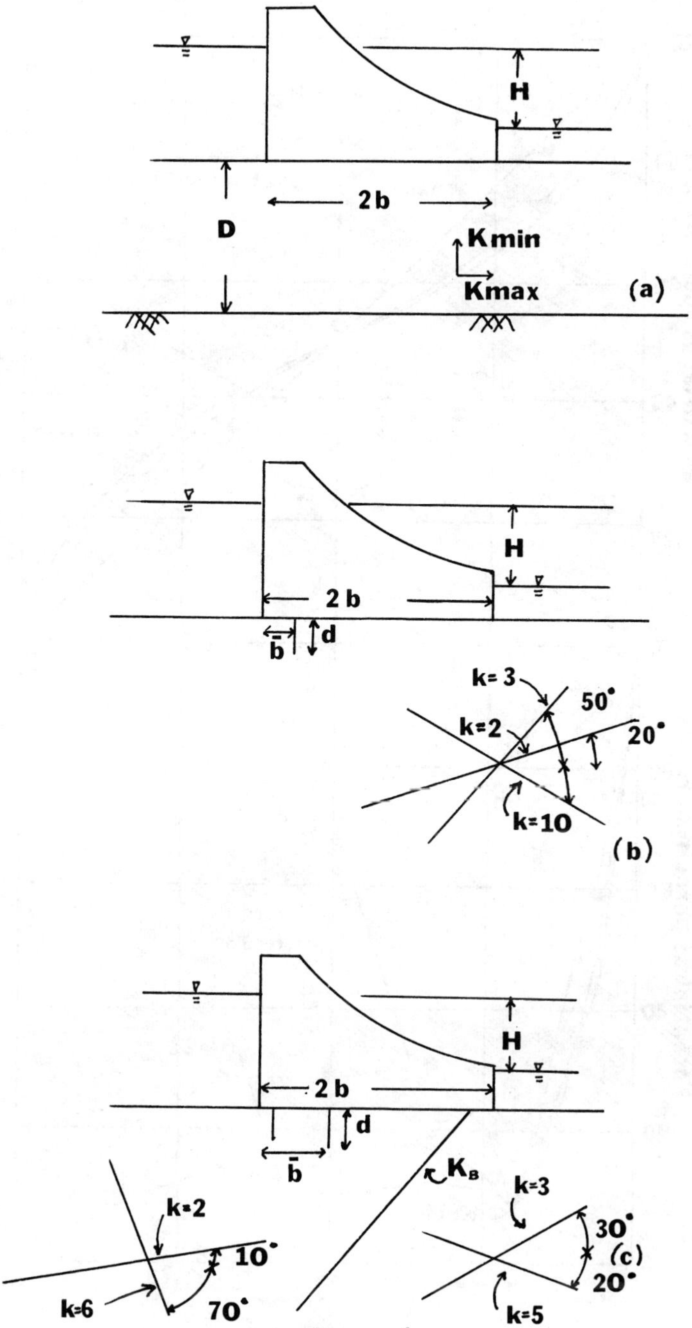
H
2b
D
Kmin
Kmax
(a)
H
2 b
b̄
d
k= 3
50°
k=2
20°
k=10
(b)
H
2 b
d
b̄
K_B
k=3
30°
(c)
20°
k=2
10°
k=6
70°
k=5

Figure 1

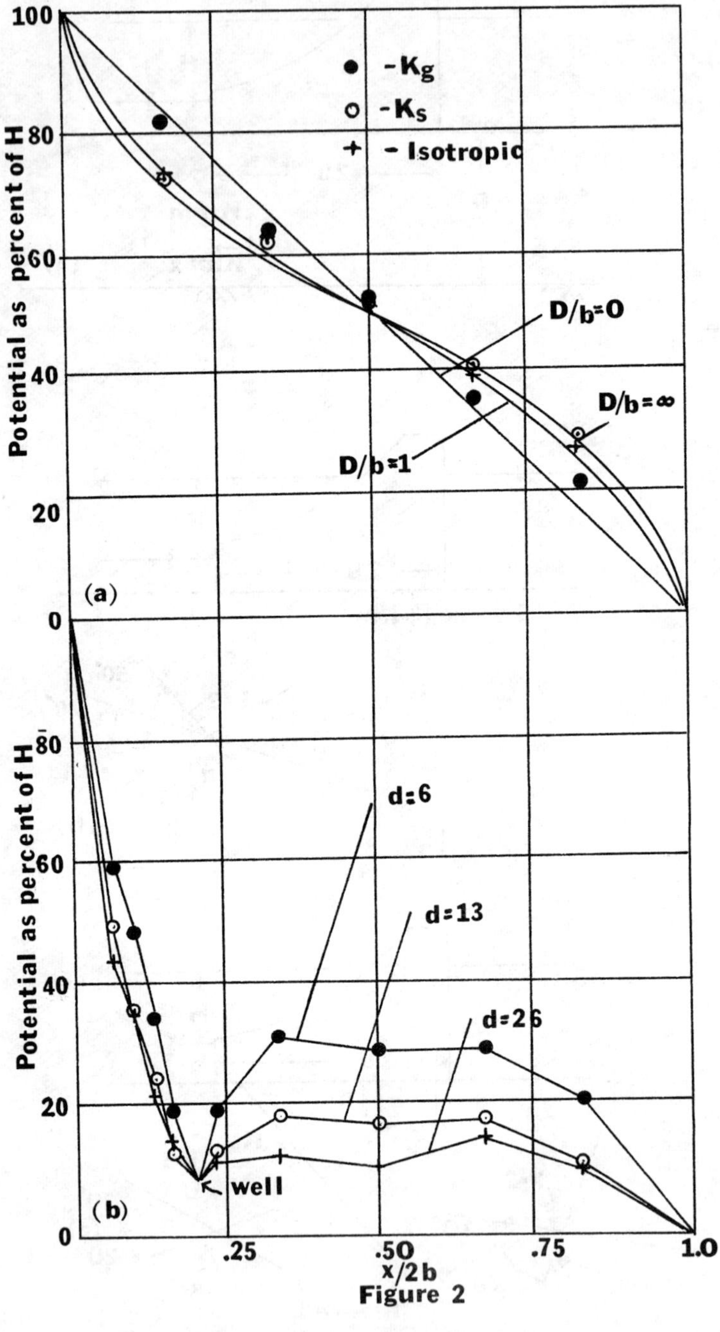
100
80
60
40
20
0
Potential as percent of H
- Kg
- Ks
- Isotropic
D/b=0
D/b=∞
D/b=1
(a)
80
60
40
20
0
Potential as percent of H
d=6
d=13
d=26
well
(b)
.25
.50
.75
1.0
x/2b

Figure 2

Solutions from Harr (3) are shown for reference in Figure 2b. The D/b ratio here would be 2. A transformation which would reduce the permeability in the direction of the resultant gradient to an equivalent isotropic medium yields a D/b ratio of 4/9. The results obtained for both permeability distributions by the continuous Monte Carlo method are shown in Figure 2a. The results indicated as for Ks are for permeability in the direction of the resultant velocity and those for Kg and for permeability in the resultant gradient direction. Greater differences in the two solutions were observed upstream and downstream of the dam.

Dam on Jointed Medium of Infinite Extent with Well. The problem as solved is shown in Figure 1b. The equivalent anisotropic medium was found to have a maximum permeability of 11.2 inclined 18 degrees to the horizontal. The minimum permeability was computed to be 3.8. The drainage was set to maintain a potential of 10 percent of H. The depth of the drainage indicated by d, was varied and the results are shown in Figure 2b.

Dam on a Non-Homogeneous Jointed Stratum of Infinite Extent With Control Measures. For all solutions shown in Figure 3, an impermeable cutoff was located a distance from the heel approximately 6 percent of the width of the dam. The depth of the cutoff was approximately 16 percent of the dam width. The boundary between the two zones is inclined 50 degrees to the horizontal, and intersects the dam at a point equal to 90 percent of the dam's width from the heel. The boundary permeability indicated by K_B is allowed to take on values

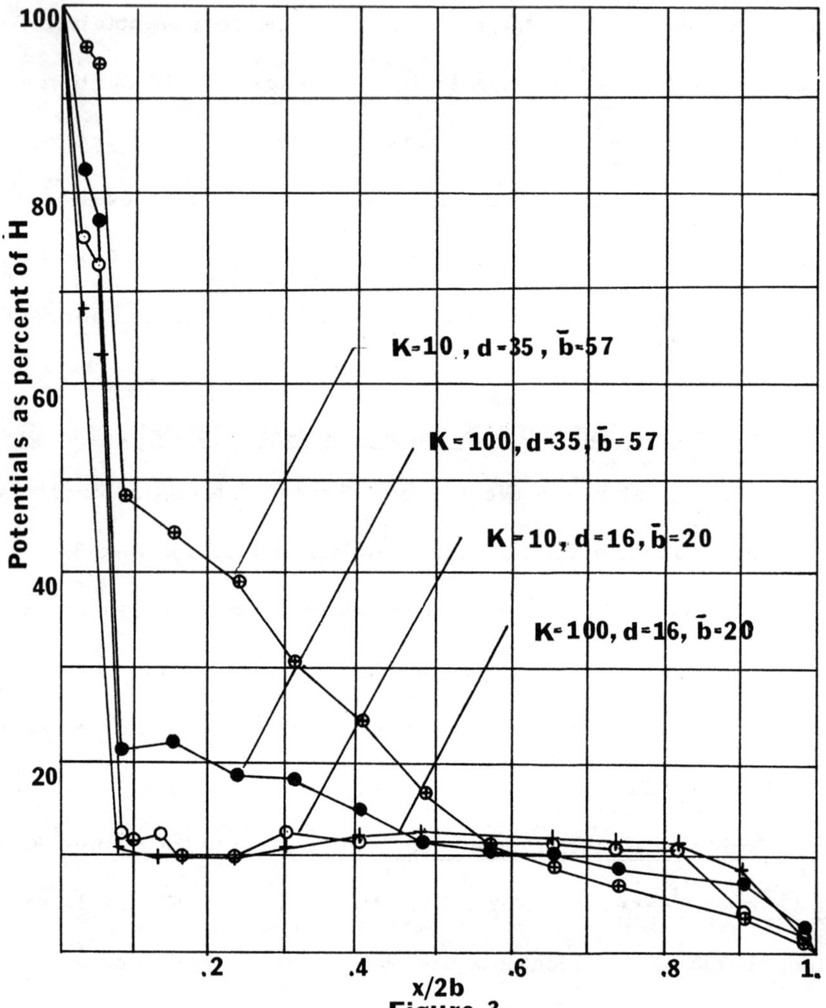

100
80
60
40
20
Potentials as percent of H
K=10 , d=35 , b̄=57
K=100, d=35, b̄=57
K=10, d=16, b̄=20
K=100, d=16, b̄=20
.2
.4
.6
.8
1.
x/2b

Figure 3

of 10 and 100. Two different depths and locations for wells are considered. The locations of the wells $\bar{b}$, are given as percent width of the dam measured from the heel. The depths of the wells d, are given as percent width of the dam. These values appear with the results in Figure 3.

DISCUSSION AND COMPARISON WITH OTHER SOLUTIONS

The problem as shown in Figure 1a was solved for both an isotropic medium and an anisotropic medium. The results obtained provide a basis for a brief discussion of the accuracy of the continuous Monte Carlo method.

The results for the isotropic case and the anisotropic case indicated by K should lie between the solutions D/b equal to 1 and D/b equal to infinity (3). All results shown are generally within 5 percent of H of the theoretical value. With a probabilistic method, larger errors are possible even though the probability becomes smaller with increasing number of trials.

The results indicated by Kg in Figure 2a are for permeability in the direction of the gradient. The differences in potentials calculated using the two definitions of permeability are apparent. However, depending on the direction chosen, the value of permeability calculated by the two definitions differs as much as 190 percent. This extreme doesn't manifest itself in Figure 2a.

In general, the reduction of truncation increases the accuracy of the result but increases solution time. The results obtained by this writer indicate that solution time is more sensitive than the accuracy.

To illustrate this point, all the results presented here were subject to a minimum of 500 walks (8). For the isotropic example, in one case, the truncation was doubled, the resulting solution time was reduced by 40 percent and the results were approximately the same. For cases illustrated in Figures 1b and 1c, no definite conclusions were reached. These problems are more complex because of the addition of wells and interior boundaries. At these points truncation must also be considered.

CONCLUSIONS

The continuous Monte Carlo method for the solution of seepage problems has been introduced along with some selected problems.

With regard to the method itself, it can be said that for two dimensional problems: 1) the continuous approach is faster than the direct method employed by previous writers (8, 15); 2) the method is extremely versatile with respect to the permeability distributions which may be handled; 3) truncation is a controlling factor in solution time. However, the optimum truncation to use for any number of completed walks is still unresolved.

The main objection to the Monte Carlo method has been its requirements for large amounts of computer time. The methods presented here reduce the time requirement depending on the truncation employed.

This writer still feels that the method is best suited to determining only a few points where the potentials are considered critical. The examples illustrated this, since only points along the base of the dam were investigated.

The unconfined case has not been considered in this paper.

LaRusso (8) has studied this case employing the finite difference Monte Carlo approach.

The application of the Monte Carlo or random walk method is not, of course, restricted to the method suggested in this paper or by previous writers (8, 15). There are other approaches, perhaps working with the joint geometry itself rather than assuming a continuous media, which might be investigated.

ACKNOWLEDGMENTS

This paper is based in part on a thesis presented by the writer at the University of Notre Dame, Notre Dame, Indiana in partial fulfillment of requirements for the degree of Master of Science in Civil Engineering, February, 1969. The writer's advisor was Ramon S. LaRusso.

REFERENCES

1. Brown, G. W., Modern Mathematics for the Engineer, McGraw-Hill Book Company, Inc., New York, New York. Vol. 1, pp. 279-301, 1956.

2. Collins, R. E., "Flow of Fluids Through Porous Materials," Reinhold Publishing Corporation. New York, New York, 1961.

3. Harr, M. E., Groundwater and Seepage, McGraw-Hill Book Company, Inc., New York, New York, 1962.

4. Hersh, R., Griego, R. J., "Brownian Motion and Potential Theory," Scientific American, Vol. 220, No. 3, pp. 66-77, March 1969.

5. Klahr, N. C., "A Monte Carlo Method for the Solution of Elliptic Partial Differential Equations," Mathematical Methods for Digital Computers, A. Ralston and H. S. Wilf, eds., John Wiley and Sons, Inc., New York, New York, 1960.

6. Krizek, R. J., "Permeability of Anisotropic Media," Highway Research Record, No. 223, pp. 72-81, 1968.

7. Krizek, R. J. and Anand, V. B., "Flow Around a Vertical Sheetpile Embedded in an Inclined Stratified Medium," Water Resources Research, Vol. 4, pp. 113-123, February 1968.

8. LaRusso, R. S., "Analysis of Seepage and Groundwater Flow by Monte Carlo Methods," Ph. D. Thesis-Department of Civil Engineering, University of Notre Dame, June 1967.

9. Leonards, G. A., "Engineering Properties of Soils," Foundation Engineering, G. A. Leonards, ed., McGraw-Hill Book Company, Inc., New York, New York, 1962.

10. LiaKopoulos, A. C., "Variation of the Permeability Tensor Ellipsoid in Homogeneous Anisotropic Soils," Water Resources Research, Vol. 1, pp. 135-141, 1965.

11. Maarsland, M., "Soil Anisotropy and Land Drainage," Drainage of Agricultural Lands, edited by James N. Luthin, pp. 216-285, American Society of Agronomy, Madison, Wisconsin, 1957.

12. Muller, M. E., "Some Continuous Monte Carlo Methods for the Dirichlet Problem," Annals of the Mathematical Statistics Society, Vol. 27, No. 3 pp. 1- 38, 1964.

13. Parsons, R. W., "Permeability of Idealized Fractured Rock," Journal of Society of Petroleum Engineers, A. I. M. M. E, Vol. 6, pp. 126-136, 1966.

14. Schvidler, M. I., "Filtration Flows in Heterogeneous Media: A Statistical Approach," Consultants Bureau, New York, New York, 1964.

15. Scott, R. F., Principles of Soil Mechanics, Addison-Wesley Publishing Company, Inc., Reading, Massachusetts, pp. 152- 156, 1963.

16. Serafim, J. L., and del Campo, A., "Interstitial Pressures on Rock Foundations of Dams," Journal of the Soil Mechanics and Foundation Division, A. S. C. E., Vol. 90, pp. 65-85, September 1968.

17. Shreider, Y. A., Methods of Statistical Testing, Elsevier Publishing Company, New York, New York pp. 1- 38, 1964.

18. Snow, D. T., "Rock Fracture Spacing, Openings, and Porosites," Journal of the Soil Mechanics and Foundation Division, A. S. C. E., Vol. 94, pp. 73-91 January 1968.

SLOPE ANALYSIS FOR EXPLOSIVE EXCAVATIONS

by

Richard H. Gates[1]

INTRODUCTION

The term "explosive excavation" is a general term which in most instances implies "cratering" or the use of large explosive charges to produce an excavation by fracturing and ejecting large volumes of soil or rock (15, 16)[2]. Fig. 1 illustrates a channel explosively excavated on the shore of Fort Peck Reservoir, Montana. As a rule of thumb, it is convenient to think of explosive excavation by chemical explosions in terms of tens of tons of explosives emplaced tens of feet in depth.

Explosive excavation must be cost-competitive to become an accepted method of excavation. As shown in Fig. 2, actual explosive excavation costs have been reduced. Based on the usable cross section, the cost for rock excavation is about $2.00 per cu yd, with many additional economies still possible.

---

[1]Major, U. S. Army, assigned as Deputy Director, U. S. Army Engineer Explosive Excavation Research Office (formerly Nuclear Cratering Group), Lawrence Radiation Laboratory, Livermore, Calif.

[2]Numerals in parentheses refer to corresponding items in Appendix—References.

FIG. 1.— EXPLOSIVELY EXCAVATED CHANNEL, FORT PECK RESERVOIR, MONTANA

A general understanding of the effects of a cratering explosion on the surrounding medium and of the nature of the various crater zones which are produced has been developed. Cratering experiments by both nuclear and chemical explosives and a typical slope analysis are discussed in Ref. 11.

INITIAL STEPS OF ANALYSIS

Classification of Media.— As in any slope stability analysis, the first step is the collection of site data. Primary considerations for classification are the ease and the expense involved in obtaining the

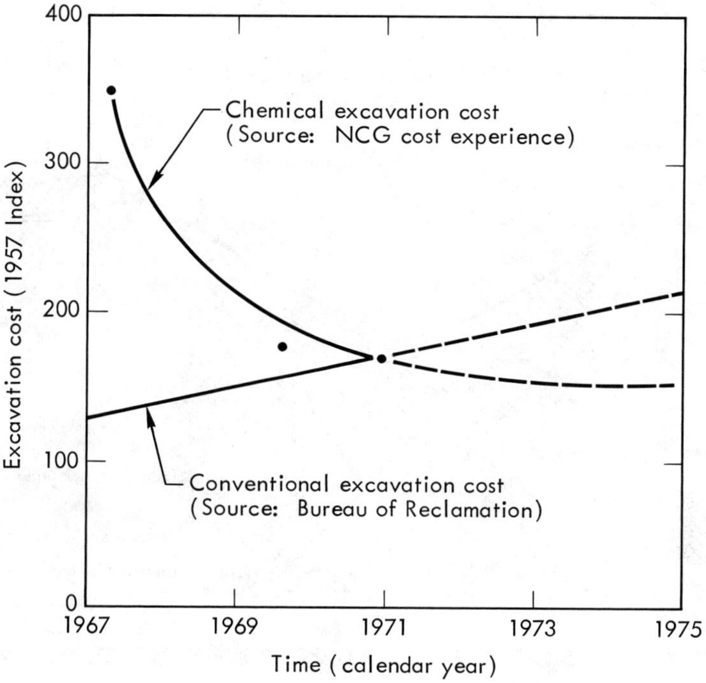

FIG. 2.— COMPARISON OF CONVENTIONAL AND EXPLOSIVE EXCAVATION COSTS

necessary data to describe adequately the material for the intended purpose.

The media must initially be classified as to whether it is common or rock excavation in accordance with Corp of Engineers Guide Specifications. One type of material, clay shale, has been set apart as a separate entity in a system development (16). This refinement results largely from the extensive studies and experiments conducted in clay shale (12, 24).

Prediction of Crater Geometry.— When explosives are placed at a depth which gives a maximum volume of excavation, a crater is

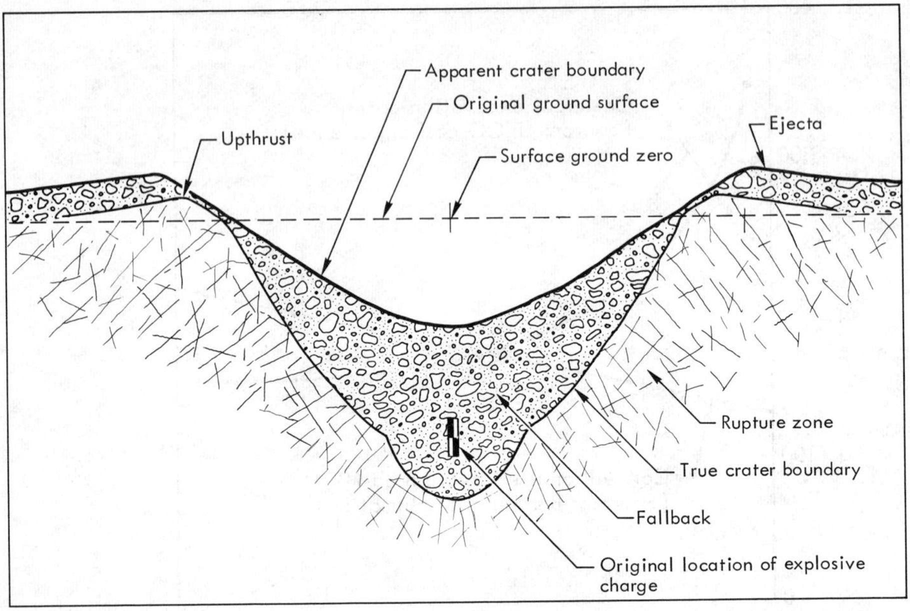

FIG. 3.— CROSS SECTION OF TYPICAL CRATER IN ROCK, SHOWING ZONES OF DISTURBANCE

produced with relatively predictable characteristics. Fig. 3 is a typical cross section of such a crater showing crater nomenclature. This figure could be considered a cross section of a row crater formed by a series of single detonations or a cross section of one such detonation. Empirical formulas have been developed that correlate crater dimensions with the magnitude of the explosive charge. The apparent, or visible, crater outlines material which may be classified into three categories: the rubble zone made up of fallback and ejecta, the rupture zone, and the intact zone. The geometry of each zone is as shown in Fig. 3.

A numerical technique has been developed by Terhune (25) to predict crater geometry in lieu of scaling from past explosions. This technique uses a digital computer to integrate the conservation laws of mass, momentum, and energy. It involves replacing the continuous spatial distribution of stress, density, velocity, etc., with a distribution defined over a grid of discrete positions (zones) in the medium. The calculations are simplified by assuming a degree of symmetry.

Two computer codes are currently used for cratering calculations. One was developed for a spherical, one-coordinate model which integrates the conservation equations written in spherical symmetry with only a radial direction of motion permitted. Another code was developed to study cylindrically symmetrical problems, such as a crater where two spatial variables are required.

At a given time, the stress, the density, the coordinates, and the particle velocity of each zone are known. The conservation-of-momentum equation provides a relationship between the applied stress field and the resulting acceleration of each zone in the grid. Accelerations acting over a small time increment produce new velocities, velocities produce displacements, displacements produce strains, and strains produce a new stress field. Time is incremented and the cycle is repeated until the mound is in freefall.

The slopes of such excavations are formed by the natural deposition of rubble after the detonation. In the creation of a row excavation, as would be needed for a canal, these slopes might be subjected once again to a dynamic loading. In some areas they would be subjected to earthquake loading. The possibility of a change to this crater geometry was studied empirically at the Nevada Test Site. Data were obtained from measurements of the effects of the seismic motions generated by the HANDLEY Event on the CABRIOLET and

SCHOONER craters (22). Accelerometers recorded the ground motions; aerial photographs and ground surveys documented the permanent ground displacements. The slopes at CABRIOLET and SCHOONER were subjected to an estimated peak radial (horizontal) acceleration of 0.74 and 0.36 g, respectively.

The analysis showed that the mode of failure of cohesionless materials, surface raveling and sloughing, did occur during the HANDLEY Event. However, the mode of failure should not be confused with a project failure, which is a mass movement of material that results in a loss of time, money, or life. Since no major changes in crater geometry occurred under severe ground acceleration, it thus appears that crater slopes are likely to remain essentially stable when subjected to large seismic motions as long as discontinuities do not cause instability in the rupture or intact zones.

Estimation of Crater Zone Stability Properties (9).— In addition to crater geometry, an analysis of crater stability requires a knowledge of the post-detonation physical and structural properties of the medium and the impact of such external factors as seepage, weathering, and seismic loading. The post-detonation properties are predicted from pre-detonation investigations of the undisturbed medium. The principles of soil mechanics are used to evaluate the properties of the rubble zone, and the principles of rock mechanics are used to evaluate the intact and rupture zones.

The material comprising the fallback and ejecta has been fractured by the explosion, broken into particles of various sizes, lifted into the air or sloughed down, and redeposited in a somewhat predictable pattern. This is crater rubble. If the detonation is in rock, the rubble is more or less loose blocks of material, the faces of which are preexisting discontinuities or blast-induced fractures. The strength analysis of the rubble zone is primarily an assessment of the stability of cohesionless rubble slopes. The factor of safety for dumped large aggregate, a situation roughly analogous to the formation of crater slopes, normally varies between 1.1 and 1.5 (20).

For other than very weak rocks, the strength of the rock mass in the rupture zone is governed by the spacing and orientation of the natural and blast-induced fractures; therefore, careful site documentation is necessary. The materials near the crater have been highly fractured by the explosion. In the rupture zone natural discontinuities are disrupted and large subsurface displacements are common (10).

Exploratory drilling in the rupture zone below the preshot ground surface has provided information on the increase in the intensity of fracturing. In the portion of the rupture zone near the true crater boundary, fracturing increases in intensity by several hundred percent as compared to the preshot in situ fracturing (Fig. 4). The majority of blast-induced fractures are alined in a direction nearly parallel to natural fractures. The effective porosity (volume of continuous voids/total volume) of the materials in the rupture zone is increased significantly as a result of the cratering detonation. This increase is due primarily to the opening of both natural and blast-induced fractures. Investigations of several craters in basalt, having a preshot effective porosity of approximately 2%, have indicated blast-induced porosities of as much as 25% near the true crater boundary (10). Fig. 4 schematically illustrates the decrease in blast-induced fracturing and effective porosity with distance from the true crater boundary.

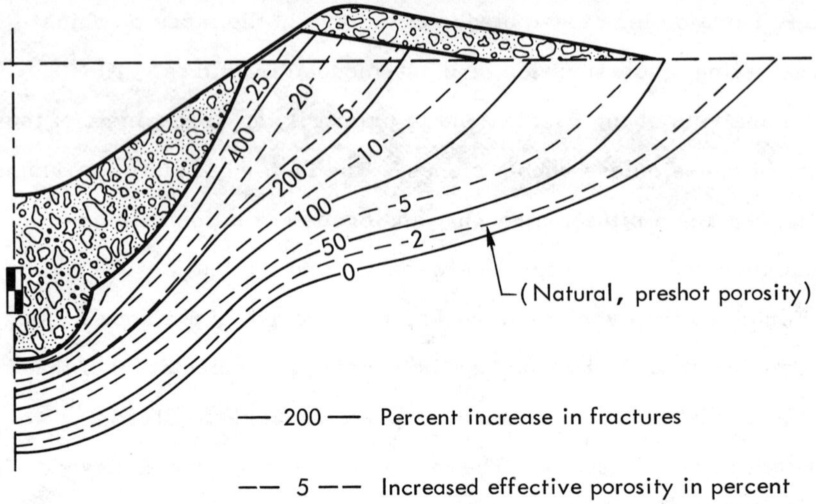

FIG. 4.— REPRESENTATIVE VALUES OF BLAST-INDUCED FRACTURING AND INCREASED POROSITY IN RUPTURE ZONE

For explosively excavated craters in rock the permeability will generally be high because of the fracturing and the material will drain easily; thus the stability of such slopes will not be affected to any great extent by seepage pressures (23).

METHODS OF ANALYSIS

Empirical Analysis.— In the empirical approach, a prediction of the performance of a crater slope is based largely upon empirical data relative to the observed performance and characteristics of comparable man-made or natural slopes. To use the empirical approach, sufficient site information must be collected so that the material in which an excavation is planned may be characterized for comparison with observations and data which have been previously gathered and catalogued in other locations. The primary advantage of this approach is that it can circumvent a lack of detailed site data and crater slope properties. Several reports deal with this method of analysis for crater slopes (12, 17, 19, 24). Don C. Banks is presenting results from Refs. 12 and 24 at this meeting.

Limit Equilibrium Analysis.— This section discusses four limit equilibrium methods: (a) the method of slices for circular slip surfaces, (b) the method of slices for composite failure surfaces, (c) the method of wedges, and (d) the method of three-dimensional wedge analysis. These methods can account for the effects of pore water pressure, seismic loadings, and long-term strength losses on the stability of a crater slope. The methods presented are two- and three-dimensional limit equilibrium analyses applicable to linear row craters of semi-infinite extent. Ref. 3 applies the first three methods to explosive excavations, and Ref. 14 presents an excellent evaluation of the fourth.

The first three methods are considered to be most applicable to those portions of a crater slope in which the slope materials behave as masses of angular, aggregated particles such as exist in the rubble zone of a crater (Fig. 3). Three-dimensional solutions for cases in which structural discontinuities may play an important role, such as in the rupture and intact zones of crater slopes in rock, are handled by the fourth method. When the fractured and intact zones behave in a two-dimensional manner, all four methods are applicable. This condition may exist when joints, fractures, and other discontinuities have strike directions which parallel the center line of the excavation or when discontinuities are randomly oriented but have such a close spacing that the mass behavior is similar to that of a relatively homogeneous material. Limit equilibrium methods of analysis have recently been developed to analyze problems in three dimensions which are explained and illustrated in Ref. 14. In rock slopes the potential system of failure surfaces already exist in the mass, but the kinematics of sliding must be checked to delineate the possible directions and surfaces on which it is physically possible for sliding to take place.

The following is an example of a two-dimensional wedge analysis. Its purpose was to evaluate whether fallback provides a buttressing effect against failure in the rupture and intact zones and to quantify the improved factor of safety. The type of failure analyzed was that of a wedge slide with the strike of the plane of sliding parallel to the slope and "daylighting" into the true crater.

The geometric configuration of the crater for analysis is shown in Fig. 5. All surfaces are considered planes. The cross section of the apparent crater has been approximated by two straight lines.

The failure surface was defined by the parameters α, β and H (Fig. 5). For each value of H, a factor of safety was calculated for various values of α and β.

The side forces between blocks are also shown in Fig. 5. The force acting between blocks 1 and 2 is denoted E_{12} and was assumed to act horizontally. E_{12} is a maximum value and is found by varying the angle η until a maximum value of E_{12} is reached.

The side force E_{23} between blocks 2 and 3 was assumed to act at an angle parallel to the line above it approximating the apparent crater. The value of E_{23} is a function of the parameter α and the weight of the wedge determined by α.

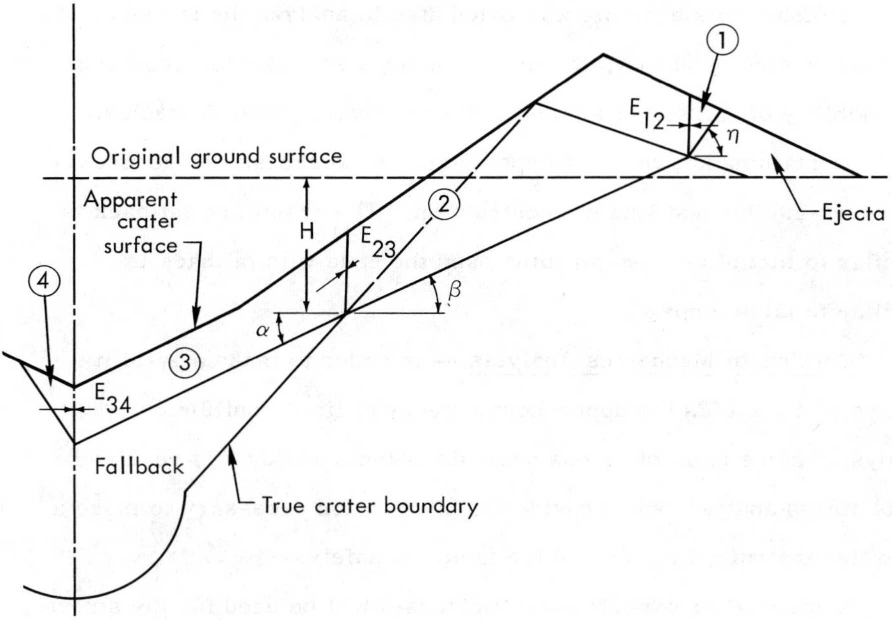

FIG. 5.— CRATER MODEL FOR LIMIT EQUILIBRIUM ANALYSIS

When the failure surface does not daylight in the crater, two passive blocks exist. The side force E_{34} was assumed to act horizontally on the boundary and can be found by the Rankine equation for a passive wedge.

For each failure surface the factor of safety was calculated for two conditions, one with fallback and one with no fallback. The analysis showed that the presence of fallback in an explosively produced crater will increase the stability of a plane of weakness in the rupture zone. The amount of increase is dependent on the dip and depth of the failure plane and the strength of the material along the plane. An increase of over 100% in the factor of safety (0.56 to 1.25) has been computed for H = 80 ft and β = 45° with the angle of internal friction along the plane of weakness equal to 30°.

Although this example was calculated to analyze the influence of fallback in slope stability, it can also be used as a tool for analyzing the stability of particular slopes. Planned craters through geologic media containing known and mapped planes of weakness can be analyzed to determine the best line of construction. The case of no fallback is similar to that of an open pit mine, and the case with fallback is similar to talus slopes.

Continuum Mechanics Analysis.— In order to make a stability analysis that avoids the upper-bound theory of limit equilibrium analysis, some form of stress analysis of the continuum is required. This stress analysis will provide the information necessary to make a stability analysis; i.e., to find the factor of safety.

Numerical or experimental methods might be used for the stress analysis. Of the numerical methods, the finite difference and the finite element methods are the most popular. Because the finite element method has been used extensively in the study of crater slope stability, this section will deal only with that method of numerical stress analysis. The finite element method, as discussed here, is a displacement method of analysis and the discussion is restricted to plane strain. Further, the emphasis will be placed on the mathematical model to represent the soil or rock.

There are limitations to the use of the finite element method. The primary limitation is the inability to determine accurately and to describe the actual complex properties of the intact rock material and the rock mass. Another limitation is the inability to model completely the inelastic behavior of rock during the loading through failure. Both of these limitations are recognized; however, many efforts in rock mechanics are directed toward improving the capabilities of the finite element method.

In the past few years the Corps of Engineers has funded many studies of the use of the finite element method for slope stability analyses (6, 7, 8, 18). The major variations studied are boundary conditions (slope angle, initial stress conditions, and methods of loading) and material properties (modulus of elasticity, Poisson's ratio, cohesion, angle of internal fraction, and unit weight). The finite element method is very versatile and can be used with different methods of modeling the stress-strain patterns of earth materials.

Some Results of Parameter Studies with Finite Element Method.— The initial state of stress at a location and the method of its application to a cut slope have been found to be most important. The initial state of stress should be applied directly to the elements at the beginning of the analysis for the most direct and probably the most realistic analysis. The vertical stress should be taken as the overburden stress, and the horizontal stress should be taken as proportional to this overburden stress with the constant of proportionality being K, the coefficient of earth pressure at rest. The value of K can be greater or less than one depending on the geologic history of the location; i.e., it need not be the value derived from elastic theory. Values less than one can be associated with normally consolidated clays and values greater than one with overconsolidated clays. The authors of both Refs. 6 and 7 found that the stresses at the toe of a slope are greater for higher values of K and the factor of safety is lower, particularly against progressive failure.

Although explosive excavations create slopes "instantly," incremental unloading or excavation is required for inelastic analyses.

This is the approach taken for the explosive excavation studied by Palmerton and Banks (21) in which horizontal layers were removed from the semi-infinite mass to create a slope modeling an explosive excavation.

In addition to the obvious relations for slope angle, cohesion, and angle of internal friction with stability, it has been pointed out by Kulhawy et al. (18) that if the technique requires a fixed Poisson's ratio, higher values of Poisson's ratio will give higher factors of safety. This relationship could be misleading. This role of Poisson's ratio is unfortunate, since it is seldom measured on intact samples and must be assumed for a rock mass. It is suggested that analyses either use a technique that modifies Poisson's ratio internally such as the method presented in Ref. 18 or several runs be made with different estimates of Poisson's ratio.

The factor of safety along assumed failure surfaces can be computed with the stresses from the finite element method; however, care should be taken to insure that the definition is consistent for both finite element and limit equilibrium analyses. When the factor of safety is defined with respect to the value of shear stress for a potential failure surface, its value from a finite element analysis is very close to that calculated by means of the best limit equilibrium analysis procedures (18).

One limitation of the finite element method was removed by Duncan and Goodman (8) when jointed rock properties were introduced into the model. Although the discontinuities would have to be fixed before the analysis was made, this limitation is satisfactory if geologic investigations have found predominate joint patterns or a particularly significant discontinuity.

Although the mathematical models to represent the stress-strain behavior of earth materials are discussed in the next section, one property of a particular model should be pointed out at this time. As mentioned above, the value selected for Poisson's ratio plays an important role in the factor of safety calculated. The inelastic technique presented by Kulhawy et al. (18) not only allowed the stress-strain curve to be simulated by a hyperbola, but also the axial strain-radial strain curve. This improvement removes the requirement to assume a value of Poisson's ratio and includes volume change characteristics.

Mathematical Models to Represent Stress-Strain Behavior of Earth Materials.— Only the imagination of the engineer limits the possible mathematical models that can be used to represent the stress-strain behavior of earth material. Although the models that have been used in connection with rock slope stability are somewhat limited, there are many models used by researchers modeling earth material to evaluate the effect on structures of ground shock from a large explosion. Some of the more sophisticated and recent are reported in Refs. 2, 4, and 5, with Ref. 4 reviewing several models.

The material properties to be modeled must be determined by extensive testing. The stress-strain curve and the shear strength envelope are frequently referred to as the constitutive equations. When Hugoniot data (measures of material compressibility as a function of very high applied loads) are also developed for the rock through its various modes of material behavior (gas, liquid, solid), the test results are called the equation-of-state. This information is particularly important when the material response from nuclear excavations (25) is modeled.

The simplest model for slope stability is the elastic assumption. Refs. 6 and 7 used this model primarily. The fact that incremental loading is not required for the elastic assumption results in a considerable saving of computer time. Although the stresses and factors of safety from this method are compatible with limit equilibrium methods, efforts continue to improve modeling of the inelastic behavior of the material.

The author (13) considered four models for the stress-strain curve, one being the elastic, two being variations of stable inelastic, and the fourth being unstable inelastic. Baker et al. (1) also applied one of the stable inelastic models to slope stability analysis. Kulhawy et al. (18) considered one stable inelastic model in addition to the elastic model.

Fig. 6 illustrates six models of stress-strain curves. All can be made to fit test results with varying degrees of accuracy. The models illustrated in Figs. 6(b), 6(c), 6(e), and 6(f) use a generalized Mises criterion. The parameters needed for this yield condition are cohesion and angle of internal friction. The hyperbolic stress-strain curve of Fig. 6(d) was first used on slopes by Kulhawy et al. (18). Three constants taken from the sample stress-strain curve to be modeled are required for this model. An additional output from these inelastic analyses is a zone in the slope that has yielded or failed. Only Palmerton and Banks (21) have applied one of these inelastic models to an explosively excavated slope. Both Kulhawy et al. (18) and Palmerton and Banks (21) applied the hyperbolic stress-strain model to the Otto Brook dam and calculated the displacements within good tolerance with the actual displacements.

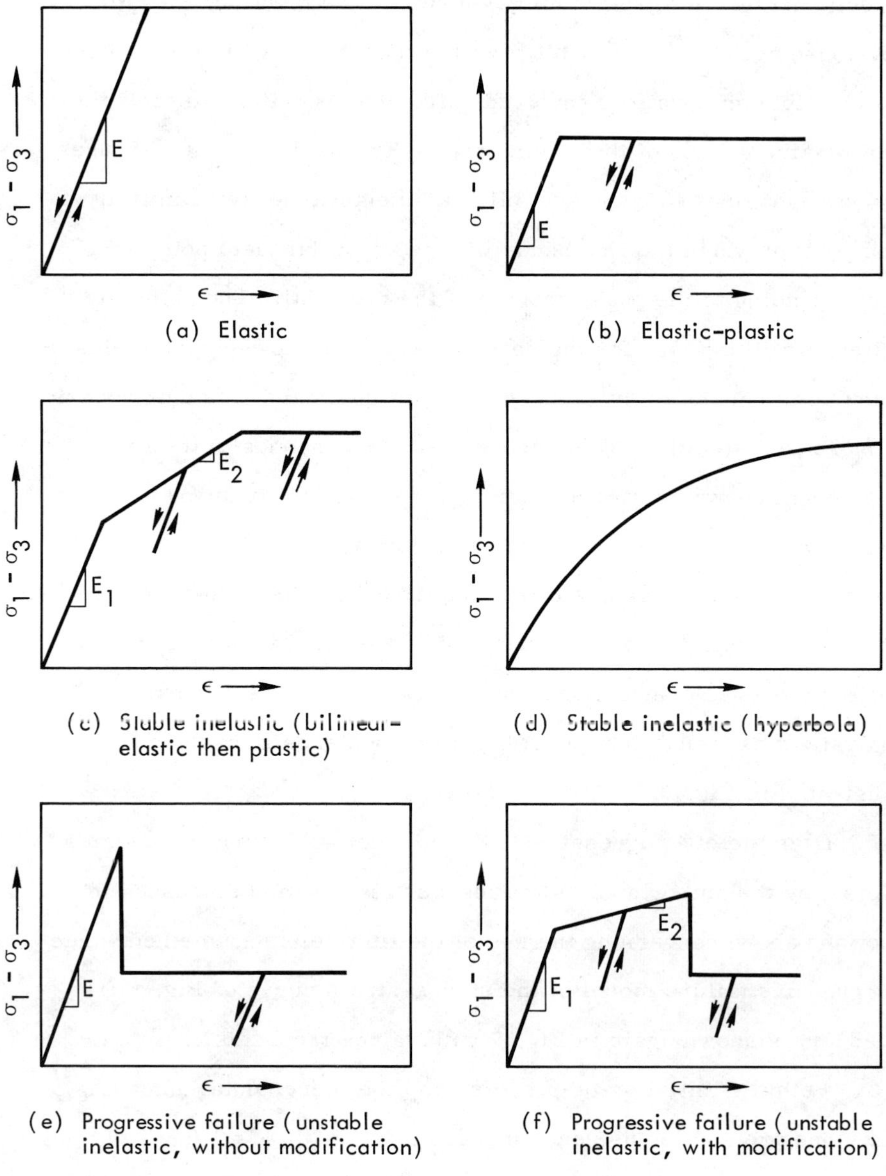

FIG. 6. — MODEL STRESS-STRAIN CURVES

Only parameter studies of a 45° slope have been made with the models of Figs. 6(b), 6(c), 6(e), and 6(f). The model of Fig. 6(f) was developed because of an inability of the model of Fig. 6(e) to balance the equilibrium equations involved, although the latter did show the progressive growth of the failure zone. The model of Fig. 6(f) uses two straight lines to represent bilinear inelastic behavior initially. The point at which the line bends will be an initial yield point corresponding to the peak strength of the material. The same yield criterion will be used for the second yield point; however, the elastic modulus is considerably reduced, and the internal friction or cohesion is increased slightly. This second yield corresponds to the reduction in strength known as strain-softening, and the stress drops to a new yield surface represented by the residual strength, usually a considerably reduced angle of internal friction with zero cohesion. At this new stress level, the material is assumed to behave in a perfectly plastic manner. Any unloading of an element during the analysis is assumed to be elastic with no hysteresis loss, as illustrated in Fig. 6.

Example of Progressive Failure Model.— A current research project by the author applies the lessons learned and discussed in the sections above concerning the use of the finite element method. The progressive failure model of the stress-strain curve of Fig. 6(f) is used and is shown again in Fig. 7 with values from this example.

Both the initial state of stress and the incremental unloading of the slope are applied as recommended above with $K = 1.0$. The strength properties selected are representative values from Ref. 24 for Cucaracha and Culebra clay shale.

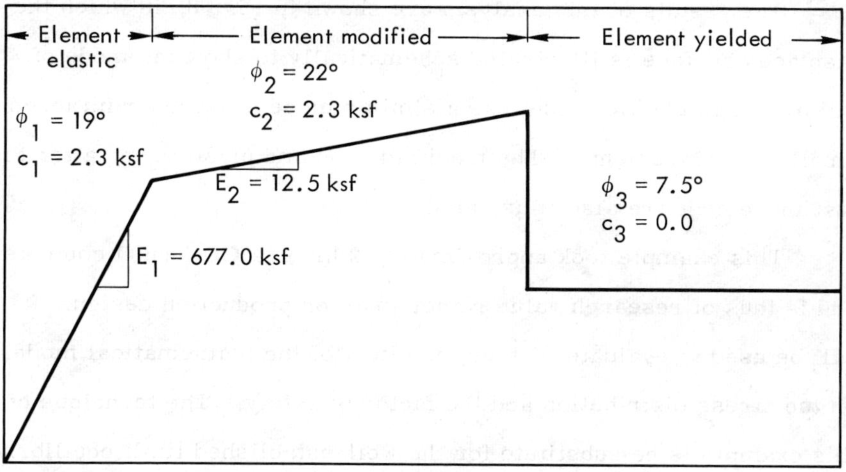

FIG. 7.— PROGRESSIVE FAILURE STRESS-STRAIN CURVE (UNSTABLE INELASTIC, WITH MODIFICATION)

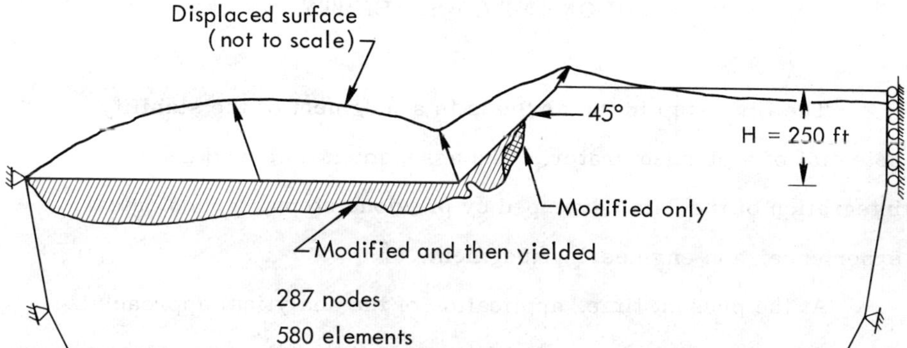

FIG. 8.— SURFACE DISPLACEMENTS AND MODIFIED AREAS OF PROGRESSIVE FAILURE ANALYSIS

The results of this analysis are shown in Fig. 8, in which the displaced surface is illustrated schematically to show the mode of failure. The elastic model has a similar mode of failure with much smaller displacements. The modified zone and plastic zone after the last increment are also illustrated.

This example took approximately 2 hr on a CDC-6600 computer and is thus of research value rather than for production design. It will be used to evaluate further the effect of the mathematical model on the stress distribution and the factor of safety. The technique of this example is no substitute for the well-established limit equilibrium methods of soil and rock mechanics at this time, but it may lead to better understanding of the effects of progressive failure.

JUDGMENT OF STABILITY

The final step in the analysis is a judgment of the stability potential of a planned crater. The assessment is based upon an integration of the data developed by previous steps, technological experience, and engineering judgment.

At the present time, application of the analytical approach is limited by several factors. To date, experimental crater slopes have experienced no significant failures which could be used to study failure geometries and backfigure material strengths. In addition, the methods required to predict precisely the parameters required for numerical analyses are not available. Although assumed ranges of parametric values might be used in lieu of precise values, calculated safety factors are very sensitive to small changes in input parameters.

SUMMARY

The assessment of crater slope stability requires specific information about crater geometry, geologic conditions, and material properties at the site, including seepage and weathering characteristics of the materials.

Fortunately there are certain general characteristics of craters and crater slopes that make them inherently stable. Fallback from dry rock media will behave as cohesionless material and may be studied by methods developed for the stability of rubble, talus, and rockfill. Factors of safety for the fallback portion of a crater slope should range between 1.1 and 1.5, and the rupture zone is buttressed by the fallback.

Rupture zone materials in rock would most likely fail along surfaces of natural or blast-induced fractures if the strength of the intact rock is significantly larger than the frictional resistance to sliding along the fractures. A major uncertainty in sample analysis and in the assessment of crater stability in general is the possibility of long-term changes in strength and seepage characteristics in response to weathering agents.

The methods of analysis of explosive excavations are continuing to improve. Limit equilibrium techniques are satisfactory for design, but the continuum mechanics techniques are helpful in the research of rock behavior.

APPENDIX.— REFERENCES

1. Baker, L. E., Sandhu, R. S., and Shieh, W. Y., "Application of Elastic-Plastic Analysis in Rock Mechanics by Finite Element Method," Rock Mechanics-Theory and Practice, Eleventh Symposium on Rock Mechanics, University of California, June 1969, Published by the Society of Mining Engineers of AIME, 1970.

2. Baladi, G. Y., and Hadala, P. F., "Ground Shock Calculation Parameter Study, Report 1, Effects of Various Nonlinear Elastic-Plastic Model Formulations," TR S-71-4, U. S. Army Engineer Waterways Experiment Station, Vicksburg, Miss., 1971.

3. Banks, D. C., "Selected Methods for Analyzing the Stability of Cratered Slopes," WES Miscellaneous Paper S-68-8, U. S. Army Engineer Waterways Experiment Station, Vicksburg, Miss., 1968.

4. Baron, M. L., "A Summary of Some Analytical Studies on Airblast Induced Ground Motions," DASA Report 2634, Prepared by Paul Weidlinger, Consulting Engineer, New York, N. Y., 1971.

5. Cherry, J. T., and Petersen, F. L., "Numerical Simulation of Stress Wave Propagation from Underground Explosions," Proceedings of Symposium on Engineering with Nuclear Explosives, American Nuclear Society and U. S. Atomic Energy Commission, Vol. 1, 1970, pp. 142-220.

6. Constantopoulos, I. V., Christian, J. T., and Whitman, R. V., "Initiation of Failure in Slopes in Overconsolidated Clays and Clay Shales," NCG TR-29, U. S. Army Engineer Nuclear Cratering Group, Livermore, Calif. (in publication).

7. Duncan, J. M., and Dunlop, P., "Slopes in Stiff-Fissured Clays and Shales," WES Contract Report S-68-4, U. S. Army Engineer Waterways Experiment Station, Vicksburg, Miss., 1968. Also published by University of California at Berkeley as Report No. TE 68-6, 1968. Also published in Journal of the Soil Mechanics and Foundation Division, ASCE, Vol. 95, No. SM2, Proc. Paper 6449, March 1969, pp. 467-492.

8. Duncan, J. M., and Goodman, R. E., "Finite Element Analyses of Slopes in Jointed Rock," WES Contract Report S-68-3, U. S. Army Engineer Waterways Experiment Station, Vicksburg, Miss., 1968.

9. "Engineering Properties of Craters and Principles of Crater Stability," IOCS Memo NCG-35, U. S. Army Engineer Nuclear Cratering Group, Livermore, Calif., 1970.

10. Fisher, P. R., "Engineering Properties of Craters—Description of Crater Zones and Site Investigation Methods," PNE 5012-I, U. S. Army Engineer Nuclear Cratering Group, Livermore, Calif., 1968.

11. Fleming, R. W., Frandsen, A. D., and LaFrenz, R. L., "Stability of Nuclear Crater Slopes in Rock," Proceedings of Symposium on Engineering With Nuclear Explosives, American Nuclear Society and U. S. Atomic Energy Commission, Vol. 2, 1970, pp. 1661-1678.

12. Fleming, R. W., Spencer, G. S., and Banks, D. C., "Empirical Study of Behavior of Clay Shale Slopes," NCG TR-15, U. S. Army Engineer Nuclear Cratering Group, Livermore, Calif., 1970.

13. Gates, R. H., "Inelastic Analysis of Slopes by the Finite Element Method," thesis presented to the University of Illinois, at

Urbana, Ill., in 1968, in partial fulfillment of the requirements for the degree of Doctor of Philosophy.

14. Hendron, A. J., Jr., Cording, E. J., and Aiyer, A. K., "Analytical and Graphical Methods for the Analysis of Slopes in Rock Masses," NCG TR-36, U. S. Army Engineer Nuclear Cratering Group, Livermore, Calif., (in publication).

15. Hughes, B. C., "Nuclear Construction Engineering Technology," NCG TR-2, U. S. Army Engineer Nuclear Cratering Group, Livermore, Calif., 1968.

16. Johnson, S. M., et al., "Explosive Excavation Technology," NCG TR-21, U. S. Army Engineer Nuclear Cratering Group, Livermore, Calif., 1971.

17. Kley, R. J., and Lutton, R. J., "A Study of Selected Rock Excavations as Related to Large Nuclear Craters," PNE 5010, U. S. Army Engineer Nuclear Cratering Group, Livermore, Calif., 1967.

18. Kulhawy, F. H., Duncan, J. M., and Seed, H. B., "Finite Element Analyses of Stress and Movements in Embankments During Construction," WES Contract Report S-69-8, U. S. Army Engineer Waterways Experiment Station, Vicksburg, Miss., 1969. Also published by University of California at Berkeley as Report No. TE-69-4, 1969. Also published in Journal of the Soil Mechanics and Foundations Division, ASCE, Vol. 96, No. SM5, as "Nonlinear Analysis of Stress and Strain in Soils," by Duncan and Chang, 1970.

19. Lutton, R. J., "Natural Rubble Slopes and Their Relevance to Crater Fallback Slopes," NCG TR-14, U. S. Army Engineer Nuclear Cratering Group, Livermore, Calif., 1969.

20. MacIver, B. N., "The Formation and Initial Stability of Slopes in Cohesionless Materials," PNE 5009, U. S. Army Engineer Waterways Experiment Station, Vicksburg, Miss., 1967.

21. Palmerton, J. B., and Banks, D. C., "Application of Finite Element Method in Determining Stability of Crater Slopes," WES Miscellaneous Paper S-71- , U. S. Army Engineer Waterways Experiment Station, Vicksburg, Miss. (in publication).

22. Shackelford, T. J., "Crater Stability Under the Influence of Large Seismic Motions," PNE 5013, U. S. Army Engineer Nuclear Cratering Group, Livermore, Calif. (in publication).

23. Sherman, W. C., and Banks, D. C., "Seepage Characteristics of Explosively Produced Craters in Soil and Rock," NCG TR-27, U. S. Army Engineer Nuclear Cratering Group, Livermore, Calif., 1970.

24. "Study of Clay Shale Slopes Along the Panama Canal, Report 1, East Culebra and West Culebra Slides and the Model Slope," IOCS Memo JAX-108, Jacksonville District, Corps of Engineers, Jacksonville, Fla., 1970.

25. Terhune, R. W., and Stubbs, T. F., "Nuclear Cratering on a Digital Computer," Proceedings of Symposium on Engineering With Nuclear Explosives, American Nuclear Society and U. S. Atomic Energy Commission, Vol. 1, 1970, pp. 334-359.

DISTRIBUTION

LRL Internal Distribution

Michael M. May

R. H. Gates 35

TID Berkeley

TID File 30

External Distribution

Division of Technical Information Extension, Oak Ridge 2

WTF/md/la

AN APPROACH TO THE STABILITY OF ROCK SLOPES BASED ON THE THEORY OF LIMITING EQUILIBRIUM WITH A MATERIAL EXHIBITING ANISOTROPIC SHEAR STRENGTH

J.E. Jennings*

INTRODUCTION

The behaviour of rock material, particularly at the lower normal stresses encountered in the slopes of open pit mines, depends almost exclusively on joints or other discontinuities in the rock. On the other hand the behaviour of soils, and in particular softer soils, depends upon the properties of the intact soil with the shear strength being almost independent of any joints which might exist in the soft soil. It can be concluded therefore that there is some boundary, related to strength of the intact material, below which the behaviour of the geological material is independent of any joints and above which its behaviour is determined almost entirely by joints in the material. This boundary is proposed at an unconfined strength of the intact material of 689,5 kN/m^2 (100 p.s.i.). If the material has a lower strength the tendency will be for it to behave as a soil with slope failures taking place on a classical cylindrical surface through intact material; joints or other pre-existing planes of weakness will have little influence. However, if the intact material has a higher strength then it behaves as a rock and slope failure will take place on planes or combinations of planes which depend on the pre-existing planes of weakness. These principles are illustrated in Fig. 1.

The figure of 689,5 kN/m^2 is important since it provides another definition of the boundary between rock and soil. It is

*Kanthack Professor of Civil Engineering, The University of the Witwatersrand, Johannesburg.

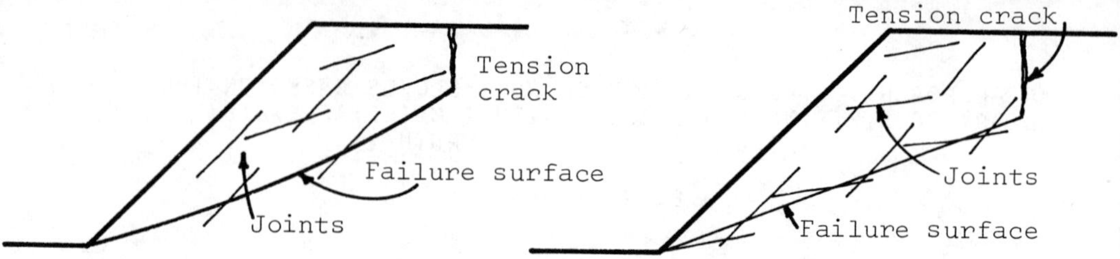

(a) Soil Slope - Rotational Failure Independent of Joints in Soil

(b) Rock Slope - Failure on Surfaces Composed of Planes which are determined by joints

Fig. 1 Comparative Behaviour of Soil and Rock Slopes

not entirely arbitrary and reference is made to Fig. 2 where physical and sensory observations in the field suggest that the boundary should lie between 276 kN/m^2 (40 p.s.i.) and 965 kN/m^2 (140 p.s.i.). Further, a rock mass which is continuously jointed in two or more directions will tend to behave as a cohesionless soil with a void ratio approaching zero; the shear strength will be grossly anisotropic, i.e. it will vary greatly as the plane under consideration rotates from the horizontal to the vertical position.

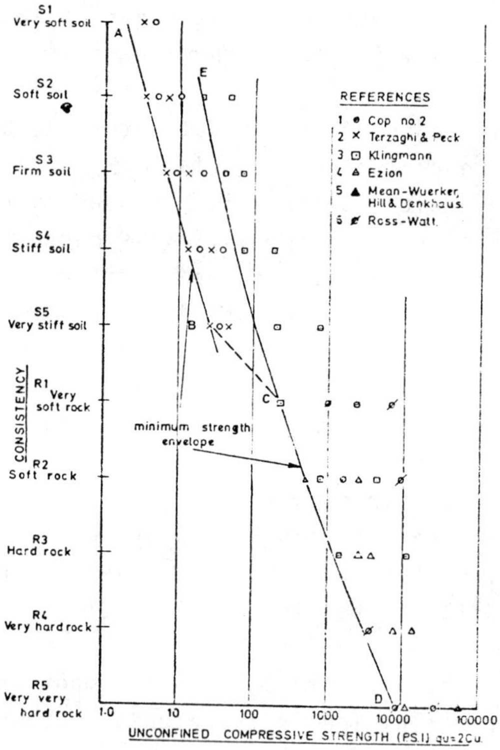

Fig. 2 Relationship between Consistency and Unconfined Compressive Strength

The slope angles which are of principle interest are those which are greater than the angle of internal friction. In a soil slope this difference depends upon the cohesion of the intact material; in a rock slope it depends upon the orientation of the discontinuities and on the measure of interlocking of the rock elements; alternatively, it depends on the extent to which the anisotropy of the material can be mobilized.

THE CONCEPT OF AN EFFECTIVE COHESION AND AN EFFECTIVE FRICTION ANGLE IN A ROCKY MATERIAL

If failure occurs in a rock slope and if this is determined principally by pre-existing joints or other planes of weakness lying in a preferential direction, the simplest mode of failure may be represented in the form shown in Fig. 3.

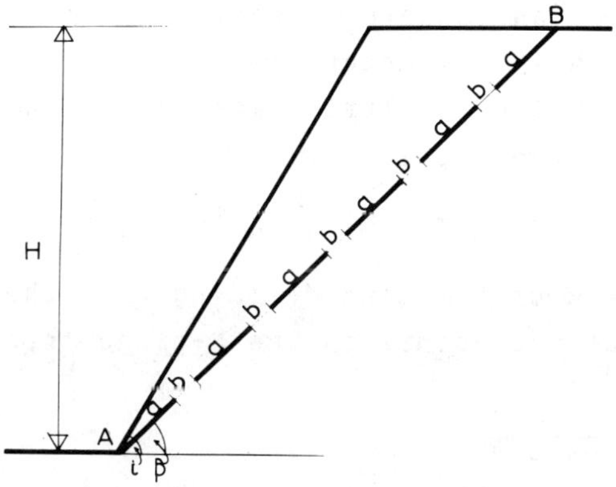

Fig. 3 Failure in a Rock Mass on a Plane defined by pre-existing Joints

The slope angle, i, is steeper than the angle of internal friction of the intact material and also of the angle of friction on the joints. The slope stands at this steeper angle only because of interlocking between joints. Alternatively, it may be considered in terms of a cohesion which exists as a result of the fact that the joints are not continuous over the whole length of the plane along which failure might possibly occur.

Let it be imagined that the material of the slope is replaced by a hypothetical material exhibiting parameters, c_a, an apparent cohesion, and ϕ_a, an apparent friction angle which taken together give the same effect as the jointed material shown in Fig. 3. Let σ_n be the average normal stress on the plane AB. Then

$$\sigma_n = \frac{\text{Weight of the Wedge of Failure}}{\text{Length of the Surface of Failure}} \cdot \cos\beta \quad \ldots \quad (1)$$

$$= \frac{\gamma H(\cot\beta - \cot i).\cos\beta}{H\ \mathrm{cosec}\ \beta}$$

The Resisting Force,

$$RF = H\ \mathrm{cosec}\ \beta(c_a + \sigma_n \tan\phi_a) \quad \ldots \quad (2)$$

If now c_m and ϕ_m and c_j and ϕ_j are respectively the Coulomb shear parameters of the intact material and joints and if Σa and Σb are respectively the cumulative lengths of the joints and intact material on the β-plane, then

$$RF = \Sigma b\ (c_m + \sigma_n \tan\phi_m) + \Sigma a\ (c_j + \sigma_n \tan\phi_j) \quad \ldots \quad (3)$$

Because $\Sigma(a) + \Sigma(b) = H\ \mathrm{cosec}\ \beta$ and defining k as the coefficient of continuity of the joints on the β-plane then

$$k = \frac{\Sigma a}{\Sigma a + \Sigma b} = \frac{\Sigma a}{H\ \mathrm{cosec}\ \beta} \quad \ldots \quad (4)$$

Equating like terms

$$c_a = (1-k)c_m + kc_j \quad \ldots \quad (5)$$

and $$\tan\phi_a = (1-k)\tan\phi_m + k\tan\phi_j \quad \ldots \quad (6)$$

Equation (3) contains a number of important implied assumptions. In most cases of limiting equilibrium the parameters c and ϕ are deduced from peak strengths of the stress:deformation curves. If one looks closely at equation (3) it will be noted that the assumption is implicitly made that the strengths of the intact material (sections "b", Fig. 3) and of the joint material (sections "a", Fig. 3) reach peak values at the same value of strain, or deformation. Much laboratory testing shows this is not true. Nevertheless, the assumption is made and the justification for it is taken from soil behaviour where it is fully accepted that the cohesional and frictional strengths are mobilized at quite different values of deformation. Further, for a soil, if the reserve of strength, as measured by the factor of safety, has a value of approximately 1,50, then the fact that there may be different factors of safety with respect to cohesion and friction is of no practical importance. This is an important facet of equation (3) and is the cause of much discussion and debate.

THE IMPORTANT JOINT FACTORS WHICH RESULT FROM THE CONCEPTS OF EFFECTIVE COHESION AND FRICTION IN A JOINTED ROCK MASS

From Fig. 3 and equations (1)-(6) it may be observed that a knowledge of the following joint properties is required:

(a) attitude of the joints with respect to the slope, i.e.
   (i) dip and dip direction of the joints;
   (ii) slope details as defined by the slope direction and slope angle;
(b) the length and width of the joints, in particular the length of the joint, as defined by "a" in Fig. 3. How do these lengths vary - are the lengths constant or do they change in a systematic fashion?
(c) the spacing between joints. Can joints be considered to lie on single planes as in Fig. 3 with a spacing "b" in between them or are the joints randomly scattered with a spacing "d" measured normally in between them?
(d) the frictional and cohesive shearing strength along the which will be determined by:
   (i) the roughness of the joint surfaces;
   (ii) gouge in the joint opening - its thickness, its shear strength and its frequency of occurrence;
(e) the cohesive and frictional shear strength of the intact rock material;

(f) any major irregularities in the surfaces of the joints, in particular waviness or change of orientation of the joints about mean values;

(g) any water pressures which may exist on the plane of potential failure and which may tend to reduce the shearing strength along the plane. Note that if water pressures are to be brought into the calculations then the c and ϕ parameters in (d) and (e) above must be considered in terms of effective stresses, i.e. total stresses normal to the failure plane minus water pressures.

If one considers the factors (a)-(g) above, it is clear that any system for collecting joint data (the joint survey) must go much further than the simple observations of dip and strike which geologists normally observe. The data must be more extensive and useful for engineering calculations of stability. Certain coarseness in these quantitative values may have to be accepted.

The factors fall into two distinct orders of importance: dip, dip direction, slope angle, slope direction, lengths of joints, spacing of joints and water pressures are of first order of importance. They lead to the calculation of quantities such as the coefficient of continuity where small errors may significantly affect the calculated factor of safety. Factors which are of second order of importance are roughness of the joint surfaces and gouge on 'opened' joints, intact material strength and waviness. Coarser evaluations are acceptable since such variations do not greatly affect the calculations.

These principles are accepted in the joint surveys described by Piteau (1970) and he gives considerable detail on the procedures used for collection of the data. The surveys are made on 'lines' where every joint which crosses the survey line is observed, this being done because the statistical analysis of the results is developed specifically for the 'line' as opposed to the 'areal' survey. The surveys are based on sound geological precepts with specific orientation towards the engineering uses to which the data will be put.

The Four Geological Propositions

The analyses proposed in Fig. 3 and by equations (1)-(6), together with any extensions of these concepts, depend upon the validity of the following four propositions:

1. Structural discontinuities, which for generality will all be referred to as joints, will exist in the rock mass as detectable features and it will be possible *to describe their properties in a quantitative way*, particularly as these relate to shear strength in the plane of the joint concerned.
2. In the rock mass definable boundaries exist, enclosing volumes which will be called *structural regions*, within which the joints occur in a systematic fashion forming one or more joint sets, each of which is designated a *design set*. For each design set the measurable joint properties are sufficiently similar to permit analysis to be performed.
3. A reliable *model* can be constructed to represent the jointing in the rock mass, i.e. sampling and analysis can be carried out on elements with reasonable expectation that similar conditions will apply to the particular portion of the rock mass under consideration.
4. Shear failure will tend to occur on *planes* or mean planes or combinations or such plane surfaces, i.e. there will be distinct anisotropy in shear strength, compelling failure on preferred planes.

In addition to the above it is also necessary to accept that water pressures in the rock mass can be measured or can be reliably estimated. This is not a general proposition of the form (1)-(4) above since such measurements can be made reasonably providing always that sufficient care is taken in the installation and observation of the water pressure measuring devices.

The idea of systematic jointing in a rock mass with several similar joint sets in a localized section is relatively easily accepted. If one examines a newly cut face and if one eliminates the man-made surfaces of fracture, such as those due to blasting which are relatively easily recognized, then it is comparatively easy to identify joint sets providing only one, two or three such sets are present. If the number of sets exceeds three, and in particular if the number is large, of the order of ten or more, then visual identification is much more difficult. One is then compelled to measure and analyze. The values of dip and dip direction are used as first measures and plots of the sterogram form (Schmidt hemispherical projection, etc.) are made. A very useful form of sorting the data is by the rectangular plot shown in Fig. 4. Further plots of the same form may be made of the data corrected for directional bias (direction of the sampling line with respect to the direction of the normal to the joint being sampled) length of sampling line, etc. These rectangular plots provide a basis for defining the design joint sets and for each set the tabulated details are as follows: structural region designation, joint set designation, dip direction, dip angle, mean joint length, maximum joint length, mean joint spacing, percentage of joints with gouge, unconfined compressive strength of intact rock, friction angle of intact rock, clean joint friction angle, cohesive strength of gouge, friction angle of gouge and percentages of joints having waviness angles greater than 5^{0} and 10^{0}.

The techniques of the joint survey and the interpretation of the observations to yield quantitative values for design have been described by Piteau (1970) and Robertson (1970) and the reader who intends applying the procedures should refer to these two authors. In general principle, however, one may summarize by saying that for the factors of first order importance such as orientation and continuity of the joints, the field observations are of a numerical form and the analyses and calculations are more sophisticated: for the factors of

Fig. 4 Rectangular Plot - Sorting of Joint Data

JOINT SURVEY FOR DE BEERS MINE
ALL JOINTS

DIRECTIONAL BIAS CORRECTED TO STANDARD LENGTH

SECTIONS ACCUMULATED 4 5 12

DIP DIRECTION ACROSS AND DIP ANGLE DOWN

	0.0	10.0	20.0	30.0	40.0	50.0	60.0	70.0	80.0	90.0	100.0	110.0	120.0	130.0	140.0	150.0	160.0	170.0	180.0
0.0	0	0	0	0	0	190	0	0	14	0	9	0	7	0	7	0	0	0	0
5.0	0	0	0	0	0	0	16	0	0	3	0	0	0	0	0	0	0	0	3
10.0	0	0	0	0	0	0	0	0	0	0	2	1	0	0	0	0	0	0	0
15.0	0	0	0	0	0	27	0	0	0	0	0	0	0	0	0	0	0	0	0
20.0	0	0	0	0	0	0	0	0	0	0	0	1	0	0	0	0	1	0	2
25.0	0	2	1	0	0	0	2	0	0	3	0	0	0	0	1	0	0	0	0
30.0	1	0	0	0	3	0	14	0	0	0	0	0	0	0	0	0	0	1	0
35.0	1	1	0	0	0	0	0	0	0	0	0	0	0	0	0	0	9	0	0
40.0	2	1	0	0	0	0	0	2	1	0	0	0	0	0	0	0	0	0	0
45.0	0	3	0	1	0	0	3	3	0	0	0	0	0	0	0	0	0	0	0
50.0	0	0	0	0	2	0	3	1	1	0	1	0	0	0	0	0	0	0	1
55.0	1	0	0	0	1	0	10	12	6	1	0	0	0	0	0	0	0	0	0
60.0	0	1	1	0	1	3	5	4	1	1	1	1	0	0	0	0	0	0	0
65.0	0	1	1	1	3	45	22	2	1	3	0	1	0	0	0	1	1	0	1
70.0	0	1	1	2	3	38	15	13	24	4	2	0	1	0	0	1	0	1	0
75.0	1	1	1	5	7	25	32	14	27	11	5	4	0	1	0	0	2	1	2
80.0	2	3	2	2	1	0	28	19	23	13	3	3	2	1	0	1	1	5	2
85.0	1	0	0	1	1	8	0	1	14	2	4	2	2	1	0	0	1	4	2

	180.0	190.0	200.0	210.0	220.0	230.0	240.0	250.0	260.0	270.0	280.0	290.0	300.0	310.0	320.0	330.0	340.0	350.0	360.0 / 0.0
85.0	2	1	4	1	1	0	2	9	5	8	0	0	0	0	0	0	1	2	1
80.0	2	3	1	2	1	0	0	1	1	1	0	1	2	0	1	0	0	2	2
75.0	2	0	2	1	1	1	0	3	4	0	1	1	0	1	1	1	1	0	1
70.0	0	0	0	0	0	0	10	0	4	0	0	0	0	1	0	0	0	0	0
65.0	1	1	0	0	0	9	2	1	0	0	0	0	0	0	0	0	0	1	0
60.0	0	0	0	0	0	0	0	2	3	0	1	0	1	0	1	1	0	0	0
55.0	0	1	0	0	0	10	0	0	0	0	1	0	0	0	0	0	0	0	0
50.0	1	0	0	0	0	0	3	1	0	0	0	0	0	0	0	0	0	2	0
45.0	0	1	1	0	0	0	0	0	0	0	0	1	0	0	0	0	0	0	0
40.0	0	0	0	0	2	4	0	0	0	0	2	0	0	0	0	0	0	0	2
35.0	0	0	1	0	0	0	13	0	5	0	0	0	0	0	0	0	1	0	1
30.0	0	0	0	2	0	0	0	0	0	1	0	0	0	0	0	0	0	0	1
25.0	0	0	0	0	6	0	17	0	0	0	1	0	0	0	0	0	0	0	0
20.0	2	0	0	0	3	0	0	0	0	0	3	0	0	0	0	0	0	1	0
15.0	0	0	0	0	0	0	7	0	9	0	0	0	0	0	0	2	0	0	0
10.0	0	0	0	0	0	0	0	0	0	0	0	0	0	0	0	0	0	0	3
5.0	3	0	4	0	0	0	0	0	0	0	0	0	0	0	0	0	0	0	0
0.0	0	0	0	0	0	0	0	0	0	0	0	0	0	0	0	0	0	0	0
90.0	0	0	0	0	0	0	0	0	0	0	3	1	2	0	0	0	1	0	1
90.0	1	1	1	1	0	0	0	8	4	6	0	0	0	0	0	0	0	0	0

second order importance such as joint friction angles, unconfined compressive strength of the rock, etc., qualitative assessments are made leading to numerical values which are coarser in character. For example, the hardness categories noted in Fig. 2 are used to assess the unconfined compressive strength of the intact rocks.

A joint survey as currently executed is a tedious operation and a simple survey on only one or two benches in a moderate mine may involve the observation of 10 000-15 000 joints - each joint involving the recording of 17 separate quantities. The mass of data is great and its handling is time-consuming, even on a high-speed computer. Attempts are being made to reduce this work and while some of these appear promising it is still too early to recommend methods other than the full joint survey described by Piteau (1970). Further, the whole statistics of interpretation is being re-examined. Difficulties have been experienced in defining the design joint sets when the data from several sets overlap. A further re-examination relates to the present method of defining joint sets on the basis of dip and dip direction only - joint lengths and joint spacing are probably of equal importance and should be taken more fully into account in the sorting of data which lead to decisions on the joint sets.

The second proposition given above relates to structural regions. Changes in jointing patterns are frequently apparent even by casual observation of freshly cut slopes. Such changes may take place sharply, as at a fault or contact, or gradually, over a somewhat ill-defined length such as a shear zone or zone of folding. A boundary to a structural region may vary from vertical to horizontal and may also be plane or curved. Because of the changes in joining in passing from one structural region to the next, the analyses must be carried out separately for each region.

Robertson (1970) has described a procedure for determining these boundaries. He has the computer plot dip and dip direction sequentially along the lengths of the sections of the line survey. At a 'sharp' boundary there are sudden changes perceived; at others the changes take place gradually over a zone within the plot. The boundaries usually coincide with the field geologists' estimates of their positions and are frequently associated with faults, contacts, dykes, shear zones or fold regions.

With further experience, close definition of the boundaries of the regions does not appear to be of great importance. What is necessary is that the engineer recognizes that jointing patterns change and hence his calculated slope angles may also change. He will probably be forced to compromise or employ the lower slope angles over two or three such changed sections. The exact position of change may not be of very great importance. This thought is associated with the attempt to reduce the work put into the current extensive joint surveys.

(a)

(b)

(c)

(d)

(e)

Fig. 5 Joint Sets and their combination in a Structural Region

(a) Joint Set A alone

(b) Joint Set B alone

(c) Joint Sets A and B superimposed

(d) Joint Set C alone

(e) Joint Sets A, B and C superimposed

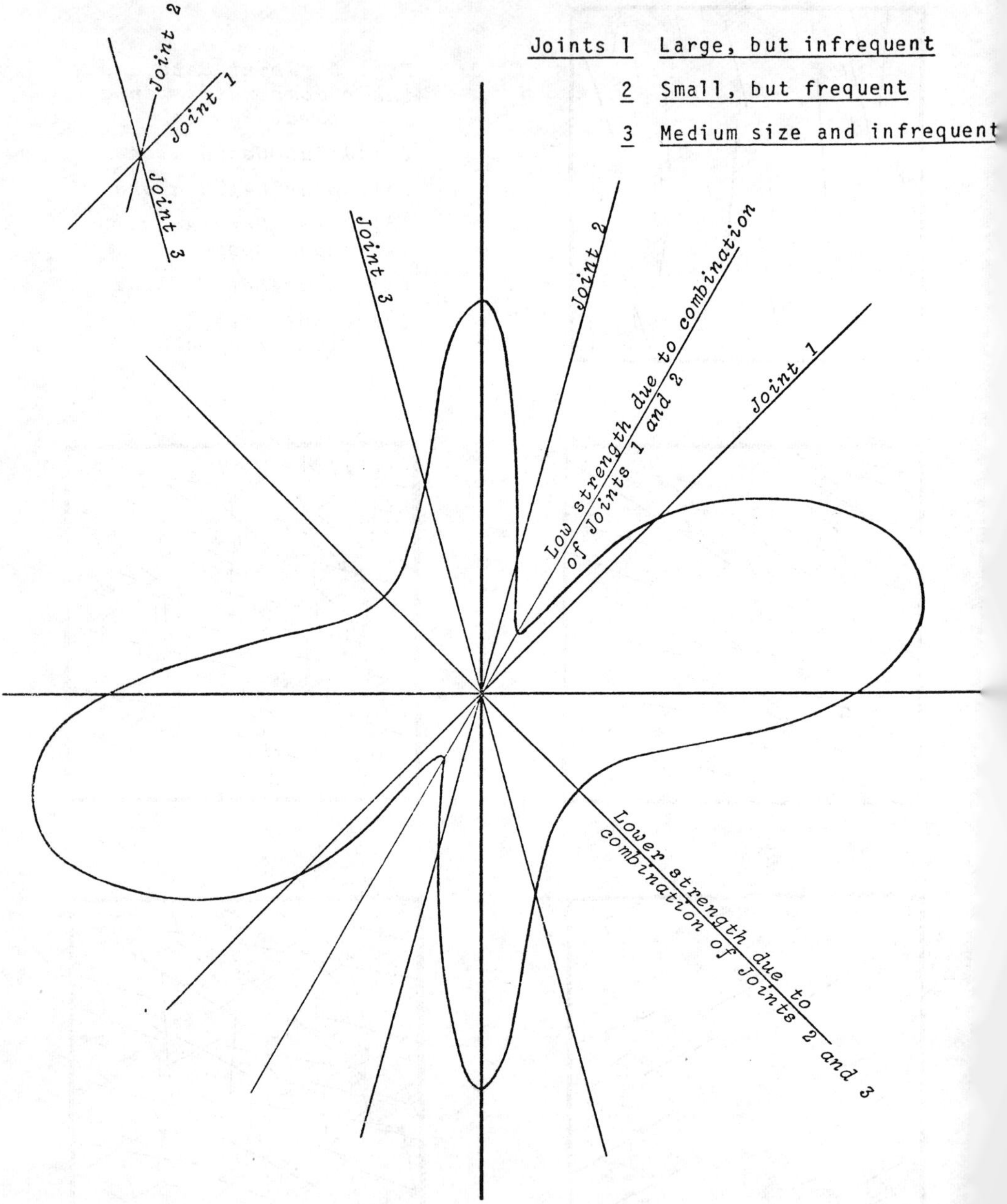

Fig. 6 Polar diagram of shear strength in jointed rock mass combination of 3 Joint Sets

The third and fourth propositions relating to a reliable mathematical model based on potential failure on planes or combinations of planes within the rock mass form the basis of the slope stability theory. The theory and the calculated results can only be good if the model reasonably represents what happens in nature where failure along planes, mean planes or combinations of planes are frequently observed in practical slopes.

MODES OF SHEAR FAILURE IN JOINTED ROCK MASSES

A jointed rock mass is highly anisotropic with respect to its shearing strength. Fig. 5 shows the progressive combination of three joint sets (data interpreted from an actual survey) and Fig. 5(e) illustrates the type of practical situation which the model must attempt to simulate. Fig. 6 illustrates the type of anisotropy which is likely to be encountered.

Before dealing with the complications demonstrated in Figs. 5 and 6 it is necessary to think in simpler conditions:

(a) if a plane of preferentially lower shear strength exists in a material then shear failure will tend to select this plane as shown in Fig. 7.

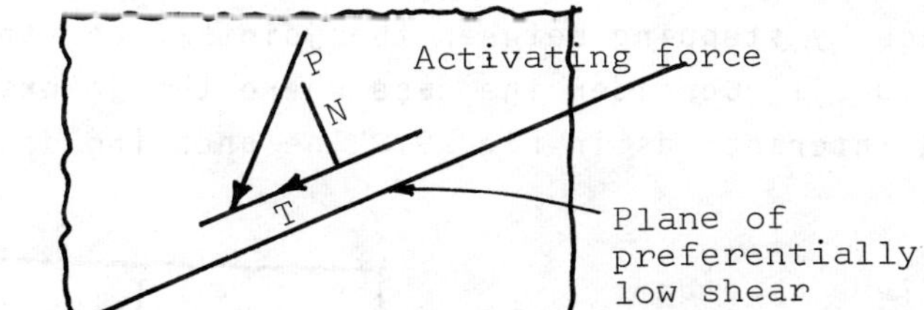

Fig. 7 Resolution of Forces for Case of a Plane Preferentially Low Shear Strength

The activating force P must be resolved into components tangential and normal to the preferential plane and if this combination gives a lower resistance than any other potential plane of failure, then the preferential plane is the critical plane of potential shear failure.

(b) if the preferential plane contains two or more low-strength discontinuities then the net shear resistance is the sum of the shear resistances on the discontinuities and on the intact material in between them as in Fig. 8.

Fig. 8 Shear along a Preferentially Weak Plane containing Finite Discontinuities

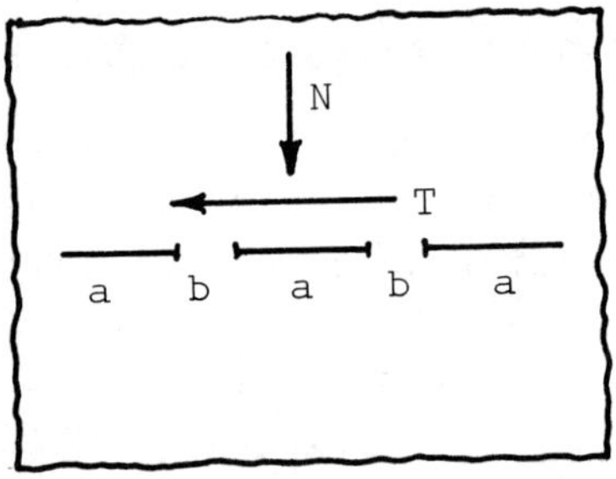

This case corresponds to Fig. 3. The concept of continuity of jointing is introduced and the shearing strength becomes a function of the cohesive and frictional parameters of the intact material and the joints all as given in equations (1)-(6).

(c) if two joint sets exist then the preferential plane need not coincide with either set and failure can take place by stepping between the joints (the stepped joint failure). Consider the case where the joints are large and intersect as in Fig. 9. The shearing displacements

Fig. 9 Shear Failure by stepping between α-joints (the main set) and ψ-joints (the secondary set) - Long Joints

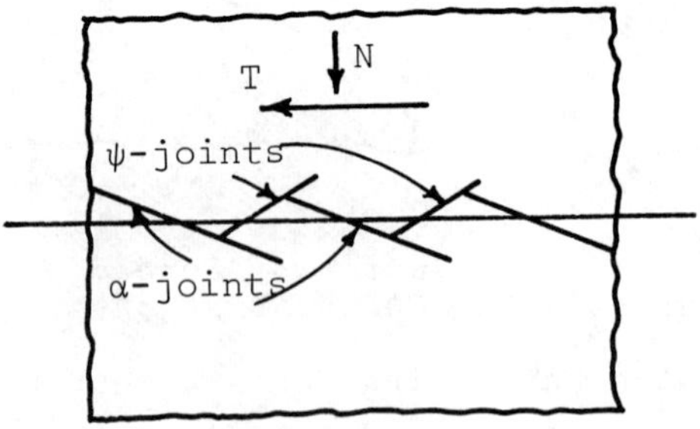

will no longer be in the direction of the (mean) preferential plane of failure but will take place in the direction of the α-joints. As shearing takes place in this direction, separation (tension) occurs on the ψ-joints. This difference in directions causes many complications in the theory.

Consider next the case where the joints are of limited length so that one part of the failure occurs on the joints and another part on intact material, as in Fig. 10. The assumption is made that the mode of failure

Fig. 10 Stepped Joint Failure - Short Joints

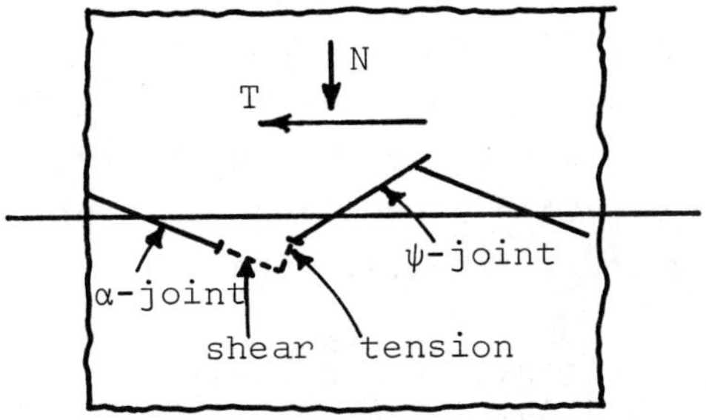

will involve a shearing along the extension of the α-joint, this plane extending until it meets the nearest ψ-joint.

The cases given above provide an underlying philosophy for the handling of the complexity of cases such as shown in Fig. 5(e). Further assumptions are required as follows:

(i) failure will occur only on two joint sets acting together to form the stepped joint mechanism. This case can often be observed in cut slopes where the joints are long and of the form shown in Fig. 9.

(ii) any line drawn through the rock mass will interesect α or ψ or joints of any other set in a systematic fashion. Depending upon the inclination of the line the number of joints intersected will be a function of the mean length of the joints, their mean spacing within the set, and the difference between the angles of the line and dip of the joints, i.e. the line through the mass may be considered as if it were a line of sampling in the joint survey.

(iii) where different numbers of α-joints and ψ-joints are intersected by the line then adjustment will be made to the lengths of the ψ-joints to deduce equivalent joints such that the number of α- and equivalent ψ-joints are equal. Further adjustment to the lengths will be made in order to deduce ψ-joints such that they fall at the bottom ends of each α-joint.

The assumptions made above are considered in detail by Jennings (1970) and persons using his methods of calculation are advised to study this paper carefully. His aim has always been to err on the conservative side when making the assumptions. Each assumption is at present the subject of further laboratory investigation.

THE EQUATIONS OF CONTINUITY AND THE RESULTING STRENGTHS FOR USE IN STABILITY CALCULATIONS

The mechanism of shear failure described above may be used directly for calculation of resisting forces in a theory of slope stability. Fig. 11 represents a slope of inclination i and height H with two design joint sets designated α and ψ the apparent angles of dip of the joints in the vector direction of movement which for the case of two-dimensional plane-strain failure is in the direction of dip of the slope. The potential plane of failure is designated β and it corresponds to the preferential planes in Figs. 7-10.

Two equations for continuity of jointing are required: one, $k_{\alpha\psi\beta}$, is for continuity in shear and another, k_t, is for continuity in tension. Both continuities apply to displacements in the α-direction. Therefore the two continuities are with respect to the projections of line AB onto the surfaces AC and BC.

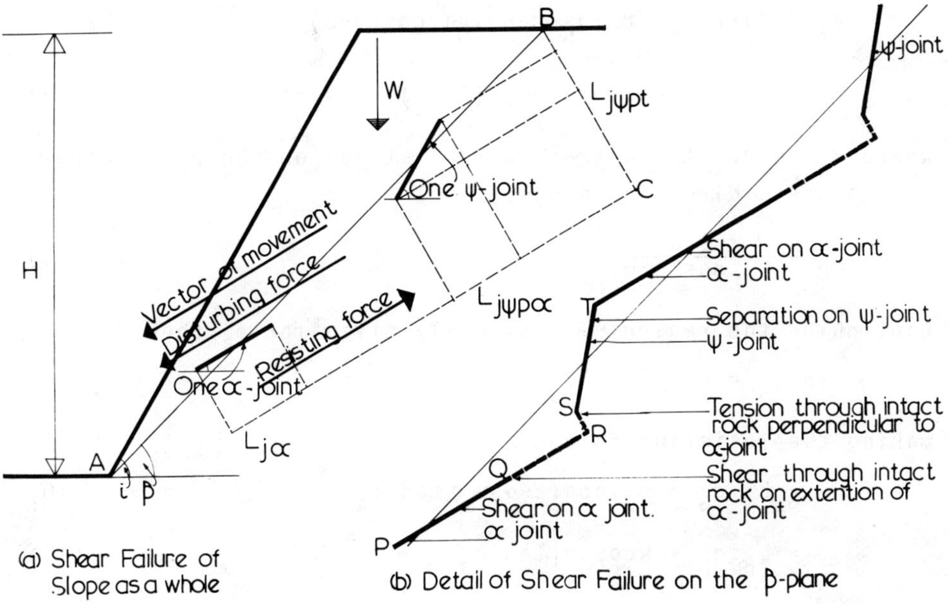

Fig. 11 Potential Failure along a Surface AB which includes pre-existing Joints

The shear continuity is made up of three parts: first, account must be taken of the possibility of a 'longest' joint of the set lying within the system; second, all α-joints intersected are projected onto AC; third, those lengths of the ψ-joints which are operative in the system are also projected onto AC. The sum of the three contributions are added together to give $k_{\alpha\psi\beta}$. Similarly, the projections of the operative lengths of the ψ-joints onto BC give the continuity of the system in tension, k_t.

It is convenient to think in terms of an effective cohesion for L_β, the length of the mean plane of failure AB. Designating this by c_β, Jennings (1970) gives

$$c_\beta = (1-k_{\alpha\psi\beta})(c_m+\sigma_n \tan \phi_m).\cos(\beta-\alpha)$$

$$+ k_{\alpha\psi\beta}.c_j.\cos(\beta-\alpha)+(1-k_t).t_m.\sin(\beta-\alpha) \qquad \ldots \quad (7)$$

where σ_n is now the average normal stress on the plane parallel to the α-direction and is

$$\sigma_n = \frac{W \cos \alpha}{L_\beta.\cos(\beta-\alpha)} \qquad \ldots \quad (8)$$

Continuing the reasoning previously given for ϕ_a then

$$\tan \phi_a = k_{\alpha\psi\beta} \tan \phi_j \qquad \ldots \quad (9)$$

making the resisting force

$$RF = L_\beta.c_\beta + L_\beta.\cos(\beta-\alpha)\sigma_n \tan \phi_a \qquad \ldots \quad (10)$$

$$= L_\beta.c_\beta + W\cos \alpha.\tan \phi_a \qquad \ldots \quad (10a)$$

The disturbing force for the potentially sliding wedge in Fig. 11 is

$$DF = W.\sin \alpha \qquad \ldots \quad (11)$$

Hence following the ordinary procedure of considering the factor of safety, F, as a number by which the cohesive and frictional strengths must be reduced (or increased when $F<1,0$) so that the slope is just on the point of failure

$$F = \frac{RF}{DF} \qquad \ldots \quad (12)$$

The factor of safety of the slope thus depends on the shear and tensile properties of the intact material, the shear parameters on the joints and the shear and tension coefficients of continuity, these latter being the quantities which define the 'interlocking' of the joints. It is this interlocking which determines the steepness of many natural cliffs and cut slopes.

Tension Cracking at the Top of Slopes

It is observed that before a slope actually fails, cracks

appear near the crest of the slope. These give the engineer warning that all is not well with the slope and the condition is particularly dangerous if the cracks become filled with water. These cracks are generally considered to be tension cracks due to the fact that active earth pressure conditions develop in the vicinity of the crest ($\sigma_H < \sigma_V$, with σ_H even becoming negative or tensile). In the ordinary theory, this negative stress is associated with a cohesion in the slope material, the depth of cracking, z_o, being taken as $2c/\gamma$, with a practical limit of one-half of the total slope height. Other and more complicated expressions can be developed for z_o and these may also take account of surcharge loading, the friction angle in the slope material, water pressures and the factor of safety as a measure of the reserve of shear strength. All such theories present great difficulties in application.

Because of complexities in setting up general expressions which take account of all possible variables, and also because the computer lends itself well to the execution of trial calculations, it is convenient to consider z_o as an independent variable and to think of the stability of the slope in two parts, namely, the stability of a lower section which slides out and an upper section which holds up. In Fig. 12 these are first, ABDC with slope angle i, critical β as β_1 and factor of safety, F_1 and, secondly, BED with slope angle 90°, critical β as β_2 and factor of safety F_2. For the crack to occur at all $F_2 > F_1$ and the appropriate value of F for the slope is then the minimum F_1. These two conditions are examined simultaneously and the values of z_o and F for the slope are determined.

There is obviously some limit to the value of z_o since active pressure conditions and hence tensile stresses cannot occur in the vicinity of the toe of the slope. Therefore the calculations are not executed for values of $z_o > \frac{3}{4}H$. An interesting result shows up in the development of 'slivers' which are

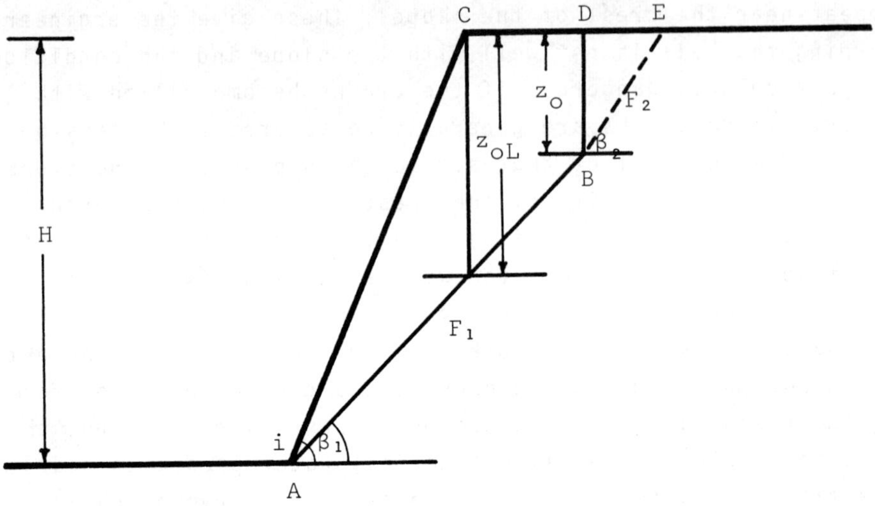

Fig. 12 The Concept of a Depth of Cracking as an independent variable in stability calculations

the relatively thin sections close to the slope which break off in the vicinity of the toe. As L_β decreases when z_o increases the first term of $k_{\alpha\psi\beta}$, dependent upon the longest α-joint, increases so that at large values of z_o, $k_{\alpha\psi\beta} \geqslant 1,0$, i.e. there is no effective cohesion component in RF (equations (7) and (10)). Failure then takes the form of a sliver of limited size. Such slivers are frequently observed in rock slopes. Slivers in the region $z_o > \frac{3}{4}H$ are ignored. For $z_o < \frac{3}{4}H$ they are noted and their relative importance is considered by the designer.

The presence of a depth of cracking, z_o, has important effects on the calculated factors of safety of a slope and it is surprising how frequently it is ignored or only given scant attention. The procedures outlined above make its inclusion a very simple matter and leave decision on its importance to the designer.

WATER PRESSURES IN THE MATERIAL OF THE SLOPE

Water pressures are of very great importance in rock slopes

and their presence may reduce the safe slope angle by as much as 20°-25°, e.g. a slope may just be safe at 45° if water pressures are absent:with full water pressure it may become unsafe at 25°. Wherever practical experience shows that the safe angle for an open pit is less than 30° one should immediately suspect the presence of water pressure. In this respect many rock slopes behave as slopes in soil with zero cohesion, a condition which applies when $k_{\alpha\psi\beta} = k_t = 1,0$.

A knowledge of the water pressures in the slope material is therefore of the greatest importance. It must be remembered that it is the *pressure* and not the *quantity of flow* which is important. Sometimes the flow will be so little that is disappears by evaporation as soon as it reaches the surface - the fact that no water flow is observed is not a certain indication that no water pressures exist in the slope material.

Water pressures within the slope material are conveniently measured by differential piezometers or by suitable 'null-flow' pressure measuring devices when the interconnection with fissure water is incomplete. The installation of piezometers and testing to confirm their reliability is described by Lyell (1970). It is emphasized that piezometers which give incorrect readings are dangerous and continued reliance on the readings is of the greatest importance. Lyell's (1970) procedures for upset tests and hydraulic fracturing to secure interconnection deserve careful attention.

Fig. 13 illustrates the principles which relate to the determination of water pressures in the slope material. If one has a knowledge of the water pressures at a number of points in an existing flow situation then it is possible (Lyell (1970) and Caldwell (1971)) to fit a flow net to these values which takes account of a uniform recharge w at the phreatic line and anisotropy of the slope material, both of which are determined by the fitting process.

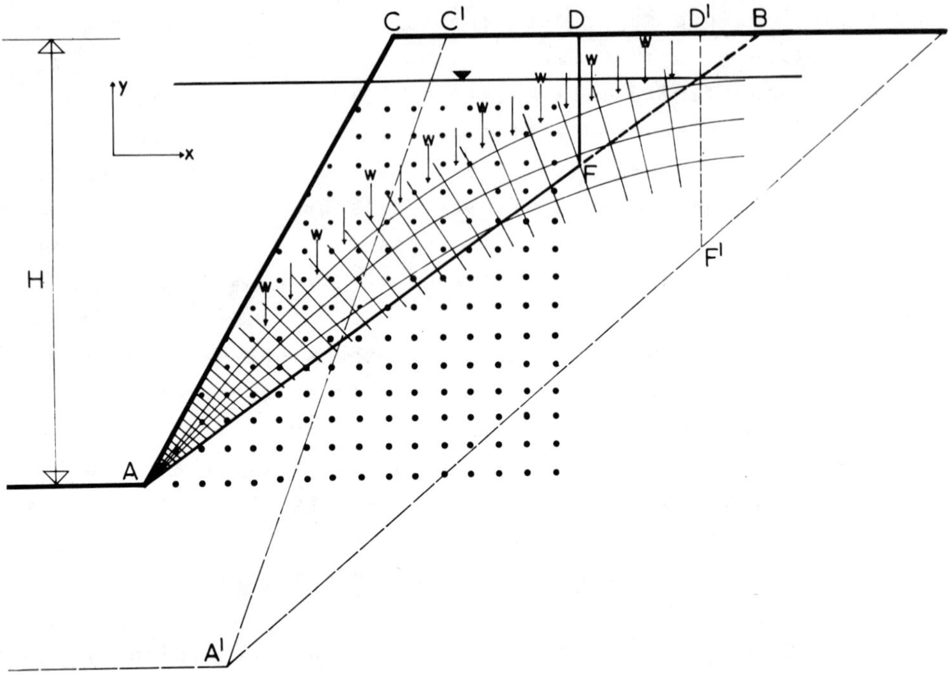

Fig. 13 Water Pressure Grid in Slope Material

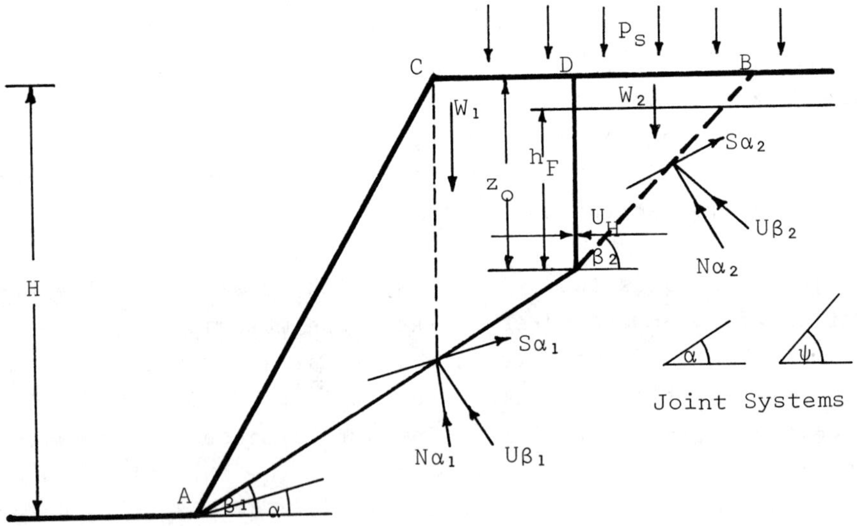

Fig. 14 Plane Failure in a Rock Slope

It is now necessary to consider the slope being cut from AC to A'C' to a new H and new i as shown in Fig. 13. This will involve a transient flow condition but again, providing the piezometers continue to give reliable readings, the new water pressure situations on any potential planes of failure can be found. These water pressures are indicated in Fig. 14.

Water pressures in tension cracks which may become filled after periods of rainfall give rise to other temporary water pressure conditions. They may alter the whole water pressure situation already described and judgement must be used in interpreting this particular transient situation as illustrated in Fig. 15.

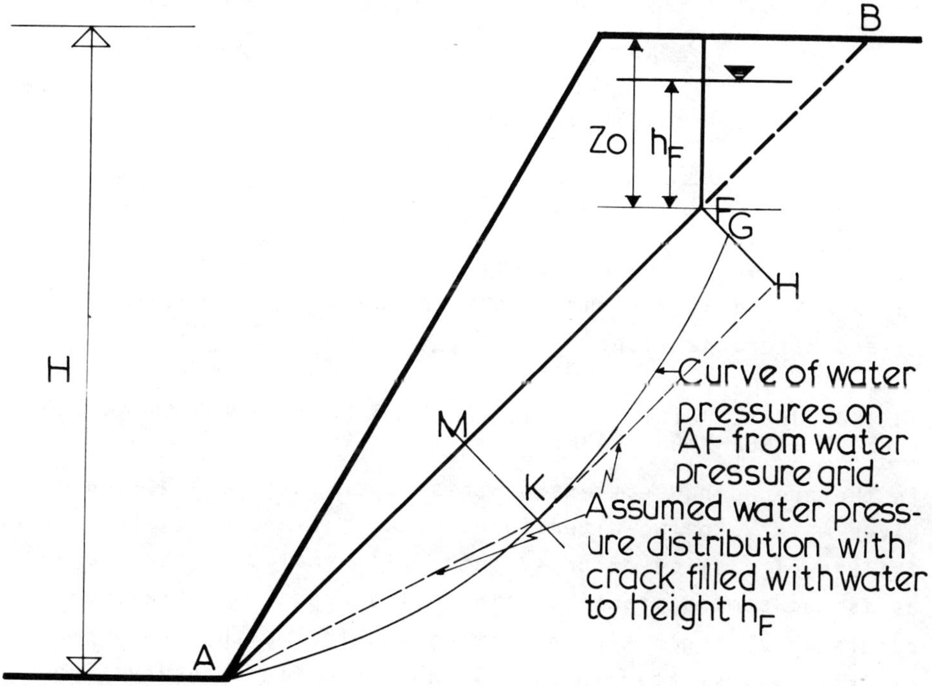

Fig. 15 Water Pressures on β-plane AF

The equations (10) and (11) now change. Using subscript 1 for the lower or sliding out wedge and subscript 2 for the upper or holding back wedge gives

$$(DF)_1 = W_1 \sin\alpha + U_H \cos\alpha + U_{\beta_1} \sin(\beta_1 - \alpha) \quad \ldots (13)$$

$$(RF)_{max.1} = c_{\beta_1}.(H - z_o)\operatorname{cosec}\beta_1 + N'_{\alpha_1} \tan\phi_{a_1} \quad \ldots (14)$$

where N'_{α_1} = force normal to the α-plane for wedge 1 reduced by the hydrostatic uplift on the α-plane

$$= W_1 \cos\alpha_1 - U_H \sin\alpha - U_{\beta_1} \cos(\beta_1 - \alpha) \quad \ldots (15)$$

with $F_1 = \dfrac{(RF)_{max.1}}{(DF)_1}$

and for the upper wedge

$$(DF)_2 = W_2 \sin\alpha - U_H \cos\alpha + U_{\beta_2} \sin(\beta_2 - \alpha) \quad \ldots (16)$$

$$(RF)_{max.2} = c_{\beta_2}.z_o \operatorname{cosec}\beta_2 + [W_2 \cos\alpha + U_H \sin\alpha - U_{\beta_2} \cos(\beta_2 - \alpha)] \tan\phi_{a_2} \quad \ldots (17)$$

with $F_2 = \dfrac{(RF)max.2}{(DF)_2}$

Solutions for F_1 and F_2 result and these are tested as before for $F_2 > F_1$ and F_1 minimum to obtain the calculated values of z_o and F for the slope.

THE CONJUGATE JOINT ZONE FAILURE OF A SLOPE - INVOLVING FRACTURE ON THREE MEAN PLANES

In the preceding sections of this paper attention has been given mainly to two-dimensional *plane failure* of a slope, i.e. failure of a slice of unit width cut normally in the slope, as is the common practice with soil slopes*. The failure occurs on a single β-plane acting together with a tension crack as illustrated in Fig. 14 and from field observation

---

* Note that most practical cases of soil slope failure occur in a spoon-shaped form. The two dimensional slice for analysis is taken from the maximum height of the spoon failure and no allowance is made for side restraints.

this appears to be the most likely two-dimensional mode when the rock mass is homogeneously jointed. It should be noted that a homogeneously jointed mass may be stratified with respect to density and hardness or shear strength. A computer program has been developed to handle this case and this can accommodate up to nine strata.

However, if the joint properties of the strata change, i.e. if we have different design joint data for the different strata it is likely that the plane failure mode will change to a mode involving failure on two, three or more mean planes of failure. One such mode is illustrated on Fig. 16 which has been given the designation "Conjugate joint zone failure mode" in which failure takes place simultaneously on the mean

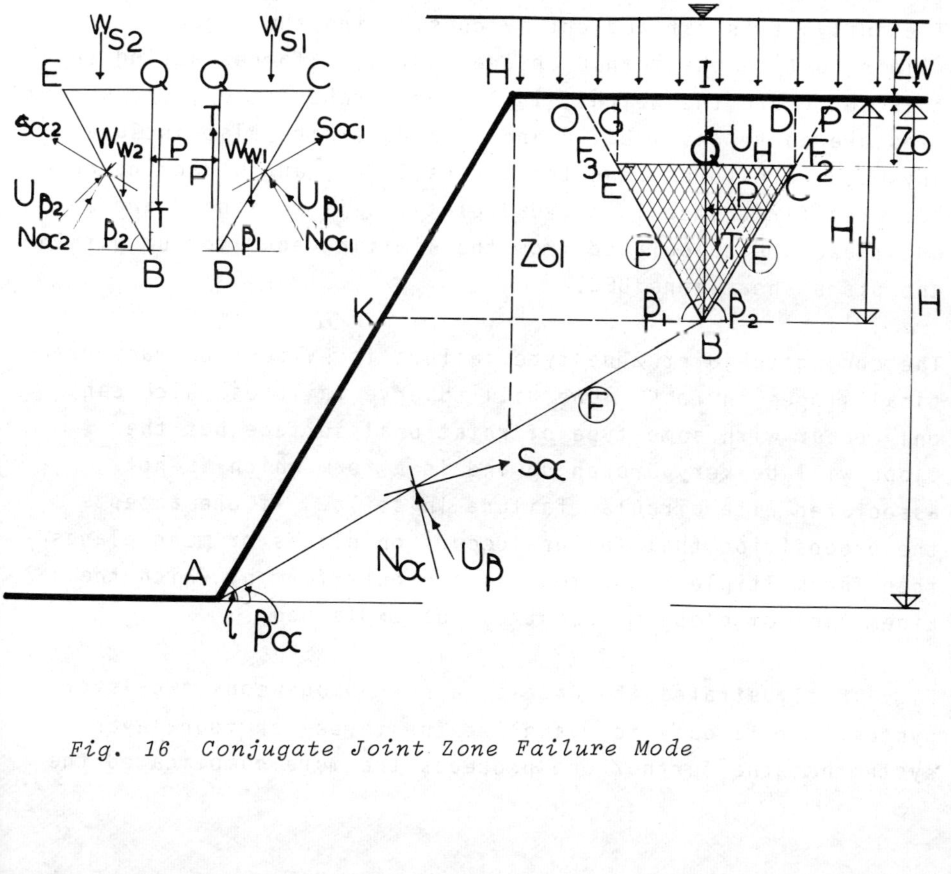

Fig. 16 Conjugate Joint Zone Failure Mode

planes defined by β_α, β_1 and β_2, together with a single associated depth of cracking z_o. The observed failure surface is ABCD and the material breaks up at the top of the slope due to zone failure in the wedge BCE. If only the upper portion of the failure surface can be seen it is possible that the mode may be confused with a circular arc failure.

Each of the surfaces AB, BC and BE may be a mean surface stepping around joint sets, an α and a ψ for each case. Failure also occurs simultaneously on planes parallel to BC and BE within the zone BCE. Vertical tension cracks occur in the zone ECDG and the wedges EGO and CDP hold up. A single factor of safety F applies to failures along ABC and ABE with F_2 and F_3 respectively for the held-up wedges CDP and EGO. As previously, z_o only has meaning when $F<F_2$ and $F<F_3$.

The analysis is carried out by considering the wedge ABQIH sliding out on the mean β_α plane, AB. The forces acting on this wedge are the weight, including surcharge, U_H, the water pressure above level EC, P and T on BQ (which also include any water pressures) and the forces S_α, N_α and U_β acting on AB. The first step in the calculations is to find P and T and these are calculated from the limiting equilibrium within the plane shear zone BEC.

The conjugate joint zone type failure is evident on many practical slopes in rock. One will observe failures which can only occur with some type of rotational surface but the slope will be very broken up and in a form which is not associated with circular failure in soils. If one accepts the proposition that failure occurs on planes or mean planes then the multiple plane mode is the only form by which the kinematics of slope movement can be explained.

Fig. 16 illustrates the case of a non-homogeneous two-layer system. It is easy to visualize the three- or four-layer system but the further one proceeds the more complicated the situation becomes. In addition, cases where failure occurs on the β_2 plane with vertical tension cracks replacing the β_1 plane in Fig. 16 are also possible modes - these have been designated the "conjugate joint block failure mode". Here again there is some field evidence to suggest that this form of failure also applies in certain cases.

These more complicated modes applying to non-homogeneously jointed strata are still under investigation. Laboratory models using steel prisms are being studied in the attempt to reproduce the modes of slope failure. To date these have shown the plane failure plus tension crack in homogeneously jointed material, the conjugate joint zone failure with tension crack with stratified materials of different jointing, and the rotational column failure as described by John (1970). More work will undoubtedly disclose other modes of failure which must be included in the analysis.

Concurrently with the model tests above, analysis of the situation shown in Fig. 16 is also proceeding. So far the equations which are used for calculating the F value have proved unstable in the computer and the cause is being sought. There is clearly a considerable field for investigation.

The Three-Dimensional Wedge Failure Mode

An important case of failure arises when the intersection of planes forms a wedge which slides out as illustrated in Fig. 17. AB and AC represent the traces of two planar features on the top surface of a slope. CD and BD are their intersections on the slope itself: point D at the toe of the slope is the point where both planar features meet. Normally the slope will be cut to depth D, or some other point just above it before failure starts to take place.

The failure of wedges has been thoroughly investigated by Wittke (1967) who proposes six basic forms ranging from simple sliding along a direction parallel to the line of intersection

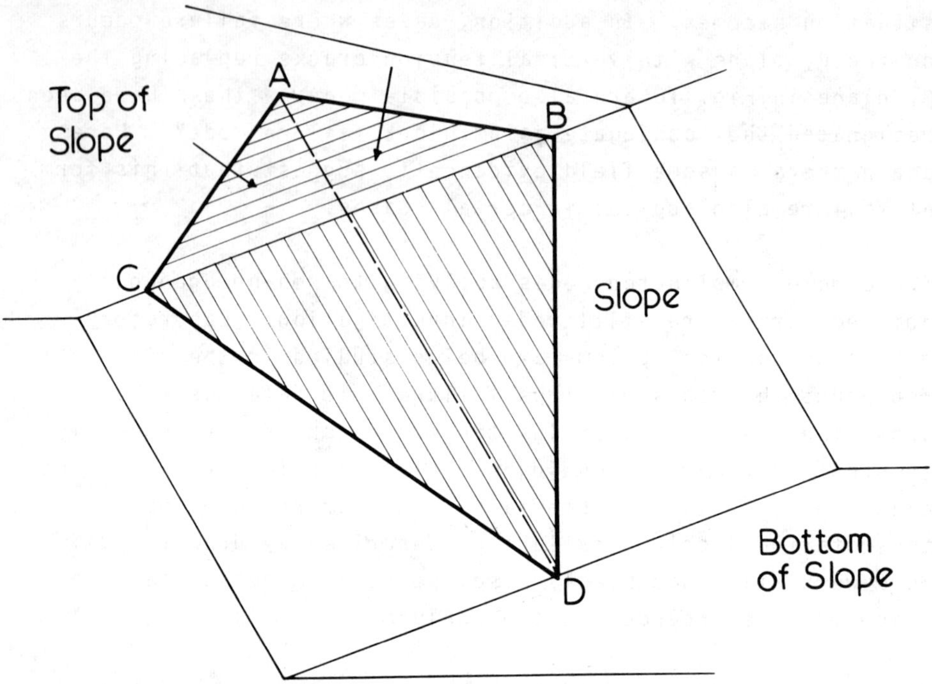

Fig. 17 The Three-Dimensional Wedge Failure Mode

of the two planes (with both planes of weakness remaining continuously in contact) through to rotation and sliding of various forms along either plane of weakness.

Wedge type failure are very common in open pit mines on heights of one or two benches (60-100 ft height) but become progressively less prevalent as the height of the slope increases. The most important wedge of great height (about 600 ft) is in the Ruth Pit, Nevada.

This type of failure appears to be associated mainly with geological planes of weakness which are at least of the same order of size as the height of the slope involved. This appears to be the reason why the frequency of this form of failure decreases as the height of the slope increases: planes of weakness of the size of bench heights (say 30-50 ft) are much more common than similar planes of 10 bench heights (say 300-500 ft).

Efforts are at present being made to place the problems of wedge failure on a computer in which the machine will examine each of the six possibilities and print out the most probable form and lowest factor of safety.

An important fact which should not be overlooked is that if both planar features are investigated separately as cases of two-dimensional failure, then it will be found that the more adversely oriented will give a factor of safety which is lower than that found for the combined plane wedge failure. Since many such planar features tend to occur as repeated, parallel features (e.g. sympathetic fault planes) this may be an important reason to pay greater attention to two-dimensional failure except in those cases where limiting values of slope angles are sought.

MONITORING OF SLOPES FOR THE CONTROL OF CONSTRUCTION

It would be very useful if field measurements could be devised which would give the engineer a measure of the safety of the slope as the excavation proceeds. This is the monitoring process and its aim should be to provide control for safety.

The movements which are measured are those which are due to excavation. Consider the excavation shown in Fig. 18 which is cut first by deepening at a flatter slope and then extended by steepening to a steeper slope, the progressive stages being shown dotted.

Consider the movements of points such as A, on the rim of the excavation, and B and C within the mass of the slope. In the initial stages of the excavation the points will tend to move upwards and inwards towards the excavation due to elastic re-

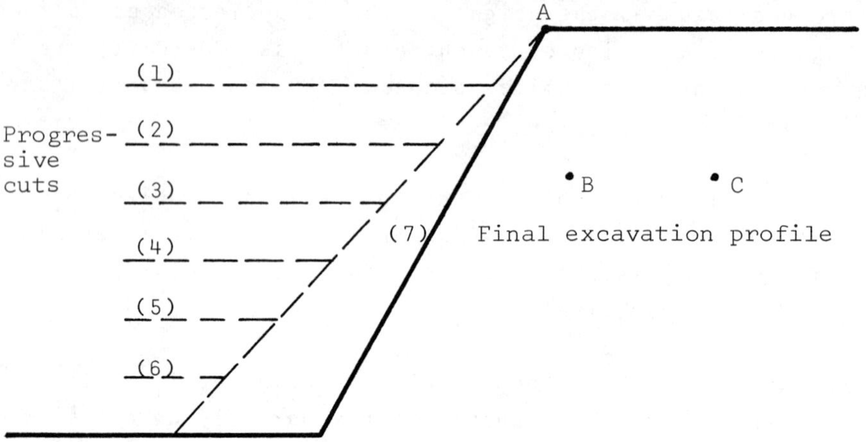

Fig. 18 Progressive Stages in an Excavation

bound resulting from the removal of load by the excavation. The points A and B being nearer the excavation will move more than point C which is more remote. Initially the movement will show linear relationship with excavation depth and will obey normal elastic theory from which the movements will be functions of the elastic constants, namely the Modulus of Elasticity, E, and Poisson's Ratio, μ. If two or more points are observed then it should be possible to calculate these constants using only the results of the movement:excavation records.

As the excavation is deepened and, in particular, when the slope is steepened in its final stages, the form of the movement:excavation curves will change. Instead of moving only upwards and inwards in a linear way, a new component of deformation due to shear strength mobilization starts to play an increasing role. This will involve downward and inward movements of points such as A and B whereas for the point C, if it is sufficiently remote and beyond the region of potential failure, the movements will continue in a linear fashion. The curves of movement of points such as A and B can be expec-

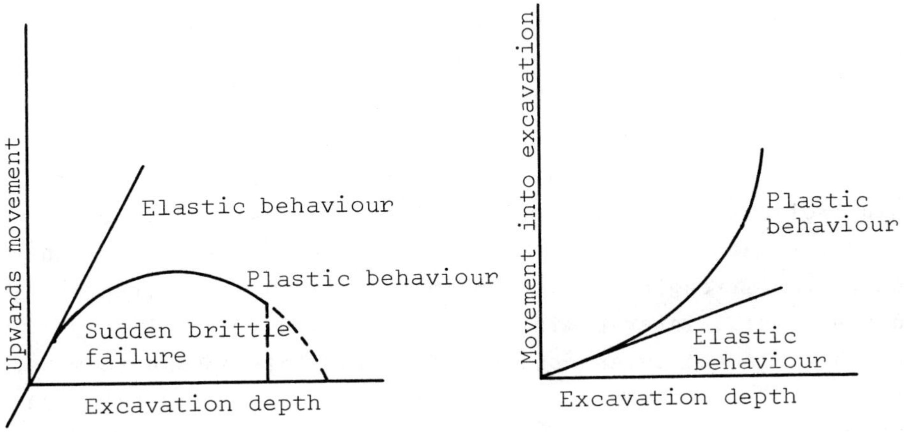

(a) Vertical Movements *(b) Horizontal Movements*

Fig. 19 Movements of Points such as A and B of Fig. 18 as Excavation Proceeds

ted to be of the forms shown in Fig. 19.

The observation of movements of points within an excavation thus provides a powerful tool for control by the engineer. The points must be carefully selected to yield maximum information and points on the rim are probably the most suitable. This monitoring is a *sine-qua-non* in all excavations which are cut to slopes which are close to the limit.

The field observations are made using modern survey techniques, employing high precision instruments and nets which are balanced by statistical methods. For vertical movements modern automatic levels are used and for horizontal movements, triangulation with good distance and angle-measuring instruments are employed. The procedures are described by Watt (1970). They are sophisticated but may be expected to give results of a high order of accuracy.

CONCLUSION

The theory for calculating the stability of rock slopes is basically an extension of the Coulomb wedge theory in earth pressure calculations. Its complication lies in the fact that rock is a material with anisotropic shear strength, certain inclinations of potential failure planes having much lower strengths than others. This is due to jointing in the rock. The strength is not determined only by the strength on the joints; interlocking with solid rock material occurs and the strength on any potential failure plane is determined partly by failure through intact rock and partly by shear on joints.

An attempt has been made to give quantitative meaning to these statements. First, four basic geological propositions are posed and the whole theory depends on their validity. Secondly, many assumptions have been made in order to make the problem tractable to calculation. Relatively insignificant consequences may result if some of these assumptions are in error but there are others such as those which concern continuity of jointing, which are of very great significance and where small errors could have serious consequences. At all stages in making assumptions attempt has been made to err on the conservative side.

Water pressures in the slope material are of major importance and the safe slope angle may be reduced to about one half its value if such water pressures are present. Prior drainage of a proposed rock slope is of the greatest importance and is probably the most important single operation that man can perform to improve his excavation conditions.

The theory has been tested against one set of rock slope conditions known to be at limiting equilibrium. More full scale tests of this type will be necessary before the basic principles can be accepted without reserve. The collection of field data and the analyses are complicated but further application will serve to simplify the procedures to a state where they can be used in ordinary engineering practice.

ACKNOWLEDGEMENTS

The theory presented in this paper has been developed over about ten years, working on open pit mines of the Anglo American Corporation of South Africa in Zambia, Swaziland and the Republic of South Africa. Grateful acknowledgement is made, first, for permission to publish relevant data and second, to the many mining engineering and geological colleagues in the Corporation with whom the author has worked but whose substantial contributions have been left without specific acknowledgement. There are, however, some persons who deserve special mention, namely Messrs. Lyell, Piteau, Robertson, Steffen and Watt, who have formed the team which undertook this project. It is difficult to trace the origins of ideas and the author acknowledges the substantial contributions these men have made. They are not alone, however: in the list of references there are many omissions only because there have not been direct references in the paper at any stage. The papers of Professor and Mrs. K. Terzaghi, Professor L. Muller, Professor D.U. Deere and Dr. K.W. John provided a background of reading against which the theory was developed. All of these authors have written extensively on the shear behaviour of rock masses on pre-existing surfaces of discontinuity. This paper builds on the foundations they have laid.

REFERENCES

Caldwell, J.A. *"Determination of Fluid Potential Distribution in Jointed Rocks"*. Unpublished thesis, University of the Witwatersrand, 1971.

Jennings, J.E. *"A Mathematical Theory for the Calculation of the Stability of Slopes in Open Cast Mines"*. Proc.Symp. Planning Open Pit Mines, S.A.Inst.Min.& Met., Sept 1970.

Lyell, K.A. *"The Interpretation of Water Pressure Factors for use in the Slope Theory".* Idem.

Piteau, D.R. *"Geological Factors significant to the Stability of Slopes cut in Rock".* Idem.

Robertson, A.MacG. *"The Interpretation of Geological Factors for use in Slope Theory".* Idem.

Watt, I.B. *"Control for Early Warning of Potential Danger in Open Pits".* Idem.

Wittke, W. *"Influence of the Shear Strength of the Joints on the Design of Prestressed Anchors to stabilize a Rock Slope".* Proc. of the Geotechnical Conference, Oslo, 1967.

****

STUDY OF CLAY SHALE SLOPES

By Don C. Banks[1]

INTRODUCTION

The U. S. Army Engineer Waterways Experiment Station (WES) has recently participated in two studies relating to the stability of clay shale slopes. The first study, sponsored by the U. S. Army Engineer Nuclear Cratering Group (NCG), concentrated on slopes found in five geologic formations or groups along the upper Missouri River valley (Fleming, Spencer, and Banks, 1970). The purposes of the study were to define characteristics typifying a potentially troublesome clay shale, to determine factors leading to instability, and to provide a basis for assessment of the long-term stability of high crater slopes in clay shale. The second study, sponsored by the Panama Canal Company (PCC) and, during the first year, by the Atlantic-Pacific Interoceanic Canal Study Commission and, during the remaining four years, by the Office, Chief of Engineers (OCE), is a continuing study to assemble information about slope stability along the present Panama Canal that will assist operation and maintenance and to develop new information that will provide a basis for more rational, economical, and reliable slope design for a new canal. During the first year of the study, the data collected and analyzed dealt mainly with the East and West Culebra slides and the Model Slope (Lutton and Banks, 1970).

---

[1]Supervisory Civ. Engr., U. S. Army Engineer Waterways Experiment Station, Vicksburg, Miss.

Previous studies by others, dealing with identification of clay shale materials, testing techniques for index and physical properties, delineation of in situ conditions and determination of factors affecting clay shale behavior, have indicated the complex nature of problems in clay shale. From studies, such as discussed in this paper, the various parameters affecting slope design are brought into focus to allow the designer to make predictions of slope behavior in other situations.

SIGNIFICANT RESULTS FROM THE EMPIRICAL STUDY OF SLOPES ALONG THE UPPER MISSOURI RIVER VALLEY

Several of the geologic formations of the Northern Great Plains are clay shales with varying reputations regarding slope stability. The five geologic units studied in the field were the Marias River formation of the Colorado group, the Claggett formation, the Bearpaw formation, three formations of the Fort Union group, and the Pierre formation, Fig. 1. The study consisted of collecting data from field mapping and office studies of aerial photographs, review of literature referring to the various geologic units, boring and sampling of selected sites, laboratory tests of selected samples, case history studies of the major dam projects in the area, and a survey of clay shale slope behavior from other geographical areas.

A brief lithological description of the five geologic units studied is shown in Table 1. From field studies, detailed topographic and surface stratigraphy were determined; some information was obtained on material characteristics from shallow samples. Additional slope data were collected from aerial photographs.

Slope Behavior

In the Claggett, Bearpaw, and Pierre formations (all marine-deposited clay shales of Late Cretaceous age), all slopes exhibited indications of failure;

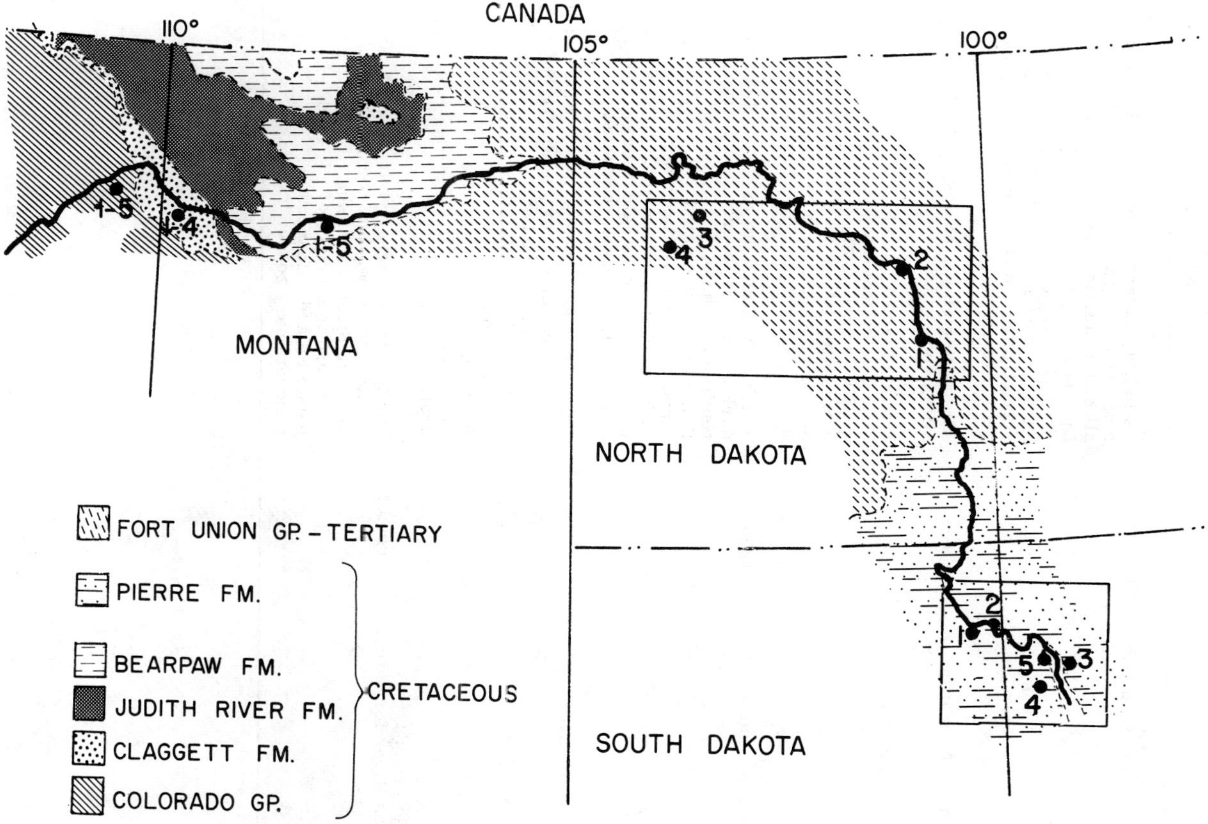

Fig. 1. Generalized geologic map showing study sites for detailed investigation.

Table 1. Lithologic description of materials studied

Unit	Approximate thickness (ft)	Description at surface	Source
Fort Union group			
Sentinel Butte formation	1150-1300	As below, except slightly, more coarse-grained overall and containing fewer lignites.	Hearn, et al. (1964)
Tongue River formation		Fine-grained sands, shaly clay, lignites, and some bentonite clays.	
Cannonball formation		Alternating clays, shales, silts, and sands.	
Pierre formation	800	Variable (see Fig. A25, Vol. 2). In general, gray bentonitic shale with scattered marly and concretionary zones.	
Montana group			
Bearpaw formation	1200	Thick unit of medium to dark gray, soft, poorly fissile clay shale. Bentonite disseminated throughout and also occurs in innumerable thin layers. Contains numerous calcareous and clay-ironstone concretions.	Erdmann (1962)
Claggett formation	450-550	Homogeneous sequence of gray to brownish-gray fissile flaky shale and bentonitic shale. Contains several bentonite layers in the lower 100 ft; sandy near top, lower 20 ft may contain scattered pebbles.	Hearn, et al. (1964)
Colorado group			
Marias River formation	800	Medium to dark-gray, silty shale, and moderately fissile siltstones containing scattered thin layers of bentonite and calcareous concretionary zones.	Erdmann (1962)

and while overall slope heights amounted to several hundred feet, inclinations were typically very low (5 to 10 deg), for example, Fig. 2. Excavations into these materials have usually exposed old failure surfaces, on which renewed sliding occurred. Slickensides were especially common in these three materials. Cementation was minimal, with the exception of parts of the Pierre. Slope inclinations in these three materials are comparable to their residual angles of internal friction. Where the Judith River sandstone formed a rigid base, certain slopes in the Bearpaw stood slightly steeper than elsewhere. Slope failures in the Colorado group, also a marine deposit of Late Cretaceous age, were relatively uncommon. Where they occurred, sliding appeared to involve a single failure, for example Fig. 3. Cementation was fairly well developed in the Colorado group; and probably as a result thereof, slope inclinations in the unit commonly were steeper than the residual angle of internal friction of the materials. Slickensides were present but not common. The Paleocene (lowermost Tertiary) Fort Union group, composed of one marine and two nonmarine formations, generally stood less steeply and showed more failures than the Colorado group, but stood steeper with fewer failures than the Claggett, Bearpaw, and Pierre formations. Notably, there were wide intraformational as well as the interformational differences within each of the five geologic units studied. Bentonitic layers with high plasticity characteristics, high swelling potential, and low residual shear strength were found in all of the units.

Laboratory Tests

Extensive laboratory testing documented in detail such properties as Atterberg limits, natural water contents, preconsolidation pressures, swelling and shrinkage behavior, and residual shear strengths. Variations in properties were determined for the different materials, but the properties of the five formations

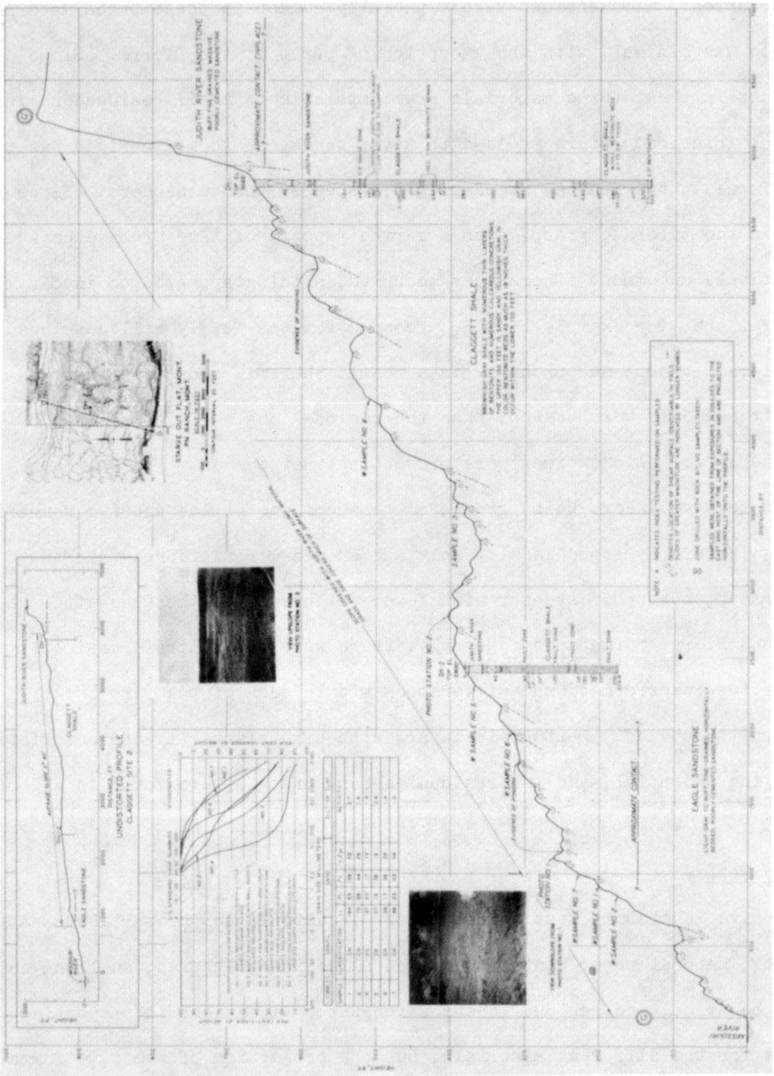

Fig. 2. Claggett site 2, geologic section G-G'.

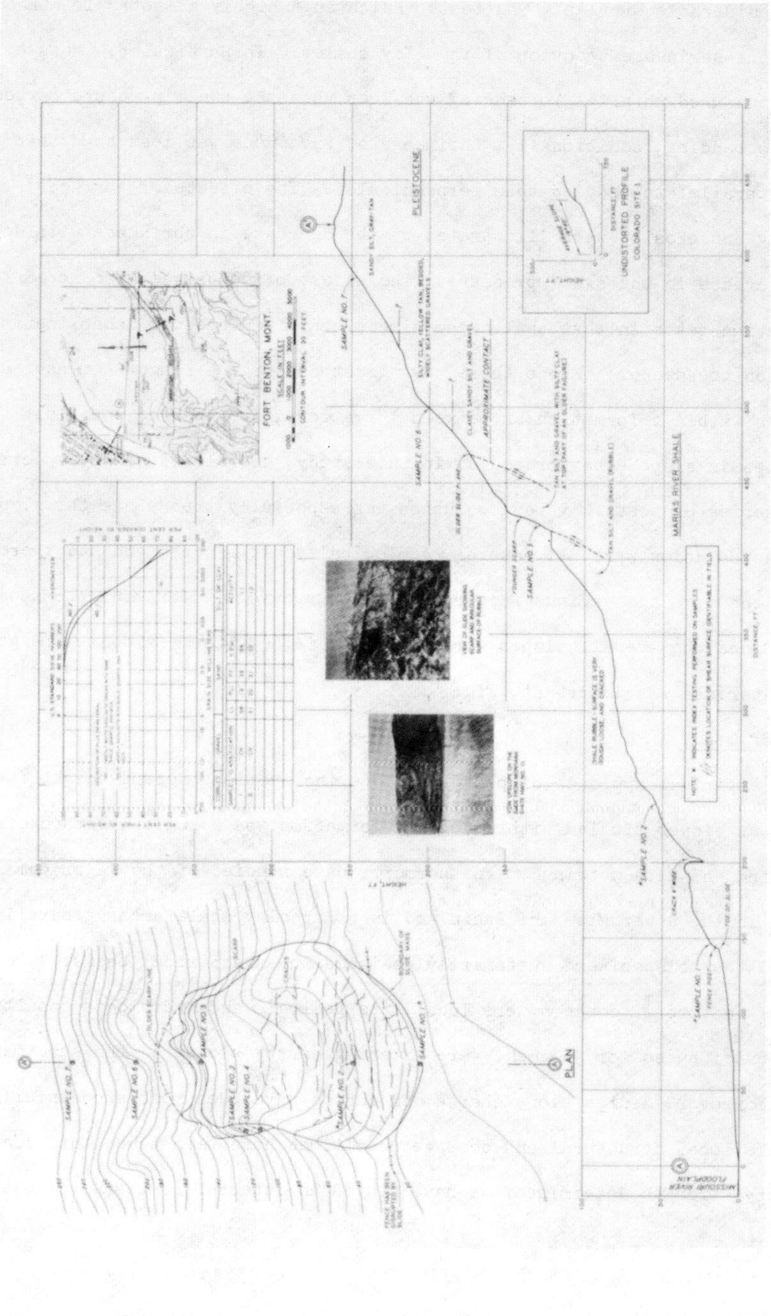

Fig. 3. Colorado site 1, geologic section A-A' and plan

showed considerable overlap. The tests indicated a highly anisotropic character to the swell-shrinkage behavior of the clay shales. In particular, the swell pressure measured parallel to bedding was less than the swell pressure perpendicular to bedding; additionally, shrinkage of specimens was less than that measured parallel to bedding than perpendicular. The effects of surface weathering and erosion were illustrated by the decrease in shrinkage with depth and an increase in the swell pressure. The effect of depth was also noted from consolidation tests in that shallow specimens exhibited steeper consolidation curves than the deeper. Since there was almost as large a range in measured properties within a formation as between formations, it appeared that only gross differences in slope behavior could, in this study, be related to any specific test or series of tests. This conclusion was especially true since the slopes under consideration were all natural, geologically old slopes, and had therefore been subjected to the various factors discussed previously for many thousands of years. Since the overall slopes were generally failed, the residual strength of the materials was important, Fig. 4.

Results of Study

The empirical approach of plotting slope inclinations versus height for the various slopes studied within a given formation was a logical method of summarizing the data obtained. Furthermore, case studies of the major dam projects in the upper Missouri Basin and in other clay shale areas generally indicated that slopes were customarily designed on an empirical basis. The principal factors of concern were local site geologic and hydrologic conditions, and the chief bases for judgment were existing nearby slopes in the same materials. Moreover, design slope charts and curves have been most successfully used where local structural and seepage conditions have been taken into account. Laboratory strength determinations have played an integral, but subordinate,

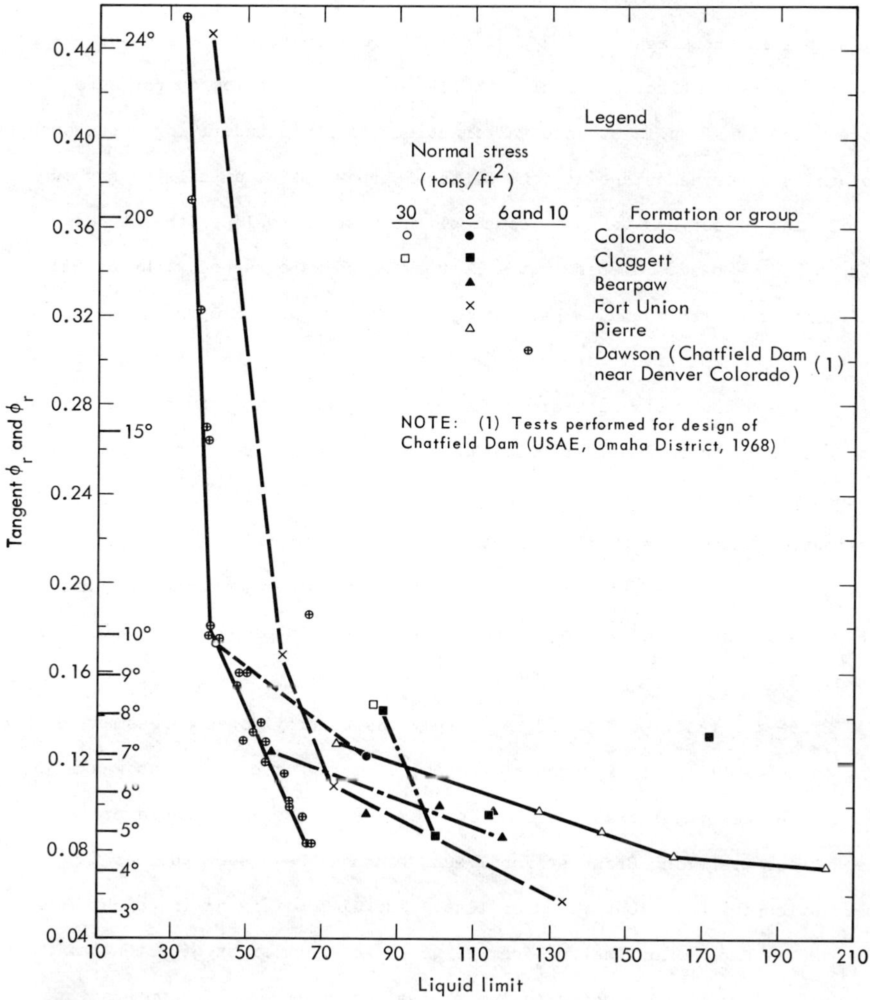

Fig. 4. Summary of residual direct shear test data.

part. Conservatism in slope design has generally been vindicated, since time-dependent phenomena have caused some slopes that stood apparently stable for varying periods of time to eventually experience some distress. The slope charts developed for the five units studied are shown in Figs. 5 through 9. The slopes in the Claggett, Bearpaw, and Pierre formations are similar. For slopes composed entirely of shale, the range in slope angles compares to the range in residual angle of internal friction. A rigid underlying base, such as the Judith River sandstone at some of the Bearpaw sites, appears to have a stabilizing effect and tends to cause slightly steeper slopes than those without such a base. Slopes in the Bearpaw which were capped with Fox Hills sandstone were also found to have steeper inclinations. Slopes in the Colorado group stand at relatively steep angles when compared to the residual friction angle. The average strength along failure surfaces must be increased from that obtained from the residual friction angle by the influence of numerous interbeds of cemented silt. Slopes in the Fort Union group stand at variable heights and inclination, depending on the formation involved. Slopes in the Sentinel Butte formation are steeper, as a group, than slopes in other formations. This behavior was thought to reflect a coarser-grained lithology in the Sentinel Butte.

Since the basic purpose of this study was to provide an assessment of the long-term stability of high crater slopes, a review of experiences was made. The only experimental crater slopes in clay shale (the Pre-Gondola project of the Nuclear Cratering Group at Fort Peck, Montana) have been subject to observation for only four years or less. The slopes varying in height from 40 to 80 ft and in inclination from 22 to 29 and show signs of potential future distress, although no massive failures have yet occurred. It appears that in the future cratered slopes for engineering purposes must be designed largely on

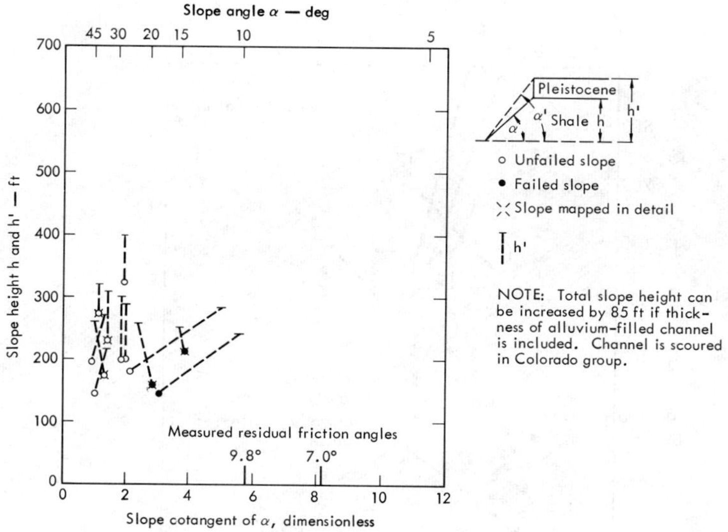

Fig. 5. Height and inclination of slopes in Colorado group.

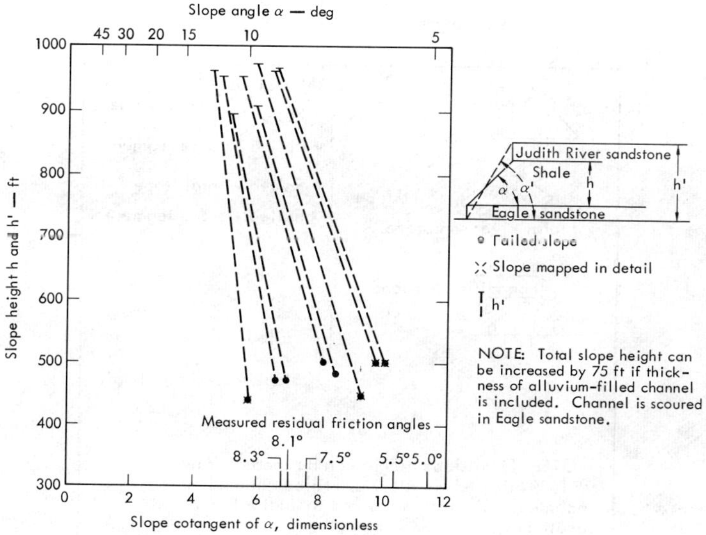

Fig. 6. Height and inclination of slopes in Claggett formation.

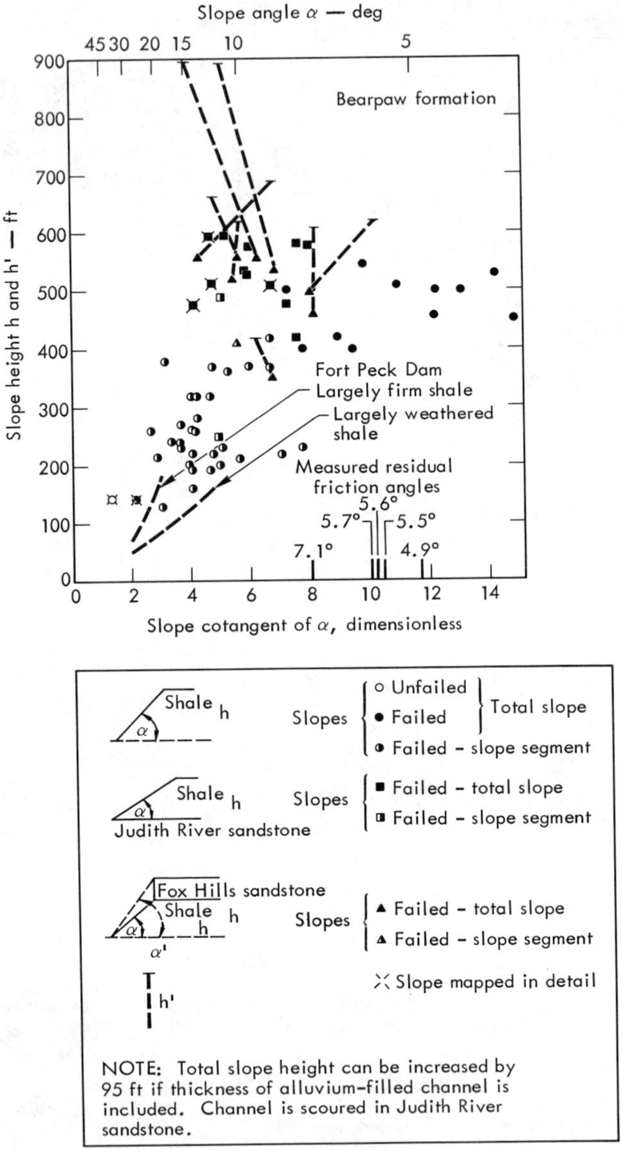

Fig. 7. Height and inclination of slopes in Bearpaw formation.

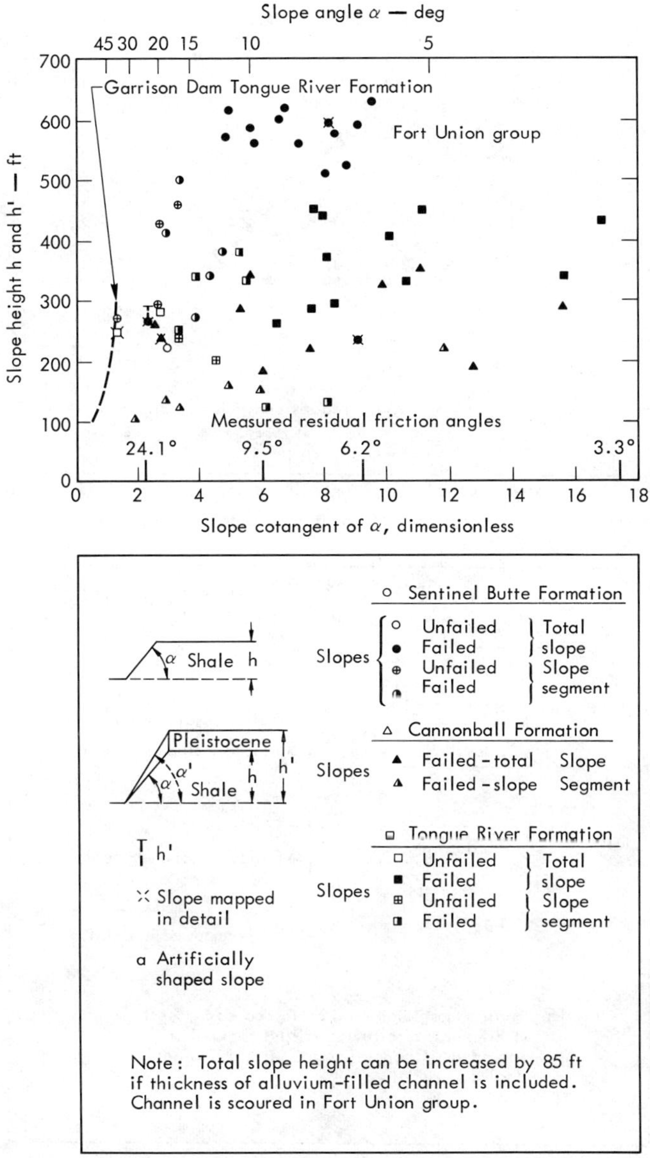

Fig. 8. Height and inclination of slopes in Fort Union group.

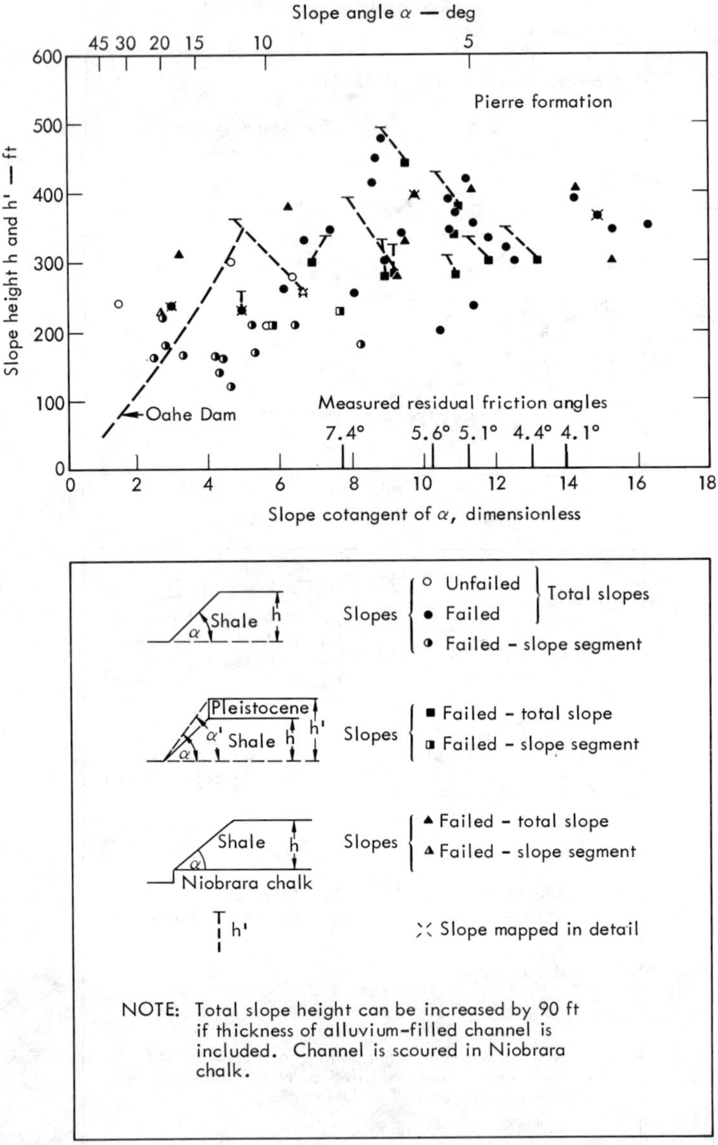

Fig. 9. Height and inclination of slopes in Pierre formation.

the basis of experience with conventionally excavated slopes, and that the design must be conservative. No experience to date indicates that it is necessary to design slopes as flat as the residual angle of internal friction of the materials, although such an inclination would be the lower limit to which slopes might ultimately adjust.

RESULTS OF STUDIES OF CLAY SHALE SLOPES ALONG THE PANAMA CANAL

The Panama Canal studies differed from the previously described work in that the slopes were constructed as part of an engineering project and data and experience records existed from which the behavior of the slopes could be reconstructed. The factors previously discussed were of course operative to varying degrees, but the failures were induced as a direct result of the construction activities. The study is a continuing project which will cover the entire 12-mile Gaillard Cut through the Isthmian Divide. The present paper covers the results of the first year's effort and concentrates on the East and West Culebra slides, and the Model Slope, Fig. 10.

Geology

The Isthmian Divide area is underlain by relatively soft Tertiary pyroclastic strata and localized masses of resistant basalt and breccia. These contrasting rocks are distinctly expressed as broad rolling topography and conical hills, respectively. A system of north-south faults and subordinate east-west fractures divides the area and is responsible for localized basaltic intrusions. North-south faults delineate a graben structure crossing the canal obliquely at the East Culebra and West Culebra slides. Bedding dips adversely toward the canal at least locally, contributing to instability in both banks. Favorable dips in the Model Slope, i.e., nearly horizontal or into the slope, may account in part for the slope's stability. The Cucaracha formation,

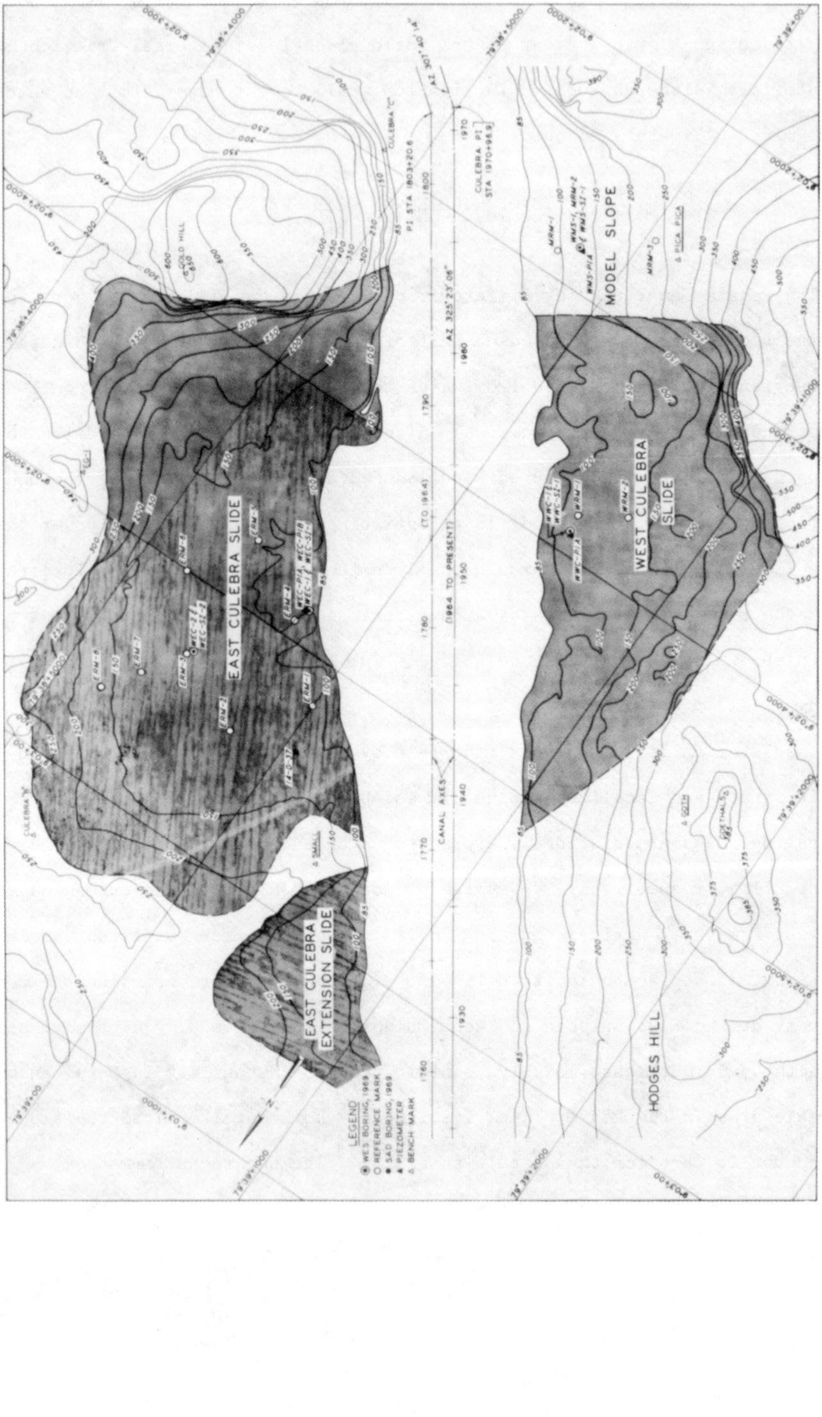

Fig. 10. General plan of study area.

composed basically of tuffaceous shale, and the underlying Culebra formation of tuffaceous shale, siltstone, and sandstone are involved in sliding. Characteristics of the Cucaracha formation that have made it particularly susceptible to sliding are: (1) an abundance of shale strata; (2) a high content of montmorillonite in the shale; and (3) well developed, small-scale fracturing in the form of slickensides in certain layers. The lower third of the Cucaracha formation seems to be a transition zone, identifiable on the basis of lithology and the presence of scattered limy concretions. The zone may be locally cemented and possibly has a slightly greater strength, but this is not corroborated by laboratory testing.

History of Sliding

When the United States assumed control in 1904, the excavation had been cut through the divide area to a depth of about 140 ft with the bottom at about el +175. Both banks of the excavation were largely overgrown with vegetation, and only vague evidence remained of surficial sliding previously experienced by the French. Subsequently, the first failures occurred in 1906-07 at sta 1777 on the west bank and sta 1789 on the east bank. Both sides were partly in residual weathered clay, but they soon evolved to deep slides.

The slides grew significantly from 1908 to 1914 by deepening in pace with excavation and progressively incorporating new blocks of ground farther back in the bank. By 1911 slides had occurred in both banks at intervals between sta 1760 and 1794. As the final excavation bottom at el +40 was reached in 1913, the East Culebra slide had attained a depth corresponding approximately with the top of the transition zone in the Cucaracha formation. Sliding was apparently taking place along one or a few favored stratigraphical horizons. Cracking of the bank usually preceded failure. Then after a few months the blocks separated by cracks gradually settled, sometimes also tilting toward the excavation and was followed by massive failure of the front block and its sliding into the excavation in a few hours or days (Goethals, 1916).

Sliding activity accelerated on the east side to a major break in 1914 almost a year after the canal was flooded for dredging. By January 1915, the West Culebra slide was divided by long cracks along geological fractures into several potentially unstable blocks. The blocks evolved to graben and horst structures as the mass crept slowly canalward. In August and September 1915, both slides converged in massive failures accompanied by upheaval of the canal bottom and blockage of the canal. An example of the reconstructed history of sliding for one section in the East Culebra slide is shown in Fig. 11.

Slow movement with occasional increased activity has continued to the present. This stage has apparently reflected a drastic deterioration of material strength since the driving force due to large volume and height are not comparable to those operating at the prime of slide activity.

Upheaval of the bottom of the excavation has been characteristic of the sliding activity since 1907. The most impressive case occurred in the culminating movement of 1915. The upheaval mass appears to correspond to a passive wedge at the front of a basically translational slide. Deep sliding surfaces have been suspected from construction days at about el -50 to -100 ft, but the more conventional picture has the sliding sufaces at el +100 to +25.

Results from Field Instrumentation - 1969

Slope Indicators. Data collected from slope indicators installed in 1969 located the present depths of sliding in the East Culebra and West Culebra slides near the bottom of the canal. These data complement interpretation of borings and give good information as to the extent of the masses now involved in sliding. Data obtained at the Model Slope installation indicated small movements at several depths which are probably closely related to the geologic structure and surface expression of cracking in the area.

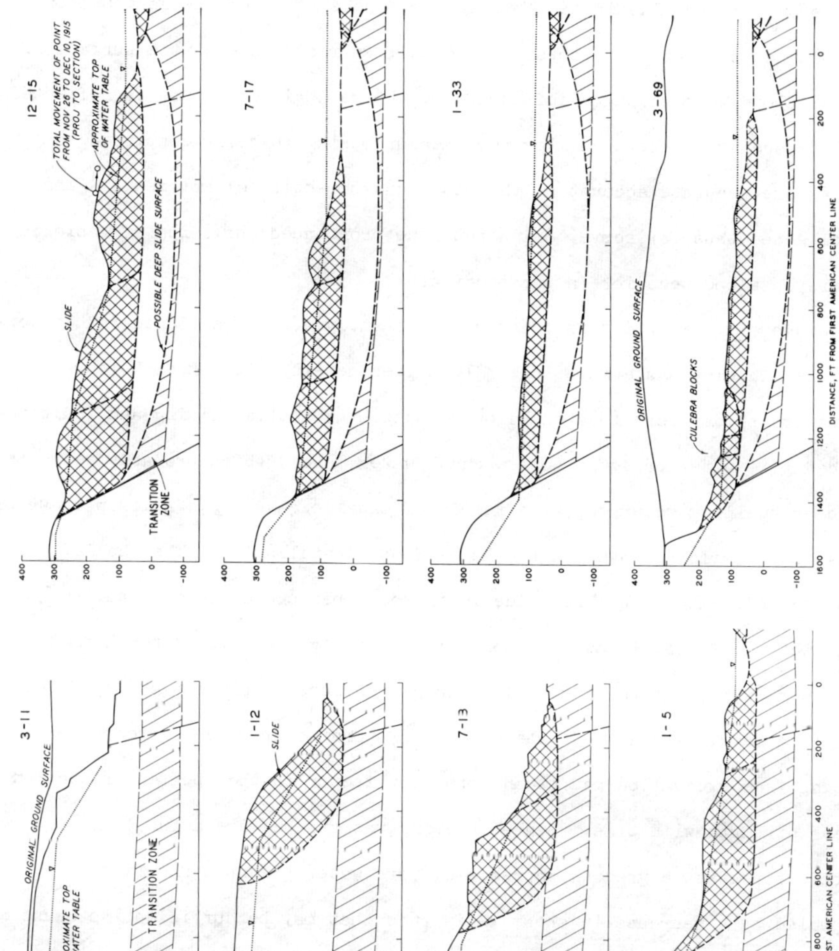

Fig. 11. Development of East Culebra slide at sta 1782+50, March 1911 to March 1969.

Surface Survey Monuments. Surface reference markers indicated continuing movements at all installations. Movements at the surface of the East Culebra and West Culebra slides were toward the canal, whereas those at the surface of the Model Slope were strongly influenced by the geologic structure as reflected by the surface crack pattern. Movement measurements are being continued to obtain a more accurate account of the rates and direction of movements. The data collected thus far serve to indicate the pronounced influence of geologic structure on slope behavior in the study area.

Piezometers. Piezometers located in the study area were placed in slickensided zones in clay shale materials. Deep piezometers in the East Culebra and West Culebra slides were located below the depth of active sliding and indicated a reduced pore water pressure (below canal water level) which probably reflects rebound or swelling occurring as a result of canal excavation. A low pore water pressure was similarly noted in the Model Slope installation. The shallow piezometer in the East Culebra slide indicated a piezometer level equal to that in the canal. Additional piezometers are being installed beneath the East Culebra slide and the Model Slope to give a more complete picture of the piezometric level with depth. The concept of pore water pressure reductions resulting from excavation may be an important factor in the apparent reduction of shear strength with time. Since shearing resistances are a function of effective stresses, a gradual increase of pore water pressure should be accompanied by a decrease in shear resistance that may be sufficient to cause failure some time after excavation. As a result of the occurrence of Cucaracha clay shales of extremely low permeability overlying Culebra sandstone, it is possible that high pore water pressures could exist in the sandstone regardless of pore water pressures in the overlying clay shale. Thus, the failures may have been influenced in part by low effective stresses and resulting low shear strengths at the clay shale-sandstone interface.

Results from Laboratory Tests

Classification Data. The laboratory classification of shale samples from the Cucaracha and Culebra formations indicated that the materials generally plotted close to the A line on the plasticity chart. The results indicated little differences in average values of the liquid and plastic limits but that exceptional values could occur at any depth. Slaking reactions were similar in the two formations with materials slaking very rapidly when air-dried and immersed in distilled water. The most important difference noted between shales in the two formations was the slight increase of materials lost in the insoluble residue test in the Culebra formation. The increase reflected the increase in marine fossils in the Culebra. A small increase was similarly noted for samples taken from the transition zone of the lower Cucaracha. From the laboratory results it was concluded that very little difference existed in the clay shale portions of the Cucaracha and Culebra formations.

Water Contents. Water contents generally ranged from 7 to 15 percent below the plastic limit; higher water contents were noted in samples taken from the slide debris. Significantly, water contents about double the average water content were determined for slickensided zones at the probable depth of sliding in the East Culebra and West Culebra slides.

Consolidation Characteristics. Preconsolidation stresses were determined to be as high as 200 tons/sq ft; preconsolidation loads from geologic evidence were estimated to be as high as 130 tons/sq ft. Swell pressures increased with depth and generally ranged from 0.8 to 1.3 times the computed effective overburden pressure assuming hydrostatic pore pressures. The expansion index averaged about 0.045, the coefficient of consolidation averaged about 0.35×10^{-4} cm^2/sec, and the coefficient of permeability averaged about 1×10^{-10} cm/sec from the unloading portion of consolidation curves between loads existing before excavation was commenced in the study area and those now existing. The extremely low value of the coefficient of permeability is corroborated by results computed from piezometer observations.

Strength Tests. The majority of previous tests conducted on Cucaracha shale samples indicated a peak drained friction angle of from 17 to 22 deg and averaging about 19 deg. The effective cohesion intercept was shown by several tests to be extremely variable, depending on thhe occurrence of slickensides, sandy seams, etc. Laboratory drained direct shear tests and slow triaxial tests indicated cohesion values of about 6 tons/sq ft; the single cohesion value from this study was 5 tons/sq ft. Residual friction angles determined in this study varied between 4.5 and 10.5 deg with an average value of approximately 7.5 deg.

Stability Analyses

East Culebra and West Culebra Slides. Cross sections of the slopes were reconstructed at selected times (see as an example Fig. 11) for stability analyses. The average strength at the time of first-time failure and the variation in average strength with time are of particular interest. Possible deep critical failure surfaces and failure surfaces corresponding to present slope indicator data were used to determine the peak effective strength, but the effect of assumed depth of sliding was small. The effect of reduced pore water pressure was also considered. Stability analyses of initial failure conditions at selected sections along the canal indicated an average effective strength envelope of $\phi' = 19$ deg and $c' \approx 0$ (Fig. 12). The friction angle thus determined compares closely with friction angles determined from test results. Analyses of failure conditions for various dates of sliding indicated that the average shear strength between the time of initial failure and the massive movements (Oct 1914 for the East Culebra slide and Aug 1915 for the West Culebra

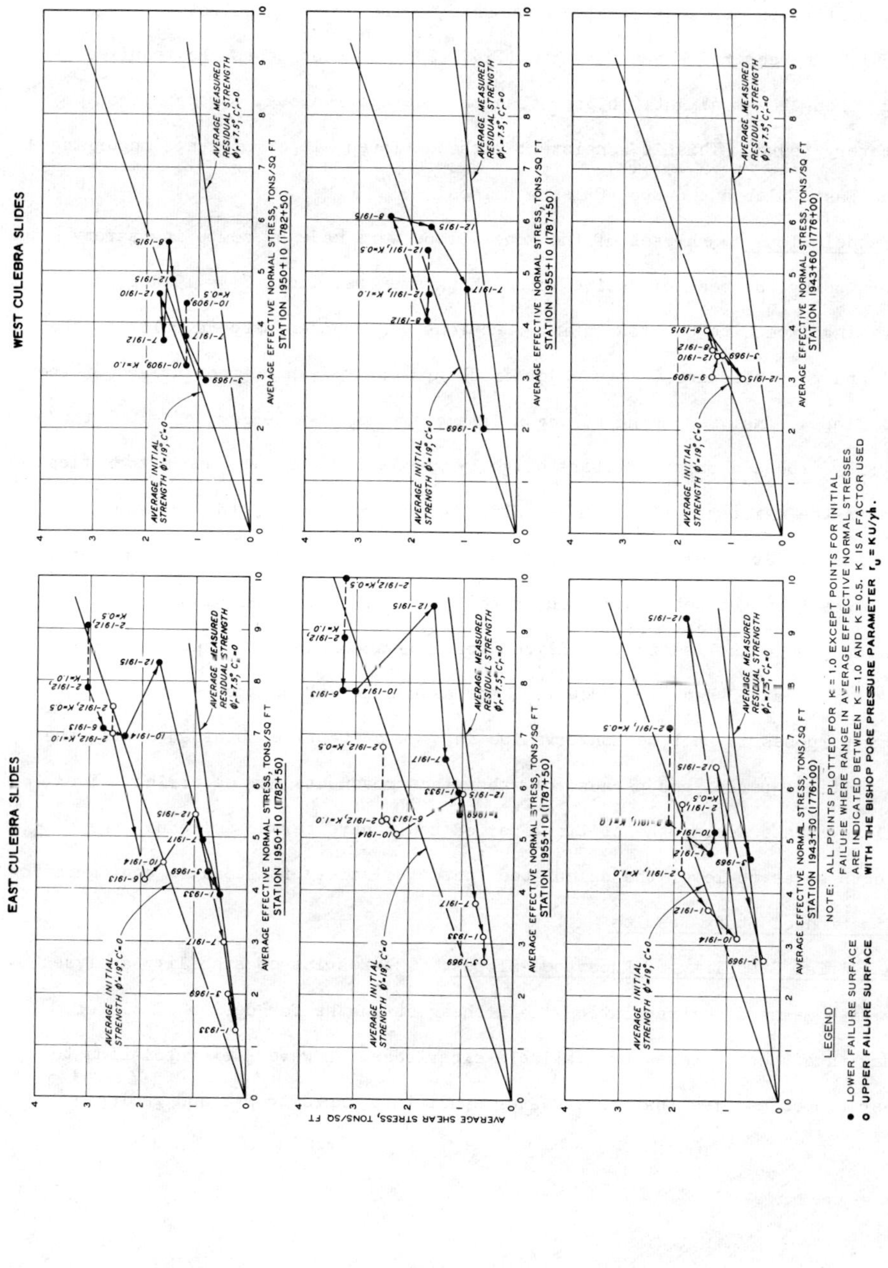

Fig. 12. Average stresses from analyses of East and West Culebra slides.

slide) remained high and near the strength envelope determined for initial failure despite the large movements that occurred. Analyses of failure conditions subsequent to the time of massive movements to the present indicate a large reduction in required shear strength (Fig. 12). In most cases the required shear strength was slightly higher than the average residual strength from laboratory tests. This is consistent with continued slope movements occurring in the East Culebra and West Culebra slides.

Model Slope. Analyses of the Model Slope were made by computing safety factors using the peak effective strength parameters, i.e., ϕ' = 19 deg and c' varying from 0 to 0.4 ton/sq ft. Sections were chosen perpendicular to the canal and perpendicular to the major crack presently seen on the surface of the Model Slope. Analyses using strength parameters of ϕ' = 19 deg, c' = 0.4 ton/sq ft generally produced safety factors slightly greater than 1.0. The most critical stage in the existence of the Model Slope was represented by the analyses of conditions in July 1917. Conditions on this date represent the canal geometry when excavation had been completed in the canal and the Model Slope had not been reduced in inclination. Analyses for 1969 conditions indicated factors of safety greater than 1.0. However, the assumption of peak strength for design purposes may not be conservative in view of results obtained during the most critical period and because of subsurface movements now occurring. Even small movements during the critical period could have served to reduce the strength so as to lower the calculated factor of safety for today's configuration.

Application of Study Results

Initial Stability of Excavated Slopes. The results of stability analyses provide a means of extrapolating field shear strengths for design of slopes higher than those covered by available experience. However, extrapolation to higher slopes assumes that groundwater conditions, anisotropy, and geologic defects and structure are similar to those of the East Culebra and West Culebra slides. However, when extrapolating to conditions where higher normal stresses will be acting, consideration should be given to possible reduction in friction angle with higher pressures as indicated by several studies and this aspect requires further investigation. While results to date must be corroborated by similar studies on other canal slope failures before they can be seriously considered for slope design, present indications are that field strengths for high (600 ft) slopes may be 25 percent higher than the previous strengths resulting from the previous studies (PCC, 1947). However, no material difference exists for slopes 300 to 400 ft high or lower. These conclusions are preliminary and will be studied further at these and other slides during subsequent phases of the study.

Decrease of Shear Strength with Time and Movement. The results obtained indicate a large decrease in shear resistance with time following initial sliding. However, the shear strength did not appear to decrease rapidly immediately following initial sliding, and the slides enlarged, with large movements, before a substantial decrease in shear strength developed. If this is corroborated by further study at the East Culebra and West Culebra and other slides, the feasibility of an observational approach to modify slope design during construction would be indicated. However, while present indications are encouraging in this respect, much additional study of the aspect is required to corroborate the findings and to establish their applicability.

SUMMARY

The reported studies have given information on the stability of clay shale slopes. The approach has been to study the geology and history of sliding, to determine the physical properties from laboratory tests, and to make stability analyses of the slopes. Geology of slopes in five different formations along the upper Missouri River were studied. The most expedient manner to view the stability characteristics of these slopes was through empirical slope charts. Design experience in the study area had used similar approaches with success when local site geologic and hydrological conditions were considered. In the Panama study, detailed records were available to reconstruct the sliding history and allow limit type analyses to be performed. Both approaches can be utilized to determine design slopes by giving proper recognition to the limitations of each.

APPENDIX I.—REFERENCES

1. Erdmann, C. E., (1962), "Geology and Mineral Resources of Missouri River Valley Between Head of Fort Peck Reservoir and Morony Dam, Montana," Department of the Interior, USGS. See also "Geology" by Erdmann, C. E., and Lemke, R. W., Appendix 3 in Joint Report on Water Resources Development for Missouri River, Fort Peck Reservoir to Vicinity of Fort Benton, Montana, USAE, Omaha District and US Bureau of Reclamation, Billings, Montana.

2. Fleming, R. W., Spencer, G. S., and Banks, D. C., (1970), "Empirical Study of the Behavior of Clay Shale Slopes," NCG Technical Report No. 15 (in two volumes), U. S. Army Engineer Nuclear Cratering Group, Livermore, Calif.

3. Goethals, G. W., (1916), "The Dry Excavation of the Panama Canal," Transactions of the International Engineering Congress, Vol 1, pp 335-385.

4. Hearn, B. C., Jr., Pecora, W. T., and Swadley, W. C., (1964), Geology of the Rattlesnake Quadrangle, Bearpaw Mountains, Blaine County, Montana, USGS Bulletin 1181-B, Washington, D. C.

5. Lutton, R. J., and Banks, D. C., (1970), "Study of Clay Shale Slopes Along the Panama Canal, Report 1: East Culebra and West Culebra Slides and the Model Slope," WES Technical Report S-70-9, U. S. Army Engineer Waterways Experiment Station, Vicksburg, Miss.

6. PCC, (1947), "Isthmian Canal Studies, Appendix 12: Slopes and Foundations," Report of the Governor of the Panama Canal, Diablo Heights, Canal Zone.

7. USAE, Omaha District, (1968), "Design Memorandum No. PC-24, Chatfield Dam and Reservoir, Embankment and Excavation," Omaha, Nebraska, December.

STABILITY STUDIES
OF A FOOTWALL SLOPE IN LAYERED COAL DEPOSIT

By C.O. Brawner,[1] M.ASCE., D.L. Pentz,[2] and J.C. Sharp[3]

INTRODUCTION

At the present time, several large open cast coal operations are in production or are being studied in the East Kootenay region of Canada's Rocky Mountains. The influence of slope stability on a proposed coal property is described in this paper. The property is located in the Kootenay Formation and exists within the area of a synclinal structure (see Fig. 1). (It is planned to develop one limb of the syncline for mining operations.) The most critical slopes in the proposed operation are in the region of the footwall, which dips at an approximate angle of 40 deg. Slopes up to 1,600 ft. in height are envisaged (see Fig. 2). This paper describes details of the overall feasibility study specifically related to the footwall stability.

The economics of the project favour the removal of all seams, except one which will underlie the proposed final footwall. The influence of this relatively weak seam on the overall footwall slope stability was studied in detail to determine whether or not the economic requirements could be realized. It is anticipated that the majority of the footwall will be defined by the bedding orientation of the coal. The degree of stabilization required to produce a stable slope was also investigated as part of the preliminary study.

---

[1]Principal, Golder, Brawner & Associates, Vancouver, B.C.
[2]Associate, Golder, Brawner & Associates, Vancouver, B.C.
[3]Senior Engineer, Golder, Brawner & Associates, Vancouver, B.C.

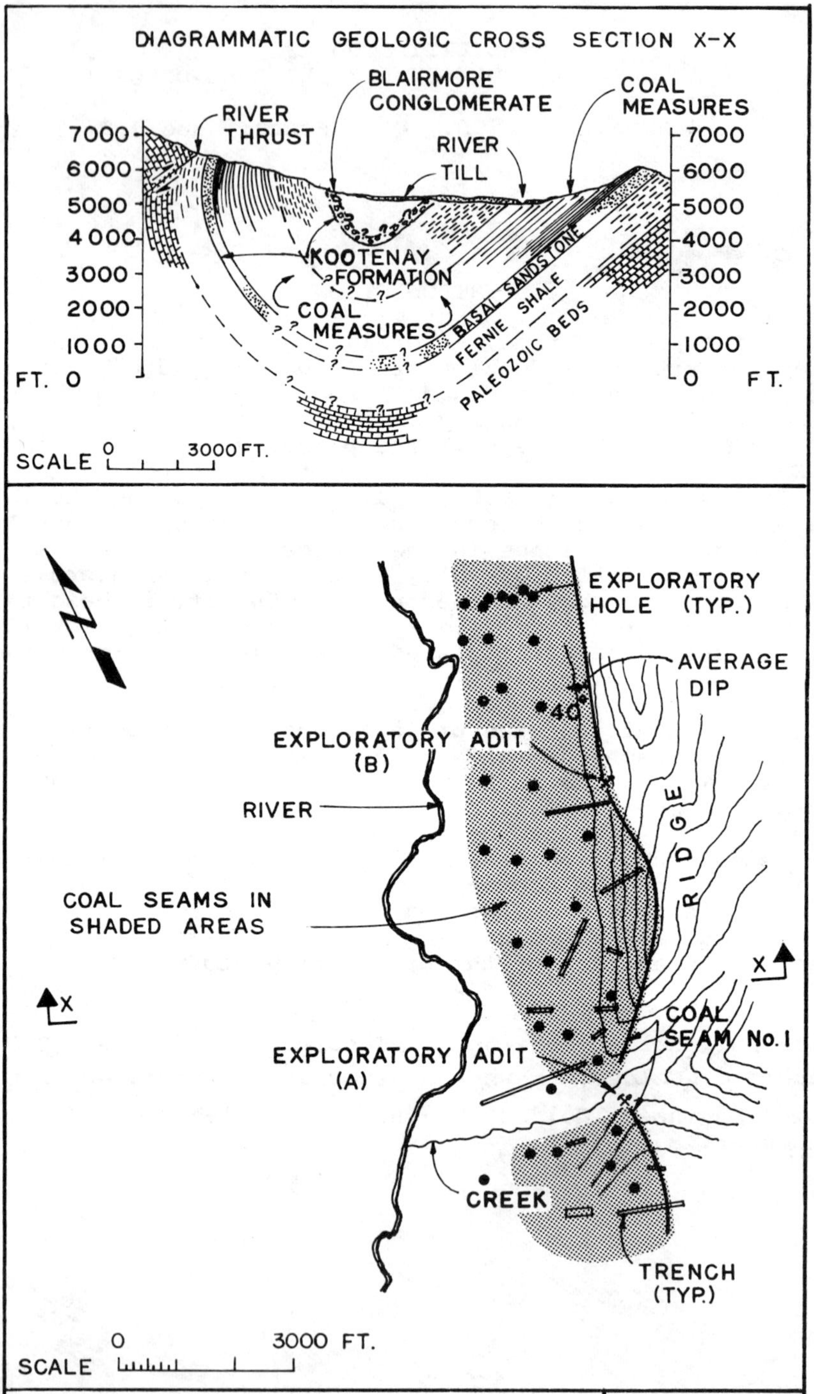

SITE PLAN | FIGURE 1

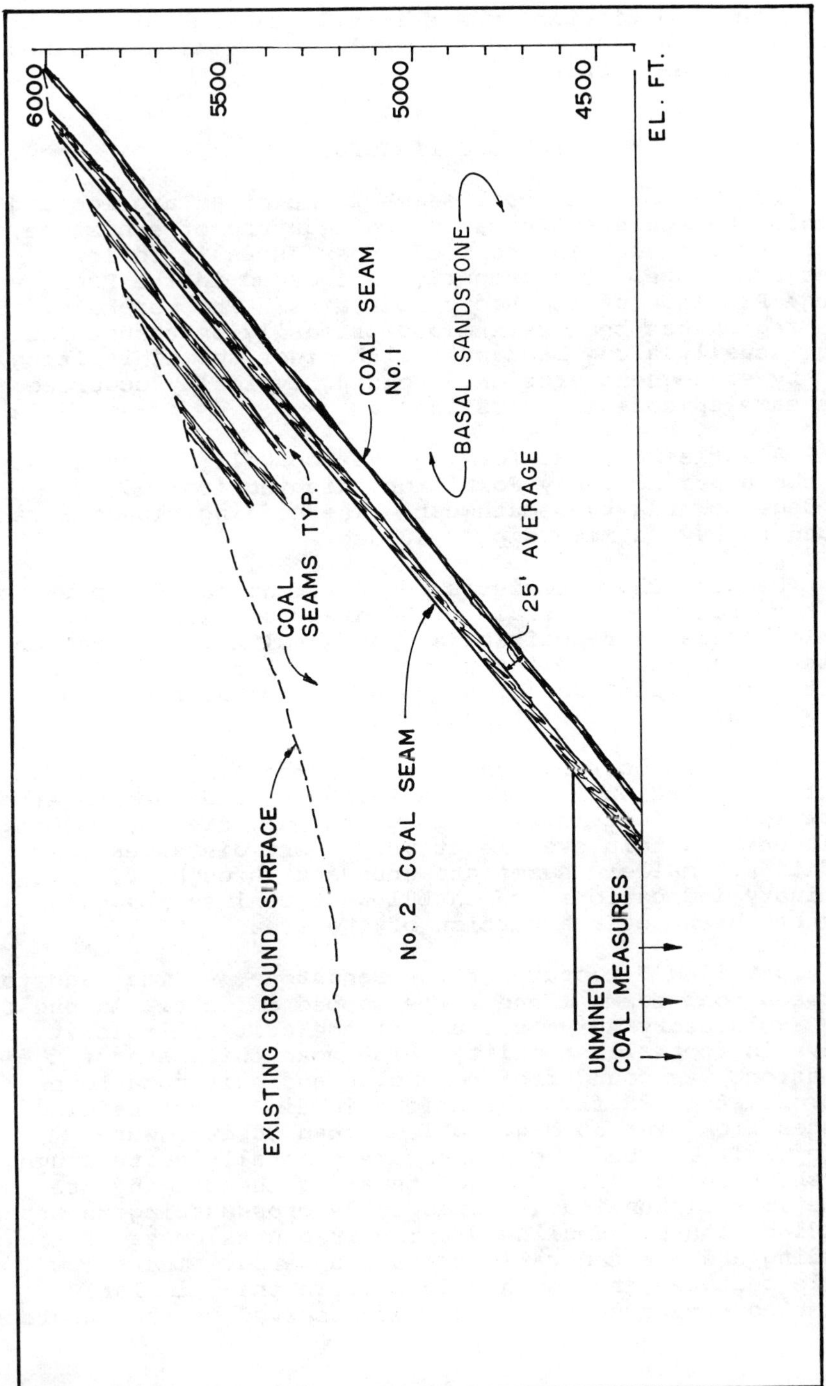

TYPICAL SECTION THROUGH MINE | FIGURE 2

The investigation was primarily centred around two exploratory adits, A and B (see Fig. 1), which provided access to the footwall.

GEOLOGIC FACTORS

General. -- The coal seams to be mined are located within the Kootenay Formation, a sequence of sandstone, siltstone, shale, and coal of upper Jurassic to lower Cretaceous Age. The property is located in the Front Range Province of the Rocky Mountains. The morphology of the region has been determined primarily by thrust faulting (usually along bedding), with minor normal faulting, and by subsequent erosion. Folding typically occurred in the same episode as thrusting.

A series of beds forming the footwall slopes exist in the upper Kootenay Formation, dipping from 35 to 50 deg. towards the southwest. The bedding planes were found to have a mean dip of 40 deg.

Bedding Characteristis. -- A variation in dip value of the bedding on a scale of hundreds of feet, which could be due to deposition and/or synclinal deformation exists. A bedding plane waviness, on a scale of 2 to 5 ft. with amplitudes of up to 6 in., produces a wide local scatter in the attitude of beds.

Borehole correlations indicate that there is a great deal of variation in both the thickness and the number of coal seams. Individual seams pinch out, divide, rejoin, thicken, and thin over relatively short distances. In addition, smaller seams are abundant throughout. Preliminary indications are that lowest coal seam (No. 1) may be absent over a portion of the area.

Detailed Structure of the Sandstone. -- The sandstone between coal seams 1 and 2 was mapped in detail in one of the Exploratory Adits because of the critical role it plays in footwall stability. The mean thickness of the sandstone was found from boreholes and adit data to be approximately 25 ft. Variations in dip of the bedding planes from over 50 deg. to less than 30 deg. were observed. The bedding planes are generally quite rough, as shown on Fig. 3. The joints along the bedding are sometimes discontinuous, frequently crosscutting to other bedding planes. Shallow dipping fractures crosscut the bedding and are generally concave upwards. These low angle features are rougher in profile than similarly oriented thrust shears that were observed in the outcrops.

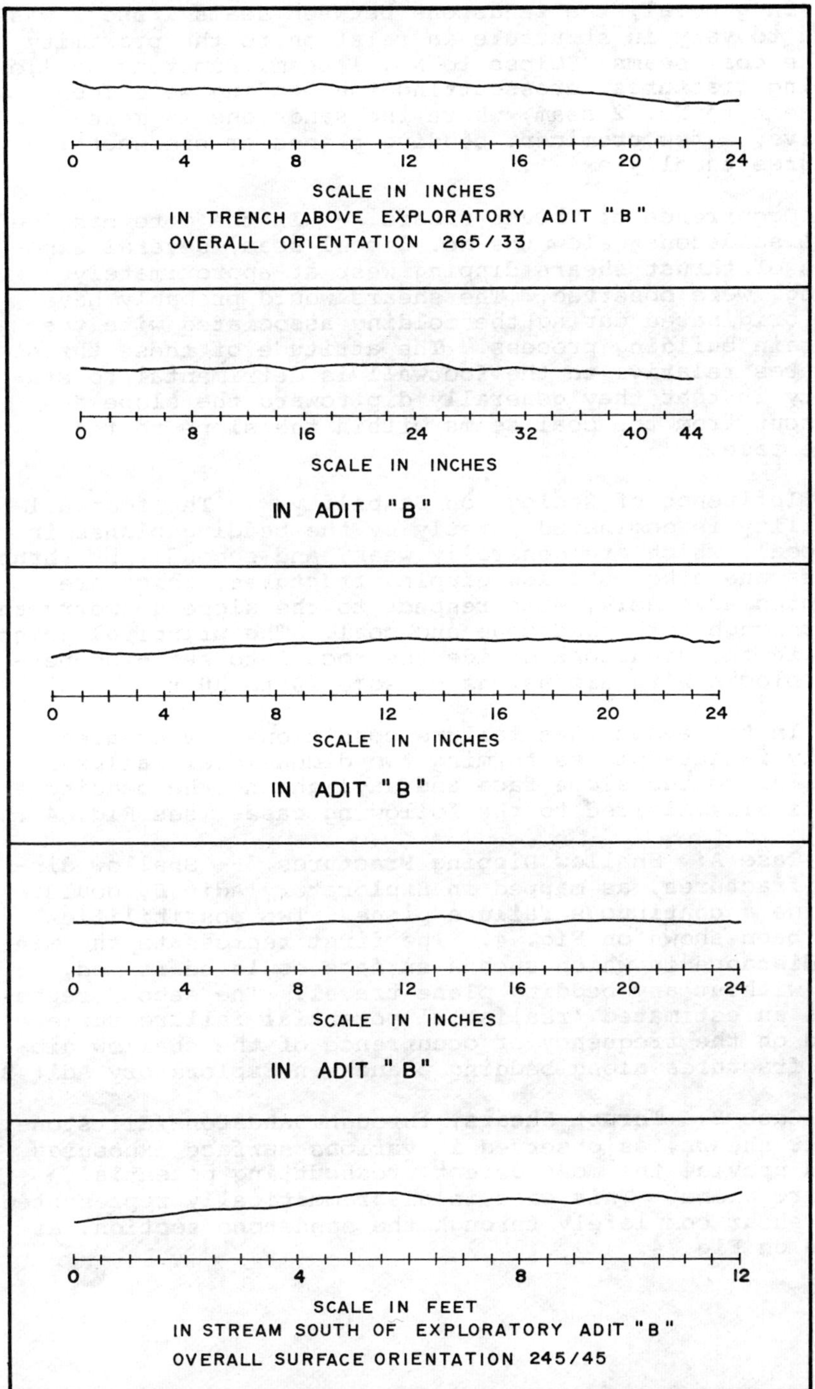

TYPICAL BEDDING PLANE PROFILES

FIGURE 3

In general, the sandstone between seams 1 and 2 was found to vary in structure in relation to the proximity of the coal seams. Close to No. 1 seam, numerous shallow dipping fractures crosscutting the bedding were found. Adjacent to No. 2 seam, where the sandstone is more massive, a few prominent bedding planes or crosscutting features usually exist.

Occurrence of Thrust Shears. -- In the more massive basal sandstone below the No. 1 coal seam, several exposures of thrust shears dipping west at approximately 20 deg. were observed. The shears would probably have been originated during the folding associated with the mountain building process. The attitude of these thrust features relative to the footwall is detrimental to stability in that they generally dip toward the slope and crosscut from the coal seams within the slope to the slope face.

Influence of Geology on Stability. -- The footwall stability is dominated firstly by the bedding planes in the coal, which are generally weak, and secondly by thrust shears and other shallow dipping fractures, which are oriented adversely, with respect to the slope geometry and cut through both sandstone and coal. The principal joint sets in the sandstone divide the rock into semi-independent blocks with dimensions of some 20 to 30 ft.

In the event that failure conditions are created, likely failure planes forming two dimensional failures parallel to the slope face and daylighting the bedding are practically limited to the following cases (see Fig. 4).

Case A: Shallow Dipping Fractures. -- Shallow dipping fractures, as mapped in Exploratory Adit B, could provide a continuous failure plane. Two possibilities have been shown on Fig. 4. The first represents the minimum distance in which such a surface could be formed, i.e. without any bedding plane travel. The second represents an estimated 'realistic' potential failure surface, based on the frequency of occurrence of the shallow dipping fractures along bedding planes in Exploratory Adit B.

Case B: Thrust Shears, through Sandstone/Siltstone. -- Thrust shears, as observed in various surface exposures would provide the most direct crosscutting potential failure plane. This case is diagrammatically represented as a shear completely through the sandstone section, as shown on Fig. 4.

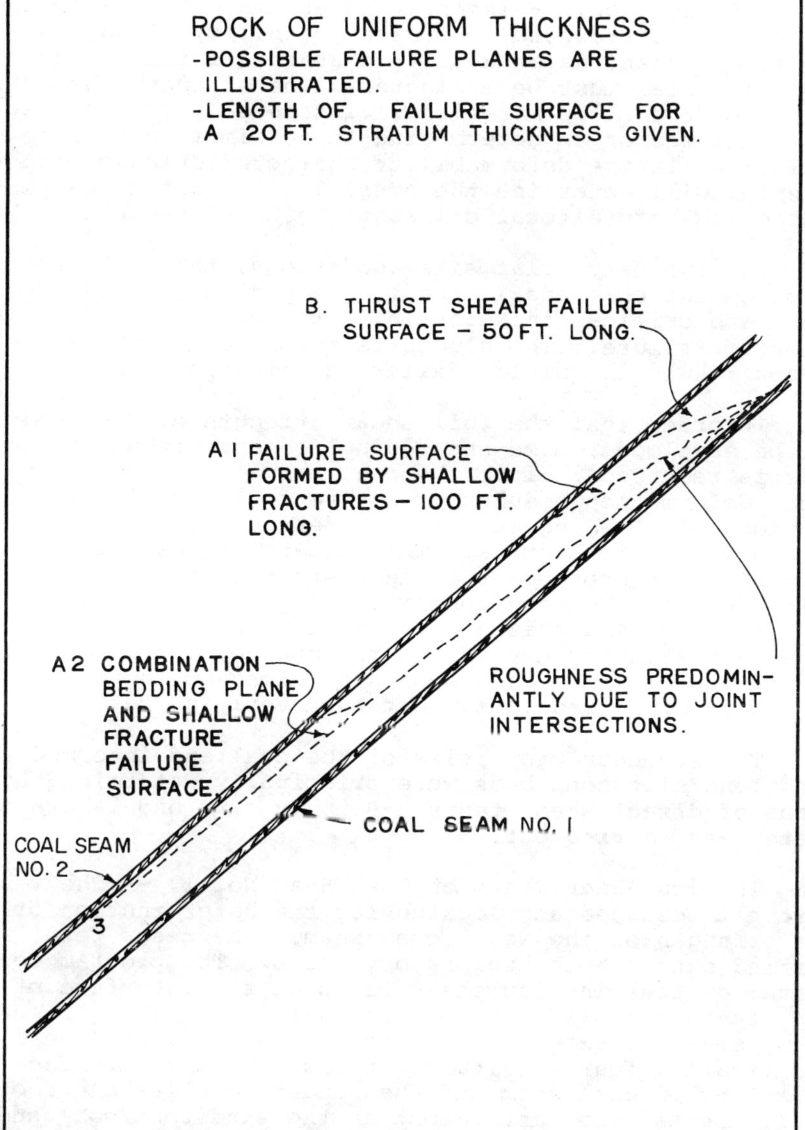

FOOTWALL FAILURE MODEL | FIGURE 4

STRENGTH OF SLOPE MATERIAL

Introduction. -- In order to estimate the safe slope angles that correspond to a given stability situation, the strength and deformability characteristics of the slope material must be evaluated. The strength characteristics of the slope material will generally be the most important factor in determining the maximum safe slope angles; while the deformability characteristics of the material will determine the overall allowable slope displacements before total collapse of the slope occurs.

For the particular site conditions, the differential displacement characteristics of the coal and the sandstone are critical in determining the stress distribution prior to failure. The alternate strong and weak layered strata result in complex failure conditions.

In order that the full shear strength of the coal may be developed, some shear displacement within the coal beds is required. Since the coal has a relatively low shear deformation modulus, relative to the sandstone, considerable loading may be transferred to the sandstone joints in critical areas. This excessive loading may be sufficient to overcome the shear strength along certain joints within the sandstone, allowing shear displacement to occur. During this displacement phase, the coal will have been steadily developing resistance, until an equilibrium phase is reached when sufficient strength will have been mobilized to support the overlying material.

The strength properties of the coal and intermediate sandstone/siltstone beds were principally determined by means of direct shear tests. Both in situ and laboratory tests were carried out.

In Situ Shear Tests on Coal Seam No. 1. -- The tests were all designed and located for the determination of the strength of the No. 1 coal seam. The tests were carried out in both exploratory adits. The program consisted of four in situ shear tests at the interface of the sandstone/siltstone hanging wall and the No. 1 coal seam, seven in situ shear tests in the No. 1 coal seam, and finally, four in situ shear tests at the interface of the No. 1 coal seam and the sandstone/siltstone footwall. Tests were carried out at the sandstone/coal and coal/sandstone contacts to determine whether planes of weakness existed at these interfaces.

Test Details - Sandstone/Coal Contact. -- The top half of the shear specimen was obtained, in the form of a 12 in. by 18 in. block of rock, by coring slots around the edges of the block down to the coal contact. The slots were prepared by drilling a series of overlapping 3 in. diameter holes using an electric coring rig and a specially fabricated template.

To provide the shear loading, steel flat jacks (typically three) were placed in series in the upper slot and grouted in place using a plaster of paris mix. The normal load was applied by means of a hydraulic loading strut and a load spreader resting on the sandstone block. The overall testing arrangement is shown on Fig. 5.

A given normal load (to represent a typical in situ stress condition) was first applied. The flat jacks were then pressurized from a hydraulic source and incremental jack pressure and deflection readings were recorded. The shear, normal, and lateral displacements of the rock block were measured using dial gauges attached to reference locations by magnetic bases (see Fig. 6).

Throughout the test, the normal load was kept constant by feeding or bleeding the hydraulic system. A steady shear loading rate was maintained at a rate of approximately 1 in. per hour.

The first test on each sample was termed an 'unsheared' test, since it was assumed that no prior shear displacement of the surfaces had occurred. The tests were all carried out 'downslope' since this mode was most representative of a potential failure condition.

On completion of the 'unsheared' test, the rock block was restored to its original position by jacking under zero normal load and tests were carried out in the 'sheared' state. Up to four 'sheared' tests were performed for each sample at various normal stresses. The tests carried out with the material in the post-sheared condition were designated sheared 1, 2, 3, etc.

Test Details - Coal/Coal. -- The in situ tests carried out on the coal were a modified form of those used on the contact. The upper half of the shear sample was prepared by jacking a specially prepared knife edge box into the prepared coal face to a depth of approximately 8 in. (slightly less than the depth of the box). As the box was pushed into the coal, the surrounding coal was gradually removed to a level corresponding to the cutting edge of the box in an attempt to minimize disturbance of the upper coal sample (see Fig. 7). After the jacking operation was completed, the void remaining in the box was grouted using a plaster of paris mix. The normal load was applied to the steel box as previously

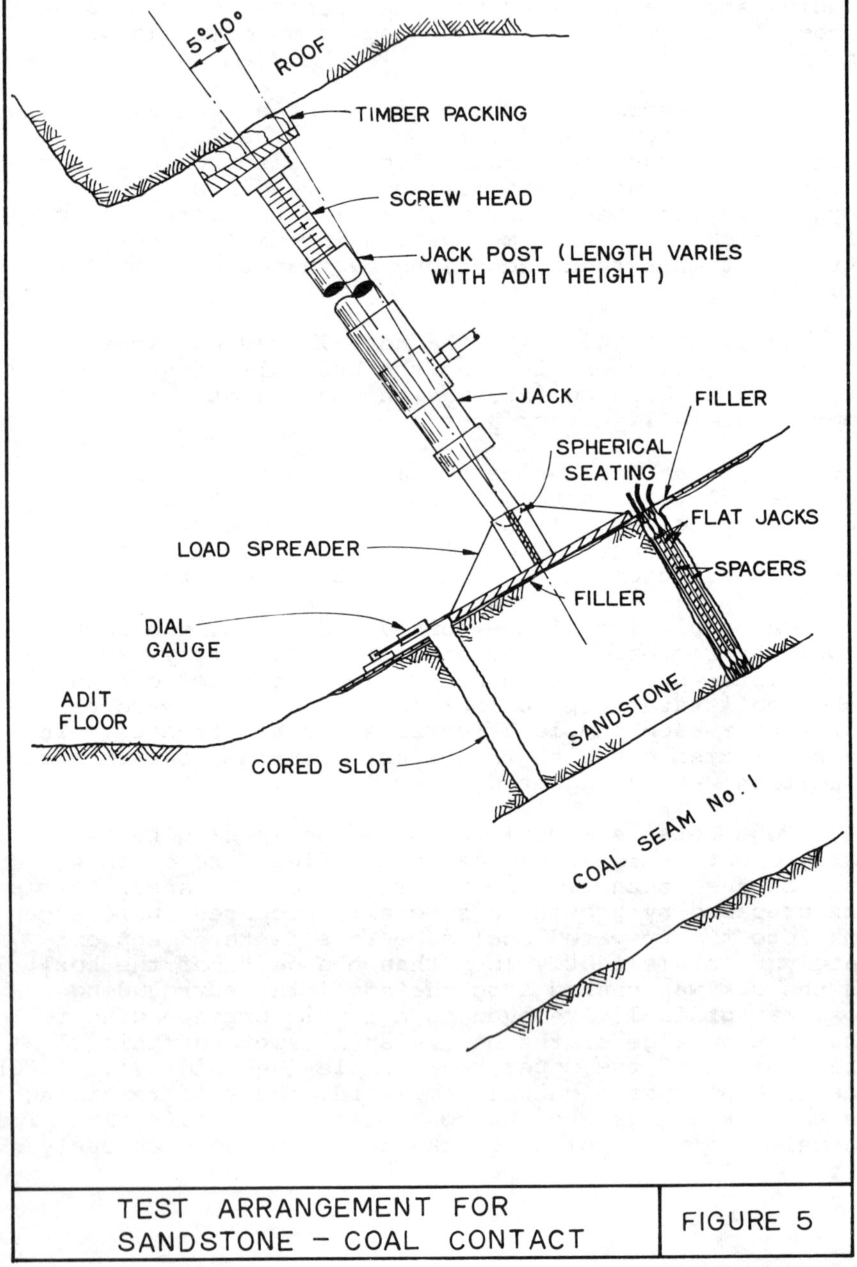

TEST ARRANGEMENT FOR SANDSTONE – COAL CONTACT

FIGURE 5

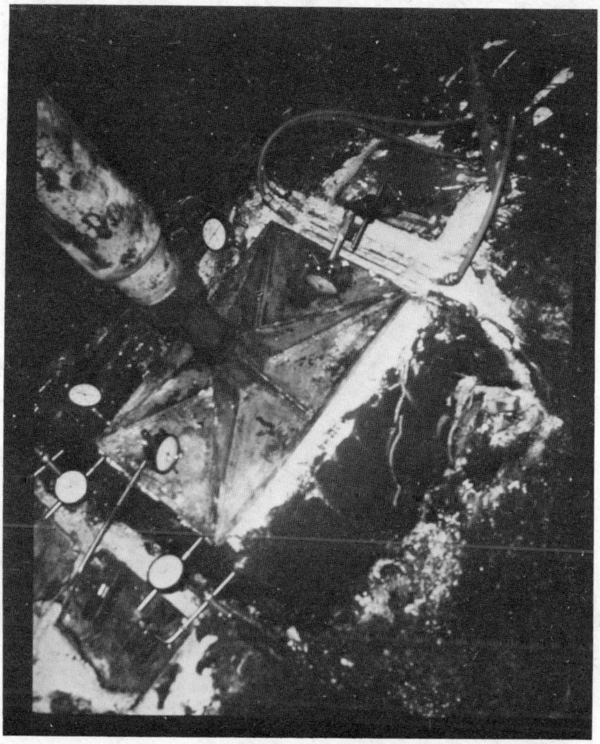

IN SITU DIRECT SHEAR TEST | FIGURE 6

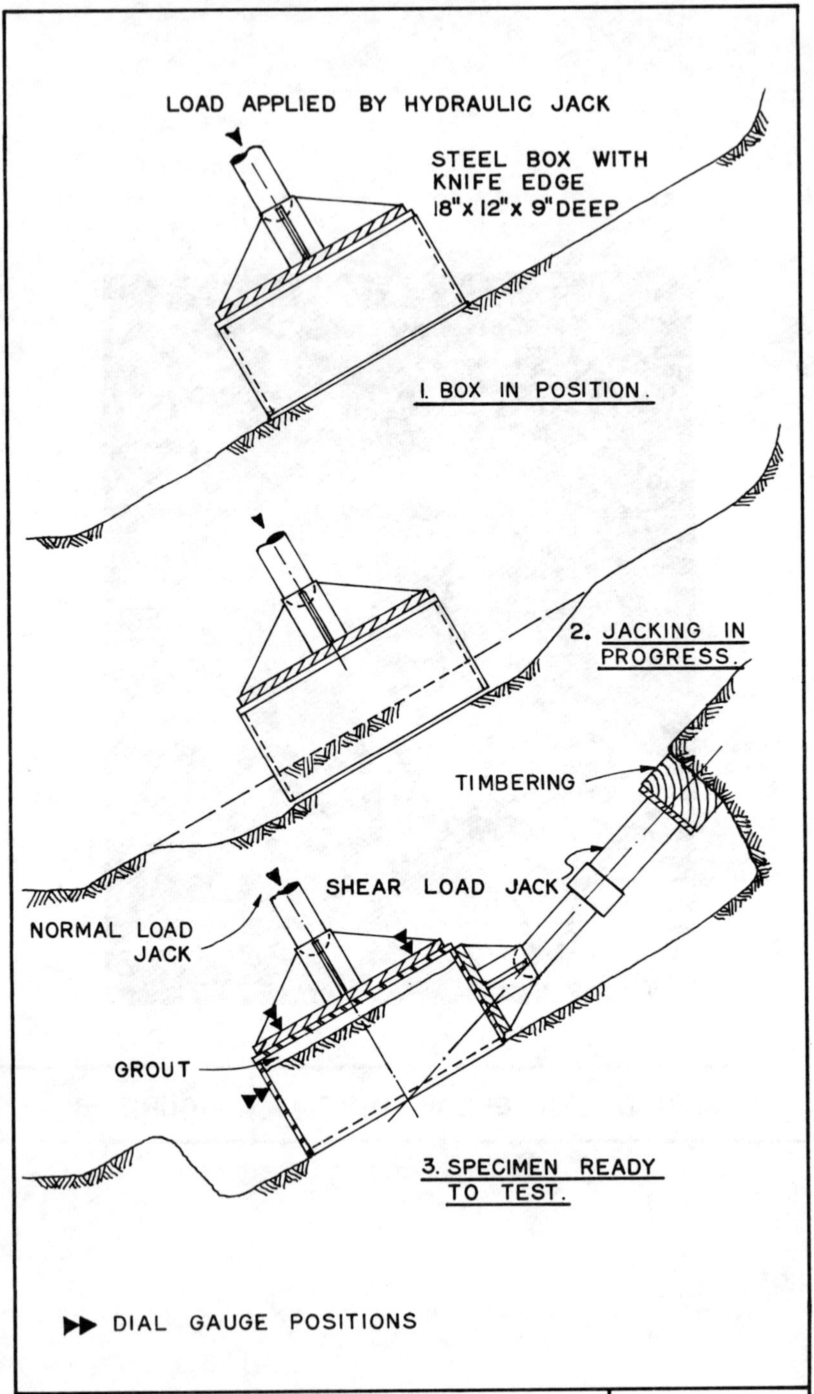

PREPARATION OF COAL SAMPLES FOR SHEAR TESTING	FIGURE 7

outlined. The shear load was applied to test specimen by a second 25-ton jack via the top end of the box. The instrumentation of the shear tests was the same as that for the sandstone/coal contact tests.

The loading rates for the tests carried out in the A Adit were the same as those used for the tests on the sandstone/coal contact. The effects of loading rate and moisture content were examined in some detail during the testing in Exploratory Adit B.

Tests in Adit B were carried out with the coal in both natural and saturated states. Saturation of the sample was carried out for approximately 12 hours prior to any loads being applied. (Following application of the normal load, a considerable quantity of water was observed to flow from the sample at the approximate shear plane location.) Load increments corresponding to approximately one-twenty-fifth of the peak load were applied at 15 minute intervals during the tests to allow any excess pore pressures to dissipate. The load was recorded both before and after each load application.

For each test set-up, an 'unsheared' test was initially performed followed by three to five 'sheared' tests. All tests were performed by shearing the upper half of the specimen downslope. The overall shear displacement during the test was of the order of 1.5 in.

Test Details - Coal/Sandstone Contact. -- The procedure for these tests was essentially the same as the one adopted for the coal/coal tests. At a given test location, the depth to the contact was accurately established and the shear box was then jacked down to within one-half inch of the surface to prevent interaction of the lower edge of the box with the rock during shearing. The remaining void at the top of the box was then grouted as before. The loading rates for these tests corresponded to those adopted for the sandstone/coal contact.

Laboratory Tests - Coal. -- Laboratory tests were conducted on bulk samples of the coal taken from the two adits. Shear tests were carried out at varying moisture contents, preconsolidation loads, and loading rates. The 25-ton capacity shear box used for these tests is shown on Fig. 8. (The sample size was approximately 18 in. by 12 in.)

The coal was placed in the shear box in loose lifts, some 2 to 3 in. in thickness, and then compacted in place. In the region of the shear plane, the coal lumps were oriented to represent their in situ attitude to a potential failure parallel to the seam. Only one test per

25 TON LABORATORY DIRECT SHEAR APPARATUS

FIGURE 8

sample was performed, the maximum travel being approximately 1 in. The normal load was applied directly to the coal via a load distribution platen. In order to minimize friction P.T.F.E. (teflon) strips were used to separate the two halves of the box during the tests.

Normal and shear stress corrections were made to compensate for the varying contact area.

Strength Evaluation of Sandstone Joints. -- Since potential slope failures would generally be controlled by the strength of the rock joints as well as the strength of the coal seams, laboratory tests were carried out to measure the shear strength of the sandstone joints that would be critical to stability.

Samples of the sandstone above the No. 1 coal seam were tested using conventional shear testing techniques. Several small scale shear tests were initially run as pilot tests. One of these tests was designed to measure the inherent frictional strength of sandstone, ϕ_p. These were supplemented by five tests on samples taken from 10 in. diameter cores.

The 10 in. diameter cores were obtained from the walls of the adits using a heavy duty coring rig (see Fig. 9). The specimens were then trimmed and mounted in the 25-ton shear box (see Fig. 8) using a plaster of paris casting.

For each specimen, one 'unsheared' and up to five 'sheared' tests were performed at varying normal stresses. The relative shear and normal displacements of the two halves of the specimen were recorded for varying shear stress values. Following each test, the quantity of 'gouge' or rock flour (resulting from grinding of the asperities during shear) formed on the surface was observed and recorded. The surface profiles of the joints were also recorded before and after testing.

Because the relatively competent sandstone is interbedded with the weaker more deformable coal seams, the overall stiffness of the failing mass will be low. It is estimated that the failure would be controlled by the shear strength of blocks having dimensions of the order of 20 to 30 ft. Owing to the effects of joint roughness and large scale irregularities, the joint strength usually increases with sample size.

The effect of joint irregularities is to increase the angle of frictional resistance above the basic frictional

CORING JOINT SAMPLES FROM ADIT WALLS

FIGURE 9

strength values.[1,2] The effect of the irregularities can be seen in the simplified roughness model shown on Fig. 10. The effective angle of shearing resistance, for no irregularities is ϕ_b. The effect of an irregularity with an inclination ϕ_i to the mean shear plane is to increase the inherent angle of shearing resistance by the amount ϕ_i to $\phi_b + \phi_i$. This phenomenon is directly related to the inclined plane effect caused by the asperities.[2]

The value of ϕ_i was estimated by measuring the path of the top half of the specimen relative to the bottom. Thus the angle of travel of the top half of the specimen relative to the mean failure plane could be determined. Using this approach, the influence of varying roughness on the strength of different samples could be ascertained.

The test results were finally corrected and modified to the full scale conditions using the joint profiles measured on site. From these data, a realistic estimate for the strength of the joint surfaces in the rock was obtained.

Mean Strength of Coal. -- Since the laboratory tests were carried out on disturbed bulk samples primarily to assess the effect of moisture variation, the results from these tests were not used in the evaluation of the design strengths. (As expected, the strengths obtained from the laboratory tests were less than the comparable in situ strengths.)

The mean in situ strengths were used for estimating the stability of the footwall slopes. It was observed from the tests that there was considerable strength variation within the coal. For slope failures of a significant magnitude, the assumption was made that the strength mobilized over large areas will be comparable to the mean strength obtained from the tests.

Even under 'sheared' test conditions (sample previously subjected to shearing) it appears that there is considerable apparent cohesion (component of shear strength that is independent of normal load) for the stress range of 40 to 180 p.s.i. The value of this apparent cohesion was found to be approximately 12.5 p.s.i. Thus from the test results for the normal stress range of 40 to 180 p.s.i. the shear strength of the coal in p.s.i.

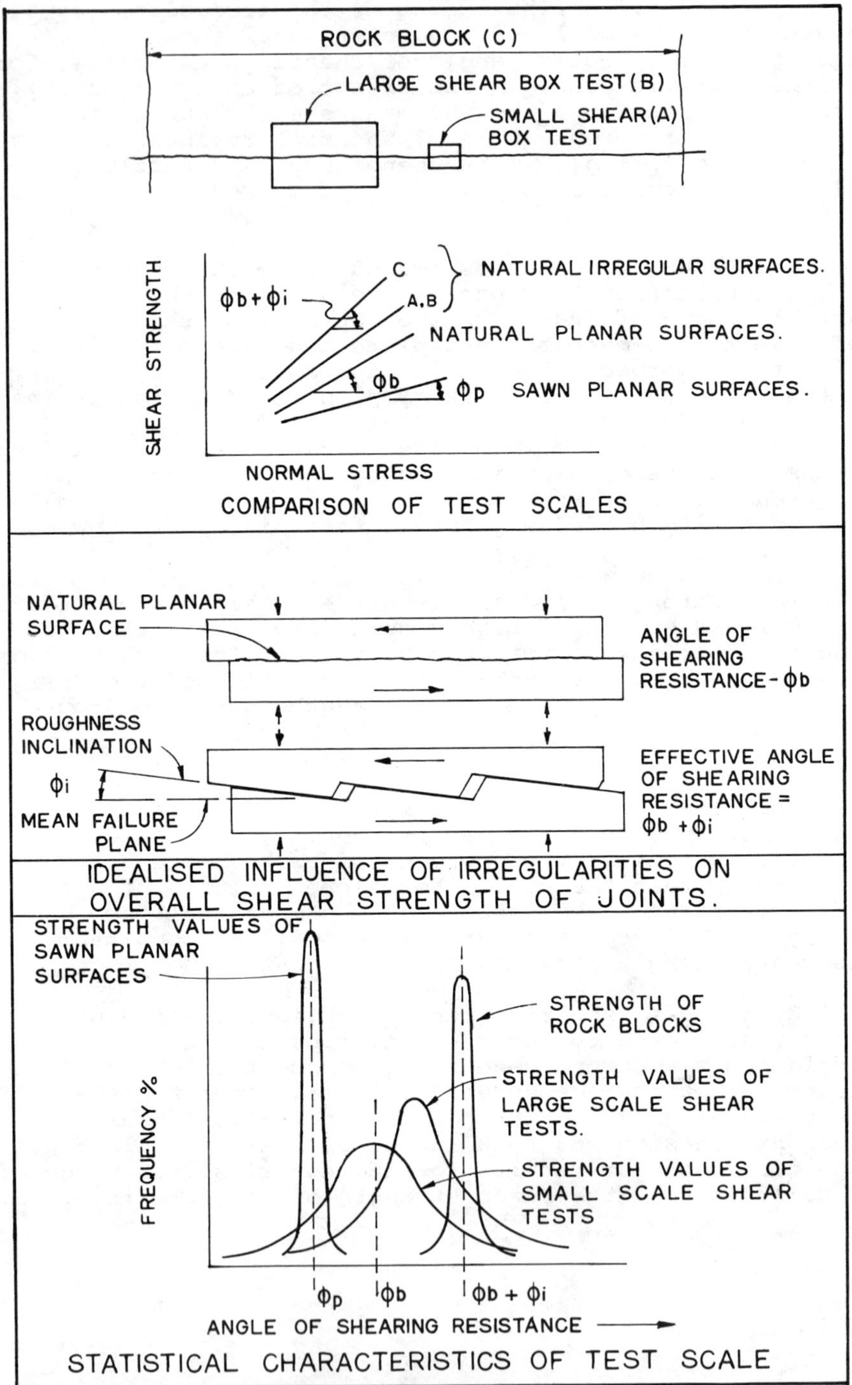

EFFECT OF TEST SCALE AND SURFACE IRREGULARITY ON STRENGTH OF ROCK JOINTS

FIGURE 10

is given by:

$$\tau_{max} = 12.5 + \sigma_n' \tan 26°$$

apparent cohesion — frictional component

where σ_n' is the effective normal stress

26° is the effective angle of shearing resistance derived from in situ tests.

Mean Design Strength Values for Sandstone. -- The shear test results were combined with the data on profiles and surfaces obtained in the field to produce mean strengths for design purposes.

Owing to deformations within the slope that would occur during excavation and which would be required to mobilize the shear strength of the coal, 'residual' strength parameters (which assume pre-shearing of the order of inches) were assumed. This assumption, to some extent, also takes into account the detrimental effect of blasting, which may cause localized overstressing leading to 'residual' strength conditions.

The average strengths from the 'sheared' tests for the sandstone joints were found to be represented by an angle of shearing resistance of 35.5 deg. (see Table I). The inherent strength of sawn planar sandstone surfaces (ϕ_p) was found to be approximately 25 deg. (angle of shearing resistance). The difference between the two strengths demonstrates clearly the influence of roughness on strength values. The major irregularities accounted for an increase in strength corresponding to an average angle of shearing resistance of 6.5 deg. for all samples. The remaining 4 deg. was accounted for by intermediate roughnesses of dimensions corresponding to several grain diameters.

Studies of the overall joint topography in the sandstone show that, for joint blocks having failure surfaces of the order of 20 to 30 ft., a lack of planarity from an assumed mean failure plane of 1 to 2 ft. could be expected. This lack of planarity is due to large scale roughnesses on the joint planes and the influence of joint intersections (shallow dipping cross joints intersecting bedding planes). This lack of planarity corresponds to an increase in the angle of shearing resistance (ϕ_i) of some 2 to 4 deg. over the average test values obtained from the shear tests.

The mean strength of actual joints was determined by adding a value of 2 deg., to account for the overall lack of planarity of large blocks, to the average test angle of shearing resistance of 35.5 deg., giving a strength value of 37.5 deg. An apparent cohesion of 3.5 p.s.i. was neglected for design purposes. The results quoted are valid for the normal stress range of 50 to 500 p.s.i., which covers the stress conditions that are liable to control significant potential failures.

Design Strength of Thrust Shears Within the Sandstone. -- It is estimated that the shear strength of the thrust shears would be considerably less than that of sandstone joints. Observations indicate that the major thrust shear features would be approximately planar owing to the significant shear displacement that is likely to have occurred during their formation. It is expected that such features would thus lack irregularity and may be infilled with sandstone gouge. The average value from the sandstone direct shear tests, minus the measured strength due to roughness, was assumed for the design strength of the shears. This corresponds to the value of 'ϕ_b' on Table I of 29 deg. It was also concluded that no cohesion should be assumed for such shears owing to their probable displacement history.

STABILITY ANALYSES

Model of Potential Footwall Failures. -- The stability of a slab of material of varying dimensions overlying No. 1 coal seam was determined to simulate footwall stability conditions. For these analyses, the following assumptions were made: (a) Slope profile approximately parallel to bedding; (b) Coal seams are approximately planar; and (c) The slope is drained.

The analyses were carried out to determine what strengths were required within the overlying sandstone to support a slab of material of length (L) and depth (D), assuming basal failure through the coal seam (see Fig. 11a). Dip values (α) for the coal seam of 35, 40, and 45 deg. were considered.

For potential failure of the main slab section shown diagrammatically on Fig. 11, the ultimate strength of the basal coal seam and the sandstone joint must be exceeded. Examination of the mode of failure and the relative strengths of the sandstone and coal showed that in general the basal coal seam reached its ultimate strength before the sandstone joint, with the result that the strength of the sandstone joint was being used to resist the downward force exerted by the slab. For any

TABLE I. -- ELK RIVER SANDSTONE RESIDUAL JOINT SHEAR STRENGTH PARAMETERS

Large Shear Box Sample Number	Small Shear Box Sample Number	Apparent Cohesion in pounds per square Inch	Measured Angle of Shearing Resistance (ϕ_r)	Geometrical Component of Shearing Resistance ϕ_i*	Basic Angle of Residual Shearing Resistance $\phi_b=\phi_r-\phi_i$
1 L		8	32.0°	2.4°	29.5°
2 L		0	32.0°	6.0°	26.0°
3 L		0	37.0°	4.6°	32.5°
4 L		0	32.0°	8.5°	23.5°
5 L		16	36.0°	3.6°	32.5°
	1 S	0	39.0°	9.4°	29.5°
	2 S	0	40.0°	10.0°	30.0°
AVERAGE VALUES		3.5	35.5°	6.5°	29.0°

*Obtained from normal displacement-shear displacement plots

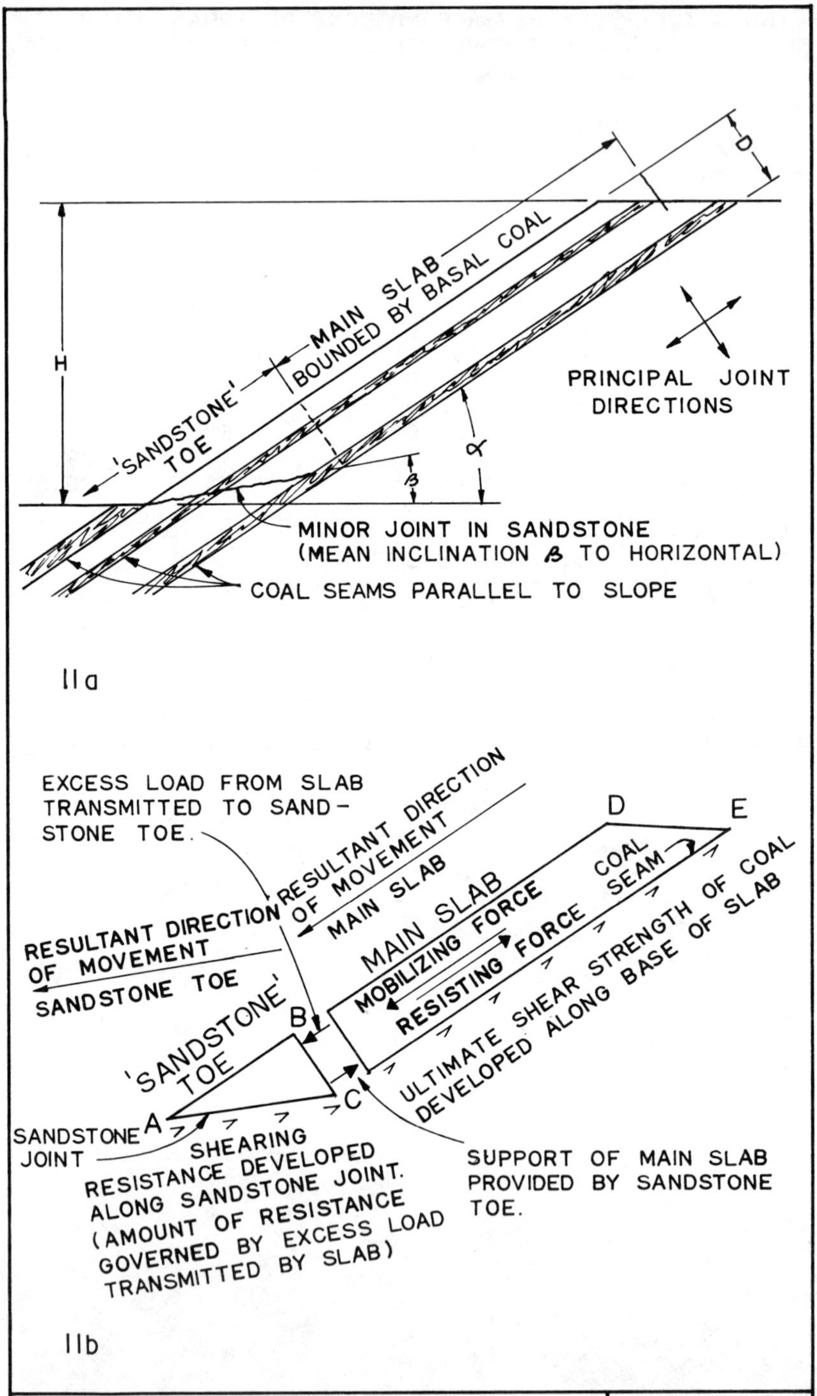

FOOTWALL FAILURE MODELS | FIGURE 11

given configuration, this downward force is the excess load, transmitted from the slab onto the sandstone toe, that the basal coal seam is unable to support. The resultant force system is shown diagrammatically on Fig. 11b. Under this configuration, the toe section ABC will be stressed in excess of the normal overburden load by the forces transmitted from the main slab.

The resultant directions of movement of the main slab section and the sandstone toe are shown on Fig. 11a. Owing to the assumed geometrical configuration, these are not coincident and shearing along the dominant joint set BC is required in the vicinity of the main slab-sandstone toe contact. The additional resistance due to this 'internal' shearing was not taken into account in the analysis, thereby rendering the analysis slightly conservative. For large movements of the main slab BCDE, considerable buckling and internal shearing would be required to conform to the failure plane. These effects in turn would provide further resistance against failure.

Stability Charts for Slope Design. -- The analyses were used to prepare design charts for the proposed slope profiles. The charts were used to plan the excavation program to ensure that unstable conditions were not developed. Because the coal derives part of its strength from frictional resistance and part from cohesion, the stability of any given slab, with respect to failure through an underlying coal seam, is a function of the average normal stresses on the failure plane and thus the slab depth. Acceptable and unacceptable slope designs, involving different slab thicknesses, were determined from the design charts. The analyses of slab stability were assumed to be representative for slab length to depth ratios greater than about 4.

Using the mean design strength parameters for coal seam No. 1, it can be shown that uniformly thick slabs of rock overlying the coal to a certain depth are stable and do not rely on strength being developed within the sandstone toe of the slope (see Fig. 14). Over this depth, shearing resistance must theoretically be developed within the sandstone cross-joints in order to support the upper rock slab.

The results are presented in terms of the angle of shearing resistance along critical joints in the sandstone (a direct function of the shear strength) that is required to be mobilized in order to support a given slab of sandstone and coal. The design strengths were obtained by applying a reduction factor of 1.1 to the mean shear strength values. For α-β values (see Fig. 11) between about 5 and 15 deg., the maximum design angle of shearing resistance is therefore approximately 35 deg. For α-β equal to about 20 deg., the equivalent value is approximately 27 deg. owing to the presence of lower strength shears at this orientation.

Since for slabs greater than a certain depth, the strength of coal seam No. 1 will be fully mobilized and additional support is required within the sandstone toe, the stability criterion has been related directly to the strength of the sandstone joints. Prior to using the main design charts, a working factor of safety with respect to failure along the relevant sandstone joint (which in turn controls the stability of the whole slab) should be established.

For any given configuration, the shear strength required to be mobilized along the sandstone joint is compared with the available resistance to determine the factor of safety. After a factor of safety has been selected, the corresponding working shear strength (ϕ'_R) for the sandstone values are obtained. This value is then assumed to be the cutoff value for a safe or unsafe stability condition.

Fig. 12 shows a plot of H and L vs. D for various values of the angle of shearing resistance ϕ'_R along potential failure surfaces through the sandstone and a slope angle of 40 deg. Four plots are given for differing values of slope failure conditions (α-β). The curves are split into two distinct regions by the value of ϕ'_R selected for a given factor of safety. Other curves for slope angles of 35 and 45 deg. were also developed.

Explanation of Stability Charts. -- Consideration is given in this example to the slope depicted in Fig. 13A, where the coal seams are at 40 deg. and the joint plane along which failure is expected has a β value of 25 deg. It is assumed that the joint plane passes through the slope toe, since this represents the worst condition. The design angle of shearing resistance (ϕ'_R) available along the joint plane is 35 deg., with no apparent cohesion; while coal seam No. 1 has a design angle of shearing resistance of 26 deg., with apparent cohesion of 1,800 p.s.i.

The feasibility of excavating this slope at various design factors of safety can be assessed by using Fig. 13B in conjunction with Fig. 13C. (Fig. 13B shows

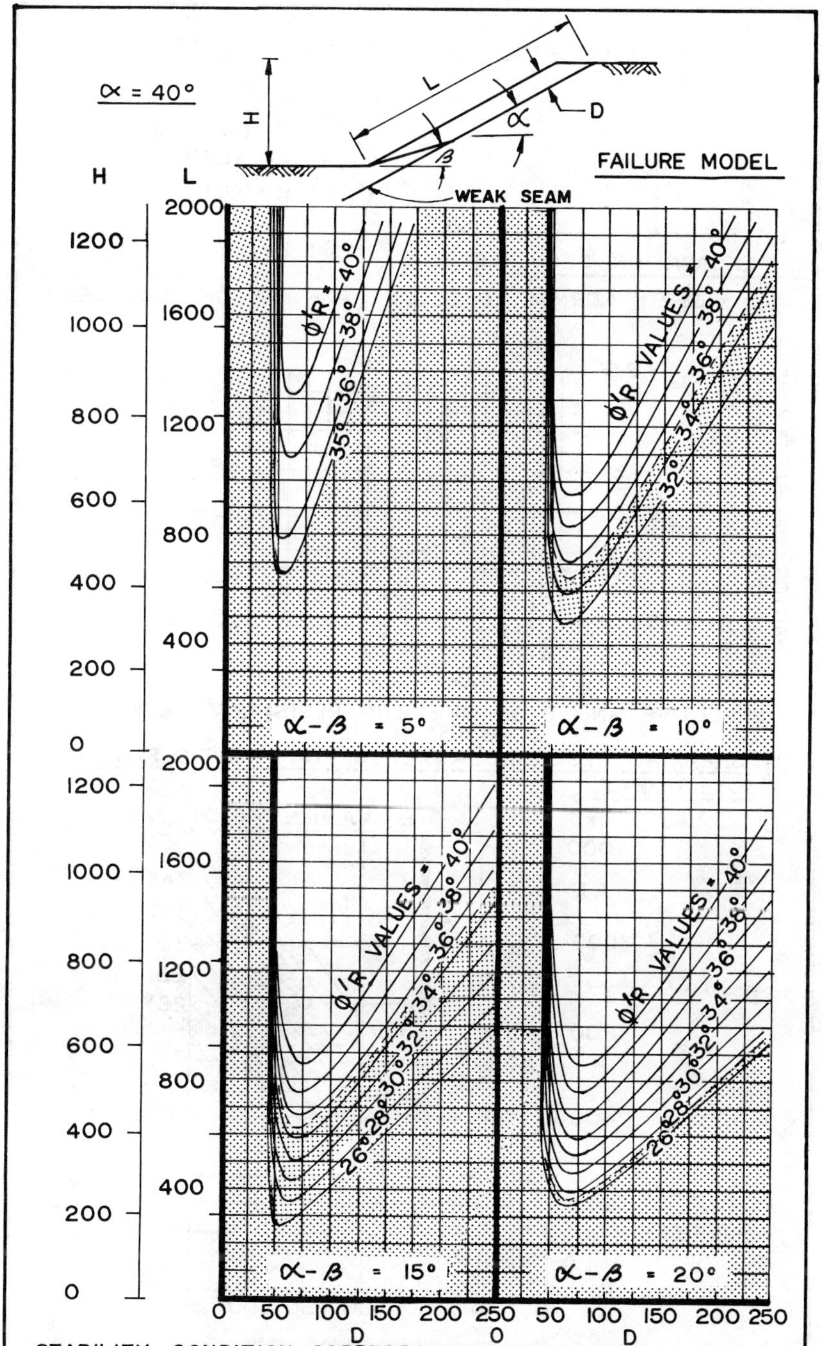

STABILITY CONDITION CORRESPONDING TO A FACTOR OF SAFETY GREATER THAN 1.0 SHOWN BY SHADED AREA.

DESIGN CURVES FOR STABILITY OF SINGLE UNIFORM THICKNESS SLAB OVERLYING COAL	FIGURE 12

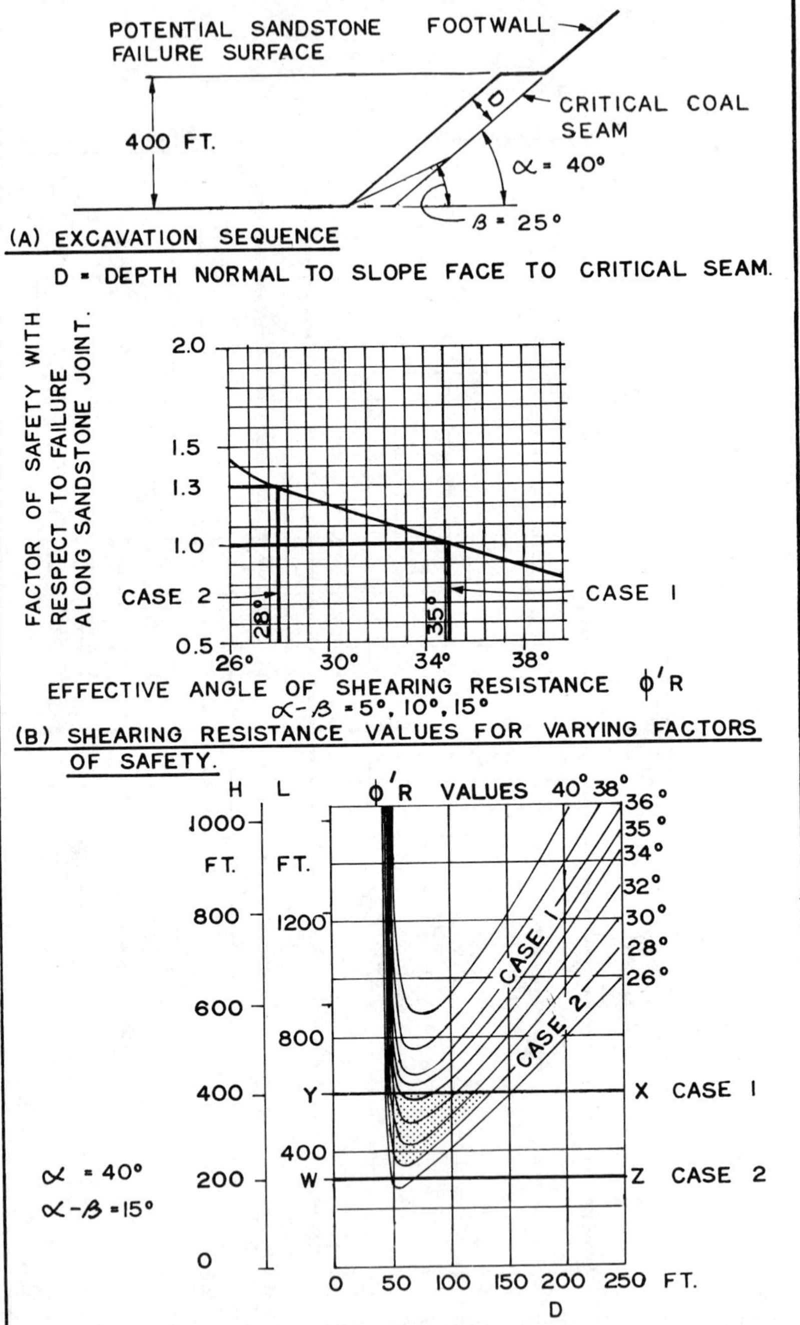

STABILITY ANALYSIS TYPICAL EXAMPLE

FIGURE 13

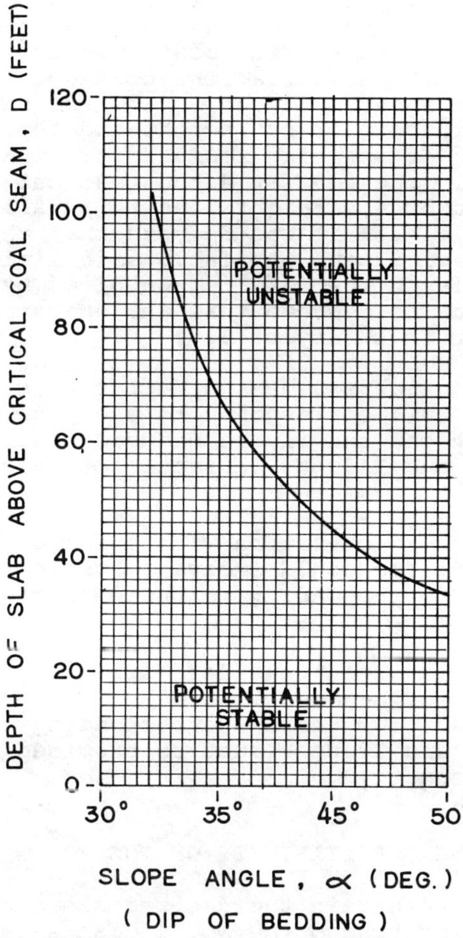

(SLAB COMPLETELY UNDERCUT NO TOE SUPPORT.)

VARIATION IN THICKNESS OF STABLE SLAB WITH SLOPE ANGLE

FIGURE 14

the relationship between the design factor of safety and the corresponding working angle of shearing resistance to be used in Fig. 13C).

Initially, (Case 1), it is required to examine the feasibility of excavating a 400 ft. high slope parallel to the bedding at an assumed design factor of safety of 1.0. Fig. 13B shows that the corresponding angle of shearing resistance is 35 deg. Fig. 13C shows the equilibrium relationship of cut height and depth of coal at various values of shearing resistance. The operation of removing all the coal at a constant cut height can be represented by a straight line XY parallel to the 'D' axis. From Fig. 13B it was deduced that the curve of ϕ'_R equal to 35 deg. is the available safe limit. Since XY lies completely in the safe region given by ϕ'_R equal to 35 deg., it can be deduced that the operation has at least a factor of safety of 1.0.

Case 2 considers a design factor of safety of 1.3. From Fig. 13B this corresponds to a ϕ'_R curve on Fig. 13C of 28 deg. It can readily be deduced that for a depth D of between 50 and 135 ft., a factor of safety less than 1.3 will result.

To produce an operation with a factor of safety of 1.3 or greater, one could excavate the slope in two 200 ft. lifts corresponding to the condition WZ on Fig. 13C.

The example given considers (for simplicity) only the condition for α-β equal 15 deg. To satisfy all conditions, the four charts corresponding to α-β equal to 5, 10, 15, and 20 deg. must be examined for the proposed slope geometry.

MAINTAINING THE STABILITY OF THE FOOTWALL BY DRAINAGE

In order to attain the desired slope profiles, water pressures acting behind the No. 1 Coal seam must be dissipated by drainage.

Where slopes are non-homogeneous, the drainage system must be adapted to suit the permeability conditions. In this case, groundwater studies indicated that the flow through the sandstone sequences was inhibited by the damming effect of the coal seams. This was attributed to the differential between the permeabilities of the coal, and the sandstone, giving rise to highly anisotropic condition within the coal sequence. A typical equipotential distribution for this type of situation, compared with a homogeneous slope, is shown on Fig. 15.

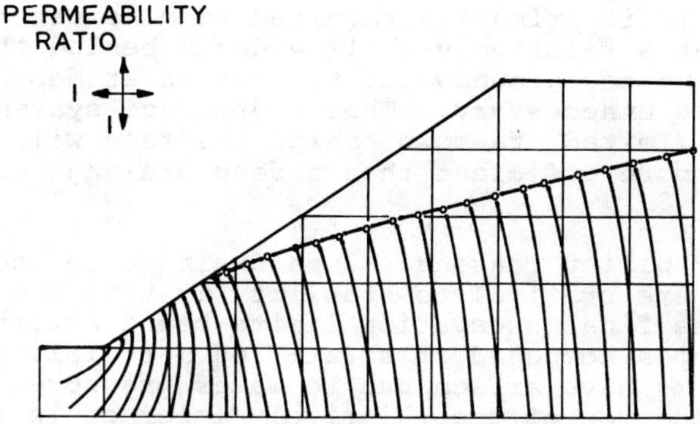

(a) ISOTROPIC ROCK SLOPE

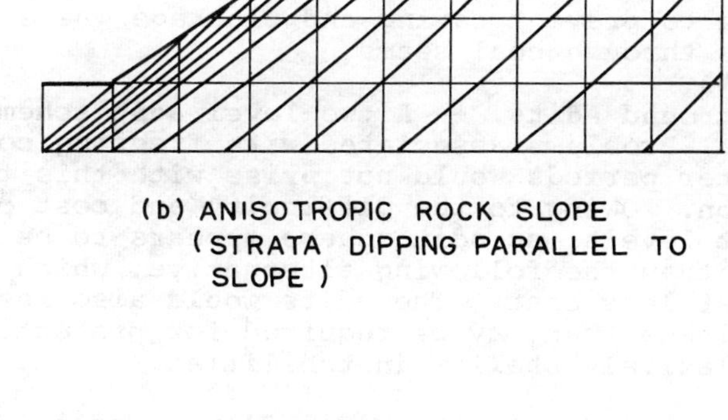

(b) ANISOTROPIC ROCK SLOPE (STRATA DIPPING PARALLEL TO SLOPE)

* LINES OF EQUAL HYDRAULIC POTENTIAL.

EQUIPOTENTIALS* FOR VARIOUS SLOPE CONDITIONS	FIGURE 15

Generally, four alternative drainage systems are available:[3] (1) Horizontal drainholes drilled from the surface; (2) underground adits within the slope; (3) wells drilled into or behind the slope; and (4) down-slope surface drainage trenches.

Since it is primarily required to dissipate water pressures at a relatively shallow depth behind the foot-wall, drawing the groundwater table down at depth behind the slope is unnecessary. Thus a drainage system that promotes a limited drawdown behind the face will be inherently more efficient than a deep drainage alternative.

For stability reasons, slope drainage behind coal seams that are critical to stability must be operative prior to the final excavation limits being attained. In similar slopes composed of stratified material, considerable problems have arisen due to water pressures building up behind the toe of the relatively impermeable seams. If excavation of material occurs such that only a small depth of cover is left above the seam, localized failure at the toe due to the effect of water pressure, may occur resulting in loss of support for the remainder of the slope.

COMPARISON OF AVAILABLE SCHEMES

Horizontal Drainholes Drilled from the Surface. -- As far as this particular site is concerned, the primary disadvantage of this method is that drainage, prior to final excavation, cannot readily be achieved. The second main problem is the collection of water from the drains and ensuring that blockage due to freezing does not occur during winter conditions. Some form of drainpipe would be required to prevent caving and blockage where the hole passes through coal seams.

Underground Adits. -- A two-level adit scheme was considered. Problems associated with freezing conditions during winter periods would not arise with this type of construction. Owing to the difficulty and cost of access to the adit levels, an adit scheme appears to be less attractive than the following alternative, which can be installed at less cost. The adits would also result in deeper drainage than may be required for protection against relatively shallow instabilities.

Wells Drilled into the Final Slope Profile (see Fig. 16). -- Owing to the relatively inexpensive cost of installation resulting from the availability of large scale drilling equipment, pumped wells appeared to

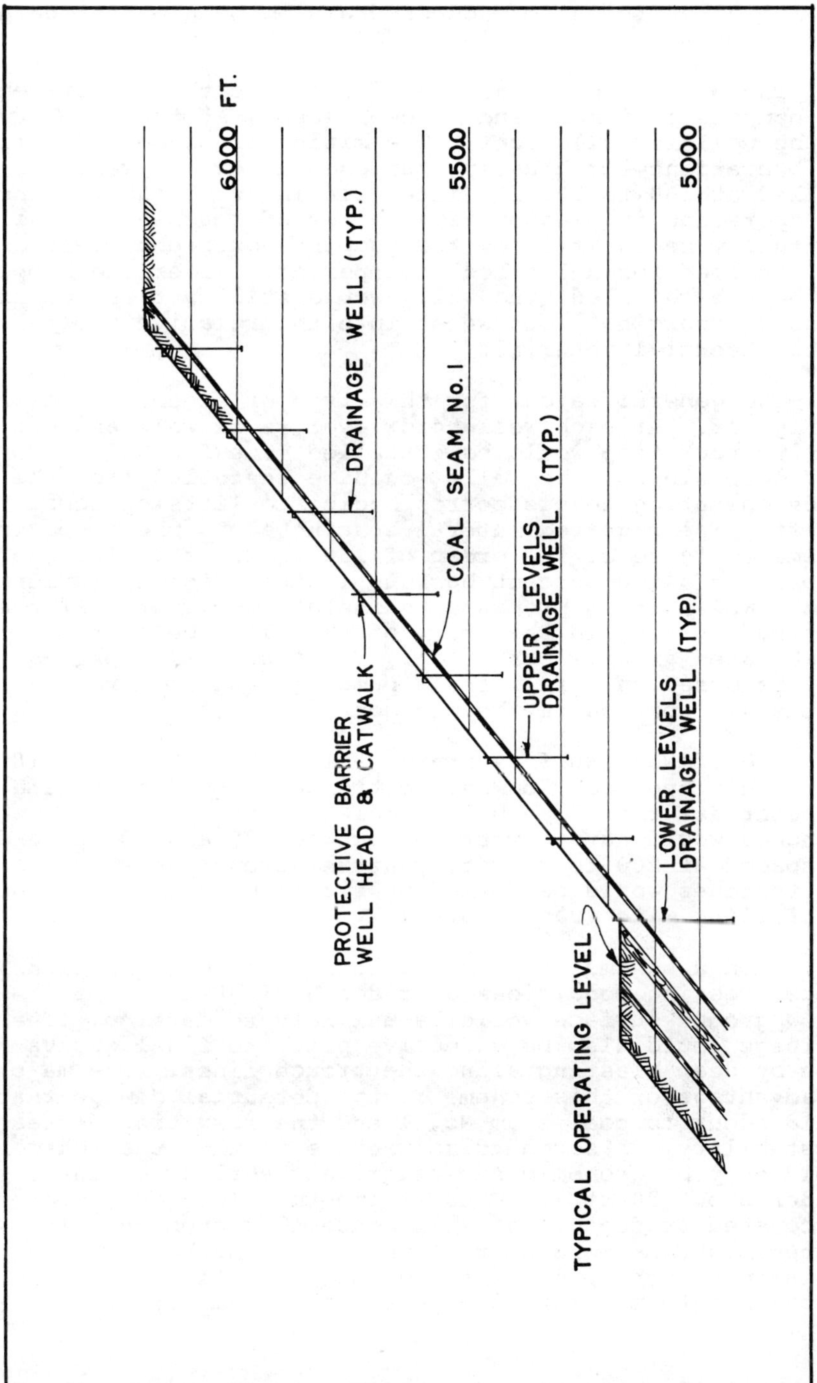

DRAINAGE SCHEME: PUMPED WELLS

FIGURE 16

provide an efficient method of draining both operating and final slopes.

The primary disadvantage of wells is the requirement for pumping equipment and a power supply at the majority of the well installations. The mining sequence and drilling operations for blasting purposes allow for wells to be drilled at 100 to 200 ft. intervals in cut height. Since the operation is planned as a series of reaches that will eventually be backfilled, the pumping equipment would only be installed for a limited time period. After the slopes had been backfilled, the wells would still act as relief wells discharging groundwater into the relatively permeable backfill material.

The general layout for this type of scheme is shown on Fig. 16. At each well head level, a catwalk and protective rock trap would be installed (see Fig. 17). For the lower slopes, the wells would be installed from the three operating levels corresponding to lifts of some 200 ft. The required lateral spacing along the slope is estimated to be of the order of 150 to 250 ft. In order to provide a 100 percent back-up in the event of failure of the well, it is proposed to install wells at half the required spacing in that part of the slope below the established groundwater level. The final well spacing will be based on a full scale pumping test at a representative location on the project.

Use of Near Surface Drainage Trenches (See Fig. 18).-- Using this approach, downslope trenches would be created to penetrate the slope below coal seam No. 1. Such trenches would have a width of between 25 and 50 ft. and be spaced at 200 to 300 ft. centres across the slope. The trenches would be formed by blasting, excavating and backfilling with graded material.

Such a scheme would be suited to winter conditions, since freezing conditions at a depth of 30 to 40 ft. below ground surface would be unlikely to develop. The drainage could also be effective prior to final excavation by deep blasting along the trench lines. The major disadvantage of this scheme is the potential damage that could occur to coal seam No. 1 and the resulting decrease in stability. A second disadvantage is that the scheme would only be economic for relatively shallow depths (under about 50 ft.). Thus if impermeable seams were discovered at depth, the well concept or even an adit system might be more attractive. On either side of the drainage trench, pre-splitting would be utilized during excavation to minimize damage to coal seam No. 1.

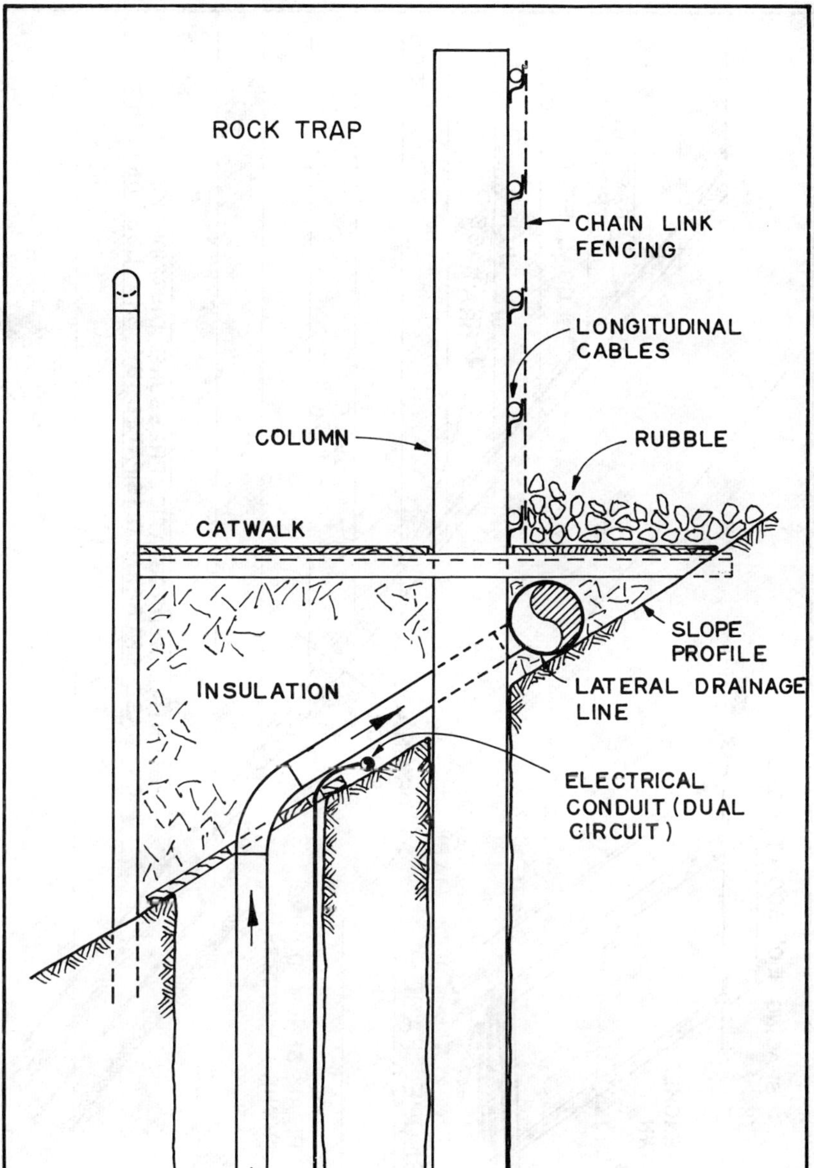

TYPICAL WELLHEAD & ROCK TRAP ARRANGEMENT — FIGURE 17

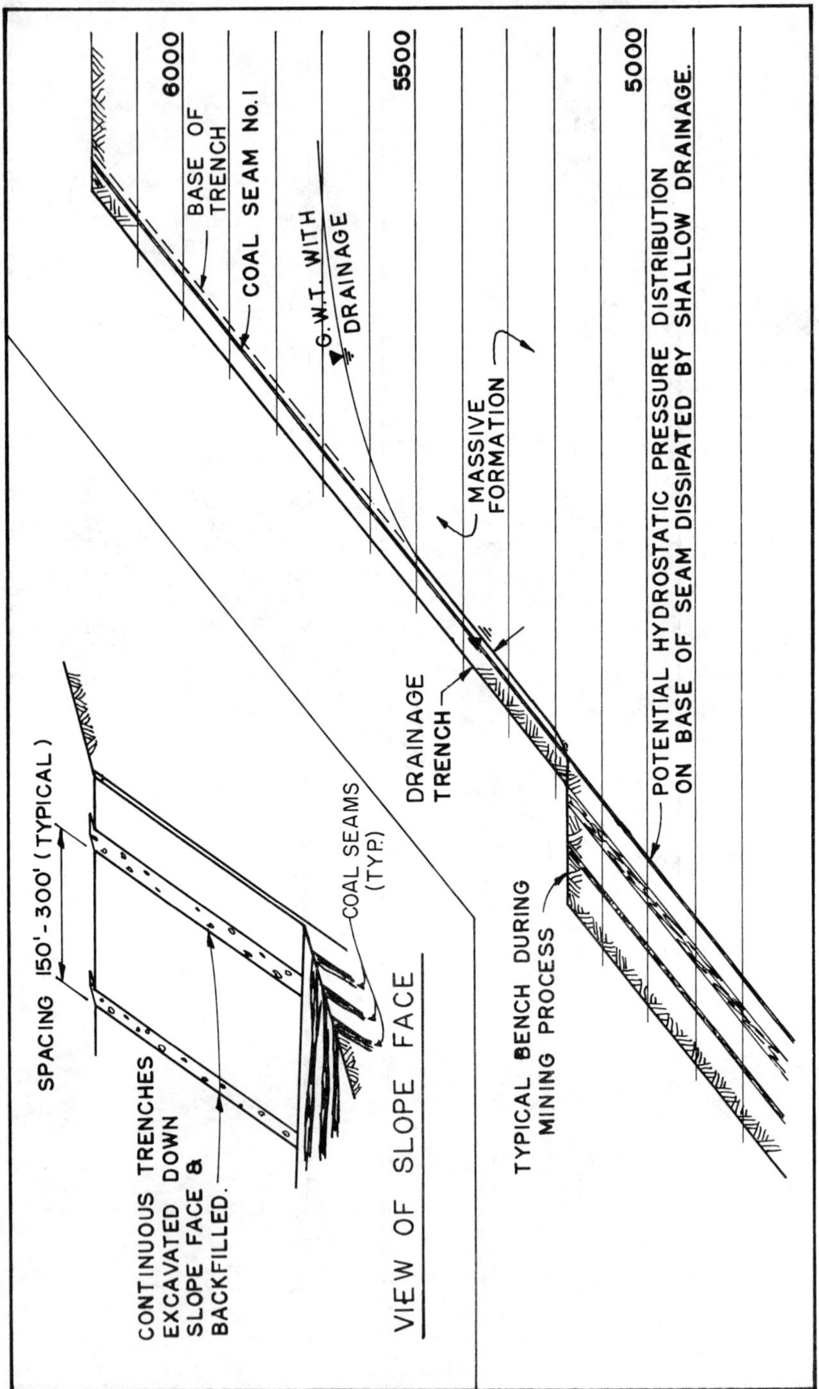

DRAINAGE SCHEME: SURFACE TRENCHES

FIGURE 18

Backfilling of the trenches could be achieved by forming buttressed arches at intervals down the trench, and backfilling up the slope behind them. The interval between arches would depend on either the angle of repose of the rock or on operational considerations. (If the angle of repose of the backfill rock is about the same as the footwall slope, the retaining arches will only be required based on operational considerations such as time of year, length of unfilled trench, etc.)

It is estimated that retaining arches could be economically constructed using segment gabions to form the arch and profiling the edges of the trench to form buttresses (see Fig. 19).

OTHER CONSIDERATIONS

As an integral part of the footwall slope design, rock traps, to retain any spalling rock and provide protection against avalanches, are proposed (see Fig. 17). The traps would be integrated with a drainage and extensometer slope monitoring system and would allow continuous access across the slope for maintenance and monitoring purposes. Because it is desirable to retain the integrity of the sandstone slab overlying coal seam No. 1, use of artificial retaining barriers in lieu of a benched slope is both economic and advantageous from an overall stability standpoint.

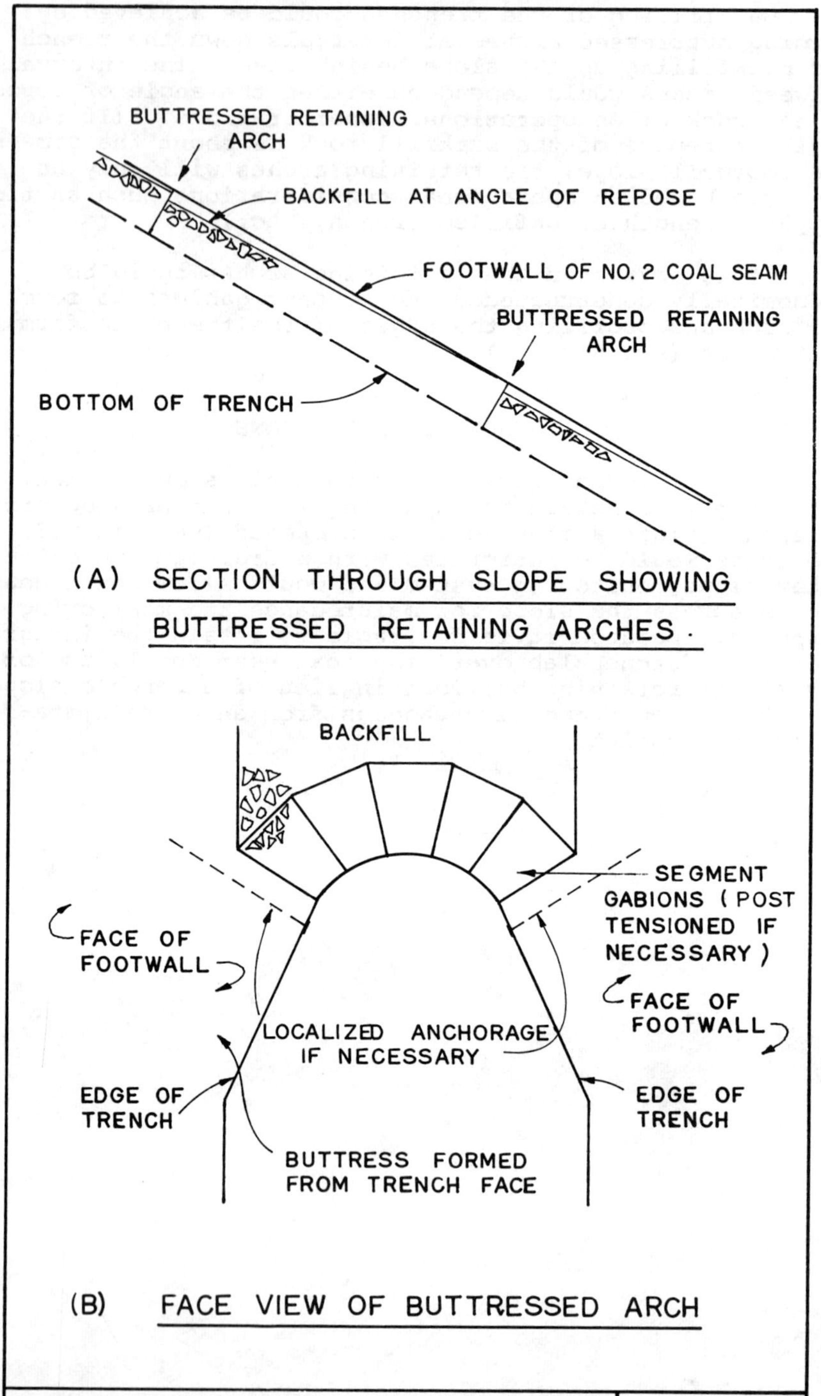

SURFACE DRAINAGE TRENCH: DETAILS OF POSSIBLE BACKFILLING OPERATION — FIGURE 19

APPENDIX 1. -- REFERENCES

1. Ripley, C.F. and Lee, K.L., "Sliding Friction Tests on Sedimentary Rock Specimens," Seventh International Congress, Large Dams, No. 4, 1961, pp. 657-671.

2. Patton, F.D., "Multiple Modes of Shear Failure in Rock," Proceedings of the First Congress on Rock Mechanics, Vol. 1, 1966, pp. 509-513.

3. Hoek, E. and Sharp, J.C., "Improving the Stability of Slopes by Drainage," Proceedings of Open Pit Mining Symposium, Johannesburg, Sept. 1970.

OPEN PIT MINE SLOPE STABILITY: THE BERKELEY PIT

by W. G. Pariseau[1] and K. Stout[2]

INTRODUCTION

The purpose of this paper is to describe the important engineering features relating to the evaluation of the stability of the Berkeley Pit a large truck haulage, open pit copper mine, owned and operated by the Anaconda Company, located near the city of Butte in southwestern Montana. Figure 1 shows the location of the city of Butte.

The Berkeley Pit is located in the east or "horsetail" area of the Butte mining district, a region of complex geology that includes multiple vein systems, numerous faults, countless fractures and several varieties of hydrothermal alteration[3]. Major structural features (widths greater than 10 feet) are related to the dominant NE (Anaconda system) and NW (Blue system) trending vein systems, the Rarus Fault (NE), and the Middle Fault system (NW). In the pit area the NE trend of the Anaconda system curves towards the east and terminates in an easterly trending en echelon pattern. The horsetail pattern is formed by SE-striking fractures that break away from segments of the en echelon vein array. Dips are steep (50-70 degrees) and southerly with the exception of the Rarus Fault which dips

---

1. Assoc. Professor, Department of Mining Engineering, Montana College of Mineral Science and Technology, Butte, Montana.
2. Professor and Head, Department of Engineering Science, Dean of the Division of Engineering, Montana College of Mineral Science and Technology, Butte, Montana.
3. Meyer, C. et al "Ore Deposits at Butte, Montana" Ore Deposits in the United States, 1933/1967 (ed. J. D. Ridge), Vol. 2., AIME, N. Y., 1968, pp. 1373-1466.

45 degrees to the northwest. Figure 2 shows the major structural features of the district which also include large igneous dikes. A zone of surface oxidation caps the ore body and extends from surface to depths of 200 to 300 feet. Several hundred feet of water-bearing alluvium overlie the eastern portion of the ore zone. Underground workings, shafts, drifts, raises, and filled stopes are extensive and account for 5 percent of the pit area[4].

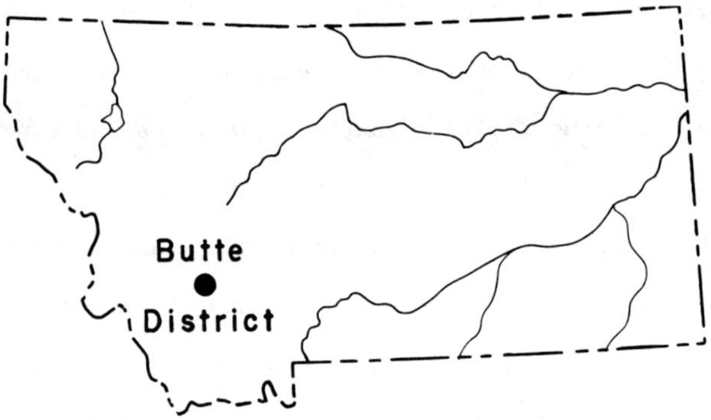

Figure 1. Location of the Berkeley Pit near Butte, Montana.

4. Goddard, C. C. Jr., "Berkeley Pit History and Geology", Mining Engineering, Vol. 11, N. 3, March 1959, pp 290-292.

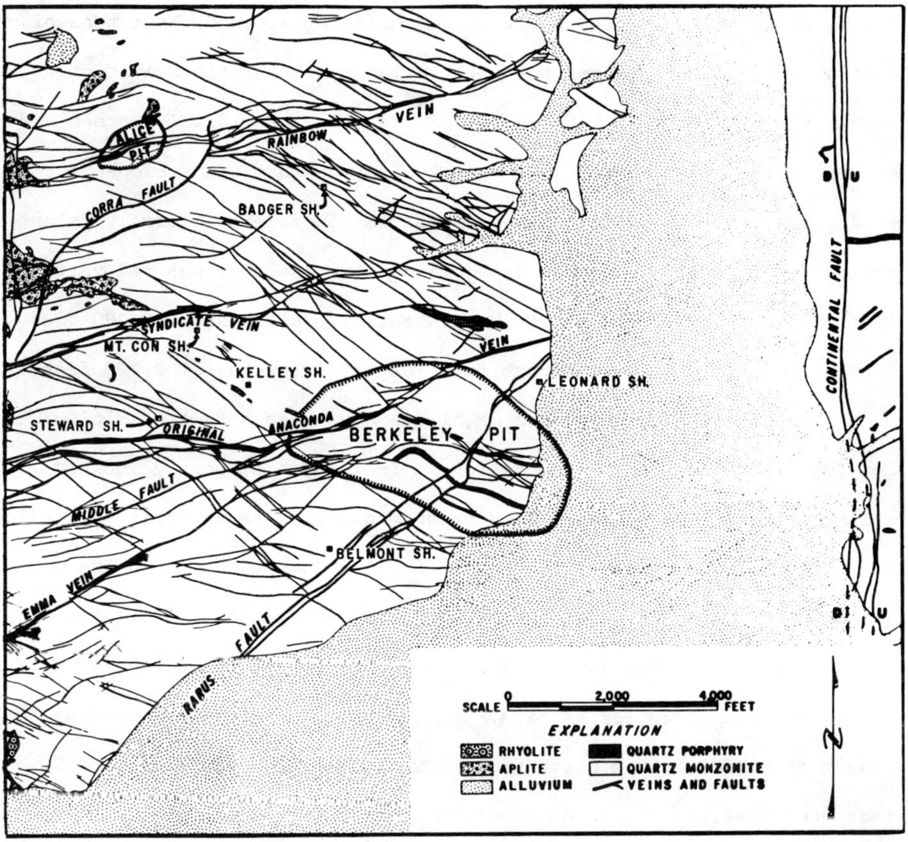

Figure 2. Areal Geology of the Butte District.

The present configuration of the Berkeley Pit, in operation since March, 1955, is that of an ovaloidal bowl approximately 6,400 feet long and 4,800 feet wide. Figure 3 shows the configuration of the Berkeley

Pit as of April, 1971. The crest of the north wall is 500 feet higher than the south rim; the west rim is approximately 400 feet higher than the east rim at the valley floor. Production of 250,000 short tons of ore and waste daily changes the pit configuration surprisingly fast. Within the pit, working benches are 33 feet high, and final benches are 67 feet high. Final berms are 33 feet wide, so that bench slopes are 63 degrees, and the overall design slope is 45 degrees. Locally, slope angles approaching 45 degrees are obtained over several hundred vertical feet. Significant portions of the present pit walls exceed 1,000 feet. Deep pit mining plans anticipate final depths in excess of 2,000 feet.

Slope failures of nuisance magnitude have been a fact of mining during the first ten years of operations. In 1965, a temporary loss of 2 million tons of ore to the mining program occurred as a result of a slide in the north wall of the pit. Slumps in alluvium developed in the eastern portion of the pit in 1966 and 1967. In 1968 another 1 million tons of ore was temporarily lost to the mining program in the north wall. In early 1969 temporary losses of 4.8 and 12.5 million tons occurred as the result of slides in the north wall. Crest displacements exceeding 20 feet were observed. A joint Anaconda-Montana Tech slope stability investigation was initiated in the Fall of 1969. This investigation is continuing.

Economics of open pit mining in general and the Berkeley Pit in particular leave little doubt as to the desirability of optimizing deep pit slopes[5]. Engineering fundamentals of the general problem should be equally clear. The problem can be posed in the following way: given

---

5. Stewart, R. M., and B. A. Kennedy, "The Role of Slope Stability in the Economics, Design and Operation of Open Pit Mines", Proc. Symp. on Stability for Open Pit Mining, Vancouver, British Columbia, Nov. 1970.

Figure 3. Present Configuration of the Berkeley Pit.

the applied loads, material properties and current pit slope geometry, first compute the stress, strain and displacement fields throught the section of interest on the basis of the known material behavior or response to load. Then assuming these predictions are reliable, evaluate the stability of the section and estimate the probability of failure. The evaluation obviously involves considerable amounts of engineering judgment. The computation involves a slightly less subjective category of questions that pertain to deep pit slope engineering technology which, despite the attention given the subject in recent years is still in its infancy, a fact attested to by the variety of seemingly competitive views towards the subject of slope stability. Seemingly competitive because enthusiastic advocates of specialized stability evaluation procedures sometimes overlook the simple fact that not all deep pit slopes are alike in detail. What is decisive in one mine may not be the case in another mine. Thus, at this juncture, there is little to be gained by formal review of the experience of others in deep pit slope engineering. References to the technical literature are made only as needed in the body of this paper.

The principal objective of the Montana Tech slope stability effort during the first phase of the investigation was the development of a reliable stability prediction and evaluation procedure. Several approaches to the problem were considered. The approach that was adopted consisted of estimating the loads applied to the slopes, determining the material properties, and measuring the actual pit geometry. Because failure was an essential feature of the problem, an inelastic analysis was pursued beyond the range of purely elastic deformation. Predictions were compared to field experience. As in every engineering problem, practical constraints relating to time and resources existed. Innovation therefore supplemented fundamentals where important questions arose to which there were no answers forthcoming.

The Berkeley Pit is mined by drilling and blasting, but contains numerous old underground workings, filled stopes and caves, and is covered by alluvium in part. The work was therefore divided into two categories composed of problems dominantly related to rock mechanics and problems dominantly related to soils mechanics. This paper describes the rock mechanics aspects of the investigation. A discussion and conclusion are included.

ROCK MECHANICS ASPECTS

One can categorically assert that the pre-mining state of stress for the Berkeley Pit was certainly not one of unit weight times depth and complete lateral restraint, nor are the rocks comprising the pit walls homogeneous, isotropic, linear and time-independent in their mechanical response to load. Real materials are complex. Simplification is therefore essential to progress. However, if a simplified analysis approximates the observed field behavior, a useful result will have been achieved. Reasonable confidence can then be placed in the application of the same procedure to problems of similar nature elsewhere. Uncertainties

---

6. Pariseau, W.G., "Influence of Topography on the Pre-mining State of Stress", 7th Canadian Symposium on Rock Mechanics, University of Alberta, Edmonton, March 1971.
7. Obert, L., R.H. Merrill and T.H. Morgan, "Borehole Deformation Gage for Determining the Stress in Mine Rock", U.S. Bureau of Mines Report of Investigation 5978, 1962.

should be explicitly recognized, of course, and appropriate degrees of risk should be attached to computed indices of stability.

Estimates of the Applied Loads

Applied loads refer to external agencies capable of causing deformation of the pit walls. Recognized static loads include those of gravitational, tectonic and thermal origin and those due to blasting. Gravitational loading was considered dominant in the analysis, tectonic and thermal loadings were not considered directly, and transient loadings were not considered at all.

The approach that was adopted then consisted in the main of computing the stress changes due to removal of material by mining. These changes were then added to the initial stresses to obtain the final stresses. Estimates of applied loads were thus reduced to considerations of specific weight and the pre-mining state of stress. Determination of the former presents no real difficulties.

Determination of the pre-mining state of stress on the basis of material models other than the elastic one presents formidable difficulties in conception and execution. Topography alone would be sufficient to produce a non-homogeneous pre-mining state of stress for the Berkeley Pit[6]. Underground workings would complicate the problem further. Moreover, the number of field tests required to define a non-homogeneous *in situ* stress field would be rather large. A full scale program of *in situ* stress measurements to determine the pre-mining state of stress thus did not appear feasible, although some experimental trials utilizing the U.S.B.M. bore hole-stress relief gage and technique were performed[7]. Unfortunately the results were inconclusive because of the tendency of the overcore to fracture before sufficient relief of the gage was obtained. Under these circumstances, the pre-mining state of stress was estimated through simulation of an erosional sequence beginning from an initially flat surface and leading to the pre-mining topography and associated stress state.

Determination of the Material Properties

The material properties of interest are dictated by the constitutive equation adopted for the analysis. In the present case, these are the specific weight and the static elastic and strength (plastic) moduli. Together with the applied loads they form the basic inputs for the computer implemented analysis of stability. Obviously much depended upon the determination of appropriate values for the rock properties. Considerations of costs and benefits indicated that large in situ test data militated against an extensive field scale testing program. The approach that was adopted was a statistical as well as a pragmatic one consisting of laboratory determination of rock properties and back-analyses of selected sections of the pit walls.

The general procedure for determining the laboratory rock properties was composed of three major steps: (1) collection of boulders at selected sample sites in the pit and transport of the samples to the Montana Tech rock mechanics laboratory, (2) coring, sawing, and lapping of the test specimens, and (3) measurement of the static elastic and plastic moduli. Usually four or five 10 to 20 pound boulders were obtained at each sample site. Some sites produced a uniform quality of rock; other sites produced a sample of mixed quality. The intent was to obtain a sample that represented the rock in the neighborhood of the sample site. Each boulder was cored with an NX-size bit (2-1/8" diameter core) until no additional core was obtainable. As many specimens as possible were slabbed from each core. Specimens of length-diamater ratio of 2 were preferable, but specimens of length-diameter ratio of 1 and disc specimens of thickness 1/2" were also cut as conditions dictated. Chips were saved for geologic study.

Elastic moduli were determined by uniaxial stress-strain tests. Peak loads were 50% of the unconfined compressive strengths during these tests. Generally the assumption of linear elasticity was supported by such tests, though not without exception. Figure 4 shows a typical stress-strain curve obtained from a laboratory test on Berkeley Pit rock. Exceptions usually occurred during tests on softer materials; occasionally these materials were too soft to core and were then tested by soil mechanics procedures. Unconfined compressive strengths were measured on core having length-diameter ratios of 1. Unconfined tensile strength was determined by the Brazil test using the 1/2 inch discs. Triaxial strength tests were carried out under confining pressures of 1,000, 3,000, and 5,000 psi. These data together with the unconfined strength data resulted in five points on the yield surface, and thus indicated the variation of the plastic moduli with confining pressure. The elastic and plastic or strength moduli were the essential inputs to the follow-on stability analysis.

As the data from the rock properties program accumulated, various efficiencies were sought and developed that expedited the requisite testing and data processing. The bulk of the data is now reduced to punch card form as it is acquired. A considerable library of information has been obtained in this way. The data library enables one to retrieve rock properties information almost immediately and in a form suitable for further processing. For example, the hierarchy of mean

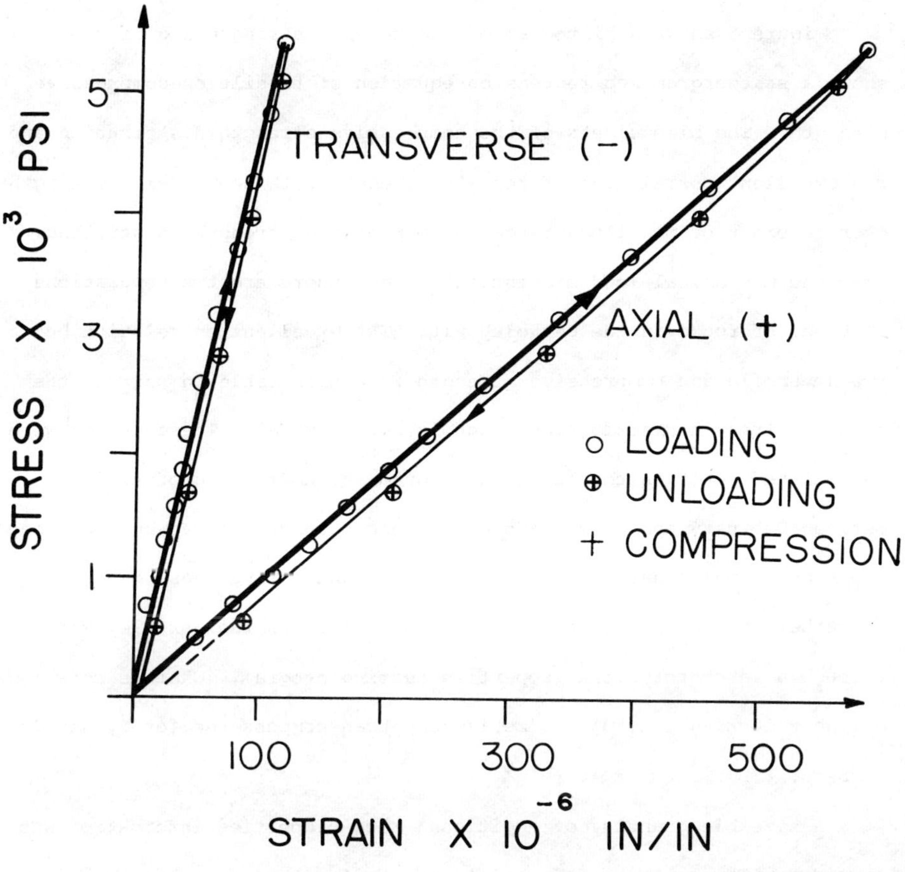

Figure 4. Typical Compressive Stress-Strain Curve of Berkeley Pit Rock Loaded to 50% of the Uniaxial Compressive Strength.

values from core to boulder to sample site can be studied in detail. The quantity of data is sufficient for various statistical questions to be posed and answered in a definite way with a quantitative proviso of risk or uncertainty.

Figure 5 shows a histogram of compressive strength and Figure 6 shows a scattergram and regression equation of tensile on compressive strength. The bimodalness of the compressive strength distribution and the excellent correlation of tensile strength with compressive strength over an order of magnitude range in compressive strength is striking. From the mechanical (and statistical) view, there are two populations or types of rocks in the Berkeley Pit. The excellent correlation between tensile and compressive strength is attributable in part to the tensile strength distribution which is also bimodal. These trends were established early in the sampling program; a quadrupling of the data between February and July changed the mean strength valves by less than 1 percent. These data clearly show that quantitative sense can be made of rather complex geotechnical situations, such as the Berkeley Pit, through a laboratory rock properties testing program. What is conceivably lacking in sample size is more than compensated for by sample numbers.

A sizeable quantity of additional rock properties information was acquired. Three items of interest concern fracture frequency, anisotropy and alteration zones. All involve geologic features and what was referred to as the geologic question. Observation of 1,946 feet of drill core resulted in a fracture frequency of 2 per foot. A control section of 400 feet was inspected by two individuals. Fracture frequencies were within 10%, so that the counting procedure was objective. The question of directional variations of material properties is currently under investigation; preliminary data indicate a lack of anisotropy on a scale of several hundred feet. It was thought that alteration zones would correlate well with strength, so that an improvement in sampling efficiency could be realized. Unfortunately, a control

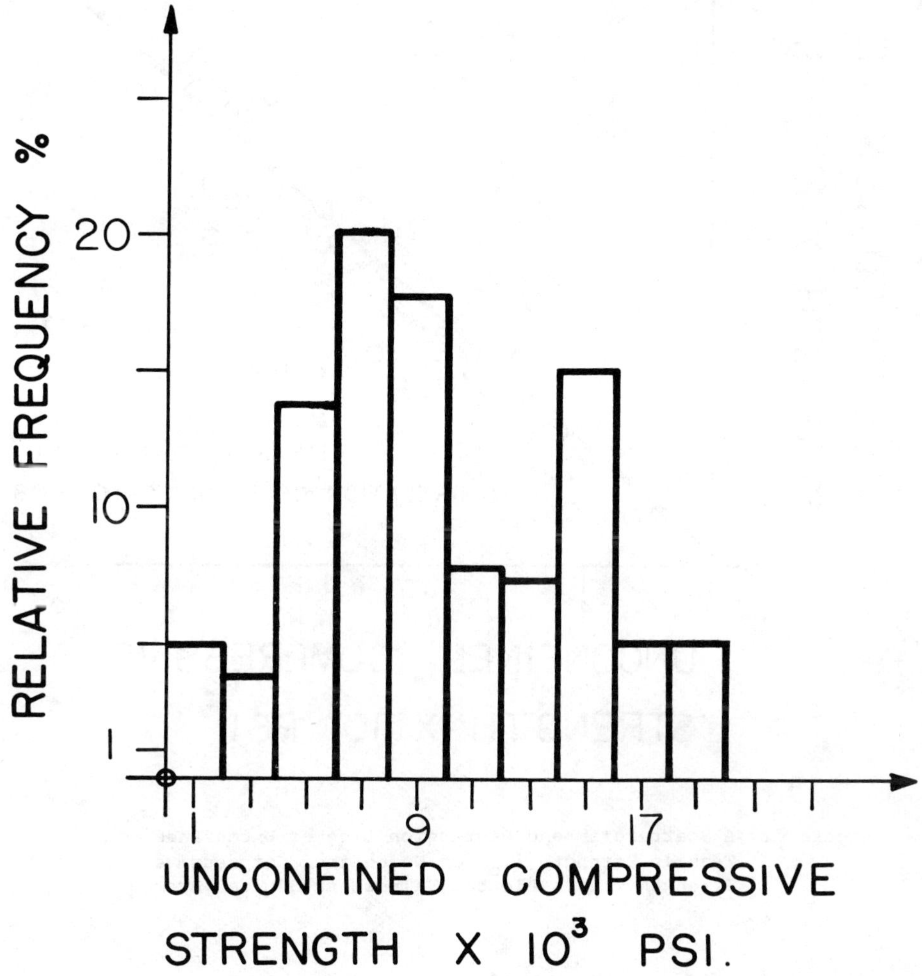

Figure 5. A Histogram of Unconfined Compressive Strength Data.

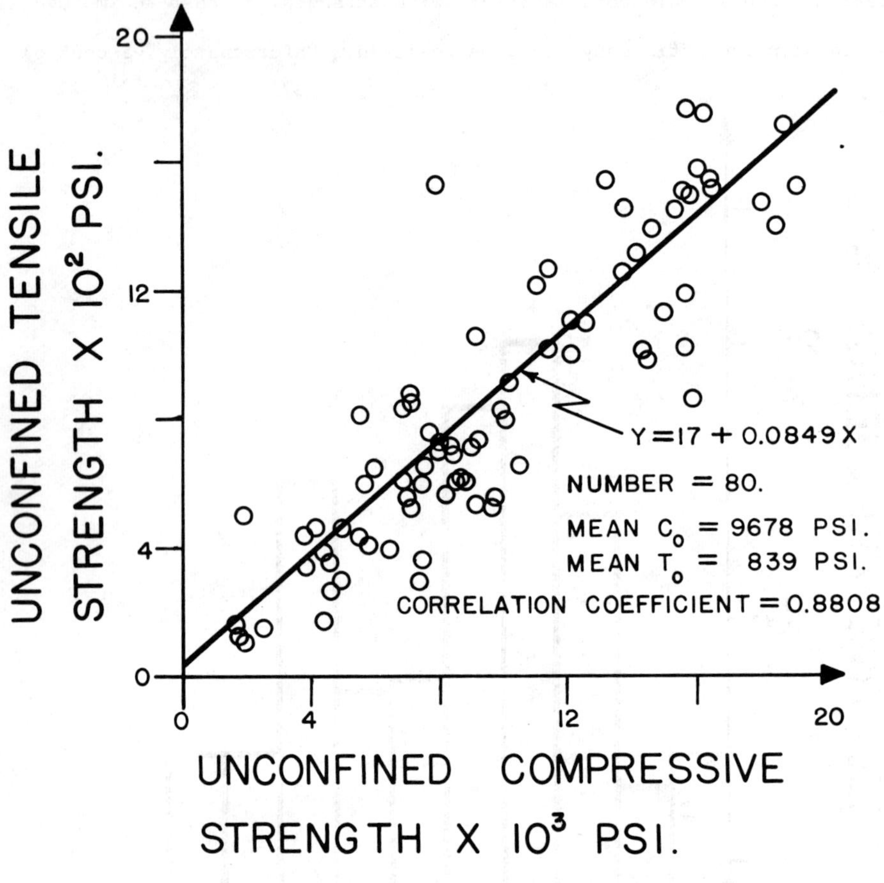

Figure 6. A Scattergram and Regression Line of Unconfined Tensile Strength (T_o) on Unconfined Compressive Strength (C_o). Plotted Values are Boulder Averages.

test showed that macroscopic inspection was not an adequate method for determining whether or not a specimen was "altered". The consensus of personnel involved in the actual properties testing, however, is that with experience one could make useful estimates of strength by inspection.

The average rock properties measured or computed on the basis of 116 boulder averages entailing approximately 1000 test are:

Unconfined Compressive Strength C_o = 9,000 psi.

Unconfined Tensile Strength T_o = 740 psi.

*Unconfined Angle of Internal Friction ϕ = 58 degrees.

**Confined Angle of Internal Friction ϕ = 37 degrees.

Average Young's Modulus E = 7.2×10^6 psi.

Average Poisson's Ratio ν = 0.22.

*Computed from C_o and T_o assuming Mohr-Coulomb failure (linear).

**Determined from triaxial tests in the range of 5-10,000 psi mean normal stress.

These properties were adjusted in extrapolation to field conditions and in the stability analysis to follow.

Analysis and Evaluation of Stability

The evaluation of stability of open pit mine slopes requires at the outset a useful measure of stability. Factor of safety as a ratio of strength to stress (each suitably defined) adequately serves the purpose. An analysis then reduced mainly to stress field computations. The stress field computations, however, can be interpreted in various ways.

All stress field computation were obtained using a finite

element program that with the exception of several minor modifications is due to Dahl[8]. In this program, the material behaves elastically-perfectly plastically and fails according to a Mohr-Coulomb condition. An associated flow rule is employed. Triangular elements are utilized throughout, and the master stiffness matrix is reassembled as required by element failures. Elements may load, fail and unload as conditions dictate. The material thus responds linearly elastically until the failure condition is satisfied. Large deformations then become possible. However, the elastic material adjacent to the plastic zone will generally restrain the plastic deformation unless the plastic zone becomes exceedingly large. Zonal heterogeneity and transverse anisotropy can be incorporated into an analysis, provided of course, that the material properties are known. Sequence of cuts and fills can be simulated as desired.

The rock in which a slope is to be cut is assumed to be initially stressed by gravity and possibly by loads of tectonic origin. Input data specifies the cut or fill profile. The computer then mines the material from an originally horizontal surface to the specified profile. An elastic analysis is first pursued; an incremental elastic-plastic calculation follows. This completes the stress field computation as well as the computations of the strain and displacement fields. In addition to printed output, punch card output of element stresses and strains and nodal point displacements is also obtained for further processing and future reference. Subsequent cuts and fills are again

---

8. Dahl, H. D., "A Finite Element Model for Anisotropic Yielding in Gravity loaded Rock", Ph.D. Thesis, The Pennsylvania State University, 1969.

computed on the basis of an initially stressed medium. The final stresses from the previous cut (or fill) serve as the initial stresses for the next cut (or fill).

A parameter study of an idealized, homogeneous, isotropic 45 degree, 500 foot deep slope was performed before using the program in a serious way. Program sensitivity for this type of problem was indicated by program run time, load increment and number of failed elements at first failure, and number of failed elements at final load level. Figure 7 shows the values of these indicators for nine combinations of elastic and plastic moduli. Young's modulus (E) of 9.45 x 10^6 psi and Poisson's ratio (ν) of 0.195 are laboratory values; a compressive strength (C_o) of 960 psi and a tensile strength (T_o) of 80 psi are one-tenth laboratory values. Figure 7 shows that the program is relatively insensitive to more than an order of magnitude change in E and to reduction of internal friction angle (ϕ) by approximately five-eights (58° to 37°). The program is sensitive to changes in Poisson's ratio and tensile strength. In retrospect these results are what one might anticipate because Poisson's ratio determines the amount of a lateral unloading (initially), and the state of stress in the majority of elements about the slope face, toe and crest is characterized by one normal stress a tension and other normal stress a compression. In the normal-shear stress plane, such states of stress are represented by Mohr circles enclosing the origin. A translation of the failure envelope towards the compression region by a reduction in tensile strength results in many more elements satisfying the yield condition.

The results of the paramenter study indicated that an order of magnitude reduction of all dimensional moduli would provide a reasonable starting set of values for back analyses of selected sections of

PLASTIC → / ELASTIC ↓	960 80	240 80	24 8	C_o *** T_o
E = 9.45 X 10^6 ν = 0.195	9.15 2 @ 0.13 39	15.10 115 @ 0.1 206	16.18 332 @ 0.1 371	Time. 1st. Fail. Final.
E = 9.45 X 10^5 ν = 0.390	10.65 2 @ 0.4 39	10.70 3 @ 0.4 75	12.14 3 @ 0.3 172	Time. 1st. Fail. Final.
E = 9.45 X 10^4 ν = 0.390	10.67 2 @ 0.4 38	10.75 3 @ 0.4 75	12.16 3 @ 0.3 172	Time. 1st. Fail. Final.

*** C_o= unconfined compressive strength, T_o= unconfined tensile strength, Time=computer run time in minutes, 1st. Fail. = the number of elements failing at the first appearance of failure and the load level at first failure (loading is applied in ten equal increments), Final is the number of elements that failed after application of the final load increment, E is Young's modulus, ν is Poisson's ratio.

Figure 7. Results of a Parameter Study Indicating Relative Sensitivity of the Computer Program to changes in Material Properties. Homogeneous 45 degree, 500 foot slope model containing 371 elements used.

the pit walls. Thus, Young's modulus, compressive strength, and tensile strength were reduced by a factor of 10 from their laboratory values for application to the field problems; Poisson's ratio and the uniaxial straint to failure were held constant. Figure 8 shows the uniaxial implications of the procedure. In effect, one moves out to a residual strength along an elastic line of greatly reduced slope (increased compliance) with increasing stress. Unloading occurs along a line of the same slope as the initial elastic portion of the "field" stress-strain curve. Physically, the reduction in strength and increase in compliance may be attributed to the presence of fractures, joints, and so forth in the field that are absent in laboratory size test specimens.

A large active slide in the north wall of the pit was selected for detailed study; three additional sections, two stable and one active, were also investigated. Figure 9 shows the element mesh used for the active slide back-analysis. A major fault zone striking sub-parallel to the slope face was represented by the inclined column of relatively small elements shown in the figure. The purpose of the back-analyses, was to bring the computer results into correspondence with observed field behavior by developing a correlation between rock properties measured in the laboratory and the rock property values required for acceptable computer results as measured against the observed field conditions. Essentially, this meant the computation of a factor of safety near unity as computed over a chain of elements. A failed element exhibits a local factor of safety of one. A factor of safety of nearly one over a chain of elements implies the existence of a zone of contiguous failed elements extending from free surface to

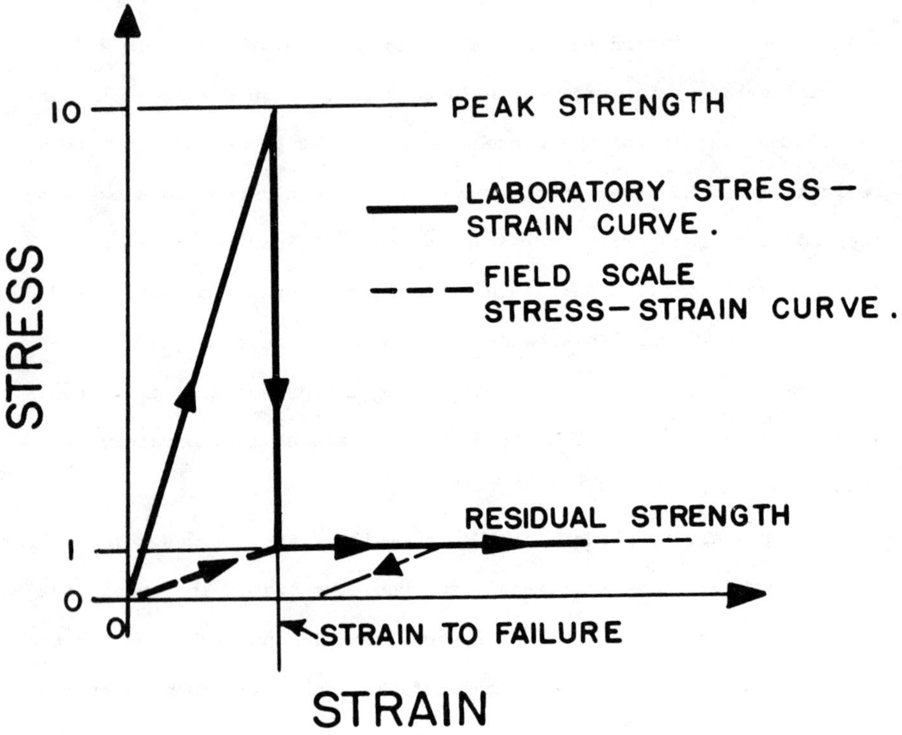

Figure 8. Schematic Representation of the Uniaxial Relationship between Laboratory and Field Stress-Strain Curves obtained through an Order of Magnitude Reduction in Dimensional Material Properties.

free surface, a toe to crest zone for example. In figure 9 the spread of the failed zone with increasing load is shown for rock properties reduced by an order of magnitude from their laboratory values. Only the upper 600 feet or so is of interest in figure 9; the lower portion of the fault zone is excluded from consideration because of its termination

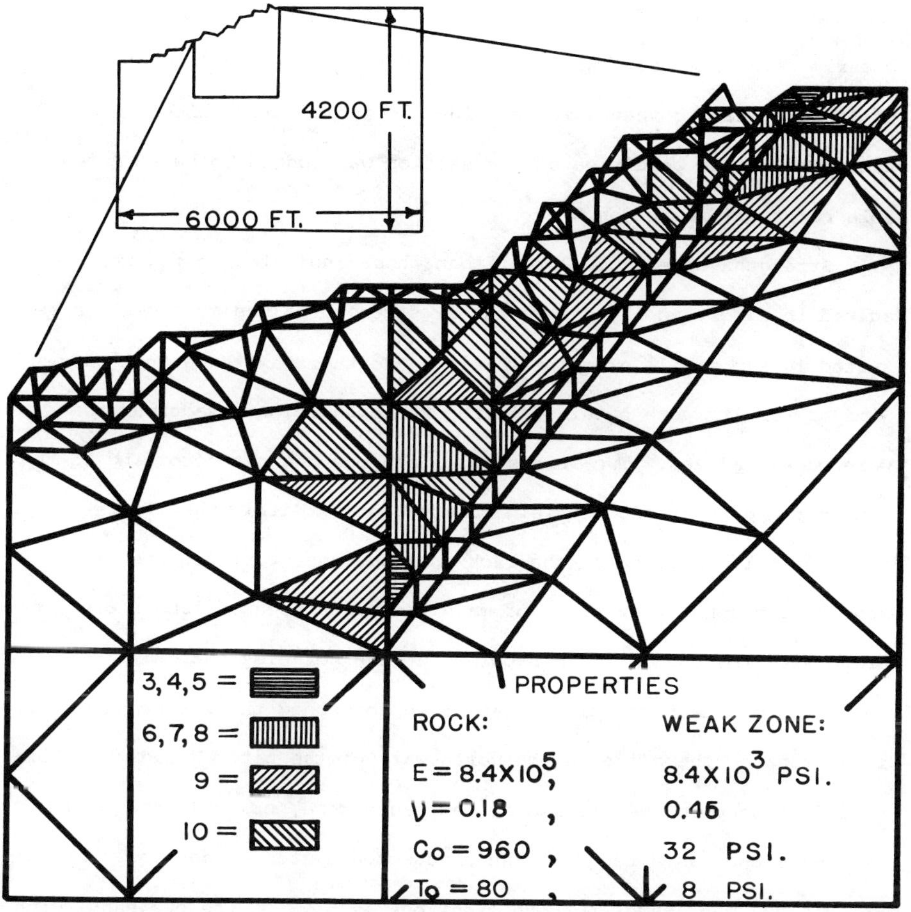

Figure 9. Results of a Finite Element Back-analysis for an Active Slide Area. Numbers indicate load level at which element failure occurred. Final load level is obtained in ten increments, and corresponds to mining to the present section profile.

in the model but not in the field. The fault zone because of its relatively high compliance behaves as a slit. Its major effect is to produce stress concentrations in the adjacent wall rock as figure 9 clearly reveals. Most failures appear during the last (tenth) increment of load indicating that a major slide was incipient. This in view of the actual field condition of the section was judged to be an acceptable computer prediction.

Strength values much larger than those indicated in figure 9 resulted in almost no element failures; strength values much smaller resulted in almost all elements failing. Neither was the actual case. Similar results were obtained in the three additional sections that were back-analyzed. Two were stable, and one was not according to the computer and according to field observations. Within the limits of the available field data, the back-analyses thus verified the dimensionally consistent order of magnitude reduction of laboratory rock properties for use in field problems at the Berkeley Pit.

Two important questions that arose concerned the influence of the fault zone on the slide in the north wall of the pit. Would the slide have developed in the absence of the fault zone, or could the slide have developed as a result of an undetected system of adversely oriented joints? The answer to both questions is no. On the one hand, without the fault zone but with dimensional rock properties still an order of magnitude less than laboratory values, only one element in the area of interest failed, whereas in fact the section was an active slide area. Moreover, it did not appear possible to lower the strengths further and still localize element failures in the area interest. On the other hand, without the major fault but with a joint of the same attitude and strength properties (as the fault) present in every element[9], almost all elements in the computer model failed, aimultaneously on joints and through the rock. Again this was not the field situation. Thus, in the present case these two plausible alternative stress field interpretations were inappropriate. There seemed little doubt that the large fault zone present in the north wall slide area was of dominant importance to the slide development. This result provided additional impetus to the mapping and delineation of major zones of weakness in the pit walls.

With the establishment of a reliable rock slope stability analysis procedure for Berkeley Pit conditions, it appeared reasonable to move from the "disaster to the design stage"[10], in order to evaluate the stability of slopes associated with the deep pit mining plans. Slopes of 45 degrees to depths in excess of 2,000 feet are currently anticipated.

Preliminary evaluations of stability for 12 potential failure surfaces shown in figure 10 were obtained for 6 conditions of structure to depths of 500, 1000, 1500, and 2000 feet. Factors of safety computed on the basis of elastic-plastic analyses for the worst-case conditions at the 1000 foot depth are shown in figure 11. Figure 11 shows that if the weakness zones are sufficiently remote from the slope face, the most dangerous failure surface is a shallow scoop-like surface in

---

9. This procedure was inspired by the "ubiquitous joint analysis" due to Goodman and described by Duncan, J. M. and R. E. Goodman, "Finite Element Analyses of Slopes in Jointed Rock", U. S. Army Engineer Waterways Experiment Station Contract Report S-68-3, February, 1968.
10. A cogent remark made originally by Professor Hoek at the Symposium of Stability for Open Pit Mining, Vancouver, British Columbia, Nov. 1970.

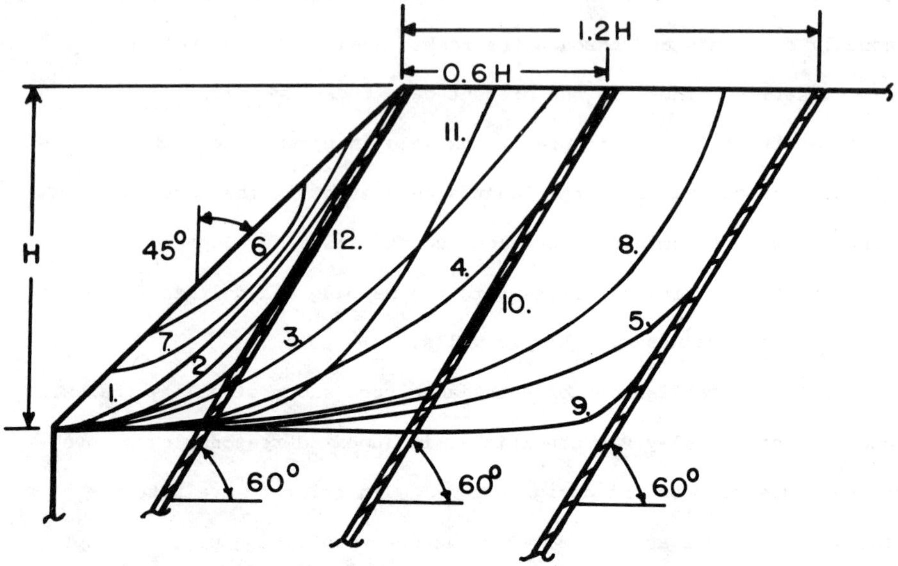

Figure 10. Locations of Potential Deep Pit Failure Surfaces and Zones of Weakness. Partial section of the Finite element model shown.

the face of the slope. This is also the case when weakness zones are absent, and represents a relatively shallow shearing along the face, otherwise the risk of large failures appeared to be high. Additional analyses indicate that even with laboratory strength values, the safety factors shown in figure 11 would increase by no more than a factor of 3. Generally the improvement of the factor of safety (with increasing strength) decreases non-linearly with depth; the improvement is four-fold at 1000 feet, three-fold at 2000 feet, and two-fold at 4000 feet. It is clear that preservation of the margin of safety that may exist will with increasing depth require a carefully engineered mining scheme to maintain and possibly reinforce the natural integrity of the pit walls.

Engineering analyses of several possibilities are being initiated.

DISCUSSION

Although the major features of the rock slope engineering investigation of the Berkeley Pit by the Montana Tech group have been described, several comments concerning the applied loads, rock properties and stability analysis are in order. Undoubtedly questions concerning some of these points have occurred to the reader, particularly if his experience has been in deep, open pit mines where conditions are unlike those in the Berkeley Pit.

Blasting which is done twice daily causes momentary acceleration of displacements measured in active slide areas. However, no catastrophic movements have resulted, so that the influence of blasting on slope stability here appears to be manifested through deterioration of the rock properties. With sufficient deterioration of the slope, a definite potential developes for triggering a large slide by a blast. If this is in fact the case, it may be feasible to account for dynamic effects by up-dating the static analyses. This approach may also be applicable to long-life stability evaluations. The alternatives of employing a genuine dynamic analysis for blast effects and of imputing a natural time scale to the material for long-life design represent a considerable increase in engineering effort which in view of the associated uncertainties may not be justified at the present time.

The great utility of the laboratory rock properties testing program is found in the back-analysis technique. Laboratory rock properties provide an independent datum from which the analyses can be initiated and referenced. In our view, without this datum, back-analyses become far too circular to be of practical value. It becomes very difficult to

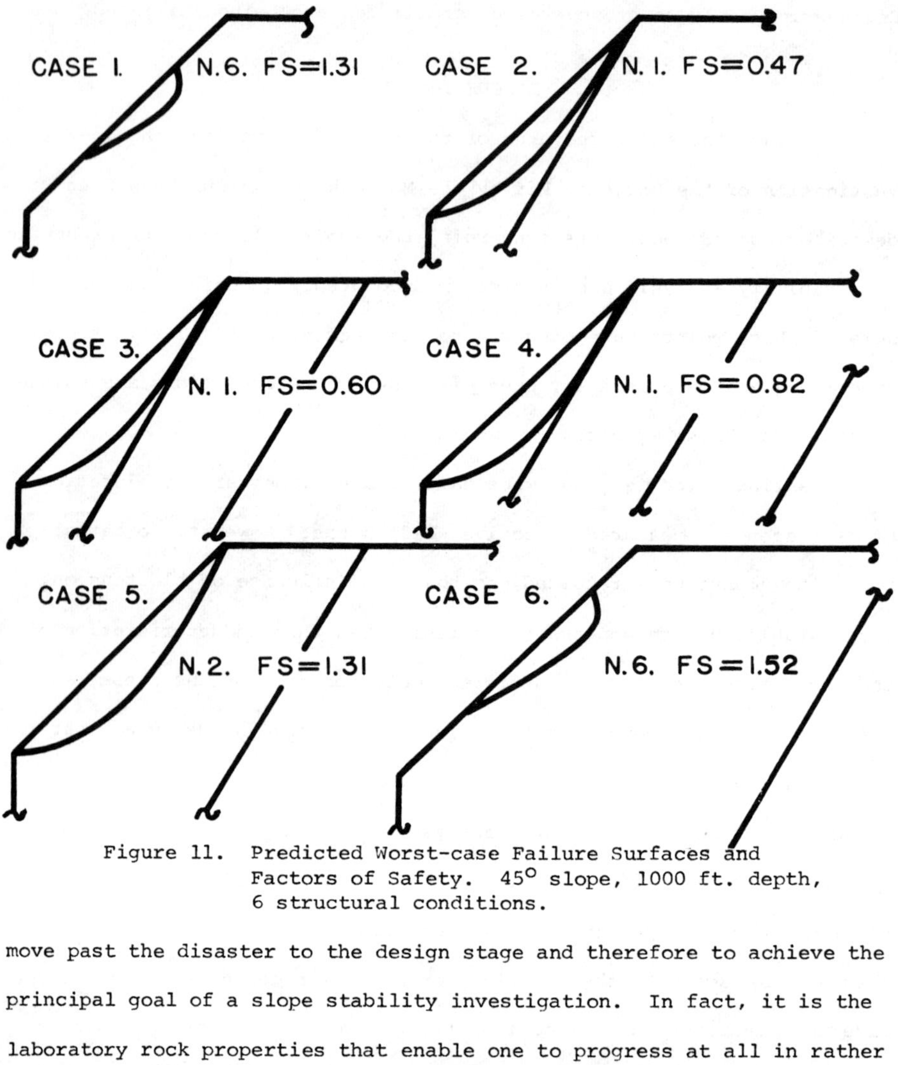

Figure 11. Predicted Worst-case Failure Surfaces and Factors of Safety. 45° slope, 1000 ft. depth, 6 structural conditions.

move past the disaster to the design stage and therefore to achieve the principal goal of a slope stability investigation. In fact, it is the laboratory rock properties that enable one to progress at all in rather complex geotechnical situations. Moreover, they enable one to make quantitative statements regarding uncertainty in the computation of safety factors. Instead of a deterministic statement that the factor of safety (F.S.) is the ratio of strength to stress, one has a probabilistic statement. The usual factor of safety becomes the mean or expected value of F.S. One is then in a position to compute probabilities of obtaining greater or lesser safety factors.

One of the essential features of slope stability analysis is the fact of permanent or plastic deformation, so that an elastic-plastic analysis was pursued. The plasticity is employed in the macroscopic or phenomenological sense, and should not be identified with the micro-view where emphasis is upon dislocation mechanisms. At the macro-level there are many mechanisms for permanent deformation including catalastic flow. Also, the direct programming of a consistent continuum theory seemed preferable to the programming artifices that have been used elsewhere to make essentially elastic finite element codes serve for the analysis of inelastic problems.

Another area worthy of further investigation concerns the behavior of rock under stress states characterized by one normal stress tension and the other compression. Little is known about the behavior or rock under such stress states. For, example, it is not known whether if fracture occurs it will be tensile or shear. Related to this area is the problem of deciding upon a failure criterion for initially stressed geologic media. The choice between a stress and strain failure criterion is no longer a false issue. Our choice was the well-known Mohr-Coulomb stress failure criterion; experimental data supported this choice.

The original parameter study designed to investigate the sensitivity of the computer progress to changes in the elastic and plastic moduli inputs was performed on a homogeneous slope. Insensitivity to more than an order of magnitude change in Young's modulus was observed. If however the slope contains several materials, the analysis can be sensitive to moduli ratios. If the moduli contrasts are large, then the sensitivity diminishes greatly, and large variations can again be tolerated without significant effect. Large contrasts were observed in the Berkeley Pit where very soft materials such as old fills, zones of intense hydrothermal alteration, and alluvium are present in addition to the igneous host rock of the Boulder Batholith.

The presence of significant amounts of soft, soil-like materials as well as the deterioration of rock properties that occurs in active slide areas suggested that traditional soil mechanics procedures would be useful. An extensive program of soil properties testing and slip-circle analysis[11] was initiated and is continuing. Unfortunately space limitations prohibit discussion of the soil mechanics aspects of the overall investigation.

CONCLUSIONS

Two strong conclusions are warranted by the investigation of slope stability in the Berkeley Pit. One is philosophical but worth stating, that is, through reliance on engineering fundamentals and prudent judgment concerning the essential features of the problem, one is able to intelligently and successfully evaluate the stability of slopes in deep, open pit mines. The second conclusion is that in the case of the Berkeley Pit, major structures (weakness zones) exert a major influence on stability. Plausible alternatives that were investigated were unable to explain the observed behavior as adequately as models containing major zones of weakness. This conclusion may appear obvious to the uninitiated. However, there are competing views concerning the reasons and mechanics

11. The authors prefer to reserve the more fashionable terminology "limit equilibrium analysis" to analyses employing the limit theorems of plasticity theory, for example, those described by Drucker, D. C., W. Prager, and H. J. Greenberg, "Extended Limit Design Theorems for Continuous Media", Q. Appl. Math., Vol. 9, pp. 381-388.

of rock slope failures. These include the "soil" view which maintains that on the scale encountered in open pit mines, a rock mass may be treated as a granular or soil medium; also included is the "jointed rock" view which almost exclusively emphasizes the importance of joints and similar discontinuities. These latter views, of course, may be applicable under conditions other than those of the Berkeley Pit.

Beyond these immediate conclusions there is an additional inference that is supported by experience. Stability analyses for deep, open pit mines may not be particularly sensitive to slope angle changes within the range of economic slope inclinations. A five degree slope angle decrease may represent a large additional stripping cost, but in terms of the load change on the slope face, the additional material removed may be insignificant relative to the original loading. Thus, the traditional habit of thinking entirely in terms of a single parameter of the problem, slope angle, may not be adequate for the design of deep, open pit mine slopes. It is hoped that the views and procedures developed herein on the basis of fundamentals and tested against actual field experience will be of some value to others interested in rock slope stability, a subject where deductions abound but facts are few.

Acknowledgments

Grateful acknowledgment is made to the Anaconda Company for permission to publish this paper and to the many student assistants, including Messrs. R. Backer and C. Roberts, to whom fell much of the necessary routine labor.

THREE-DIMENSIONAL APPROACH FOR DESIGN OF CUTS IN JOINTED ROCK

By Francois E. Heuzé[1], A. M. ASCE and Richard E. Goodman[2], M. ASCE

I. INTRODUCTION

Rock cuts often have a high enough value, because of their potential for damaging installations or their repair cost, that a rational design procedure is warranted. This paper discusses a three step-procedure for designing cuts in jointed hard rock.

1. Recognition of the attributes of the rock mass and of the scale and geometry of the cutting. The discontinuities are classified as single or multiple features because they will be handled differently.

2. Evaluation and presentation of all quantities that must be input in analyses. One must integrate results from geological mapping, mechanical testing, and flow studies with cost data.

3. Selection of relevant tools for three dimensional study from those available to the designer: physical models, finite element or other mathematical analog models, stereographic projection, and vector analysis. The tools are assembled in a design sequence with full appreciation for their assumptions.

Since the imput involves ill-defined quantities -- such as rock and joint properties; locations, orientation, spacing, and character of planar discontinuities; water regimes; and cost -- the procedure must provide

[1]Lecturer and Assistant Research Engineer, Geological Engineering, University of California, Berkeley.

[2]Associate Professor, Geological Engineering, University of California, Berkeley.

for sensitivity analysis of parameters.

Selected elements of the design procedure developed herein are applied, for illustration purpose, to an actual design: the high cuts for the left spillway at Auburn Dam, California (Heuzé and Goodman - 1971a, 1971b).

II. THE NATURE OF THE ROCK SLOPE DESIGN PROBLEM

The excavation of a rock slope involves unloading of an originally buried rock mass. The wide range of behavior possibilities or rock cuttings reflects the broad spectrum in rock mass characteristics, and it is therefore appropriate to review rock mass classifications. Rock masses can be classified with respect to their character, with respect to their modes of failure when excavated and with respect to techniques available to analyze and/or improve their stability.

1. Rock Mass Characteristics

The properties of a given site for rock excavation depend on the combined effect of rock and discontinuities. Classification of the various combinations of rock and discontinuity types on a functional basis has been proposed by many workers. Coates (1964) and Muller (1964) offered general purpose behavioral groupings for rock masses. Terzaghi (1946), Laufer (1958), Talobre (1967) and Obert and Duvall (1967) offered rock mass classification for tunneling. More recently such a scheme was presented by Duncan and Goodman (1968) that considers specifically the problem of surface excavations in rock. The major division is between continuous, pseudo-continuous and discontinuous masses based entirely on the relative importance of discontinuities with respect to stability. The different types are illustrated in Figure 1 with examples of related rock types and include: a) Continuous masses: 1. strong homogeneous, 2. weak homogeneous and terraced rocks, 3. pseudo-

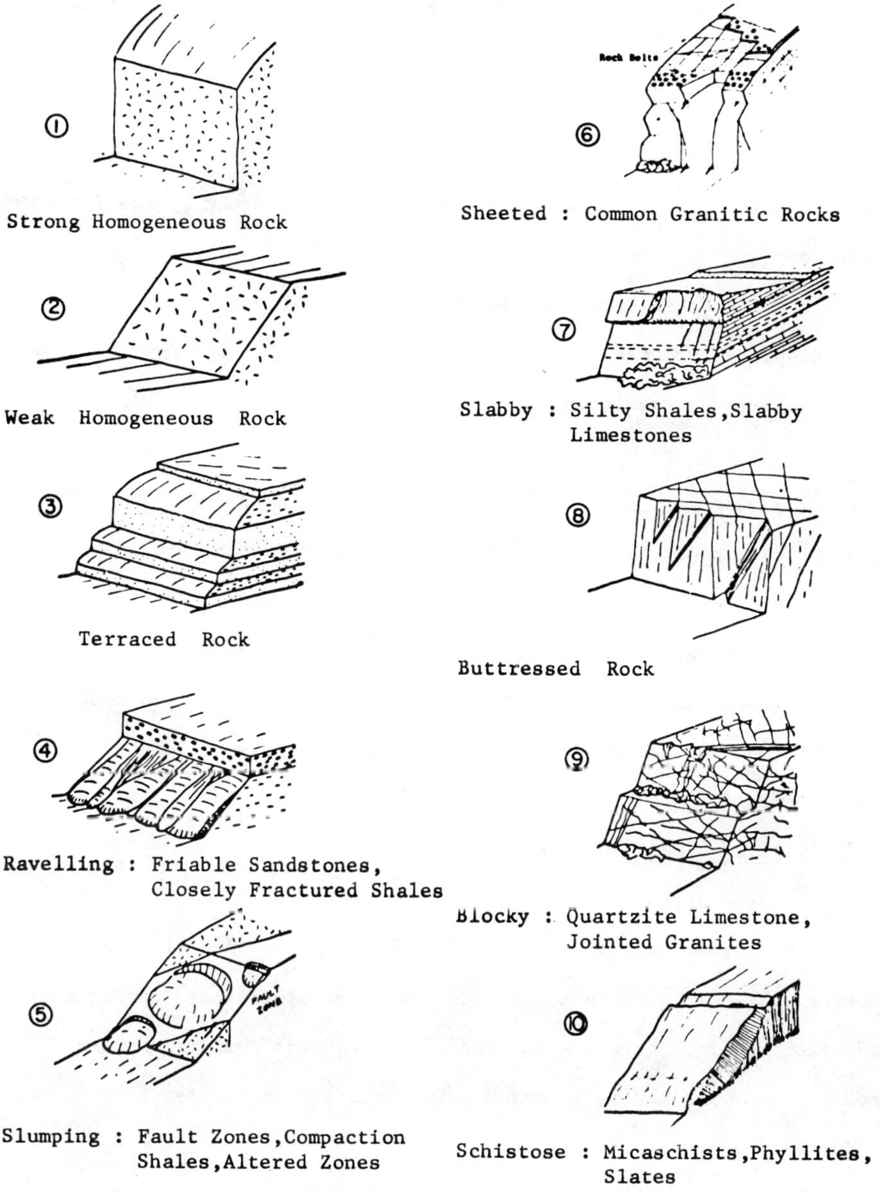

Figure 1 : Classification of Rock Masses for Surface Excavations

1-5 : Continuous , 6-10 : Discontinuous

continuous masses, 4. ravelling, and 5. slumping rock; b) discontinuous: 6. sheeted, 7. slabby, 8. blocky 9. buttressed, and 10. schistose rock.

2. Rock Masses Failures

Failure of slopes in rock masses can be classified in one of the following categories:

a. sliding on existing planes of weakness with negligible deformation of the displaced volumes

b. rotation or overturning of rock blocks

c. sliding on planes of weakness with substantial deformation of the sliding mass

d. formation of failure surfaces within homogeneous materials in zones where effective stresses exceed available strength

The first three modes are primarily conditioned by properties of the discontinuities and can be expected to prevail in hard jointed rock. The fourth may be applicable in large cuttings in closely jointed rock, which becomes essentially statistically homogeneous.

3. Analysis of Slopes in Rock

There are three different approaches to slope design in rock:

a. The stress analysis approach which includes: description of existing stresses; description of the system - material and configuration; prediction of final stress state and displacements; comparison with strength; analysis may be repeated using modified properties to simulate nonlinear behavior.

b. The limit equilibrium approach which includes: description of the system - material and configuration; examination of admissible slip modes; estimation of net driving forces on trial slip planes; and comparison with net resisting forces on the slip planes.

c. Empirical approach which includes: description of the system - material and configuration; classification into a behavioral type; appraisal and design based on previous experience; and adjustments based on readings of monitoring devices and observations. This article is concerned with the problem of an efficient design, thus deals with the first two approaches only.

4. High Cuts in Jointed Rock: Design Approach

The problem dealt with here concerns the stability of high cuts in sheeted, slabby, buttressed or blocky rock. The modes of failure one must account for are sliding, rotation and overturning of rock blocks bounded by discontinuities; the blocks may or may not be deformable. Accordingly, one shall make use primarily of a limit equilibrium approach; it is only when block deformability are significant that a stress analysis need be used. This can be the case when very high cuts are considered.

III. DEFINITION AND ANALYSIS OF THE INPUT FOR DESIGN

The input of the design is composed of:

1. The results of geological mapping and petrographic studies. This provides information as to the extent, spatial arrangement and character of the different rock formation and discontinuities.

2. The results of testing for mechanical properties which reveal gross properties of the rock mass such as degree of weathering and also provide quantitative measurement of the strength and deformability of the rock components and of the in situ stresses.

3. The results of flow studies: the presence of fluids in the rock mass may alter the properties of the discontinuities and modify the overall state of stress or the local systems of forces.

4. Finally, approximate cost data are generally available for

estimate of expenses related to excavation, drainage, support/reinforcement and grouting - generally absent will be the cost of failure of part or all of the structure.

1. Results of Geological Mapping

Emphasis must be placed on efficient use of the often considerable amount of data available from the geologist(s). In the case of slope design in jointed hard rock, relevant data will be chiefly the strike, dip, spacing and extent of discontinuities (joints, faults, dikes, ...). From these data, the designer must select the scale of the blocks that he must consider in his analysis; he must also determine whether they are formed by unique discontinuities, or multiple, penetrative discontinuities.

The extent and spacing of the geological features must be compared to the scale of the proposed works. Figure 2 presents a classification of discontinuities according to scale after Duncan and Goodman (1968). The scale gives a clue as to the general character of the discontinuities as extension fractures are generally clean with matching surfaces whereas shear fractures tend to be filled with weathered material.

It is proposed here that a distinction be made between single features defined as those discontinuities that can be mapped and accounted for individually, and multiple features, i.e. those that are essentially ubiquitous. Multiple features can be grouped in sets for which only overall attitude and spacing are obtained. The former will be major faults and dikes, the latter will be ubiquitous joint sets. It is feasible to analyze the stability of all blocks formed by intersections of single features, but one can only do sample analysis of block formed by the full complement of multiple features. Typical single rock volumes are represented on Figure 3.

Whereas single features can generally be defined in space in a unique

Name		Scale of Observation	Typical Spacing	Main Modes of Origin
Macro and Microfissures		a lab sample	<0.1 in.	alteration and extension fracturing
(may be taken as a rock block characteristic)				
Partings; bedding joints; cleavage foliation	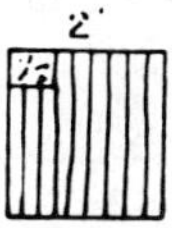	an in situ test block	0.1" to 2"	extension fracturing
Joints (A) Seams (B)	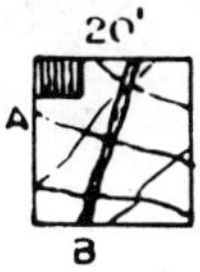	an exploratory excavation	2" to 20 feet	extension fracturing and shear failure
Minor faults, crushed zones; sheared zones		a completed cutting	20 feet to 200 feet	shear failure
Major faults		a mountain range	>200 feet	shear failure

Figure 2 ; Classification of Discontinuities According to Scale

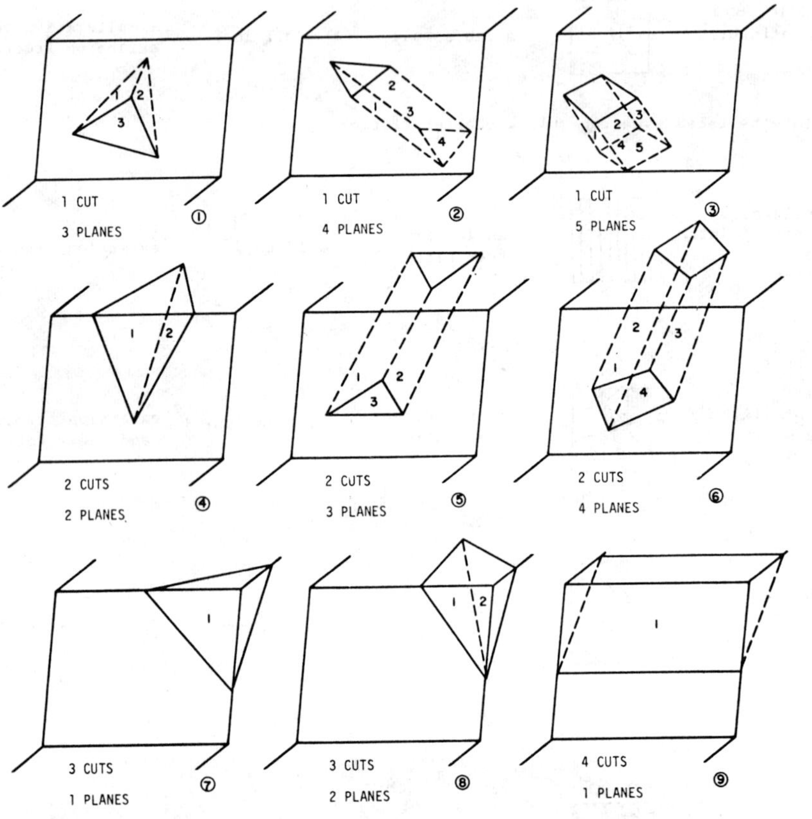

Figure 3 : Examples of Single Rock Volumes

manner, observations of joint sets involve imprecision; data reveal only spatial distributions about central tendencies. It is proposed here that these joint set data be evaluated in two steps:

a. Determination of the most significant joint sets by relative weighting of each set within the joint population.

b. For each set retained, determination of the strike, dip, and spacing values for design, based on all observations.

The efficiency of such a procedure in making raw data amenable to analysis will be demonstrated in the case of the Auburn Dam left spillway site. In this example, the site was subdivided into several geological "blocks", owing to the complexity of the geology and to the presence of major faults acting as boundaries to regions where scatter of observations could be reduced. Nevertheless the range of values on joint sets remained wide in each block as shown in Table 1. Subdivision of the sites in somewhat homogeneous regions is a satisfactory way of handling the inhomogeneity of the rock mass and of presenting the geological input data to the designer. Such a procedure can be rationally based on prominent sub-region boundaries (e.g. major faults) or on the results of statistical evaluations (Da Silveira et al., 1966).

a. Determination of primary joint sets. Altogether one notes the presence of seven joint sets on the site. They will be evaluated in terms of two indices:

1. The average of the normalized number of observations, defined as the average over all blocks, of the ratio

$$J = \frac{\text{number of observations for a set}}{\text{total number of observations}}$$

2. The average of the percent areal distribution (A) defined as the ratio of tunnel, drift, or borehole length over which a set was observed, to the total tunnel, drill or borehole length at a given site.

Table 1: Summary of Joint Set Data - Auburn Dam Left Abutment

Block	Joint Set	Strike		Dip		Spacing Average (ft)	Length Average (ft)	Number of Observations
		Ave.	Range	Ave.	Range			
C	A-1	N 60 W	180	13 SW	40	1.6	8.0	20
	A-2	N 52 E	10	7 SE	5	1.5	7.5	2
	B	N 80 E	63	88 NW	48	1.7	3.6	36
	D-1	N 67 W	45	47 SW	47	1.7	7.5	26
	D-2	N 72 E	44	46 NW	51	0.9	7.0	14
	E	N 53 E	89	41 NW	53	1.6	4.7	21
	G	N 62 W	70	12 NE	20	1.3	9.5	4
D	A-1	N 90 E	50	15 S	27	1.8	-	2
	B	N 55 E	63	89 NW	43	1.6	2.3	10
	D-1	N 54 W	57	45 SW	50	1.6	7.7	7
	D-2	N 82 E	33	57 SE	23	1.3	-	2
	E	N 40 E	81	36 NW	30	1.6	6.0	9
E	A-1	N 72 W	20	10 SW	24	1.9	-	3
	A-2	N 70 E	20	15 SE	20	1.5	-	3
	B	N 63 E	67	83 NW	50	2.1	2.5	21
	D-1	N 61 W	83	38 SW	53	1.1	7.7	16
	D-2	N 72 E	32	55 SE	48	1.3	-	3
	E	N 42 E	71	43 NW	44	1.7	8.6	15
F	A-1	N 82 W	17	31 SW	15	1.6	-	1
	B	N 82 E	10	81 SE	10	0.6	3.0	1
	D-1	N 62 W	60	50 SW	38	1.0	8.5	4
	E	N 70 E	58	38 NW	32	0.9	6.5	3
	G	N 45 W	33	45 NE	32	0.9	7.0	3
G	A-1	N 58 W	68	13 SW	13	0.8	9.5	2
	B	N 58 E	50	84 NW	64	0.9	4.9	7
	D-1	N 42 W	47	38 SW	22	2.0	-	2
	D-2	N 69 E	25	47 SE	26	1.1	15.0	2
	E	N 53 E	60	24 NW	30	0.5	14.2	2
	G	N 42 W	20	42 NE	20	0.7	4.0	1

Sets occurring infrequently are not retained as ubiquitous features. This is the case for sets A-2 and G of Table 1. For the remaining 5 sets the two indices above are computed (Tables 2 and 3). Examination of Tables 2 and 3 reveals that the joint sets fall into two groups: Group 1; Sets B, D-1 and E and Group 2; Sets A-1 and D-2. Group 1 is more important than Group 2 since both indices A and N are larger. Accordingly we shall retain joint sets B, D-1 and E for further analysis. Note that the primary joint sets number three; this, according to Terzaghi (1962), is a characteristics of a large variety of rock masses. The selection of design values for the 3 sets above must consider the joint properties. An infrequent set marked by great weakness must undoubtedly be heavily weighted for design purposes. However, if it is so scarce in occurrence that a weighting becomes difficult it should then be considered as a family of individuals (single features). Single features are not subject to the procedure developed here and can generally be input directly in the analysis as discussed later.

Table 2. Normalized Number of Observations

(from data of Table 1)

Set	Blocks					$\bar{J}$ (average)
	C	D	E J=	F	G	
A-1	17	7	5	8	13	10.0
B	31	33	36	8	13	31.2
D-1	22	23	28	34	13	24.0
D-2	12	7	5	25	13	12.4
E	18	30	26	25	13	22.4
						100.0

$$\frac{[\bar{J}(A\text{-}1) + \bar{J}(D\text{-}2)]}{2} \Big/ \frac{[\bar{J}(B) + \bar{J}(D\text{-}1) + \bar{J}(E)]}{3} = 43\%$$

Table 3. Percent Areal Distribution (A)

(from data on Table 1)

Set	Blocks					$\bar{A}$
	C	D	E A=	F	G	
A-1	27	22	10	31	31	24
B	27	77	26	46	58	49
D-1	28	46	30	71	16	38
D-2	7	28	12	63	23	23
E	27	73	25	51	29	41

$$\frac{[\bar{A}(A-1) + \bar{A}\ (D-2)]}{2} \Big/ \frac{[\bar{A}\ (B) + \bar{A}\ (D-1) + \bar{A}(E)]}{3} = 55\%$$

b. <u>Selection of primary joint sets design values (strike and dip)</u>

Assuming that observed strike and dip values for a joint set are distributed, one must decide which percentage of the observations shall be retained. It is suggested here that this percentage be no less than 65 to 68%; this corresponds, in a normal distribution, to observations whose values are within one standard deviation from the mean. An additional consideration must be made of the absolute value of the design range thus retained. It is proposed here that design "girdles" of joint sets be no more than about 30° in strike and dip ranges. It is suggested also that the percent of population be expanded rather than the range decreased; this makes the design values utilized more representative.

<u>Example</u> - Consider the following data excerpted from Table 1 (Auburn left spillway, joint sets of block D):

Table 4: Values for all Observations Joint Sets B, D-1, E

Set	Strike	Dip
B	N 68 E ± 32	90 ± 21
D-1	N 41 W ± 28	40 SW ± 25
E	N 22 E ± 40	36 NW ± 15

Ranges of design values are then chosen with the two above criteria in mind. The result is as follows:

Table 5: Design Values for Joint Sets B, D-1, E

Set	Strike			Dip		
	Average	Range	%	Average	Range	%
B	N 68 E	± 8	80	90	± 10	80
D-1	N 41 W	± 11	79	40 SW	± 8	79
E	N 22 E	± 14	88	36 NW	± 6	88

The overall effect of such a procedure is best illustrated by the stereographic plots of Figures 4a to 4d. If there were no scatter, each of the three sets would be represented by a unique strike and dip which would show as a great circle on the stereo plot (Figure 4a). This very conveniently, would limit the directions of plane intersection to three lines, appearing as points in the projection. Unfortunately, we often start (as in this example) with a large number of sets and a large scatter. The possible areas of joint set intersection will then show as the unmanageable shaded area in Figure 4b. A good knowledge of directions of joint sets intersections is critical in the design considered here, since these define primary directions of failure in the rock mass. The next step shown on Figure 4c consists in retaining only the primary sets, and the final step produces the design graph of Figure 4d. As discussed further it is then

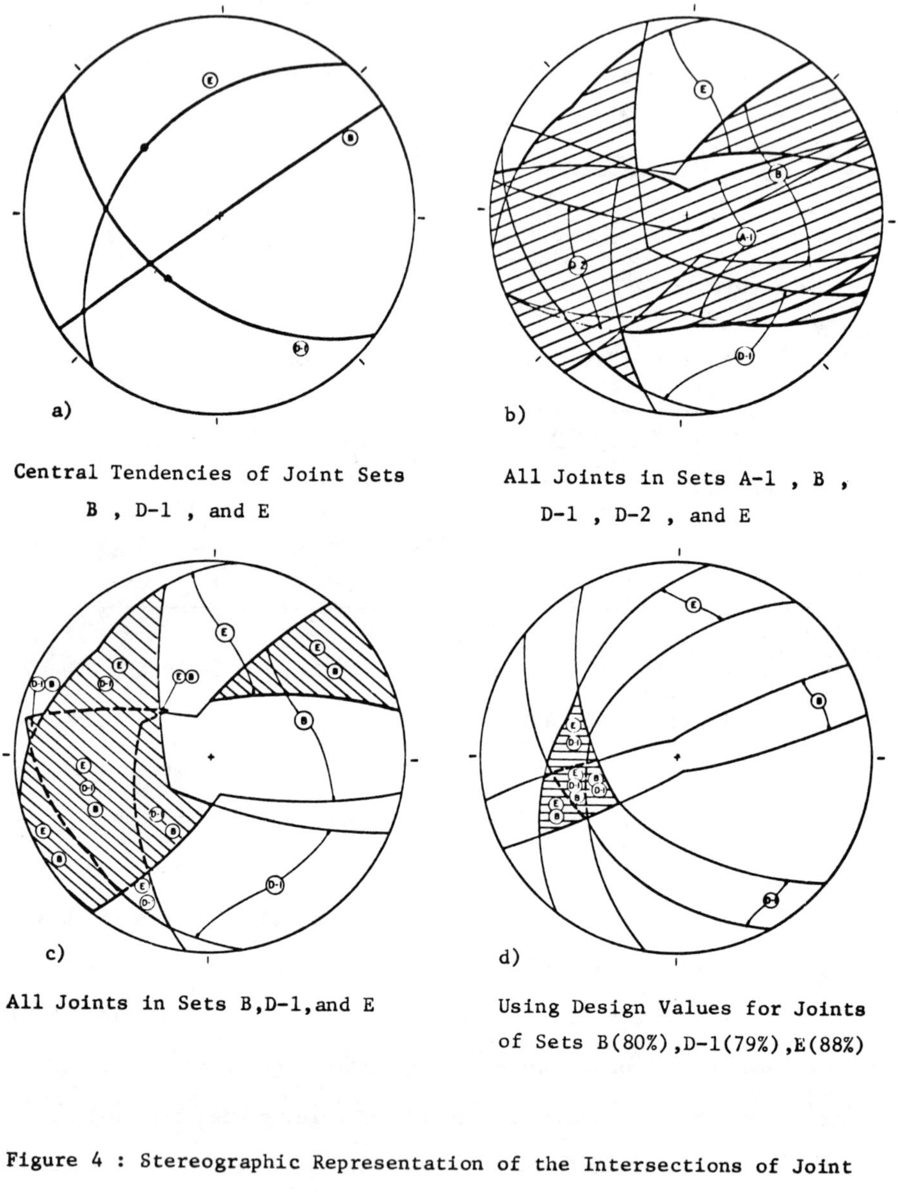

a) Central Tendencies of Joint Sets B , D-1 , and E

b) All Joints in Sets A-1 , B , D-1 , D-2 , and E

c) All Joints in Sets B,D-1,and E

d) Using Design Values for Joints of Sets B(80%),D-1(79%),E(88%)

Figure 4 : Stereographic Representation of the Intersections of Joint Sets Girdles - Auburn Dam,Left Abutment,Block D .

possible to come to realistic conclusions concerning the kinematics of the problem.

c. Determination of design values for joint spacing. An exact knowledge of joint spacing may not be as critical for two reasons:

1. Joint sets form rock blocks of different sizes. However, before their exact size is known one can do valuable kinematic analysis knowing only the directions (strike and dip) of limiting planes.

2. In addition, of all the possible rock volumes found, only a fraction will be entirely defined by joint sets. Many others will have fault boundaries or mixed boundaries (faults and joints). The distance between faults usually far exceeds joint spacings. At the limit, the analysis may only consider a discrete number of major discontinuities and treat the intervening medium as randomly fractured for which overall equivalent strength and deformability properties will be assigned or otherwise obtained. Nevertheless, the respective scaling of joint spacing and structure size remains one of the preliminary design requirements.

d. Summary. The following steps were proposed for the evaluation of geological mapping for engineering purpose:

1. Determine scale of geological features vs. scale of planned excavation.

2. Distinguish between single and multiple geological features.

3. Characterize multiple features for design:

a. Determine the primary joint sets: (i) Use normalized number of observations; (ii) Use normalized percent areal distribution; (iii) Introduce joint properties in the weighting.

b. Select design values for strike and dip of primary sets: (i) An arbitrary minimum percent is retained (suggest 2/3); (ii) An arbitrary maximum range is admissible (suggest 30°).

c. Select design values for spacing.

2. Results of Rock Testing In-Situ

Geomechanical field testing is conducted to provide information on the strength, deformability, and in-situ state of stress of the rock mass. The discontinuities are emphasized here as they tend to be the weakest and most deformable components of the system. Each test has its own theory and method for data reduction. However the following must always be kept in mind to insure the quality of the testing program from planning to conclusion (Heuzé - 1971):

a. When measuring the strength of discontinuities, normal load values must be in the range of those expected in-situ if tests are conducted in the laboratory. The effect of water pressure and water action must be evaluated.

b. The measurement of in-situ stresses can be subject to very significant errors if, for example, the deformability of the rock mass is not determined accurately (Heuzé, 1971). This is particularly true of methods using deformation gages where errors in modulus (E) will be directly carried into the stress values. Abnormally high values of stresses may indicate a particular influence of faults or other discontinuities on the stress field, or the presence of tectonic activity (near surface st stresses in outcrops). In either case the excavation may encounter stability problems. As an example, large rebound in surface excavation into a highly stressed bedded formation caused openings along bedding planes and grouting of the site proved difficult. In addition to the absolute magnitude of the stress field one must also consider its anisotropy; this is particularly relevant to the choice of orientation for excavations.

c. There are two prerequisites to a meaningful deformability test in jointed rock: first, that the scale of the test be adequate when compared to the spacing of fractures so that an overall deformability picture can be obtained; and second, that the load ranges utilized be those of the projected structure. In the case of a high cut for example, the toe of the slope shall be loaded by the dead weight of the overburden to an extent that is relatively easy to evaluate. Finally, the contrast in deformability between rock blocks and discontinuities will decide whether a "rigid block" type of analysis is adequate or if one must account for the deformability of the blocks by Finite Element analysis, for example. (Mahtab and Goodman, 1970, St. John, 1971.) To summarize, when evaluating the quality of a geomechanical testing program, attention focuses on the following points:

1. scale of the tests vs. spacing of discontinuities
2. scale of the tests vs. scale of the structure
3. load range of the tests vs. working stresses of the structure
4. "abnormality" of stress magnitude and/or orientation
5. deformability contrast of discontinuities vs. rock blocks.

3. <u>Results of Flow Studies</u>

The purpose of such studies, in the problem at hand, is to determine the spatial distribution of the fluids and the values of potentials in order to assess the magnitude of flow forces acting on the structure and the possible physico-chemical influence on the discontinuities. In contrast to flow in porous media, flow in fractured rock masses saw very little systematic investigations until recent years. Answers are now being developed in the following areas.

a. <u>Basic laws of fluid percolation in fractured media</u>. It will be generally assumed that

1. All fluid flow is along open fractures (Snow, 1968). Matrix

permeability is generally negligible when compared to fracture conductivity, Wilson and Witherspoon (1970).

2. The flow regime is laminar (Louis - 1969, 1970). Turbulent flow will occur only when abnormally large fractures are found in the vicinity of abnormally high gradient.

3. The rock is inert to the fluid.

b. Distribution and geometry of fractures: According to Snow (1968) all fractures with apertures in excess of 35 microns are potential water conduits. It is thus very unlikely that a numerical approach to fracture flow could include all discontinuities. Most of them will be analyzed through statistical evaluation of water tests (Snow, 1969, 1970).

c. Fluid conductivity of geological discontinuities: Fluid conductivity is very sensitive to fracture aperture (Louis, 1969). A continuum approach to this question is acceptable only when the fracture spacing is quite small (1/1000, Louis) when compared to the scale of the structure. In other circumstances individual determinations must be made by means of in-situ probes (Louis and Maini, 1970).

d. Flow patterns in fractured media: When the continuum approach is permissible, anisotropic solutions for the potential distributions are obtained by the Finite Element Method (Zienkiewicz and Cheung, 1967). Analog models (Sharp 1970a, 1970b) also provide satisfactory answers. In the modelling of flow through a discontinuum, two groups have made most of the contributions to date: Louis and Wittke (1967) and Louis (1969) on one hand using graphical and numerical three dimensional methods, modelled flow in networks of discontinuities; on the other hand, Wilson and Witherspoon (1970) using the Finite Element Method include rock matrix as well as fractures in their analysis; they also account (in two dimensions) for varying apertures and interference at fracture intersection.

Both groups however neglect two significant factors: (i) the deformability of the fractures and (ii) the coupling between fluid pressures and rock stresses. These two features are in turn included in Noorishad and Witherspoon's two-dimensional Finite Element solution (1971). They do not foresee major difficuties in extending this work to 3 dimensions.

e. Flow forces: Wittke and Louis (1966) and later Louis (1969) made the first attempts to systematically calculate flow forces per unit volume in jointed rock. They distinguish "bouyant forces" and shear forces. It is assumed that the flow gradient is fairly constant across a rock element. Since the treatment involves integration around the surface of the rock element, one obtains the net flow forces acting on the unit volume of rock. Consequently their evaluation reflects the compressive force due to the absolute mean pressure acting in any region of the system. On the contrary, Noorishad and Witherspoon (1971) do not require assumptions of constant gradients, and obtain the total effect of the absolute pressure on any part of the system.

f. Effect of grouting on flow pattern: Grouting is carried out to reduce water losses, or to reduce uplift pressures or to decrease exit flow gradient, or any combination of the three. The reduction in fluid conductivity of fractures could readily be handled by the techniques of d. above if the extent of the ground percolation were known (Nonveiller, 1970). In the present state of the art (Wittke, 1970) it is more likely that studies will be conducted in a parametric manner to determine which fractures are critical with respect to grouting, and what decrease in conductivity is required for an acceptable flow pattern (Wilson and Witherspoon, 1970 - Noorishad and Witherspoon, 1971).

g. Effect of drainage on flow pattern: This problem is indeed more amenable to direct treatment since the designer can set the location and nature of the drainage feature, and readily determine its effect by one of the techniques mentioned in d. Very pertinent examples concerning optimum location of drainage galleries for rock slopes are presented by Sharp (1970a, 1970b).

h. Summary: The analysis of fluid flow phenomena in jointed rock masses is quite less sophisticated and comprehensive than mechanical analyses of the rock components themselves. At such an early stage in the state-of-the-art the intent is not here to present a procedure for obtaining flow regime data. Instead, reference was made to several lines of research, and their merits were discussed. In a logical order, fluid flow effects will be determined through the knowledge of: basic laws of percolation in rock masses; fracture distribution and aperture, fracture conductivity, and the ensuing use of graphical or numerical methods.

4. Evaluation of Cost Data

Fairly accurate cost data are generally available regarding the main options open to the designer for modifying his preliminary cut choice; these are: additional excavation, drainage (drill holes, tunnels) grouting and rock reinforcement (bolts, anchors, tendons). Reorienting or re-positioning the cuts may also involve special cost. The following remarks are proposed to help in integrating cost data within a general design sequence:

a. Remedial measures such as mentioned above are considered only after the choice has narrowed to a reasonable number of options. These will be the cases of marginal stability where a definite trade-off between various options can be studied.

b. Within the same option, various choices may involve large variations in cost with little difference in efficiency; for example it is not so much the amount of steel that its proper orientation with respect to failure surfaces which will matter. Likewise a large number of drainage holes may not be as efficient as a single drainage tunnel of comparable or even smaller cost. The hydraulic efficiency of such features in discontinuous media is not quite amenable to empirical determinations. It must be the object of fully three-dimensional investigations.

c. It is proposed here that individual cost functions of the various types of design improvements be expressed in terms of a common parameter: the increase in strength or the decrease in applied forces along the critical failure surface. Figure 5 illustrates this concept where the chosen parameter is an equivalent angle of friction on a plane of failure. It follows that the total cost function can then be easily optimized by standard operational techniques.

IV. THREE DIMENSIONAL STABILITY ANALYSIS OF ROCK SLOPE

1. The Steps of the Analysis

To be stated simply, the stability analysis portion of the total design sequence will involve the following four steps

a. Determination of volumes to be analysed which implies identification of the real intersections and of the positive half normals. -- To be termed real, the intersection of two planes of discontinuities must lie inside the rock mass, and to be positive the half normal to the face of a rock volume must be directed towards the outside of this volume --.

b. Application of kinematic stability tests: the blocking and tapering tests defined below.

c. For those volumes passing the above tests, determination of the system of active (weight, seismicity, thrust of structures, water forces)

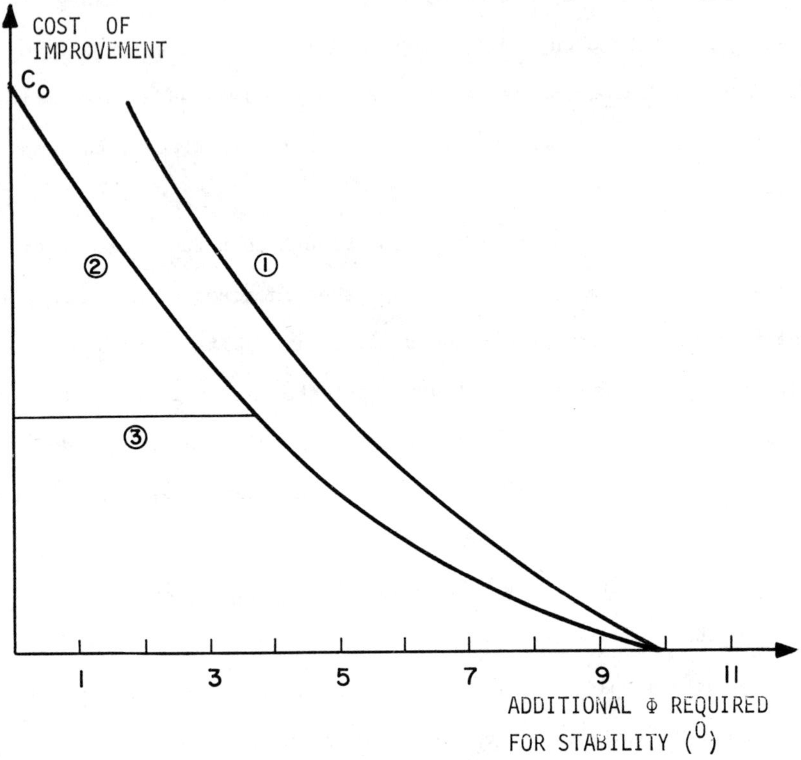

IF FUNCTION 1 : NO SINGLE IMPROVEMENT WILL GIVE STABILITY

IF FUNCTION 2 : STABILITY PROGRESSIVELY OBTAINED AT TOTAL COST C_0

FUNCTION 3 CAN BE ENCOUNTERED WHEN A KINEMATIC CONSTRAINT,SUCH AS A BERM,IS INTRODUCED IN THE COURSE OF THE IMPROVEMENT PROCESS.

Figure 5 : Possible Shapes of Cost Functions for Improvements to Design and passive (cohesive and frictional) forces.

d. Then, application of one or more of the three dimensional stability analysis methods - mentioned below.

2. <u>Cuts and Rock Volumes</u>

The designer is now ready to choose preliminary cuts in the rock mass, considering the purpose of the excavation and known geological factors. These cuts, together with geological discontinuities, will delineate rock volumes. Following the recognition of single features and multiple features it is proposed here to distinguish between single volumes -- formed by single features, and for which a unique size is given --, and multiple volumes -- formed by a combination of single and multiple features -- In this case no unique size is defined for the blocks. It is still possible to analyze their stability but to a lesser extent than for the former type since several parameters of the analysis depend upon size: cohesive strength, relative magnitude of flow forces, weight and seismic forces. On the other hand frictional strength is independent of size and so is the ratio of weight to seismic force where using a pseudo-static analysis. It follows that the analysis of multiple volumes will essentially be of a kinematic nature. Single volumes will also be referred to as <u>fault volumes</u>, and multiple volumes will also be termed <u>mixed volumes</u> if defined by faults and joints or <u>joint volumes</u> if entirely defined by joint sets. The preliminary choice of cut orientation (strike and dip) can be greatly helped by the use of stereographic plots of the proposed cut and known geological discontinuities at the proposed site of the cut. This is illustrated in the case of a particular cut for the Auburn spillway. Figure 6 shows the discontinuities and the real intersections. One can now make an evaluation of any proposed cut in terms of the number of <u>daylighting intersections</u> it creates

-- an intersection will daylight in the lower hemisphere if its plunge is shallower than the dip of the cut while the opposite is true in the upper hemisphere. To put it in non numerical terms, this intersection is piercing through the plane of the cut slope, towards the outside of the rock mass. Table 6 gives the number of such intersections, which are possible directions of blocks failure, for various strikes and dips of proposed cut. The average of such number for various strikes only is then shown on Figure 7; this figure shows vividly how adequate orientation of the cut, if compatible with the purpose of the project, can drastically reduce the number of possible failure events.

Table 6: Number of Real Intersections as a Function of Cut Strike and Dip for the Data of Figure 6

Strike		Dip 30	Dip 45	Dip 60	$\bar{M} = [M_{30} + M_{45} + M_{60}]/3$
0	Easterly Dip	13	18	18	16
30		15	19	20	18
60		16	22	26	21
90		18	34	42	31
120	Westerly Dip	32	47	48	42
150		31	39	41	37
180		28	34	36	33
210		17	30	34	27
240		9	10	16	12
270		2	3	3	3
300	East. Dip	5	7	8	7
330		8	9	13	10

3. The Kinematic Tests: Blocking and Tapering

When considering all possible block shapes and positions it can happen

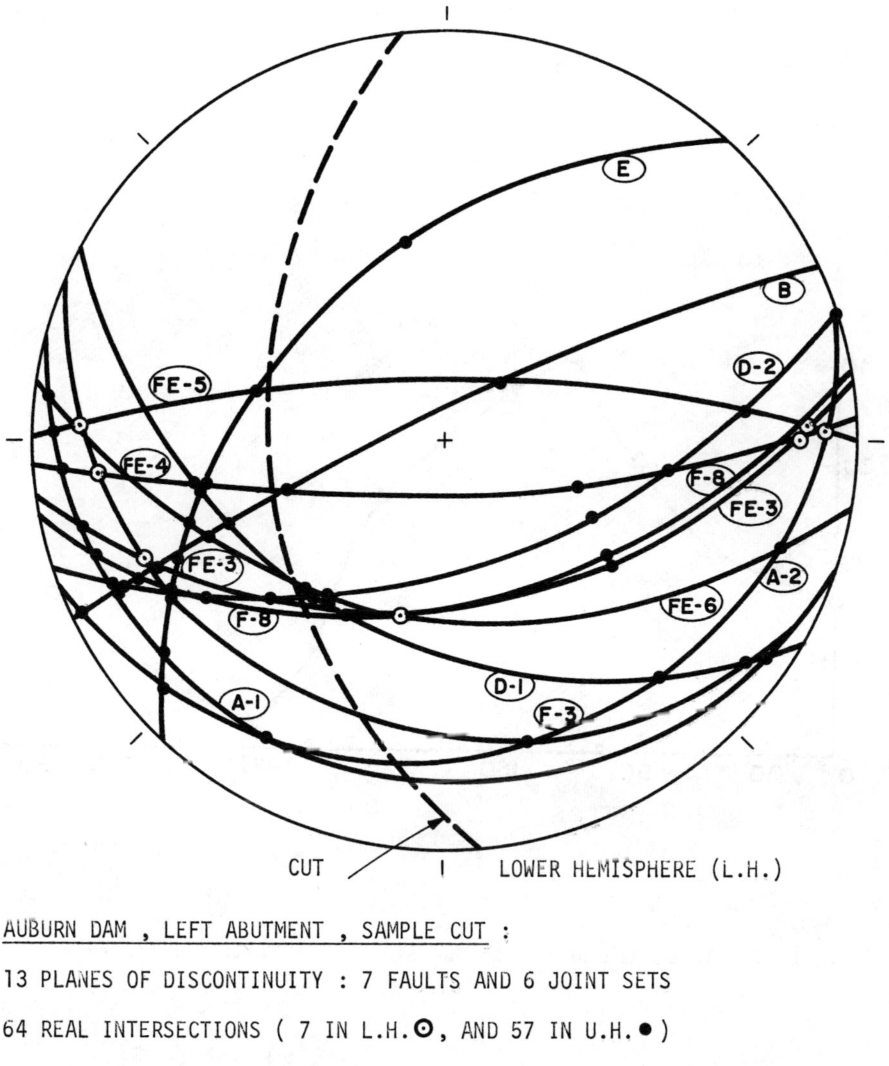

AUBURN DAM , LEFT ABUTMENT , SAMPLE CUT :

13 PLANES OF DISCONTINUITY : 7 FAULTS AND 6 JOINT SETS

64 REAL INTERSECTIONS (7 IN L.H. ⊙, AND 57 IN U.H. ●)

Figure 6 : Typical Cut in Densely Faulted Rock Mass

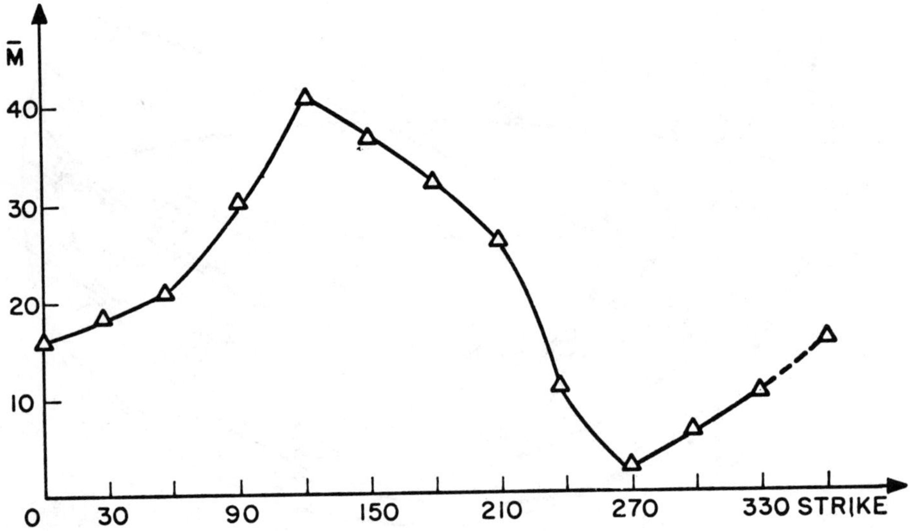

Figure 7 : Variation of $\bar{M}$ With Cut Strike, for Faults of Figure 6

however that certain modes of sliding are in fact kinematically impossible: The wedge cannot move in this mode because it would crush adjacent block(s) in the process, and this is not admissible with a rigid block assumption. When all real intersections are determined and the general resultant of forces R is such that it could create failure, two additional constraints must also be satisfied for movement to take place:

Constraint 1: Plane slide: A plane slide is kinematically possible with respect to a given cut if the intersections adjacent to the plane are possible failure directions.

Constraint 2: Intersection slide: An intersection slide is kinematically possible with respect to a given cut if this intersection is not blocked (i.e. daylights) and if the adjacent faces are not tapered. Indeed this is to be evaluated for both directions corresponding to the line of intersection (i.e. in both upper and lower hemisphere - U.H. and L.H.)

a. Blocking. An intersection will be blocked with respect to a cut if it is steeper than the cut slope in the lower hemisphere or flatter than the cut slope in the upper hemisphere. Steeper means an algebraically greater plunge counted >o downward from the horizontal plane. Although an intersection is flatter than the slope (in L.H.), it may still be blocked if it is deep inside the rock mass. (This last case cannot indeed be resolved for multiple volumes). Figure 8a shows an example of blocked intersection for a 2-plane wedge. Figures 8b and 8c illustrate respectively, also for a 2-plane wedge, the stereographic representations of a daylighting and a blocked intersection.

b. Tapering. Simple tests for tapering are presented for two different cases, according to whether the rock volume has 3 supporting planes or more than 3 planes. Case of 3 planes -- any of the following conditions will create tapering:

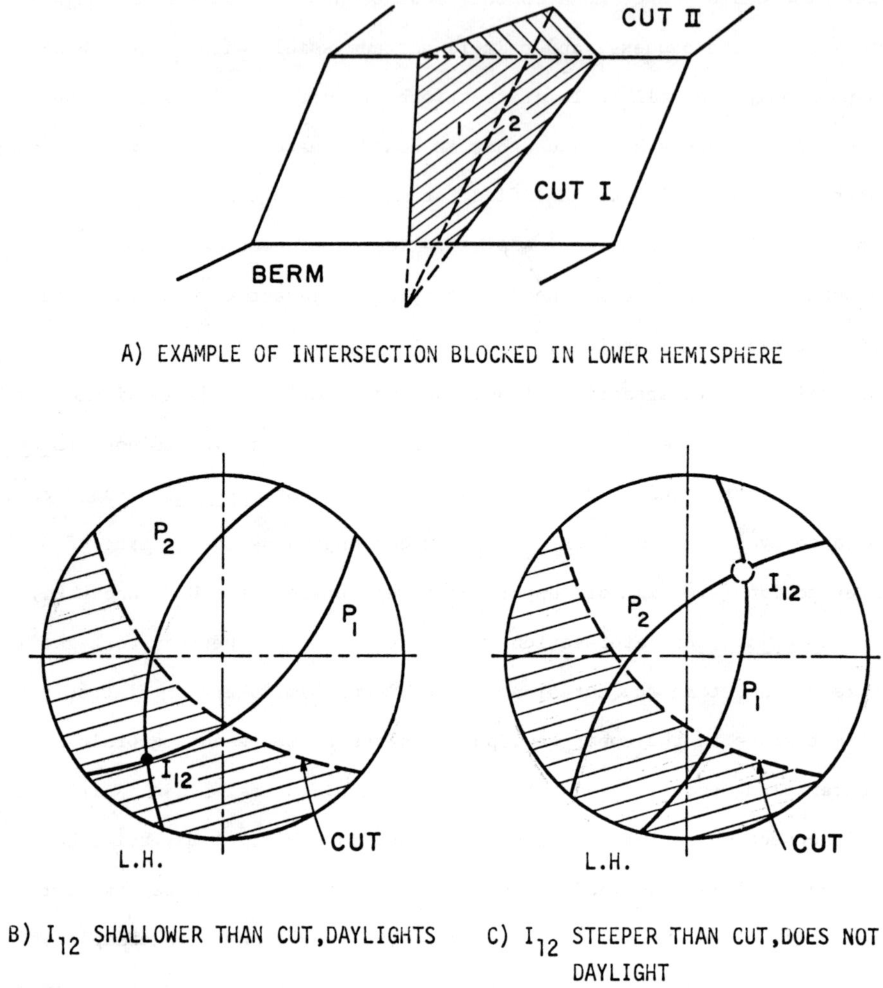

Figure 8 : Stereographic Representation of Daylighting and Blocked Intersections

(1) The apex of the wedge is outside the rock mass.

(2) The area of a section of the wedge parallel to the cut and inside the rock mass is greater than the area exposed on the cut.

(3) Any intersection is simultaneously higher in the slope and more steeply inclined than one or both of the other intersections.

Case of 4 planes or more: Tapering can only be created by non-adjacent faces, and the criterion is as follows: Two non-adjacent faces of a wedge are tapered with respect to a given cut if the angle between their outward normals, measured in the half-space outside the cut, is less than 180°.

Figure 9 shows examples of tapered wedges with 3 or 4 planes. Figure 10 analyzes in stereographic projection one of these wedges; it shows the real intersections, whether or not they daylight, and which faces are tapered.

The advantage of performing these kinematic tests lies in their simplicity and in the fact that they do not require knowledge of the systems of forces involved in the stability analysis. When in presence of a large number of potentially unstable rock volumes their use may drastically reduce the number of cases for which further stability analysis is needed (Blocking: Panet and Struillou, 1970. Tapering: Heuzé and Goodman, 1971b)

c. Application to multiple volumes, mixed modes. The most general case one will encounter is one of multiple volumes and mixed modes. In the case of single volumes, Figure 6, one could determine uniquely all intersections; this is no longer possible here and the designer is in presence of a problem typified in Figure 11 (Auburn Dam, Block E). Several principal joint sets intersect in two girdles (hatched areas) and a number of single features (faults), also create mixed intersections.

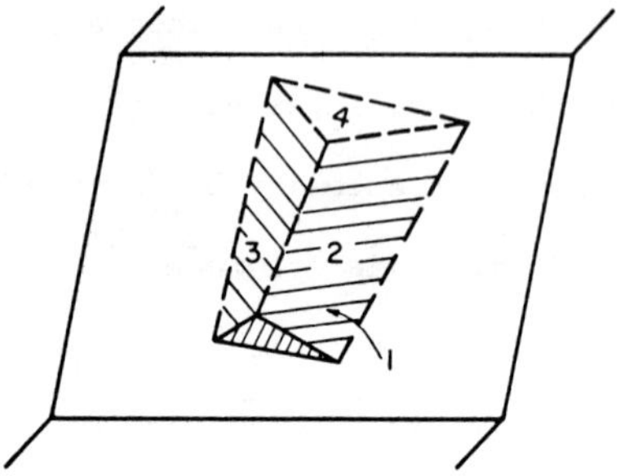

TAPERED W/R TO FACES 1,2,3

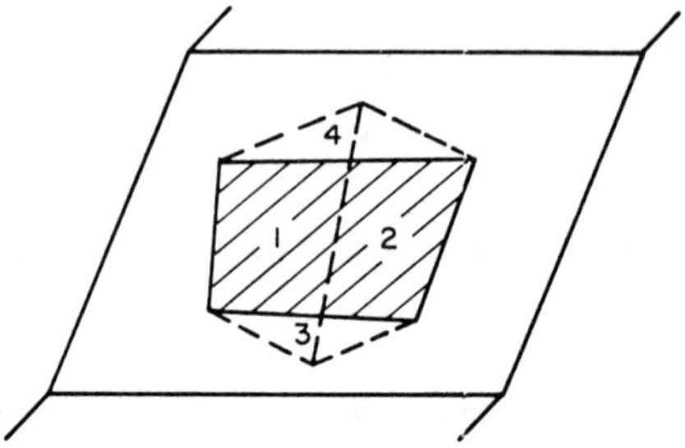

TAPERED W/R TO FACES 3,4

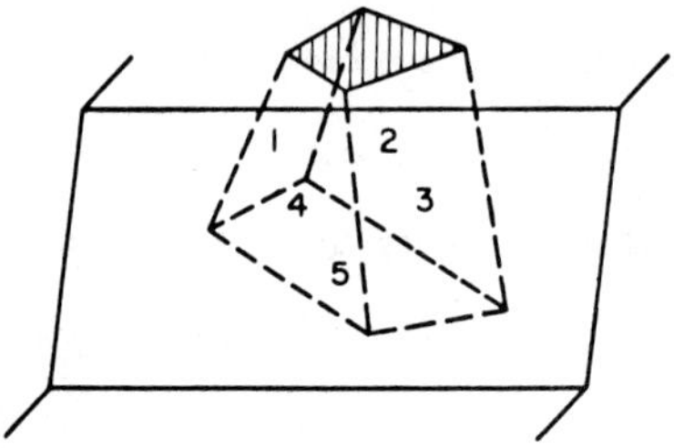

TAPERED W/R TO FACES 1,3 & 2,4

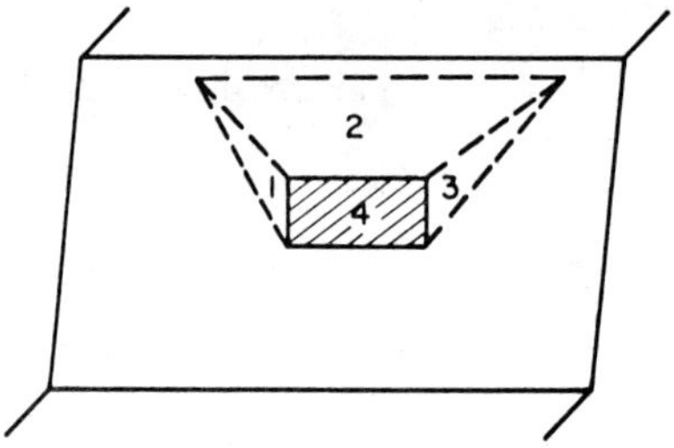

TAPERED W/R TO FACES 1,3

Figure 9 : Examples of Tapered Wedges

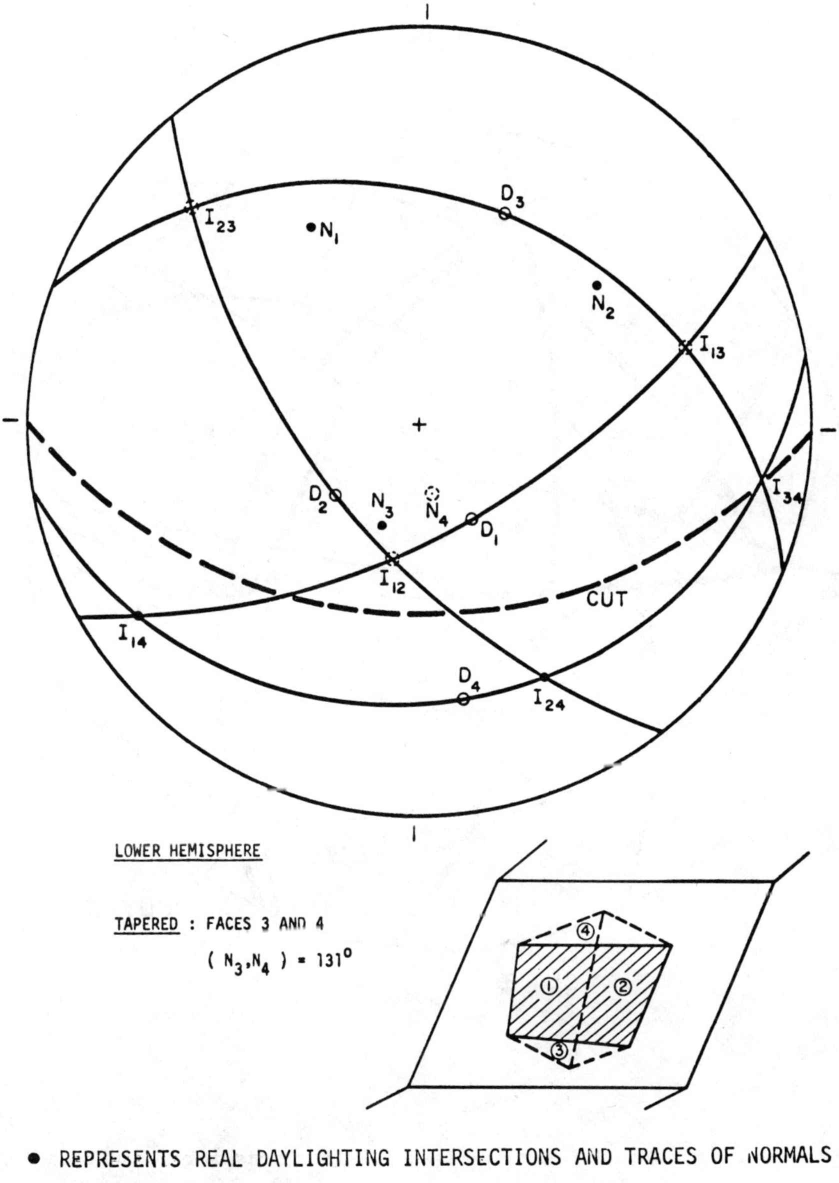

● REPRESENTS REAL DAYLIGHTING INTERSECTIONS AND TRACES OF NORMALS IN LOWER HEMISPHERE

⊙ REPRESENTS OTHER REAL INTERSECTIONS AND TRACES OF NORMALS IN U.H.

o REPRESENTS TRUE DIP VECTORS OF THE PLANES OF DISCONTINUITY

Figure 10 : Stereographic Plot of a 4-Plane Tapered Wedge

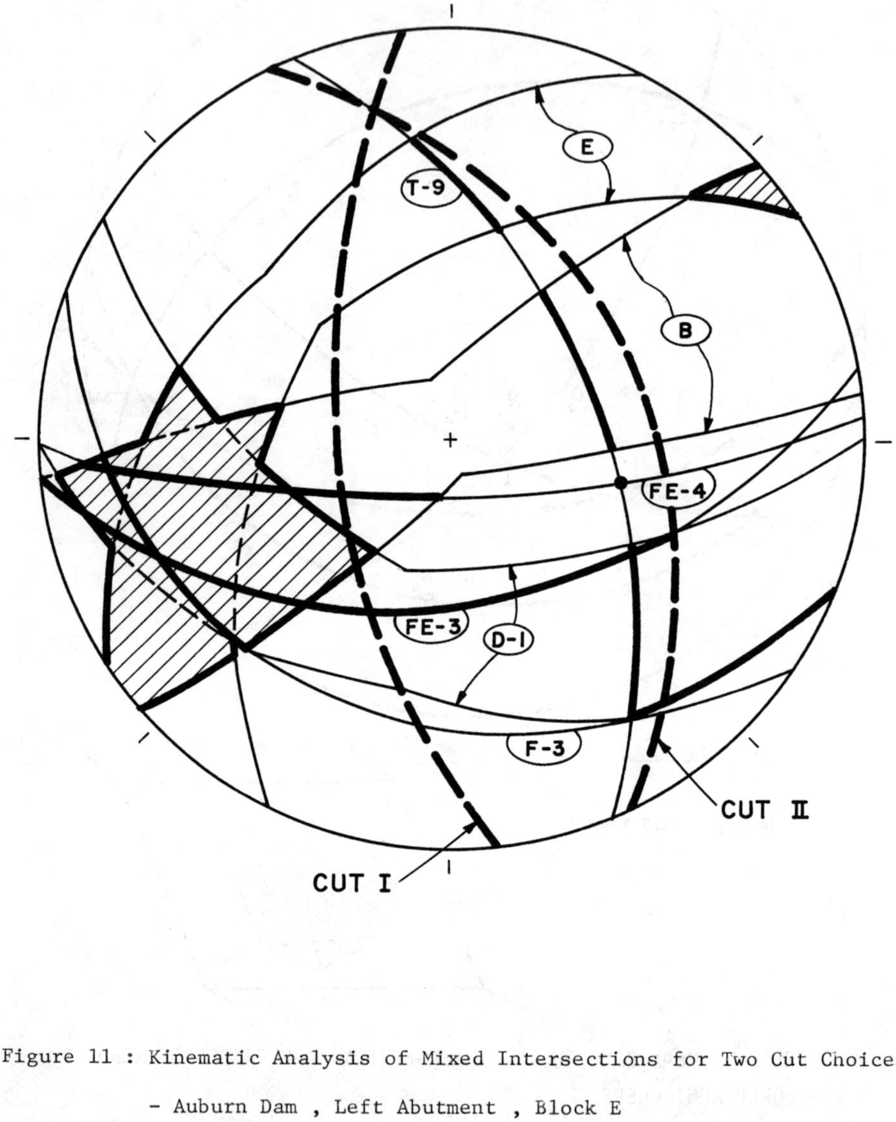

Figure 11 : Kinematic Analysis of Mixed Intersections for Two Cut Choices - Auburn Dam , Left Abutment , Block E

The locus of the trace of these intersections is shown in heavy lines where the faults intersect the girdles of joint sets. Then one can quite precisely, although qualitatively, evaluate the respective merits of proposed cuts. As illustrated here, cut I would create a large number of possibly unstable (daylighting) intersections of all three faults with all three joint sets, whereas cut II would only create daylighting intersections between fault F-3 and joint set D-1, with plunges smaller than about 12°.

4. The System of Forces

Any complete analysis will include the following set of forces:

a. The weight of the rock blocks

b. The thrust of adjacent structures

c. Seismic forces: if a stress analysis is used (Clough and Chopra, 1966) the model can be dynamic. In other cases a pseudo-static approach will be taken which equates the shock to a force proportional to the weight of the rock volume.

d. Water forces: methods to determine flow patterns including static and dynamic water forces were discussed above. This information can be used as well in stress or limit equilibrium approaches. The need for using coupled stress-flow solutions is again emphasized here (Noorishad and Witherspoon, 1970 - Dubois and Noorishad, 1971).

5. The Three-Dimensional Methods and Models

The two approaches to stability analysis, stress and limit equilibrium, include a number of techniques which are briefly listed here as a more complete treatment is beyond the scope of this article and can be found in published reports (Heuzé and Goodman, 1971a).

a. The stress approach: Two techniques are mentioned: 1. Photoelasticity (Agarwal, 1964) and 2. Finite Element (Mahtab and Goodman,

1970 - St. John, 1971). Their use is particularly warranted when large rock blocks are deformable to a significant degree so that the behavior of the rock mass is no longer clearly governed by the discontinuities. Except in weak rocks the blocks can be considered to behave in a linear elastic manner, whereas severe non linearities exist at the level of the joints.

b. The limit equilibrium approach: It can utilize the following tools:

1. Inert physical models and materials (Fumagalli, 1968 - Stimpson, 1970). Most of the information they provide is generally of a qualitative nature because it is very hard to satisfy scaling requirements for all important variables, simultaneously. However, they remain very valuable for the conduct of kinematic studies (Goodman and Erguvanli, 1971).

2. Analytical methods of vector operations (Wittke, 1965 - Goodman and Taylor, 1966). Their main advantage is their ability to handle, in addition to sliding modes, the rotational and toppling modes of block failures.

3. Stereographic projection (Londe, Vigier, Vormeringer, 1969, 1970 - John, 1968, 1970 - Goodman, 1969). This is a very powerful technique of three dimensional analysis for the sliding stability of rock slopes. Frictional and cohesive strength of the discontinuities are accounted for; all forces can be modelled including water pressure which can be graphically varied parametrically.

c. Summary. The interrelationship of methods and inputs discussed above is summarized in Table 7.

Table 7: Character of Design Methods for Slopes in Jointed Rock

	Limit Equilibrium	Stress Analysis
Examples	Inert physical models Analytical models (vectors) Stereographic projections	Photoelastic models Finite Element methods
Applies to	Single volumes or Multiple volumes (Must be non deformable)	Single volumes and their boundaries (All components may be deformable)
Systems of Forces	Weight as a centroid force Thrust as an external force Seismicity as a static equivalent Water action as external forces	Weight as a ubiquitous force Thrust as boundary loading Seismicity as a dynamic action with appropriate modes Water forces coupled with other modes of loading

IV. PROPOSED DESIGN SEQUENCE

To summarize, a design procedure will be proposed now. Options available at different stages are discussed together with simplifications in some special cases.

1. The Design Procedure

This logical sequence for optimal design of high cuts in jointed hard rock integrates the tools and methods discussed above as follows:

a. Analysis of geological data

1. Consider location and scale of project
2. Determine single vs. multiple geologic features
3. Select design values of geometry and strength for discontinuities, corresponding to a desired confidence level. $(N_1)^*$

---

*Numbers N_1, N_2 ... represent options along the design sequence and are discussed below.

b. Selection of preliminary cuts (3 stages)

Stage 1

4. Determine all real intersections in the vicinity of the proposed project. This is a bigger number than for any particular cut, but a lesser number than all possible intersections appearing on a stereo-plot of geological data.

5. Establish the relationship between number of such intersections and cut orientation (strike and dip). An example is given on Figures 6 and 7 and in Table 6.

6. From 1. and 5. select a number of preliminary cuts, stage 1. For each cut stage 1:

7. Define all single and multiple volumes.

8. For all volumes perform complete kinematic tests.

9. Determine total number of possible failures for the cut.

Stage 2

10. Select an aribtrary maximum number acceptable in 9. (N_2), and a range of dips for design (N_3). (N_2) and (N_3) determine the stage 2 cuts. For each cut stage 2

11. Determine the flow regime with minimum drainage features (N_4) - coupled flow-stress Finite Element solutions are preferred here.

12. For each single volume related to the cut

a. Map a complete stereoplot including all forces; water pressures (U's) are maximum and correspond to results of 11. This takes care of sliding modes.

b. Conduct vector analysis of rotational failures.

c. Evaluate stiffness ratios (R) and apertures (α) of wedge dihedra.

d. For any R, above an arbitrary limit and α below another limit. (N_5), execute a Finite Element stability analysis of the wedge.

13. For multiple volumes related to the cut

a. Conduct a stereographic stability analysis to the limited extent possible in this case (N_6); friction can be accounted for but not cohesion.

b. Define the maximum block size for which instability will be accepted (N_7). This means that for each multiple volume, a complete analysis of the type 12. above will be performed on different block sizes in order to relate size to instability.

Stage 3

14. Determine the cuts of limit stability with the maximum U,s of point 11 and the maximum size of point 13.b. This limit stability will be expressed in terms of one or two friction angles ϕ_o, ϕ'_o and limit water pressures U_{io}.

15. Select the cuts on either side of the limit stability within specified ranges of ϕ'_s and U's (N_8). They are stage 3 cuts.

c. Final cut selection

The final, optimal, cut is chosen from stage 3 cuts.

16. Establish, for each cut, the individual cost functions of the various methods of improvement and the costs for a given degree of stability with the initial confidence levels. It is again suggested that degrees of stability be expressed in terms of excess degrees of friction, or equivalent friction, available.

17. Optimize the total cost function for each cut according to a given criterion or combination thereof (dollars, environment, time, ...) and select the most advantageous design.

18. Compare optimum designs for various confidence levels. The choice of a new level of confidence requires going over steps 1 through 17 of the procedure.

2. Remarks on Available Design Options

The procedure presented above is meant to be a logical sequence of events. It entails for final, numerical results, the making of decisions with respect to numerical constraints and values. The design options N_1, N_2, etc. are as follows:

N_1: represents the level of confidence one desires. It is directly related to the statistical evaluation of geological data. The choice of N_1 will be based in large part upon costs of failure or partial failure of the structure.

N_2: is an internal cut-off number for which no value is to be proposed that would apply in every case.

N_3: is a dip range which can generally be rather wide (30 to 60°).

N_4: represents the assumptions on what the minimum drainage features will be. It varies with the project.

N_5: relates to the influence of joint stiffness and wedge apertures. Typical limits here could be suggested as R = 100 and = 120°.

N_6: a finite number of face orientations will be selected for each multiple volume; 2 values per face, for a typical tetrahedral wedge, is a number that can be handled with appropriate automated techniques.

N_7: This is to distinguish between ravelling and partial failure. A lower limit will be set for cuts above installations, or with traffic underneath them, for example.

N_8: These are internal cut-off numbers again. Acceptable values for stage 3 cuts could be within a $\Delta\phi = 5^o$ and $\Delta U = + 20\%$ to $- 40\%$ of the limit stability cut.

3. Remarks on Some Special Cases

The proposed procedure is easily adaptable to cases where partial information only is available before decisions, or where the data do not enable complete resolution of unknowns without additional assumptions. Assuming for example that only three systems of joints are present and no faults can be recognized, the analysis will rely upon kinematic tests and maximum unstable block size before proceeding to the block by block investigation. A more thorough exploration of the girdle of orientation can also be conducted in this case.

It can also happen that water forces are especially difficult to evaluate. One will then look for upper values and express all final results in a parametric manner - U's are the parameters - until more information is gained. Also, when the slope is known to be naturally drained and if no external load is present, including seismic shocks, the determination of final stability is greatly simplified. Conversely one may assume complete drainage to obtain a first evaluation of the problem at hand. This type of approach was used in the case of the Auburn Dam left spillway.

V. CONCLUSIONS

All preceeding discussions lead to the following conclusions:

1. The problem of analyzing the stability of high cuts in jointed hard rocks has little in common with that of slopes in a continuous medium.

2. Only three dimensional techniques should be considered as realistic tools of investigations.

3. Such techniques are presently available and can indeed be refined. They were discussed and integrated in a logical design sequence.

4. A major improvement in the series of tools available consists in the introduction of integrated solutions which account for the coupled flow-conductivity-stresses in jointed rock masses.

5. The level of confidence attached to the final results is the same as that attributed to the input. In this respect it can be argued

for example that multiple joint system are not always the product of independent phenomena and can accordingly be regrouped for statistical analysis. i.e. in a 3 set-system the choice of 85% of each set's population yields an 85% confidence level and not $(85\%)^3 = 52\%$.

6. The amount of resources devoted to a particular analysis is dependent upon the choice of numerical parameters (N_i) as discussed above.

7. Considering the number of variables involved in the problem it is strongly recommended that such a procedure be automated for maximum efficiency.

* * * *

ACKNOWLEDGMENTS

The research forming the basis for this article was supported by United States Bureau of Reclamation Contract No. 14-06-D-6990. The authors are indebted to the Bureau's personnel in Denver and in the Auburn field office for their cooperation.

APPENDIX I: REFERENCES

1. Clough, R. W., and Chopra, A. A. (1966) "Earthquake Stress Analysis in Earth Dams", J. of Eng. Mech. Div. ASCE, v. 92, SM2, pp. 197-212.

2. Coates, D. F. (1964) "Classification of Rocks for Rock Mechanics" Int. J. Rock Mech. Min. Sci., v. 1, n. 3.

3. da Silveira, A. F., Rodrigues, F. P., Grossmann, N. F., and Mendes, F. M. (1966) "Quantitative Characterization of the Geometric Parameter of Jointing in Rock Masses", Proc. 1st Congr. Int. Soc. Rock Mech., vol. 1, n. 2.21, Lisbon, Sept.

4. Dubois, J. and Noorishad, J. (1971) "A Coupled Flow-Stress, Two Dimensional Finite Element Program" Internal Report, Geological Engineering, University of California, Berkeley, Calif.

5. Duncan, J. M. and Goodman, R. E. (1968) "Finite Element Analysis of Slopes in Jointed Rock" Report to U. S. Army Corps of Engineers, Vicksburg, Miss., Contract No. DACW39-67-C-0091 by Geotechnical Engineering, University of California, Berkeley.

6. Fumagalli, E., (1968) "Model Simulation of Rock Mechanics Problems" Ch. 11 in Rock Mechanics in Engineering Practice, Zienkiewicz and Stagg Ed.

7. Goodman, R. E. and Taylor, R. L. (1966) "Methods of Analysis for Rock Slopes and Abutments: A Review of Recent Developments" Proc. 8th Symp. Rock Mech.: Failure and Breakage of Rock, Fairhurst Ed. AIME, pp. 303-320.

8. Goodman, R. E. (1969) "Discussion of Graphical Stability Analysis of Slopes in Jointed Rock" by Klaus John, Proc. ASCE, v. 95, n. SM2, March.

9. Goodman, R. E. and Erguvanli, K. (1971) "Applications of Models to Engineering Geology for Rock Excavations" submitted for publication to the Bulletin of Ass. of Eng. Geol.

10. Heuzé, F. E. (1971) "Sources of Errors in Rock Mechanics Field Measurements and Related Solutions" Int. J. Rock Mech. Min. Sci. (in press).

11. Heuzé, F. E. and Goodman, R. E. (1971a) "A Design Procedure for High Cuts in Jointed Hard Rocks - Three-Dimensional Solutions - Final Report to U. S. Bureau of Reclamation, Denver, Contract 14-06-D-6990, Vol. I, from Geological Engineering, University of California, Berkeley.

12. Heuzé, F. E. and Goodman, R. E. (1971b) "Three-Dimensional Stability Studies of the Auburn Dam Spillway Project" Report to U. S. Bureau of Reclamation, Denver from Geological Engineering, University of California, Berkeley, Contract 14-06-D 6990, Vol. II.

13. John, K. W. (1968) "Graphical Stability Analysis of Slopes in Jointed Rock" Proc. ASCE J. Soil Mech. Found. Div. v. 94, SM2, March.

14. John, K. W. (1970) "Three Dimensional Stability Analyses of Slopes in Jointed Rock" Paper presented at the Johannesburg Symposium on Rock Mechanics, South Africa, Sept.

15. Laufer, H. (1958) "Gebergsklassifizierung fur den Stollenbau" Geologie und Bauwesen, v. 24, n. 1.

16. Londe, P., Vigier, G., and Vormeringer, R. (1969) "Stability of Rock Slopes, A Three Dimensional Study" Proc. ASCE J. Soil Mech. and Found. Div., v. 95, SM7, Jan.

17. Londe, P., Vigier, G. and Vormeringer, R. (1970) "Stability of Rock Slopes - Graphical Methods" Proc. ASCE, v. 96, n. SM4, July.

18. Louis, C. (1969) "A Study of Groundwater Flow in Jointed Rock and Its Influence on the Stability of Rock Masses" Imperial College,

London, Rock Mech. Res. Report Sept.

19. Louis, C. (1970) "Ecoulement à Trois Dimensions dans les Roches Fissurées", Revue Industrie Minérale, Special Issue, July.

20. Louis, C. (1970) "Determination of In-Situ Hydraulic Parameters in Jointed Rock" Proc. 2nd Cong. Int. Soc. Rock Mech. v. 1, Theme 1, n. 32, Belgrade, Sept.

21. Mahtab, A. and Goodman, R. E. (1970) "Three-Dimensional Finite Element Analysis of Jointed Rock Slopes", Proc. 2nd Cong. Int. Soc. Rock Mech., Belgrade, v. 3, n. 7-12.

22. Muller, L. (1964) "The Stability of Rock Slopes and the Effect of Water on the Same", Int. J. Rock Mech. Min. Sci., v. 1.

23. Nonveiller, E. (1970) "A Rational Approach to the Design of Grout Curtains" Proc. 2nd Cong. Int. Soc. Rock Mech., Belgrade, v. 2, paper 6-13.

24. Noorishad, J. and Witherspoon, P. A. (1971) "Finite Element Analysis of Rock Mass Behavior Under Coupled Flow-Stress Action" Geological Engineering Report, University of California, Berkeley.

25. Obert, L. and Duvall, W. I. (1967) "Rock Mechanics and the Design of Structures in Rock", Wiley.

26. Panet, M. and Struillou, R. (1970) "Méthodes d'Etude de la Stabilité des Déblais Rocheux pour les Projets de Routes et d'Autoroutes" Proc. 2nd Cong. Int. Soc. Rock Mech., v. 3, Theme 7, n. 9, Belgrade, Sept.

27. Sharp, J. C. (1970a) "Fluid Flow Through Fissured Media" Ph. D., Dissertation, Imperial College, London, Sept.

28. Sharp, J. C. (1970b) "Drainage Characteristics of Subsurface Galleries" Proc. 2nd Cong. Int. Soc. Rock Mech., v. 3, Theme 6, n. 10, Belgrade, Sept.

29. Snow, D. T. (1968) "Rock Fracture Spacings, Openings and Porosities" J. Soil Mech. Found. Div. ASCE, SM7, v. 94, Jan.

30. Snow, D. T. (1969) "Anisotropic Permeability of Fractured Media" Water Resources Research, v. 5, n. 6, Dec.

31. Snow, D. T. (1970) "The Frequency and Apertures of Fractures in Rock" Int. Journ. Rock Mech. Min. Science, v. 7, n. 1.

32. St. John, C. M. (1971) "Three-Dimensional Analysis of Jointed Rock Slopes", preprint for Symposium Int. Soc. Rock Mech., Nancy, France, Oct. 4-6.

33. Stimpson, B. (1970) "Modelling Materials for Engineering Rock Mechanics", Int. Journ. Rock Mech. Min. Sci., v. 7, n. 1, January.

34. Talobre, J. A. (1967) "La Mécanique des Roches" 2nd Ed., Dunod.

35. Terzaghi, K. (1946) "Introduction to Tunnel Geology" Rock Tunnelling with Steel Supports, Proctor and White Ed., Commercial Shearing and Stamping Co., Youngstown, Ohio (reprinted 1966).

36. Wilson, C. R. and Witherspoon, P. A. (1970) "An Investigation of Laminar Flow in Fractured Porous Rock" Report 70-6 Geotechnical Engineering, Univ. of California, Berkeley.

37. Wittke, W. (1965) "Methods to Analyze the Stability of Rock Slopes, With and Without Additional Loading", Felsmechanik und Ingenieurgeologie Supplementum II, Springer, Vienna, pp. 52-79.

38. Wittke, W. and Louis, C. (1966) "Determination of the Influence of Ground Water Flow on the Stability of Slopes and Structures in Jointed Rock", Proc. 1st Congr. Int. Soc. Rock Mech, Lisbon, v. II, n. 6.14, pp. 201-206.

39. Wittke, W. (1970) "General Report of Theme 6: Improvement of the Properties of Rock Masses", Proc. 2nd Int. Cong. Rock Mech., Belgrade vol. 3.

40. Zienkiewicz, O. C., and Cheung, Y. K. (1967) "The Finite Element Method in Structural and Continuum Mechanics" McGraw-Hill.

APPENDIX II: NOTATIONS

The following symbols are used in this paper:

$\bar{J}$: Normalized number of joint set observations

$\bar{M}$: Weighted number of real daylighting intersections as a function of cut strike

N_i : A numerical option in the design procedure

R : Stiffness ratio of rock wedge

U : Water pressure

U_{io} : Limit water pressure

α : Wedge aperture

ϕ_o, ϕ'_o : Limit friction angles

TOCKS ISLAND PROJECT SPILLWAY ROCK MECHANICS STUDIES

A. Depman, K. Dodds* and D. Parrillio****

**U. S. Corps of Engineers
Philadelphia District
Philadelphia, Pennsylvania 19106

*Foundation Sciences, Inc.
520 S. W. Sixth Avenue
Portland, Oregon 97204

ABSTRACT

The Tocks Island Project is located on the Delaware River near the Delaware Water gap. The rock mechanics studies reported in the paper were made for the design of the approximately 400-foot high spillway cut slope. Bedrock in the area is the shale, siltstone, sandstone, quartzite, and argillite of the High Falls Formation of Silurian age. Slope stability problems centered on weak zones approximately parallel to the slope of the hillside. Rock mechanics studies for the project include; 14 flat jack tests; 12, 200-ton plate loading tests; 2, 200-ton sustained load tests; 86 in situ seismic measurements of rock dynamic elastic properties; 27 flat jack measurements of rock static elastic properties; and 21 seismic measurements on rock cores of the dynamic elastic properties. The results of these tests establish that the in situ stress in the rock, at the site of the spillway cut, is about 1000 psi; that the stress generally follows the rock structure; and that it can be accounted for by the weight of the overburden. The studies confirm laboratory measures of the coefficient of friction along the weak zones and indicate a stable cut slope following the main weak zone.

I. GENERAL SITE CONDITIONS

Introduction--The Tocks Island Project is the largest of several authorized projects in the Delaware River watershed, covering eastern Pennsylvania, lower New York, and western New Jersey. The project was authorized by the Flood Control Act of 1962 to be constructed substantially in accordance with the recommendations of the Chief of Engineers to meet the basin needs for flood control, water supply, recreation and hydroelectric power.

The dam site is located on the Delaware River about 5 miles above the Delaware Water Gap and 7 miles northeast of Stroudsburg, Pennsylvania. When in operation, the dam will control one-third of the entire Delaware River Basin drainage area.

The dam will impound a lake 37 miles long extending to nearby Port Jervis, New York. Both features form an integral part of the Delaware Water Gap National Recreation Area authorized by Congress to be developed for outdoor recreational use in the population-swollen megalopolis which forms the Northeastern corner of the United States.

The dam will be constructed of about 7 million cubic yards of earth and rock fill with a central impervious core and will have a top elevation 160 feet higher than the existing stream bed.

The presently planned dam will be composed of three major features:

1. 2400-foot long embankment section connected to a spillway section;

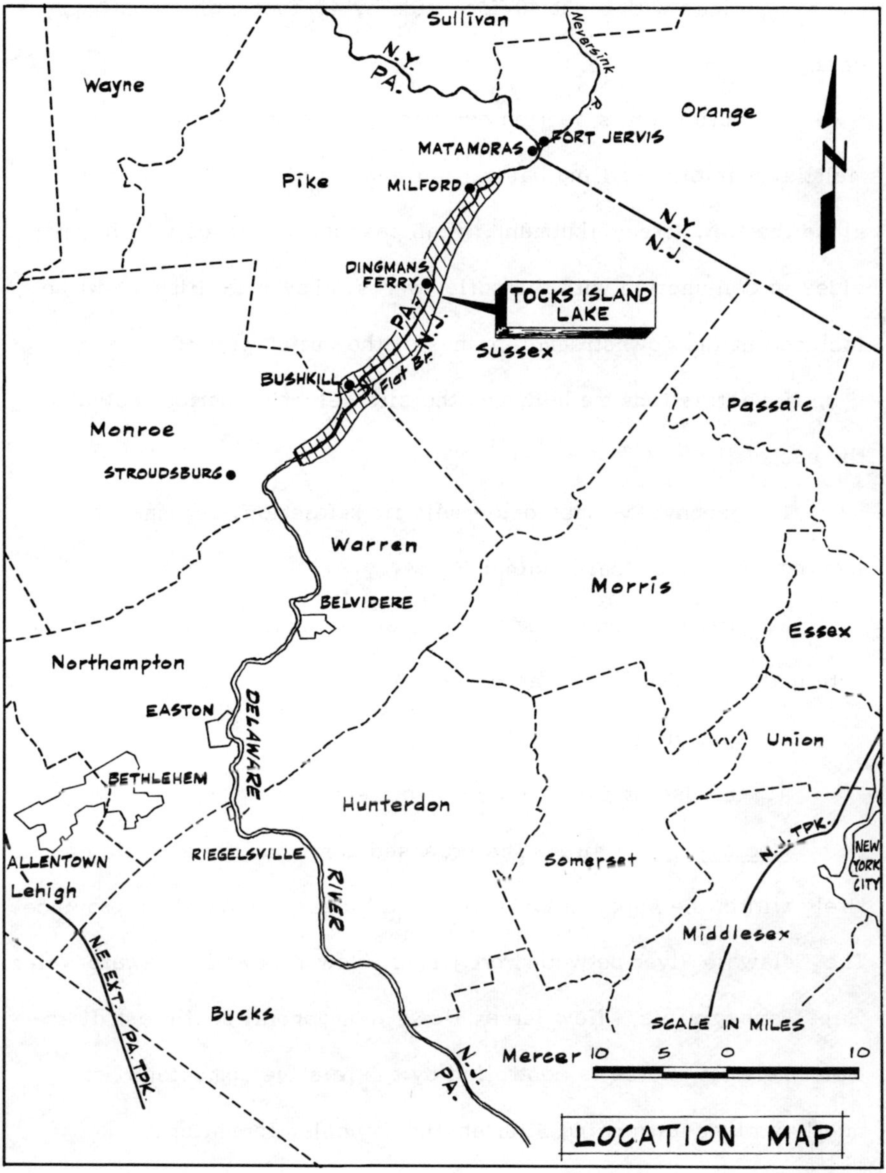
Sullivan
Neversink R.
N.Y.
PA.
Wayne
Orange
FORT JERVIS
MATAMORAS
Pike
MILFORD
N.Y.
N.J.
DINGMANS FERRY
TOCKS ISLAND LAKE
PA.
N.J.
Flat Br.
Sussex
BUSHKILL
Passaic
Monroe
STROUDSBURG
Warren
Morris
BELVIDERE
Essex
Northampton
EASTON
DELAWARE
Union
BETHLEHEM
Hunterdon
N.J. TPK.
NEW YORK CITY
ALLENTOWN
RIEGELSVILLE
Somerset
Lehigh
RIVER
Middlesex
NE. EXT. PA. TPK.
Bucks
SCALE IN MILES
Mercer
10 5 0 10
N.J.
PA.
LOCATION MAP

Figure 1

2. 383-foot wide concrete spillway located in the left abutment, equipped with eight 40-foot wide by 40-foot high operating gates;

3. outlet works section comprised of an intake tower, with multilevel inlets, stilling basin, and tunnels also located in rock along the New Jersey abutment. Fish passage facilities will be provided in conjunction with the outlet works. Power facilities will be included at the downstream terminus of the outlet works.

Investigations made during the site selection period included the use of:

1. geophysical methods - seismic refraction, gamma-ray logging, and electrical logging;

2. borings - including BX, NX, and 4-inch and $5\frac{1}{2}$-inch coring, with undisturbed sampling as needed;

3. test pits; and

4. the use of the NX borehole camera.

Geology of the site--The proposed dam and lake area lie entirely within the Appalachian Valley and Ridge physiographic province. The Delaware River between Port Jervis, New York and Delaware Water Gap, Pennsylvania - New Jersey flows in a general southwest direction where its course is controlled by the relative competency of nearly northeast-trending Silurian and Devonian formations.

At the dam site, stream bed occurs near elevation 300 feet MSL datum. The left abutment is formed by the steep western slope of Kittatinny Mountain which crests at about elevation 1500. Bedrock of the Silurian High Falls Formation is at or within 10 feet of the surface in most areas, but there are some zones where overburden is as deep as 32 feet. From exposures on the left abutment the bedrock surface plunges steeply below the river to depths of 125 to 140 feet, where the High Falls Formation is in gradational contact with the more calcareous Poxono Island Formation. The bedrock surface rises gently below the right abutment where the slope steepens and structurally complex Silurian and Devonian limestones, shales, siltstones, sandstones and dolomites outcrop near elevation 450. The materials above bedrock beneath the river and the right abutment are a series of glacially derived deposits of Pleistocene age which are capped by a thin veneer of alluvium.

The High Falls Formation in the left abutment consists of a gradational series of shales, siltstones, sandstones, quartzites and argillites representing depositional environments ranging from low gradient fluvial to tidal flat.

Structurally, the left abutment forms the southeast limb of a broad southwest-plunging asymmetrical syncline with a southeast-dipping axial plane. Bedding strikes NE-SW parallel to the ridge and dips to the NW from 0-50 degrees with the dip steepening toward the river. The one exception to this within the spillway area occurs in the dip reversals associated with the closed fold just landward of the spillway. The northeast striking cleavage is well developed in finer grain-

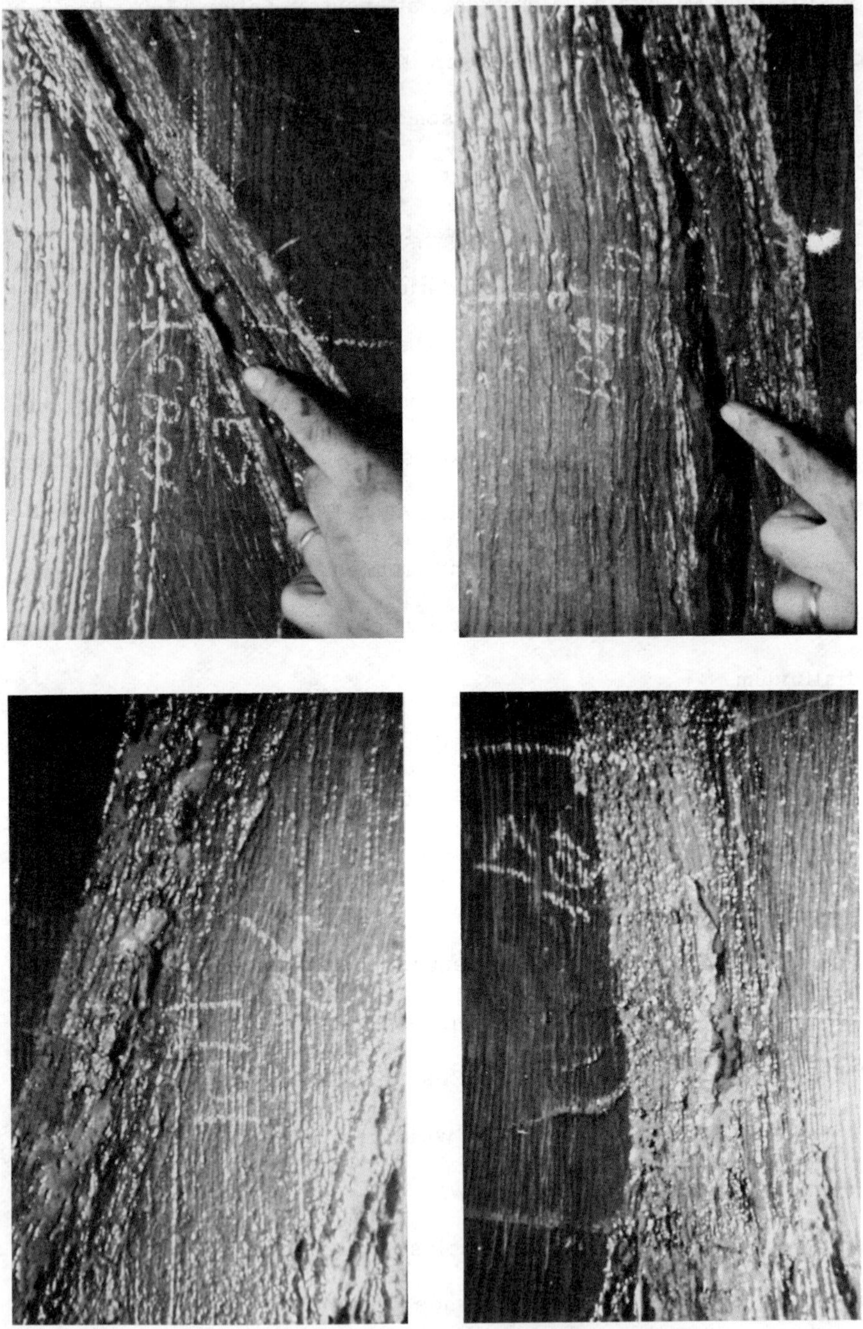

Zone 12 as encountered in the walls of 36" Calyx Hole.

Figure 2

ed beds and dips 45 to 80 degrees to the SE.

Deformation in the abutment is less intense than in older rocks of the region and may represent only a gravitational sliding into the Appalachian basin to the west generated by tectonic forces further east.

Two prominent joint sets are well developed. Both strike approximately parallel to the bedding (N62° - 74°E). One set, dipping to the southeast generally parallel to cleavage, becomes tight or incipient and rare with depth. Joint surfaces are irregular to slightly uneven. Dips average 55 degrees but vary from 30 to 80 degrees with steeper dips predominating to the southeast.

Another joint set dips to the northwest at an average of 52 degrees, but ranges from 15 to 85 degrees. This set usually has smooth surfaces and prominent vertical and lateral extent. Joint planes are usually steeper than bedding planes, but there are many zones where both dip coincidentally.

A third joint strikes approximately north-south with an average dip of 79 degrees to the east, but ranges from 65 degrees NE to 85 degrees SW.

Overall, joint spacing ranges from 0.5 feet to 10 feet and generally increases with depth. Partly open joints were found down to the elevation of the lowest structures, as evidenced by staining, but are rare below 250 feet. Based on geometric analysis, detached blocks of any size are statistically uncommon at depth.

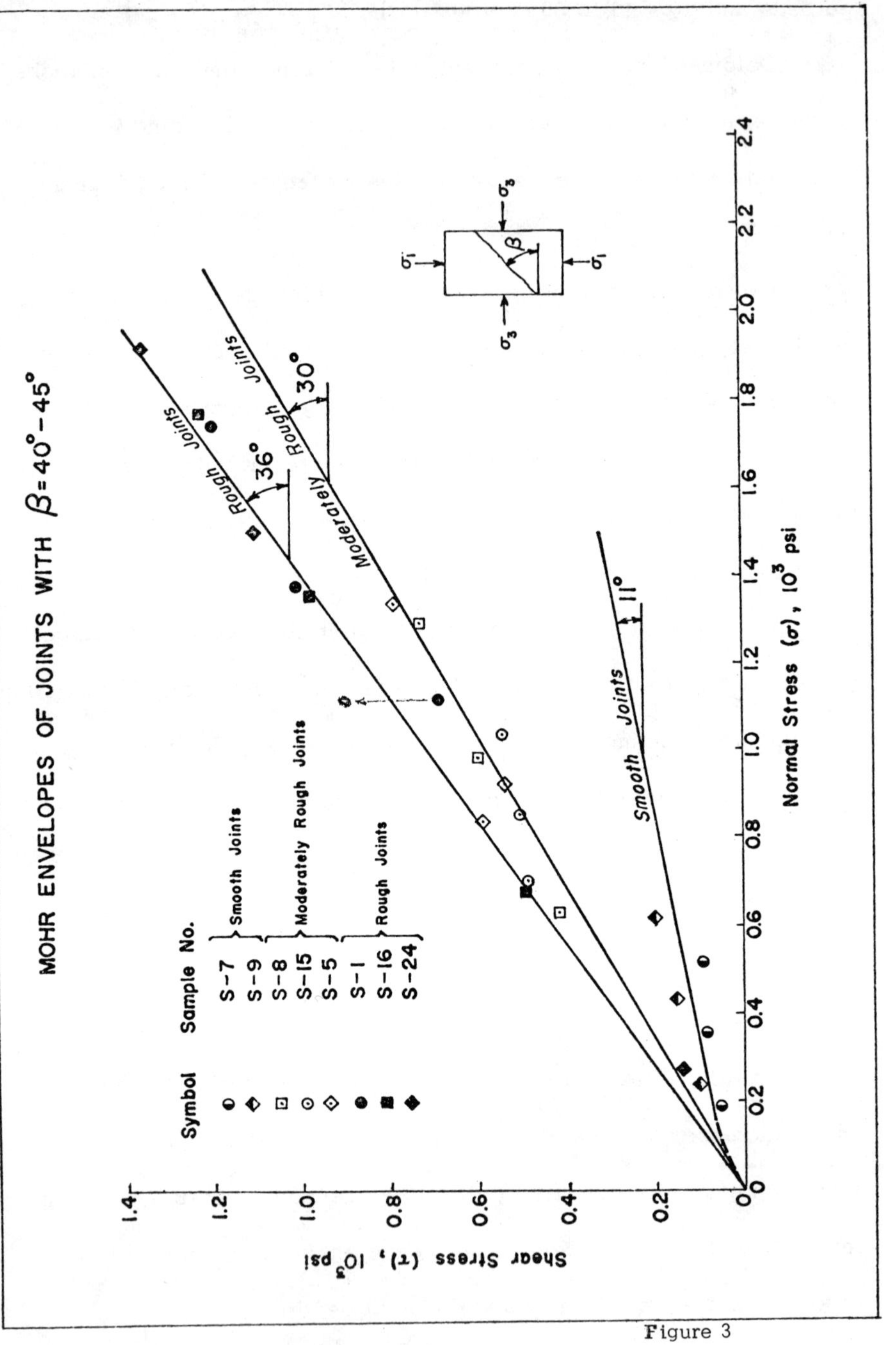
MOHR ENVELOPES OF JOINTS WITH β = 40° - 45°
Symbol
Sample No.
S-7
S-9
Smooth Joints
S-8
S-15
S-5
Moderately Rough Joints
S-1
S-16
S-24
Rough Joints
Rough Joints
36°
Moderately Rough Joints
30°
Smooth Joints
11°
σ1
σ3
β
Shear Stress (τ), 10³ psi
Normal Stress (σ), 10³ psi
0
0.2
0.4
0.6
0.8
1.0
1.2
1.4
1.6
1.8
2.0
2.2
2.4

Figure 3

Laboratory tests to determine the frictional characteristics of open joints with varying degrees of roughness showed that with joint angles as steep as 35 degrees from the horizontal, all failures occurred in intact rock regardless of joint surface conditions. As joint angles increased, friction angles were controlled by the degree of roughness present. Cohesion values were attributed to shearing of asperities on the joint surfaces.

II. SPILLWAY DESIGN AND NEED FOR DETAILED STUDIES

Initial investigations in the area were directed mainly at the location of the embankment on glacially derived foundation materials. A boring program initiated to develop this information revealed unfavorable conditions for a right abutment spillway and outlet works. Accordingly, in the final stages of site development, tentative locations for structures in the left abutment were investigated.

During these investigations it became evident that persistent core loss zones were being experienced in borings made throughout the left abutment area. In the analyses of the slope stability, those zones appeared to be controlling rock features which required further definition and evaluation prior to any final location or orientation of left abutment structures, and the investigations of the zones needed to be of a type that would permit in situ observation and study. As a result, in addition to conventional core borings, four 36-inch calyx holes were drilled. One calyx hole 96 feet deep was drilled on the riverward side of the proposed spillway crest structure to investigate weak rock zones which

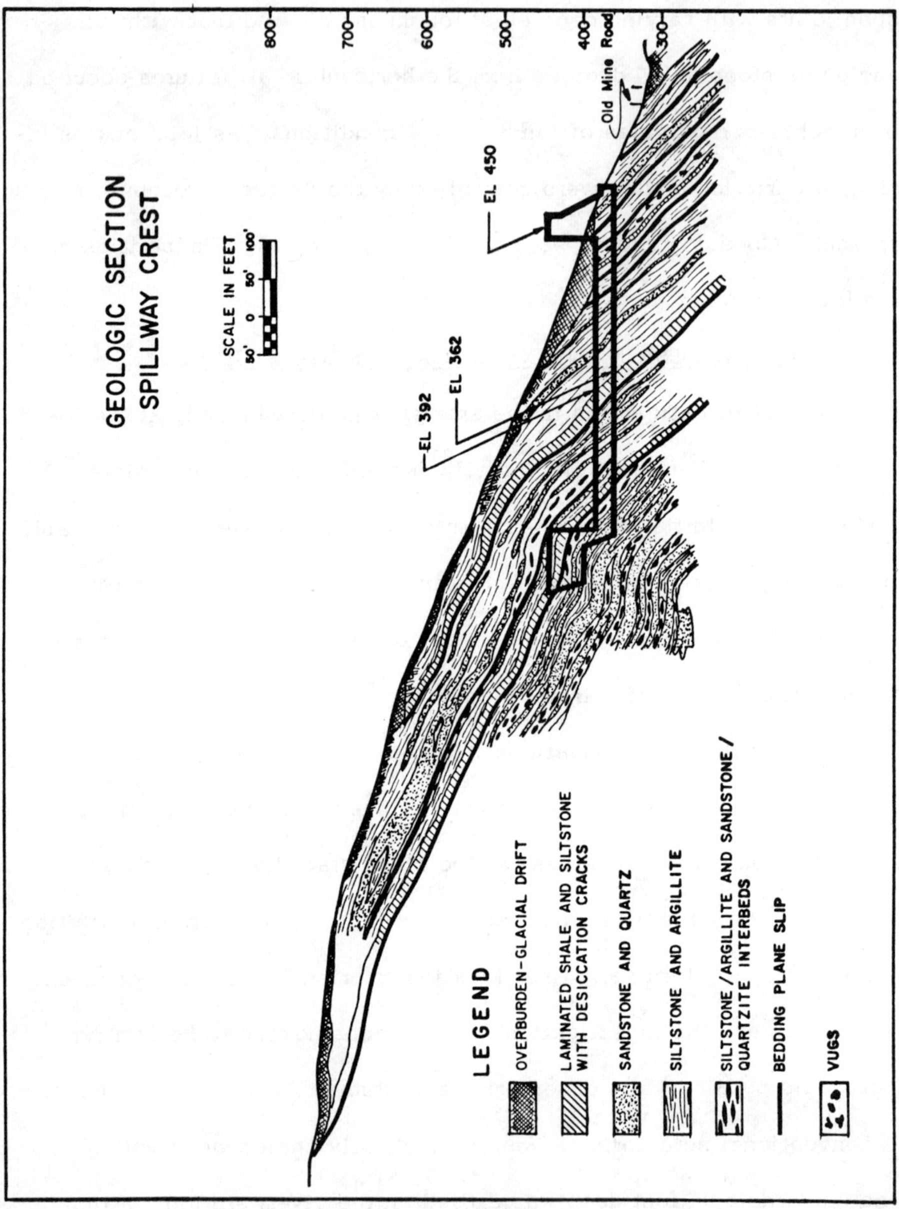
GEOLOGIC SECTION
SPILLWAY CREST
SCALE IN FEET
50' 0 50' 100'
800
700
600
500
400
300
Old Mine Road
EL 450
EL 392
EL 362
LEGEND
OVERBURDEN-GLACIAL DRIFT
LAMINATED SHALE AND SILTSTONE WITH DESICCATION CRACKS
SANDSTONE AND QUARTZ
SILTSTONE AND ARGILLITE
SILTSTONE / ARGILLITE AND SANDSTONE / QUARTZITE INTERBEDS
BEDDING PLANE SLIP
VUGS

Figure 4

would be encountered in excavation and would be used as foundation material for the massive right spillway wall. The other three calyx holes, 290, 184, and 134 feet deep respectively, were located in the steep slopes landward of the approach channel, intake structure, and powerhouse complex.

As a result of these explorations, several weak zones were found and are classified structurally as bedding plane slips. In general, they occur in lithologic units consisting of laminated siltstones and shales, exhibiting such features as desiccation cracks, varying degrees of weathering, clay seams, slickensided surfaces and/or mineral replacement. For the sake of location in an otherwise repetitious stratigraphic sequence, the three major zones of significance have been identified by number and are described as follows:

Zone 10 is the uppermost bedding plane slip influencing the spillway design and is a unit of laminated shale with desiccation cracks. The bottom foot of this zone is highly weathered and contains a layer of friable skeletal quartz-manganese and slickensides. A thin seam of clay is also present. These same conditions were found in all borings where this stratigraphic interval was encountered. It must be considered a continuous bedding plane slip throughout the left abutment. This unit will be encountered in landward cuts for both the intake and discharge channels, where it will be dipping into the excavation. It will form a continuous trace across the approach channel, spillway, stilling basin and discharge channel.

Zone 11 is a bedding plane slip which was positively identified in ten of the boreholes. It was also encountered in calyx holes DTC-1, 3 and 4. The areal distribution of the holes in which Zone 11 was identified show that it is not a single continuous feature. Indications of its presence may have been obliterated by poor recovery for the stratigraphic interval in which this bedding plane slip should have been present. Where encountered, this bedding plane slip is at the contact between a siltstone above and distinctly different lithology below; in a sedimentary section with features indicative of slickensides and is at least partly healed by quartz.

Zone 12 is the stratigraphically lowest major bedding plane slip encountered within the proposed excavation and is most critical to design of slopes. It is a series of laminated shales and siltstones with desiccation cracks. Bedding plane slips close to the base and associated porous quartz-manganese zones are always present. The zone of slickensides, although only 0.1 to 0.2 feet thick, show a variation in direction of about 40 degrees, indicating that movement was not in one direction or necessarily all at the same time. In fact, the first movement may have occurred before complete consolidation of the sediment, while the last movement may have occurred as a result of rebound associated with removal of a 10,000-foot thick iceload at the end of Pleistocene times. Where encountered in this area, Zone 12 is within the boundary of fluctuating water tables. It may have been a significant passageway for groundwater, as witnessed by various degrees of weathering. This condition also contributed to the partial solution of the calcareous beds in the immediate area of the bedding plane slips, subsequent deposition of the manganese, and resultant porous nature of this zone.

The significant features in this zone are the slickensided laminated bedrock, the associated maganese-quartz replacement zones and the secondary weathering. In addition to the many individual, closely spaced bedding plane slips, there is a hairline to $\frac{1}{4}$-inch gray clay seam.

A 6-foot section of 36-inch calyx core with Zone 12 in the middle was rockbolted and overcored. The layer of quartz-manganese was 0.2 feet thick, highly pervious, friable, but intact around the entire circumference. In all borings and explorations, a bedding plane slip occurred at this stratigraphic feature. Accordingly, Zone 12 must be considered a continuous plane throughout the left abutment.

Other bedding plane slips have been noted below Zone 12, but they are not considered to have any appreciable lateral extent. No weathering is associated with these features and they are usually partly or completely healed with quartz. In many cases, the quartz cement in these zones was stronger than the enclosing parent rock, as evidenced by breaks in calyx cores.

In order to facilitate further studies, a 5-foot by 7-foot exploratory adit, 600 feet in length, was constructed at the base of the proposed excavation for the spillway crest structure with a 100-foot downstream crosscut along Zone 12 and another 37-foot long downstream crosscut landward of the spillway and in the crown rock of a proposed diversion tunnel.

In addition, tests for shear strength of materials in Zone 12 were performed on material obtained from 4-inch core, and overcored and rock-bolted 36-inch calyx core. Also, remolded tests on clays from this zone in the exploratory adit were performed at the same time. Piezometers were installed in the 4-inch holes to monitor pressures in Zone 12.

III. ROCK MECHANICS STUDIES

The studies to be described were the initial work in the evaluation of the in situ rock conditions for the spillway cut. They had as their purpose: establishing the general deformation properties of the rock, measuring the in situ stress level and pattern, and evaluating the probable range for the strength parameters of the rock along recognized critical weak surfaces.

The rock mechanics tests completed as part of this study include: fourteen flat jack measurements of residual stress, twelve 200-ton plate loading tests and two 200-ton sustained load tests to measure the static viscoelastic properties of the rock, eighty-six in situ seismic measurements of the static viscoelastic properties of the rock, and twenty-one seismic measurements of the dynamic elastic properties of rock cores.

The layout of the tests is shown on Figure 5. The pattern of tests developed was designed to produce some knowledge of the rock on both sides of Zone 12; that is, to establish if the presence of Zone 12

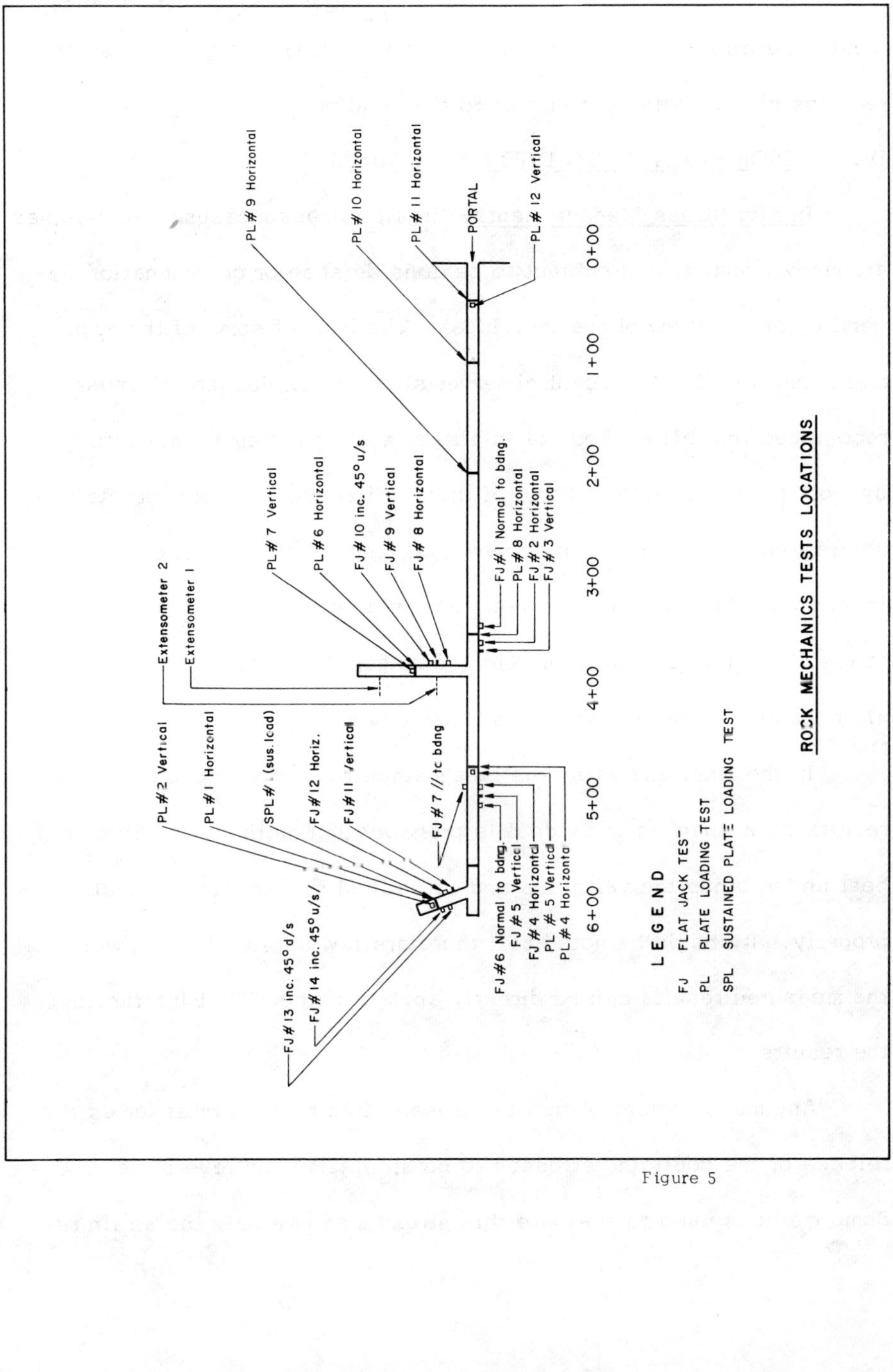
ROCK MECHANICS TESTS LOCATIONS
PORTAL
0+00
1+00
2+00
3+00
4+00
5+00
6+00
PL # 9 Horizontal
PL # 10 Horizontal
PL # 11 Horizontal
PL # 12 Vertical
Extensometer 2
Extensometer 1
PL # 7 Vertical
PL # 6 Horizontal
FJ # 10 inc. 45° u/s
FJ # 9 Vertical
FJ # 8 Horizontal
FJ # 1 Normal to bdng.
PL # 8 Horizontal
FJ # 2 Horizontal
FJ # 3 Vertical
PL # 2 Vertical
PL # 1 Horizontal
SPL # 1 (sus. load)
FJ # 12 Horiz.
FJ # 11 Vertical
FJ # 7 // to bdng
FJ # 13 inc. 45° d/s
FJ # 14 inc. 45° u/s
FJ # 6 Normal to bdng
FJ # 5 Vertical
FJ # 4 Horizontal
PL # 5 Vertical
PL # 4 Horizontal
LEGEND
FJ FLAT JACK TEST
PL PLATE LOADING TEST
SPL SUSTAINED PLATE LOADING TEST

Figure 5

had an appreciable influence on the in situ stresses and the physical properties of the rock. Plate loading tests were made in mutually perpendicular directions and flat jack tests in mutually perpendicular directions plus parallel and normal to the bedding.

IV. DESCRIPTION OF STUDIES AND RESULTS

In Situ Stress Measurements-- In situ stress measuring techniques are many, varied, and subject to zealous defense or condemnation depending on the view of the individual. The heat of some of the arguments can leave an innocent observer slightly nonplussed. It must be recognized that all methods of in situ stress measurement are affected by more than one factor. Some of these influences can be: stress concentrations around the opening where the test is made, stress relief, inability to make all measurements at a single location when analysis of results requires this assumption, changes in elastic modulus of rock after the test is made, variations in rock fabric.

In the final analysis, the test method that will produce the best results for a specific individual is probably that method for which he best understands the variables and can adjust the measured results properly. It should be noted that there are few applications in which the measured results can be directly applied without first interpreting the results.

Any measurement of in situ stresses in a rock mass assumes the release of the confining stresses to be an elastic and reversible process. Some methods used to evaluate this stress measure only the strain resulting from the elastic recovery of the material when the in situ stress is removed, and rely on a calculation based on the material's modulus of elasticity to determine that stress. Other methods measure the resulting strain and then reapply a controlled stress to force the material back to its original position. This reapplied stress is considered to be equivalent to the in situ stress, (Dodds, 1970). Both approaches have limitations, but because of the difficulty in accurately determining the actual modulus of elasticity, the second method (flat jack measurement) is considered the more reliable and thus was chosen.

Summarized, the flat jack test method consists of: a. establishing very accurate measuring points on a prepared rock surface, b. relieving the confinement on the rock by cutting a slot between the reference points and allowing the rock to move into the slot, c. forcing the rock back to its original position with the flat jack, and recording the force required; which is then assumed to be approximately equal to the stress of the rock normal to the plane of the slot. Details of the testing procedure can be found in Dodds, 1966. Figure 6 shows a flat jack installed in the rock and Figure 7 shows the rock slot prior to the installation of the jack.

For this study, flat jack tests to measure the residual stress in the rock were made at fourteen locations with the flat jacks at different orientations on the walls of the adit. Above (river side) Zone 12, three flat jack tests were made on the walls of the main adit at approximately station 3+50, and three flat jack tests in the first crosscut along Zone

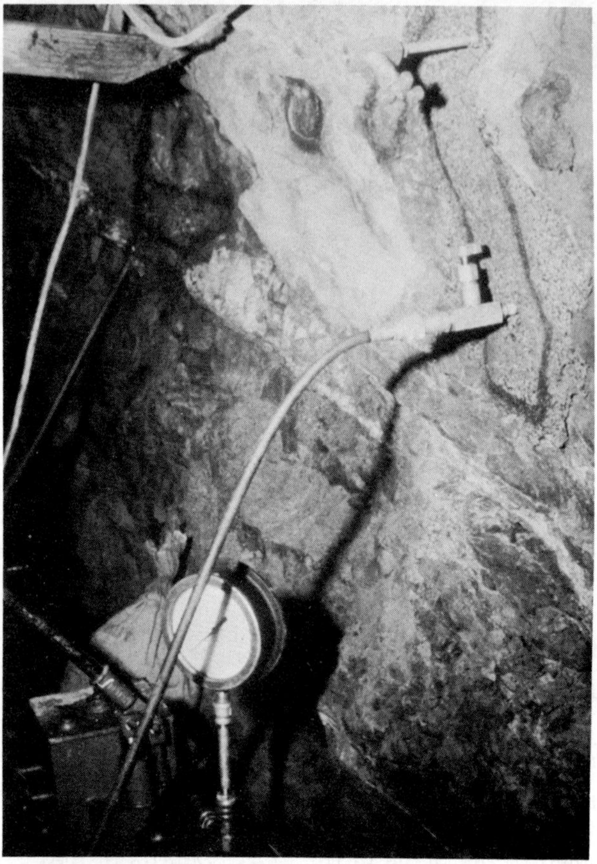

Flat jack test equipment installed.

Figure 6

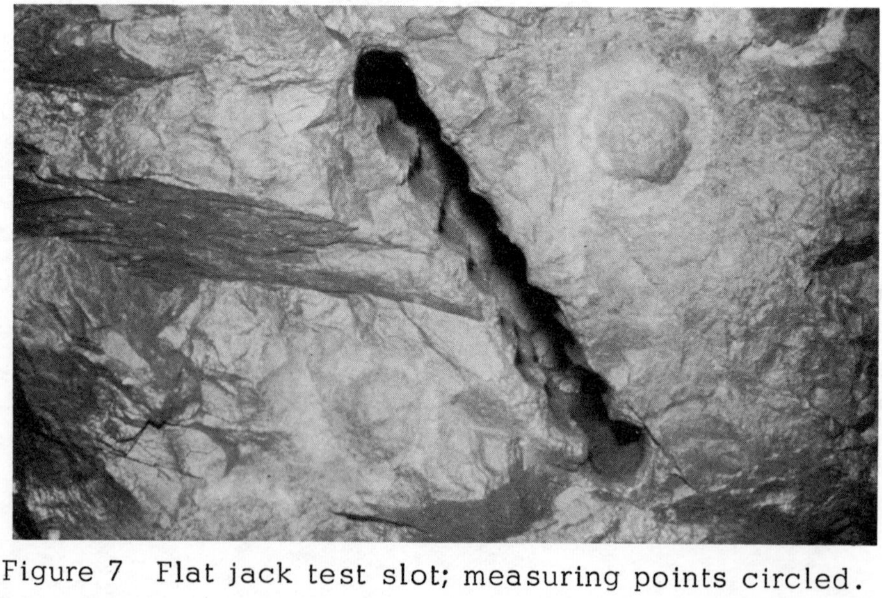

Figure 7 Flat jack test slot; measuring points circled.

12. Below (land side) Zone 12, flat jack tests were carried out in four locations at approximately station 4+85 and four locations at the second crosscut at station 5+80, approximately 30 feet from the intersection of the crosscut with the main adit. No measurements were made of stress intensity or direction in the rock mass away from the openings. The data collected are sufficient to define the general stress intensities at the surface of the openings, and to make reliable estimates of the general undisturbed stress field.

Results of the Flat Jack Tests--The results of the flat jack tests are given on Table I and Figure 8. The results of these tests indicate that most of the stress in the rock can be accounted for by gravity load. However, the distribution of the stress in the rock is controlled by the geologic structures (bedding and shear zones).

1. The results of the stress measurements made between Zone 12 and the surface indicate an undisturbed maximum stress level of about 1,000 psi. This stress is directed almost parallel to Zone 12 and is attributed to the weight of the entire rock mass between Zone 12 and its intersection with the surface.

2. Flat jack tests 4,5,6, and 7 were made toward the hill side of Zone 12. These tests show a maximum adit surface stress level of 525, which is only 50 per cent of that on the other side of Zone 12. These tests show a greater amount of inelastic strain than tests in other areas, which suggests that the results are affected by stress relief. As no measurements of stress or strain behind the adit surface were

TABLE I
FLAT JACK TESTS

FLAT JACK TESTS	JACK POSITION	CANCELLATION PRESSURE	MOVEMENT AT CANCELLATION	"E"
		psi	inches	$x10^6$ psi
1-Top	Perpendicular			
1-Bottom	to Bedding 37°	1950	.0028	5.07
2-Top	Horizontal	650	.0024	1.97
2-Bottom		850	.0028	2.21
3-Left	Vertical	1350	.0017	5.78
3-Right		1350	.0022	4.47
4-Top	Horizontal	500	.0012	3.03
4-Bottom		500	.0010	3.65
5-Left	Vertical	575	.0023	1.82
5-Right		350	.0014	1.82
6-Top	Perpendicular	250	.0020	.91
6-Bottom	to Bedding 77°	250	.0016	1.14
7-Top	Parallel	350	.0011	2.32
7-Bottom	to Bedding	800	.0028	2.09
8-Top	Horizontal	600	.0012	3.64
8-Bottom		650	.0014	3.38
9-Left	Vertical	1000	.0009	8.08
9-Right		650	.0011	4.30
10-Top	Inclined 45°	500	.0004	9.1
10-Bottom	downstream	500	.0005	7.28
11-Left	Vertical	(4200)	.0025	(12.2)
11-Right		(4130)	.0026	(11.5)
12-Top	Horizontal	500	.0024	1.52
12-Bottom		600	.0011	3.97
13-Top	Inclined 45°	580	.0010	4.22
13-Bottom	downstream	400	.0007	4.16
14-Left	Inclined 45°	600	.0007	6.24
14-Right	downstream	1100	.0012	6.67

( ) = Values not used for analysis

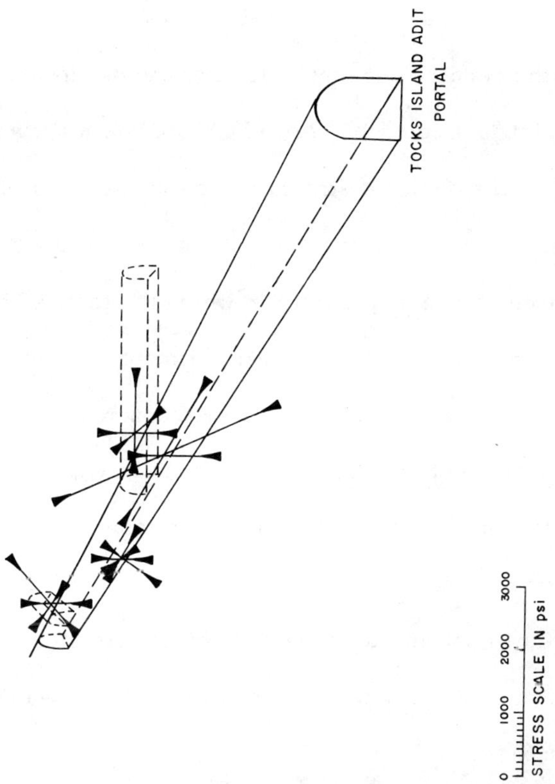

FLAT JACK TESTS, STRESSES NORMAL TO PLANE OF JACKS

Figure 8

included as part of this study, no accurate evaluation of the effects of stress relief on the test results can be made. However, it is estimated that stress relief has lowered the measured stress values in this area by between 15 and 35 percent.

The maximum stress measured in the area again appears to be controlled by the bedding because it follows the secondary folding.

3. Flat jack tests 11, 12, 13 and 14 again show a stress field controlled by the geologic structure, see Figure 8. Test 11 appears to be affected by its proximity to the intersection of the two adits and by the high strength of the quartzite in which the test is located; both facts tend to increase the measured stress. Except as a demonstration of the fact that higher stresses are carried by the more competent rock, the results of test FJ-11 are not used in the analyses.

The maximum stress level in this series of tests is 850 psi and its direction is generally parallel to the bedding. This series of flat jack tests does not appear to be appreciably affected by stress relief and so the results are taken as generally indicative of the stress field existing in the rock.

Rock Viscoelastic Properties--The rock's static and dynamic viscoelastic properties were measured by plate loading tests, sustained load tests, flat jack tests, and seismic velocity measurements.

Plate Loading Tests--Plate loading tests measure the in situ viscoelastic characteristics of a rock under controlled loading conditions. The results of the tests define the elastic and inelastic behavior of the

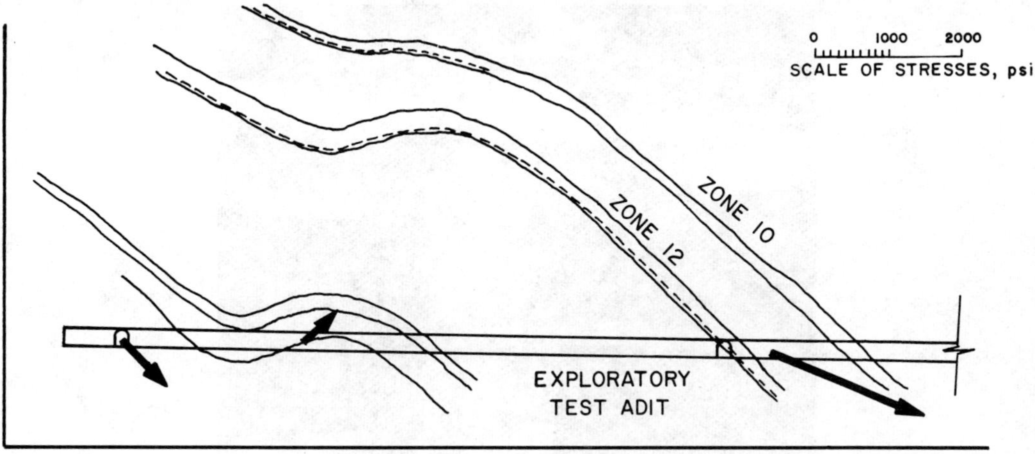

Figure 9. The relationship between measured maximum stresses and geologic structure is shown by arrows.

material and are essential in evaluating the long-term reaction of rock to load. The rigid die-type plate loading test is an averaging type of test, in that it mechanically averages the variable properties of the material under the plate. The most important of these variables are: change in rock type or mineral composition, joints and other fractures, and the degree of alteration. Tests were made at twelve locations giving twenty-four complete tests, as measurements are made at both ends of the loading assembly.

The equipment used for the tests consisted of a 200-ton hydraulic ram and pump, 3-inch thick steel bearing plates of 100 in^2 area, dial gauges reading to .0001 inch located each 120 degrees around each bearing plate, and the required support assembly. The equipment assembled and ready to test is shown on Figure 10.

Figure 10. Equipment assembled for vertical plate loading test.

1. The beam support cradle is set up and leveled by means of the adjusting screws in the legs.

2. Sections of beams, the ram, and bearing plates are placed on the cradle and bolted together until the sum of their length is 2 to 4 inches less than the distance betwen the rock surfaces.

3. The assembly is checked to make sure it is absolutely level.

4. A form is built around each bearing plate and a sulphur cap is poured.

5. The gauge-holding post and brackets are installed, being sure that there is no contact between this equipment and the test beam assembly.

Each test consisted of four cycles of loading and unloading the bearing plates and recording the rock deflections. Ten readings of deflection were made during each loading and unloading cycle with the load held constant at the maximum and minimum load of each cycle for 15 minutes to allow short-term creep measurements to be made. The cycles of one test consisted of: 50-ton maximum load in increments of 5 tons, 100-ton maximum load in increments of 10 tons, 150-ton maximum load in increments of 15 tons, and 200-ton maximum load in increments of 20 tons. One complete test required eighty loading and unloading steps and was carried out in 6 to 8 hours.

The results of the tests are given on Table II and a typical stress-strain curve is shown on Figure 11.

Three moduli of deformation are given on Table II: the working modulus (E), the secant modulus (E_S), and the recovery modulus (E_R).

The working modulus is a tangent modulus taken at a point on the stress-strain curve which in our opinion best represents the behavior of the material. The secant modulus is computed using maximum stress and strain values. A secant modulus greatly lower than the working modulus usually indicates open fractures in the material, the closing of which results in increased initial strain, or it indicates plastic deformation. The recovery modulus is a secant modulus taken on the recovery curve. Its value increases as the elastic response of the material decreases. The ratio E/E_R is a measure of the elastic recovery of the rock. A perfectly elastic material would have an E/E_R of 1.

TABLE II

PLATE LOADING
TESTS

Plate Loading Test No.	Moduli of Deformation			Percent Elastic Recovery		E/E_R	Type of Rock
	$Ex10^6$	E_Sx10^6	E_Rx10^6	Total	Last Cycle		
PL-1 Ram	1.00	1.00	2.20	50	73	.45	Quartzite
PL-1 Butt	0.80	0.67	2.61	26	68	.31	Quartzite
PL-2 Ram	0.42	0.28	0.72	39	74	.58	Quartzite
PL-2 Butt	3.20	3.20	15.30	18	28	.21	Quartzite
PL-3 Ram	1.59	1.50	3.39	44	73	.47	Sdy. Argillite
PL-3 Butt	1.08	0.98	3.62	30	90	.30	Sdy. Argillite
PL-4 Ram	0.94	0.86	2.90	34	86	.32	Argillite
PL-4 Butt	0.48	0.48	0.76	62	86	.63	Argillite
PL-5 Ram	---------No values due to failure of bottom pad----------						
PL-5 Butt	5.40	5.40	9.93	44	51	.61	Quartzite
PL-6 Ram	3.60	3.60	VHV	17	100	---	Arg. sis
PL-6 Butt	3.27	3.27	VHV	0	0	---	Argillite
PL-7 Ram	0.455	0.455	0.453	70	88	1.00	----------
PL-7 Butt	1.55	1.37	5.08	25	58	0.31	Mot. Arg. sis
PL-8 Ram	2.00	1.39	3.91	37	88	.51	Quartzite
PL-8 Butt	1.55	1.09	3.58	30	71	.43	Qtzitic ss
PL-9 Ram	1.31	0.94	2.20	42	80	.59	Siltstone
PL-9 Butt	2.70	2.20	4.03	50	86	.67	Siltstone
PL-10 Ram	0.54	0.43	1.71	23	54	.32	Mot. sis
PL-10 Butt	0.85	0.68	1.80	42	76	.47	Mot. sis
PL-11 Ram	0.21	0.17	2.20	25	60	.10	Mot. sis
PL-11 Butt	0.52	0.47	1.20	38	71	.43	Mot. sis
PL-12 Ram	Test not valid due to failure of bottom pad during first cycle						
PL-12 Butt							Mot. sis

E = working modulus of deformation
E_S = secant modulus of deformation
E_R = recovery modulus of deformation

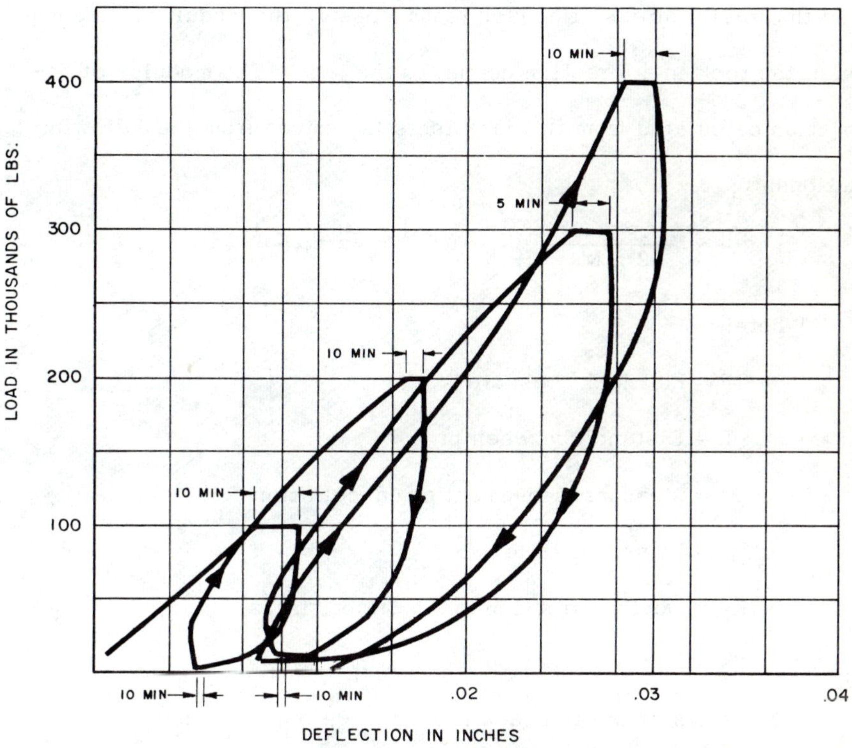

PLATE LOADING TEST CURVE

Figure 11

Generally, an E/E_R value of more than .85 is considered satisfactory.

Moduli of Deformation from Flat Jack Tests-- In addition to measuring the in situ stress, flat jack tests measure the moduli of deformation of the rock in a direction normal to the jack. The modulus of deformation calculated from flat jack tests is derived from the following relationship:

$$E = \frac{P\,L}{2\,\pi\,M}\,(K_1 - \mu K_1{}^1 + K_2 - \mu K_2{}^1)$$

Where:

P = Flat jack pressure

L = Distance between pins

M = Measured movement at cancellation

μ = Poisson's ratio

K_1 = Depth correction factor major stress

K_2 = Depth correction factor minor stress

The results of the flat jack moduli of deformation tests are given on Table I and a sample stress-strain curve is shown on Figure 12.

In Situ Dynamic Modulus of Elasticity Measurements-- Dynamic moduli of elasticity were measured in situ by measuring the seismic velocity of a shock wave through the rock using a seismic timer capable of recording wave travel time to .1 millisecond.

The formula used for modulu of elasticity calculations is:

$$E = \frac{\rho V_s^2 \left[\left(\frac{V_p}{V_s}\right)^2 - 4 \right]}{\left(\frac{V_p}{V_s}\right)^2 - 1}$$

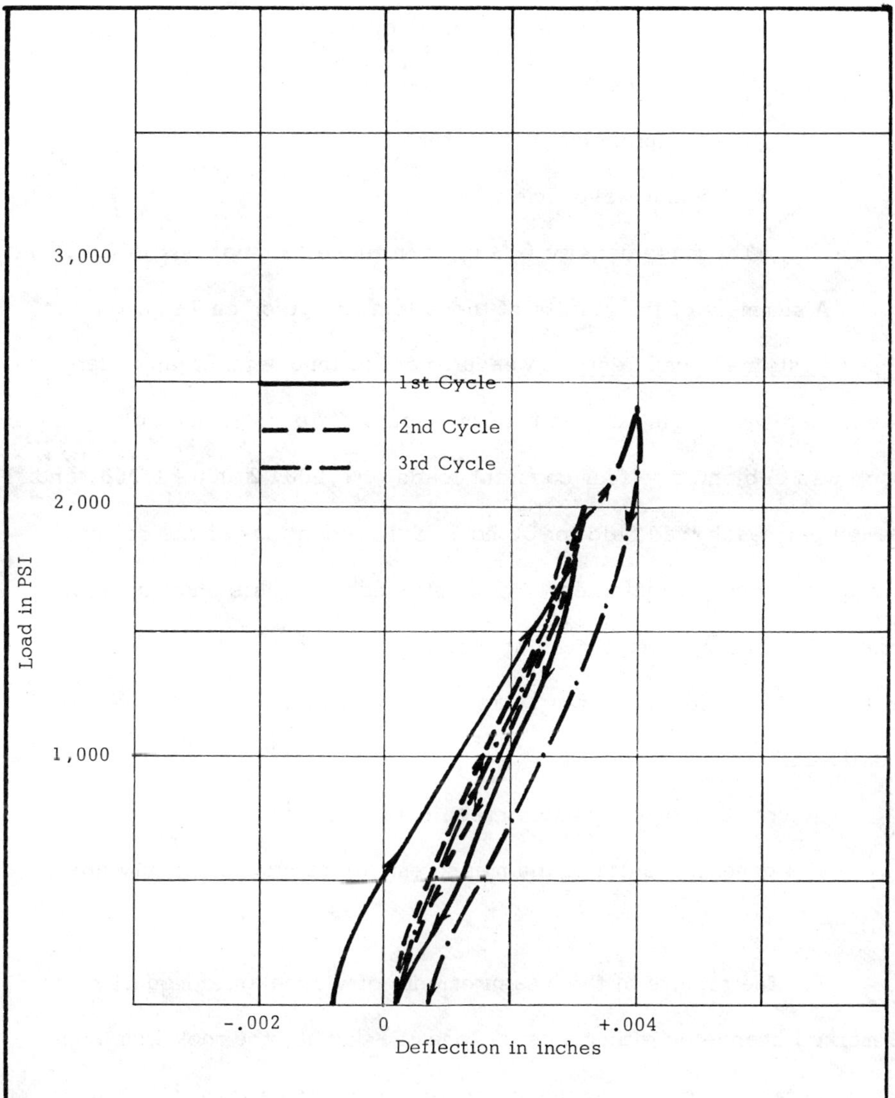

Flat Jack Test Curve

Figure 12

Where:

E = Young's modulus of elasticity

ρ = Mass density

V_p = Compressional wave time

V_p = Shear wave time

The mass density (ρ) used in these calculations is 2.7.

A summary of the results of these tests is given on Table IV.

Sustained Load Tests-- Measurement of long-term creep under constant load was carried out on both crosscuts of Tocks Island adit. The rock was subjected to four constant loads of 1,000, 2,000, 3,000, and 3,800 psi, each load held for 36 hours with recordings of the deflections made every 10 minutes. The sustained load tests show no accelerating secondary creep, even at loads of 3,800 psi. The viscosity values determined and shown on Table III also indicate very high viscosity on both tests.

Figure 13 shows a sample creep curve for these tests.

The general results of the measurements of viscoelasticity are as follows:

1. The results of the measurements of viscoelastic moduli show a marked change on either side of Zone 12; that is, the rock from Zone 12 to surface, eliminating the weathered zone, has a generally higher E value than the rock to the hill side of Zone 12. This can be due to various reasons, such as more stress relief in the rock to the hill side of Zone 12 or an increase in modulus on the surface side of Zone 12 be-

TABLE III

SUSTAINED PLATE LOADING TESTS

Test	Pressure (psi)	Movement (inches)	Time (hours)	Time (minutes)	Creep Rate (minutes)	Viscosity (poises)
1-Ram	1000					
1-Ram	2000					
1-Ram	3000	.0004	23	1380	2.8×10^7	4.5×10^{16}
1-Ram	3800	.0005	17	1020	4.9×10^7	3.2×10^{16}
1-Butt	1000					
1-Butt	2000					
1-Butt	3000	.0003	20	1200	2.5×10^7	4.9×10^{16}
1-Butt	3800	.0012	22	1320	9.0×10^7	1.8×10^{16}
2-Ram	1000	.0003	17	1020	2.9×10^7	1.4×10^{16}
2-Ram	2000	.0002	16	960	2.0×10^7	4.2×10^{16}
2-Ram	3000	.0002	17	1020	1.9×10^7	6.5×10^{16}
2-Ram	3800	.0002	23	1380	1.4×10^7	11.4×10^{16}
2-Butt	1000	0	36	2160	0	0
2-Butt	2000	.00015	16	960	1.5×10^7	5.5×10^{16}
2-Butt	3000	.0002	26	1560	1.2×10^7	10.4×10^{16}
2-Butt	3800	.0001	36	2160	0.4×10^7	37.4×10^{16}

$$\text{Viscosity} = \frac{d\epsilon/\text{sec}}{6.9 \times 10^4} \quad \text{(pressure in psi)}$$

cause the higher in situ stress there acts as an additional confining pressure, which increases the modulus. At this time, we feel that the second reason probably has the most influence.

2. Design "elastic constants" for sound, hard rock types should be E = 2.8×10^6 psi, and a Poisson's ratio of .21.

3. Design "elastic constants" for weak, weathered rock should be E = 0.6×10^6 psi, and a Poisson's ratio of .4.

TABLE IV

Comparison of "E" Values on Both Sides of Zone 12

	Above Zone 12	Below Zone 12
Plate Loading Tests		
Main Straight Adit	1.9×10^6	2.7×10^6
Cross Cut	3.5×10^6	2.1×10^6
Flat Jack Tests		
Main Straight Adit	3.9×10^6	2.3×10^6
Crosscut	6.0×10^6	4.5×10^6
Seismic Tests		
Main Straight Adit	3.7×10^6	
	2.0×10^6	1.4×10^6
Crosscut	7.2×10^6	4.6×10^6

The rock mechanics data collected during the study were utilized in the analyses of the potential stability of the cut slopes at various back slope angles.

The analyses were based on the following assumptions of the physical conditions at the site.

1. The stability of the excvated slope will be primarily controlled by resistance to sliding along Zone 12.

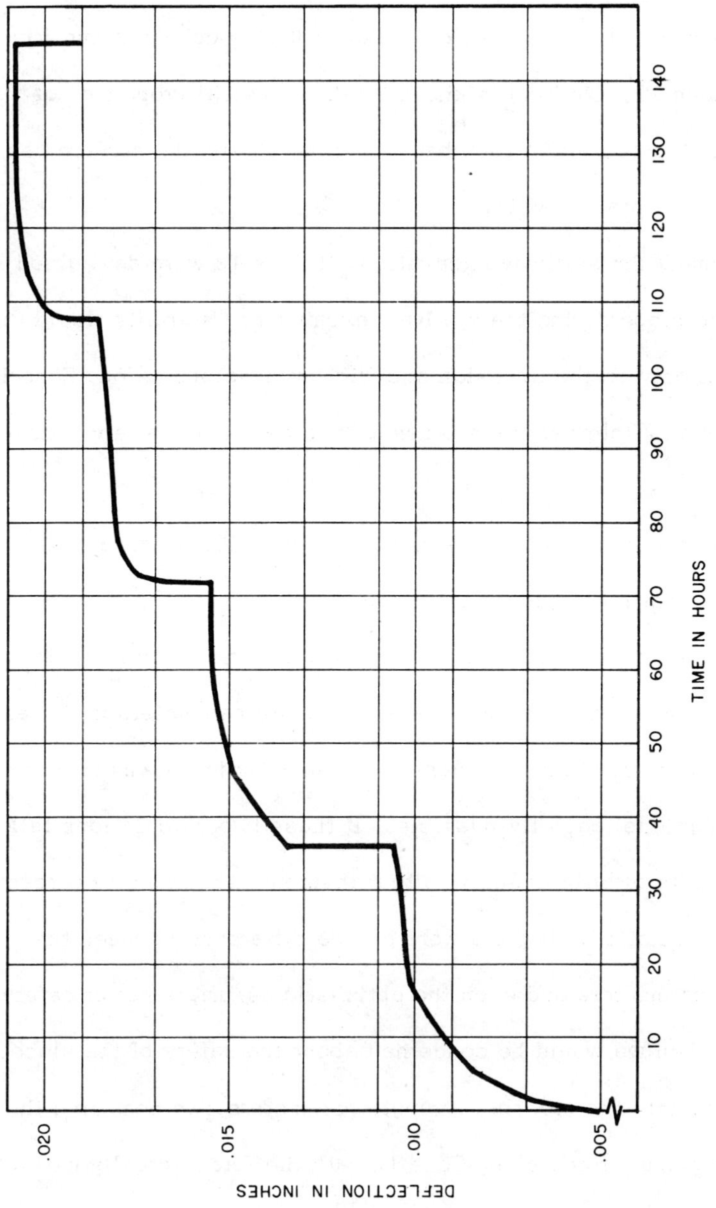

SUSTAINED PLATE LOADING TEST CURVE

Figure 13

2. The present stability of the hillside is due to the shear resistance of the rock between Zone 12 and Zone 10 plus cohesion and friction along Zone 12. Nothing is known of the physical properties of Zone 10, and the section of rock between Zone 10 and the surface has been ignored for this analysis.

3. During the spillway excavation, if Zone 12 were daylighted at the toe of the slope during the spillway excavation, stability would then be dependent only on the cohesion and friction developed along Zone 12.

4. Zone 12 intersects the present hillside surface at approximately elevation 760.

NOTE: This is as yet unproven because of a thick Pleistocene drift cover near projected outcrop.

To attempt to evaluate the significance of Zone 12 on hillside stability, we first analyzed the apparent strength parameters from measured data then calculated the necessary strength using the most optimistic assumptions logically possible. If these two calculations fell within the limits of error of the in situ measurements (about 30 percent) we would be satisfied. If there were a wide difference between the "measured" parameters and even the optimistic assumptions for safety factor of 1, then we would be concerned about the safety of the slope.

To calculate the probable strength parameters, we compared the approximate driving force along Zone 12 with the force actually measured by the flat jacks using the following assumptions:

a. dimensions as shown on Figure 14.

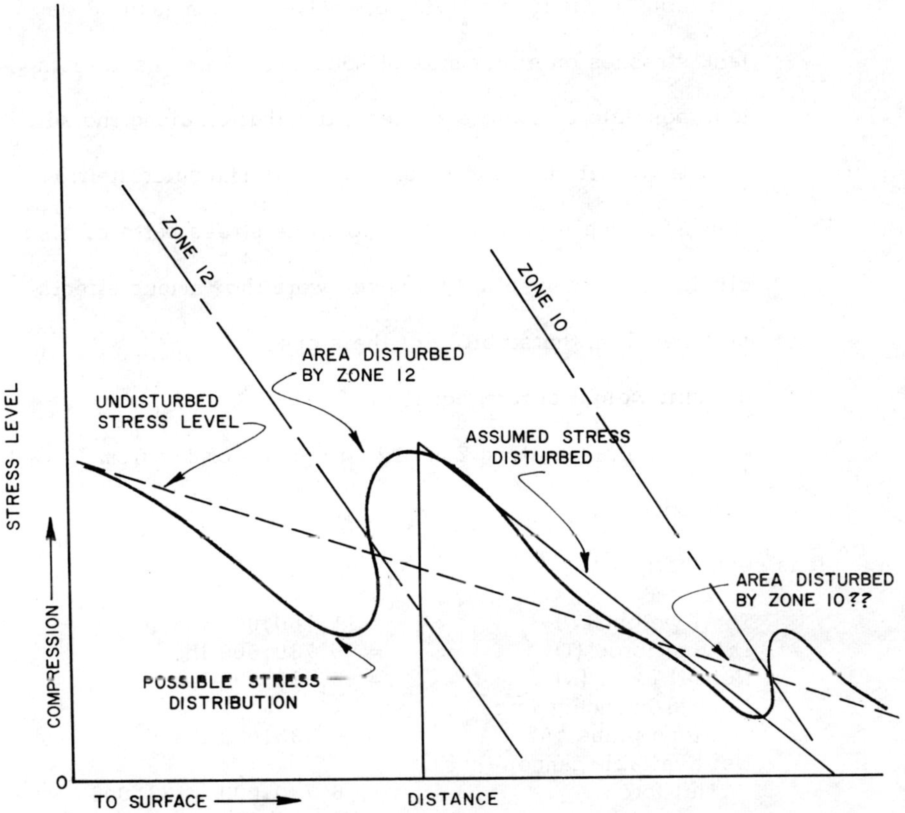

FIGURE 14 — ASSUMED STRESS DISTRIBUTION ALONG THE EXPLORATORY ADIT.

b. triangular distribution of neutral boundary forces with maximum head equal to 50 feet,

c. triangular distribution of stress across plane "A", (This is based upon the only piece of measured information available at the time of this paper; i.e., the measured flat jack stresses on either side of Zone 12.) From these tests it is possible to assume a stress distribution along the adit similar to that shown in Figure 14. The triangular distribution is a simplification of this possible stress pattern. Also eliminated, for simplicity's sake, were the various effects of Zone 10 on the stability of the slope.

d. unit weight of rock equal to 170 lb/ft^3,

e. failure plane along Zone 12 with angle of 37° from horizontal.

it is seen that;

Total weight (W)	= 11,180,000 lbs
Driving force (T)	= 6,730,000 lbs
Normal force (N)	= 8,930,000 lbs
Neutral boundary force along plane "A"	= 133,000 lbs
Net available tangential force	= 6,730,000 - 133,000
	= 6,600,000 lbs

Corresponding pressure from available tangential force at flat jack test location number 1 = 1,080 psi

Measured pressure at flat jack test location number 1 = 1,005 psi[1]

(1) The stress has been corrected for the effects of the adit on the stress measured by the flat jack.

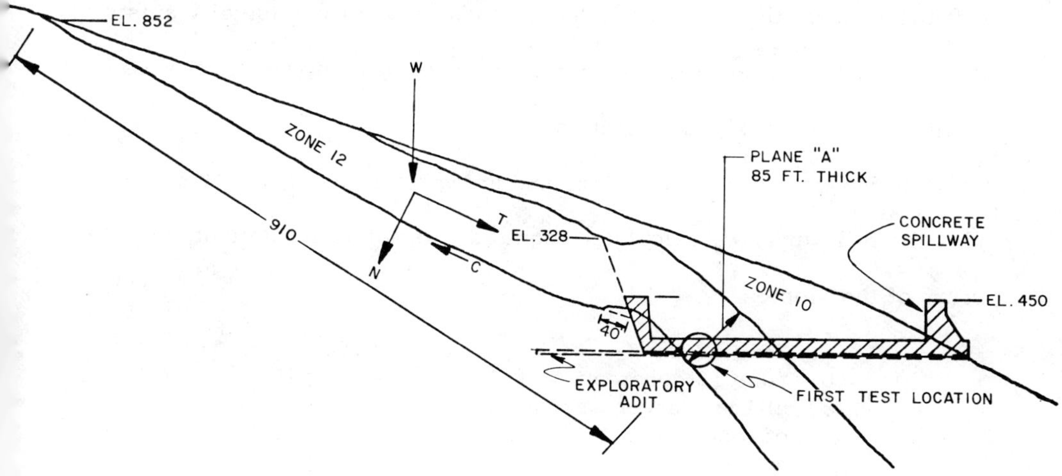

Figure 15. Calculation of approximate driving force along Zone 12 compared to measured pressure by flat jacks.

The difference of 75 psi between the measured and the available pressure, within the accuracy of the assumptions, is for all practical purposes negligible; suggesting that at the present, the total weight of the slope is resting on the plane "A" and receiving very little assistance from any shear strength or cohesion along Zone 12. This conclusion compares closely with the results of the laboratory tests on the fracture filling material (0 cohesion 11° Ø) and further indicates the strong possibility that the clay fracture filling is distributed over most of the fracture surface with little or no rock-to-rock contact present.

The calculations following are made to define the strength parameters required for a stability factor of 1 for the slope. We do not mean to suggest that a tan Ø of .58 or cohesion of 14 psi actually exists along Zone 12, nor are these assumptions related in any way to the preceding calculation of Ø from test data.

In the light of the foregoing, and using the same assumptions plus the following, we calculated required parameters for delicate equilibrium of the slope:

a.	Zone 12 tan Ø	= .58
b.	Zone 12 cohesion	= 14 psi
c.	ultimate shear strength of rock in dip reversal, average from core lab test[1]	= 188 psi

Calculating the safety factor by the sliding wedge method:

W, T, N, being the same as in point 4 above,

Neutral boundary force along Zone 12	= 1,420 000 lbs
Resisting force	= CL+N tan Ø + dip reversal strength
	= 1,840,000+4,350,000 +1,080,000
	= 7,270,000 lbs
Safety factor	= 7,720,000/6,730,000= 1.08

Considering that the strength assumed for this calculation is the maximum possible (C = 14 psi, Ø = 30°), it is concluded:

a. The stability of the slope above Zone 12 is, at best, marginal and problems could be encountered during construction even before the excavation intersects Zone 12.

b. A slope of $\frac{1}{4}$ to 1 is the best to mobilize the additional resistance due to the bedding dip reversal of Zone 12 shown on geologic sections. Laying back the slope further will not give additional safety

---

(1) MRD Report No. 65/760 dated August 1966

and, in fact, will increase the likelihood of failure.

c. Some additional treatment for the cut was considered. Choices were: complete removal of all rock down to Zone 12; drainage of the slope to control pore pressures; reinforcement of the slope with pretensioned cables; use of gated spillway to reduce excavation of rock in front of Zone 12.

d. Studies were made, for cost comparison purposes, on the possibility of using high-tension cable reinforcement to raise the safety factor of the rock mass above Zone 12. Using the results of the laboratory tests on the clay seam of Zone 12 with C = 0 and tan ϕ = .18534, it would require twelve 2-inch diameter high-strength cable reinforcements per lineal foot of width of rock mass as shown on the general layout, Figure 16.

V. APPLICATION OF STUDIES TO DESIGN

From information obtained during those studies, several general conclusions were drawn and have been applied to structural design.

1. Many of the lithologic features are contemporaneous with deposition and are not deletereous to the structural characteristics of the rock.

2. The numerous bedding plane slickensides developed in the lower portion of the High Falls formation are of limited extent because of the depositional geometry of these surfaces. They probably represent major adjustments to regional tectonics and are, in part, prelithification in age.

3. Bedding plane slips in the upper portion of the High Falls are extensive, post-folding in age and were probably caused by folding.

4. Southeast dipping joints become tight or incipient and rare with depth and rarely transect different lithologies. Although two other joint sets with different orientations exist, detached blocks of any size are statistically uncommon at depth.

The spillway location selected is that which gives the best foundation conditions throughout its length without substantially increasing the height of maximum cuts and resulting excavation quantities. Locations up or downstream would not affect conditions appreciably since this direction is parallel to the trace of all major geologic features. The proposed location minimizes the influence of discontinuities and weathered zones while providing a suitable foundation for the right spillway wall. Figure 17 shows the relationship of the proposed spillway cut to geologic discontinuities from the spillway crest to the stilling basin.

The stability of any slope which intersects Zone 12 would be controlled by the shear resistance of this bedding plane slip. Because of the condition and geometry of this zone, all material in slopes above Zone 12 will be removed up to the outcrop of this material. Slopes of $\frac{1}{2}$ to 1, or flatter, will be used below Zone 12. Benches every 50 feet in elevation in the landward side of the approach channel were provided to catch raveled material and give an added factor of safety for this high cut in the critical location with respect to the intake channel.

Because any spillway location within the Zone 12 outcrop area

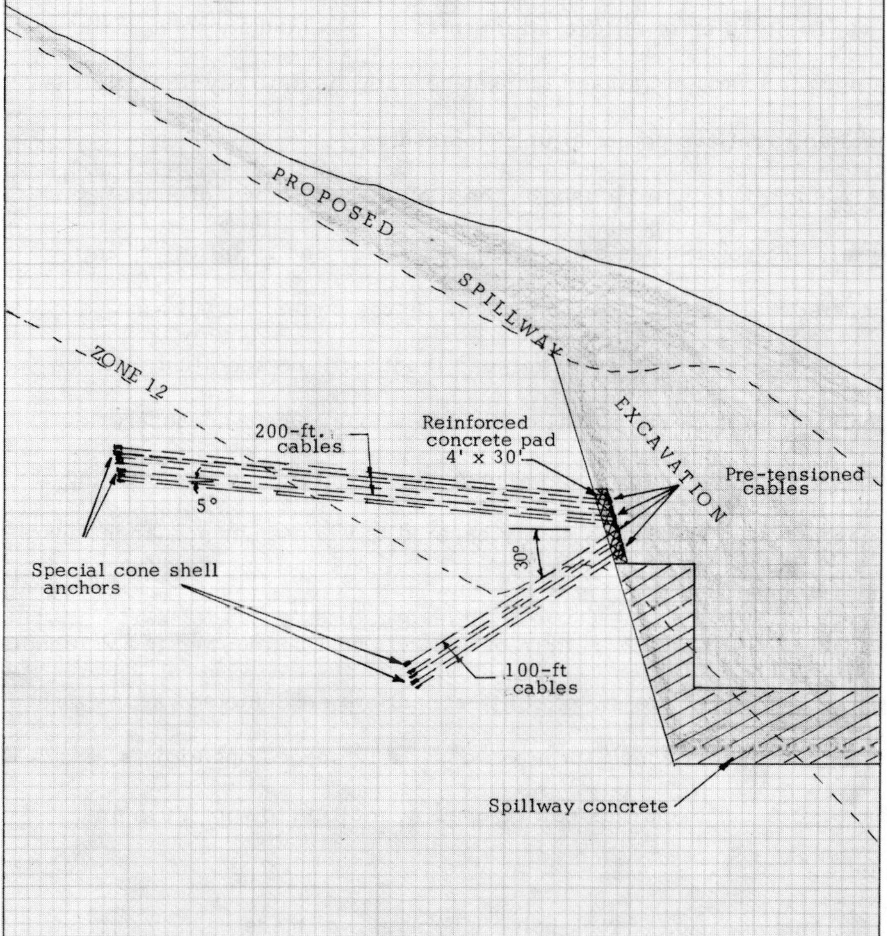

Suggested tension cable treatment of Zone 12
to increase safety factor of slope against sliding
to 1.15

Figure 16

would have the same top of slope, the proposed location was moved 60 feet landward to improve foundation conditions. This did not appreciably increase excavation quantities.

Drain holes to relieve lateral pressures behind anchored walls will be provided, and a drainage blanket beneath the spillway slab will be required to relieve potential uplift pressures.

Rock bolts will be used in the high slopes, crown of tunnels, portal, and powerhouse excavations for safety and stability and in anchored walls as part of the design. In the portal, powerhouse, and anchored walls a general pattern and length will be specified, but the program will be flexible to allow modifications, where necessary, as excavation progresses. Safety mesh will be provided where raveling of steep slopes will provide a safety hazared to workmen or permanent operating personnel.

No specific pattern of supports will be required in the tunnels. If areas which require structural treatment are encountered, such as shear zones, weathered zones, or areas of closely spaced open joints, a system of rock bolts and ties can be used in addition to wire mesh and gunite.

An evaluation of rock bolt anchorages was made on calyx core from DTC3 at 45 feet and DTC1 at 133 feet. Results have shown it possible to obtain sufficient anchorage with expansion type shells to develop full strength of the bolt. Bolt lengths should be staggered where a close spacing is required to preclude the possibility of opening continuous tensile cracks at the depth of the stressed expansion shell anchor. Bolting will be installed as soon as possible after excavation.

In the excavation of slopes, presplitting techniques will be used wherever possible, and provision will be made in contract specifications for experimentation to develop these techniques in different types of rock. All precautions will be taken to ensure that foundation rock is disturbed as little as possible, especially adjacent to structures.

REFERENCES

1. Depman, A. J. and Parrillo, D. G. - Geology of Tocks Island Area and Its Engineering Significance: Geology of Selected Areas in New Jersey and Pennsylvania. GSA Guidebook, 1969.

2. U. S. Army Corps of Engineers, Missouri River Division Laboratory. Tests for Strength Characteristics of Rock, MRD Laboratory No. 65/760, 1966.

3. U. S. Army Corps of Engineers, Missouri River Division Laboratory. Frictional Resistance of Jointed Rock Under Low Confining Pressure, MRD Laboratory No. 67/537, 1967.

4. U. S. Army Corps of Engineers, Missouri River Division Laboratory. Shear Strength Tests of Shear Zone Materials at Tocks Island Spillway Cut, MRD Laboratory No. 68/481, 1969.

5. U. S. Army Corps of Engineers, Missouri River Division Laboratory. Evaluation of Rock Bolt Anchorages, MRD Laboratory No. 68/396, 1969, pp. 86-94.

6. Dodds, Donald J., 1970, Flat Jack Tests; Special Publication, MRD Laboratory, U. S. Army Corps of Engineers.

7. Dodds, R. Kenneth, 1970, In Situ Shear Strength Testing of Rock: in Special Procedures for Testing Soil and Rock for Engineering Purposes; STP 479, American Society for Testing and Materials.

THE SLIDE AT BRILLIANT CUT

James V. Hamel[1]

INTRODUCTION

On March 20, 1941, a rock mass of over 110,000 cu. yds. volume slumped from the side of Brilliant Cut in Pittsburgh, Pennsylvania. The toe of this rock mass displaced three sets of railroad tracks and a train was derailed. The failure was apparently triggered by water pressures which existed in the slope because natural drainage outlets in the slope face were blocked with ice. The rear of the failure mass was defined by a vertical joint at the crest of the cut slope. This joint had begun to open several years prior to the slide. Approximately 70% of the failure surface passed through layers of soft clay shale and indurated clay at the base of the cut. (Indurated clay is a designation used by the Ohio River Division, U.S. Army Corps of Engineers, and certain other organizations for massive, slickensided claystone.)

This slide is described in detail and then analyzed as a large scale natural shear test. The set of effective stress Mohr-Coulomb shear strength parameters required in the basal formations for limiting equilibrium of the failure mass was calculated with the Morgenstern-Price method of stability analysis. These

---

[1] Assistant Professor of Civil Engineering, South Dakota School of Mines and Technology, Rapid City, South Dakota 57701

calculated strength parameters are compared with those measured by various investigators for indurated clays and clay shales similar to those in the basal formations at Brilliant. The roles of peak and residual shear strengths in the Brilliant slide are also discussed.

The following description of the Brilliant slide was constructed largely from information given by Ackenheil (1954). The 1940 and 1941 photographs of Brilliant Cut and additional unpublished information on the geology and geometry of the slide were furnished by Ferguson (1968).

DESCRIPTION OF THE SLIDE

Location and Geology

Brilliant Cut is located at the intersection of Washington Boulevard and Allegheny River Boulevard in Pittsburgh, Pennsylvania. Fig. 1 is a location map. The cut is at the nose of a hill within the right of way of the Penn-Central (formerly Pennsylvania) Railroad. This hill lies at the junction of an abandoned river valley and the present valley of the Allegheny River. A plan of the slide site is shown in Fig. 2.

The rocks at Brilliant are cyclic sedimentary rocks of the Conemaugh Formation (mid-Pennsylvanian age) of western Pennsylvania, West Virginia, and eastern Ohio. These rocks have been described in detail by Philbrick (1953, 1960). They are essentially flat-lying and they consist of repeated beds of sandstone, shale, and indurated clay (claystone) with occasional thin beds of limestone and coal. There is considerable vertical and lateral variation in the sandstones and shales. The shales in particular grade vertically into sandy and silty shales, carbonaceous shales, shaly sandstones and siltstones.

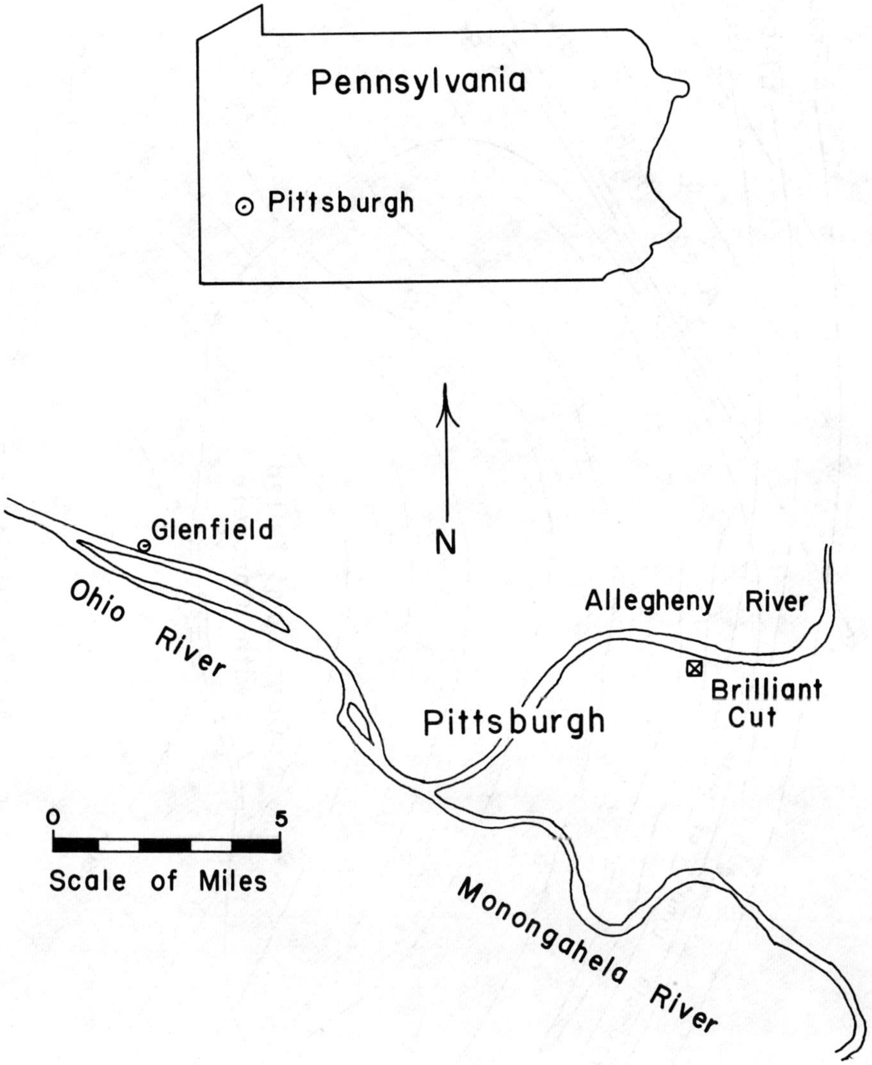

FIG. 1.--LOCATION OF BRILLIANT CUT

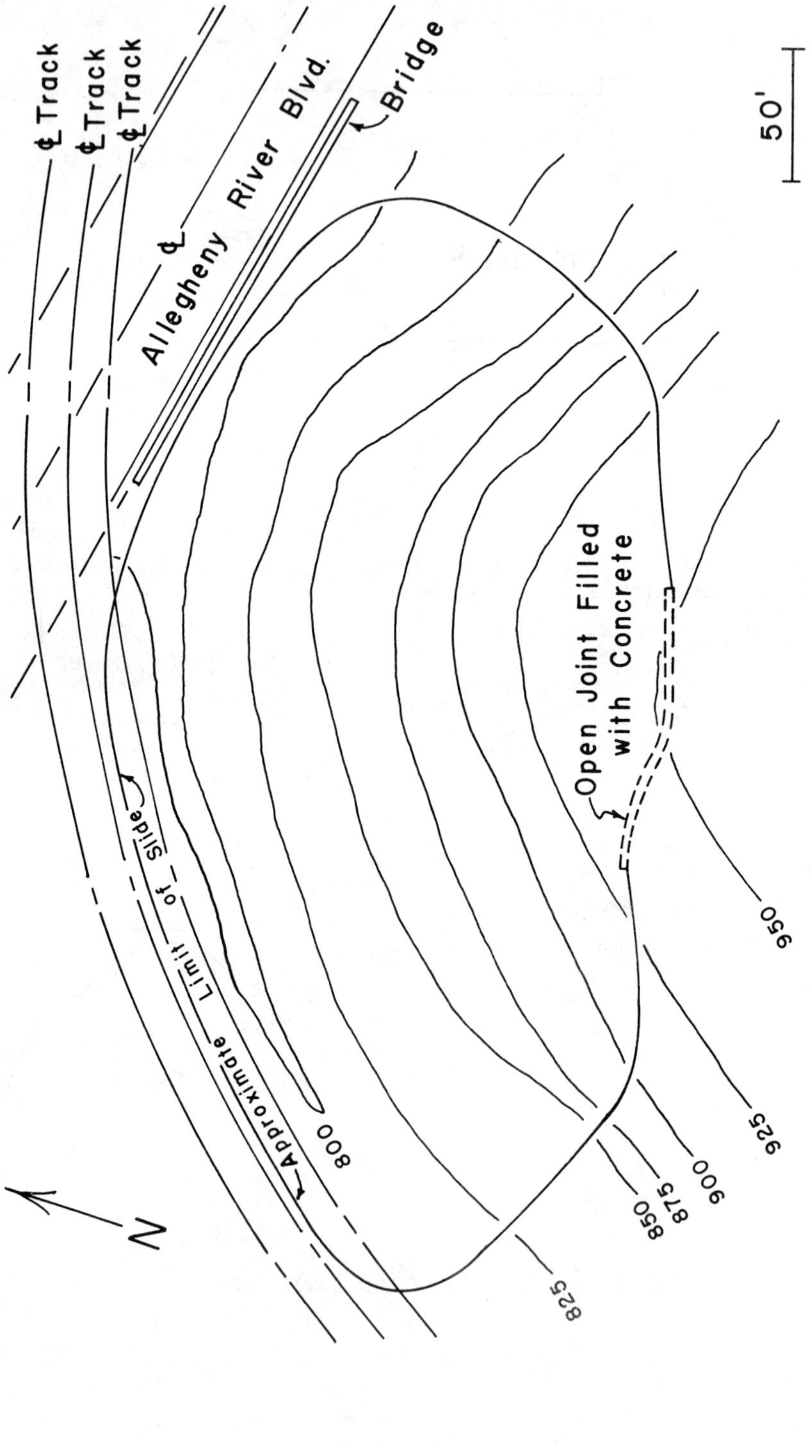

FIG. 2.--PLAN OF BRILLIANT CUT

The sandstones, coals, limestones, and sometimes the sandy shales are the more permeable beds of such as sequence due to jointing. The indurated clays, claystones, and silty or clayey shales are generally quite impermeable. Perched water tables are common where these impermeable beds underlie the more permeable beds.

Fig. 3 shows a generalized section through Brilliant Cut before the slide and Fig. 4 shows a detailed stratigraphic column. This geologic information was obtained by the Pittsburgh District, U.S. Army Corps of Engineers, in 1940-41 from surface outcrops and from borings made after the slide.

The beds overlying the Duquesne Coal (elevation 880) have numerous vertical joints generally parallel to the face of the slope. These joints are attributed to stress relief in the valley walls during and after downcutting by the rivers which formed the valleys. The mechanism of this valley wall stress relief and joint formation has been discussed by Ferguson (1967).

<u>Background</u>

The railroad tracks were originally constructed in the early 1900's with a sharp 15° curve requiring only a slight sidehill cut at Brilliant. According to Scharff (1920), a slide of unspecified size occurred in 1904 during this original construction. Very little information on this slide is available today. Scharff stated that it involved motion of fill and underlying shale and that it threatened the Brilliant pumping station of the Pittsburgh water department which was under construction on the bank of the Allegheny River at that time. This 1904 slide was probably of appreciable size because the water pumping station is located about 300 ft. north of the present Allegheny River Boulevard.

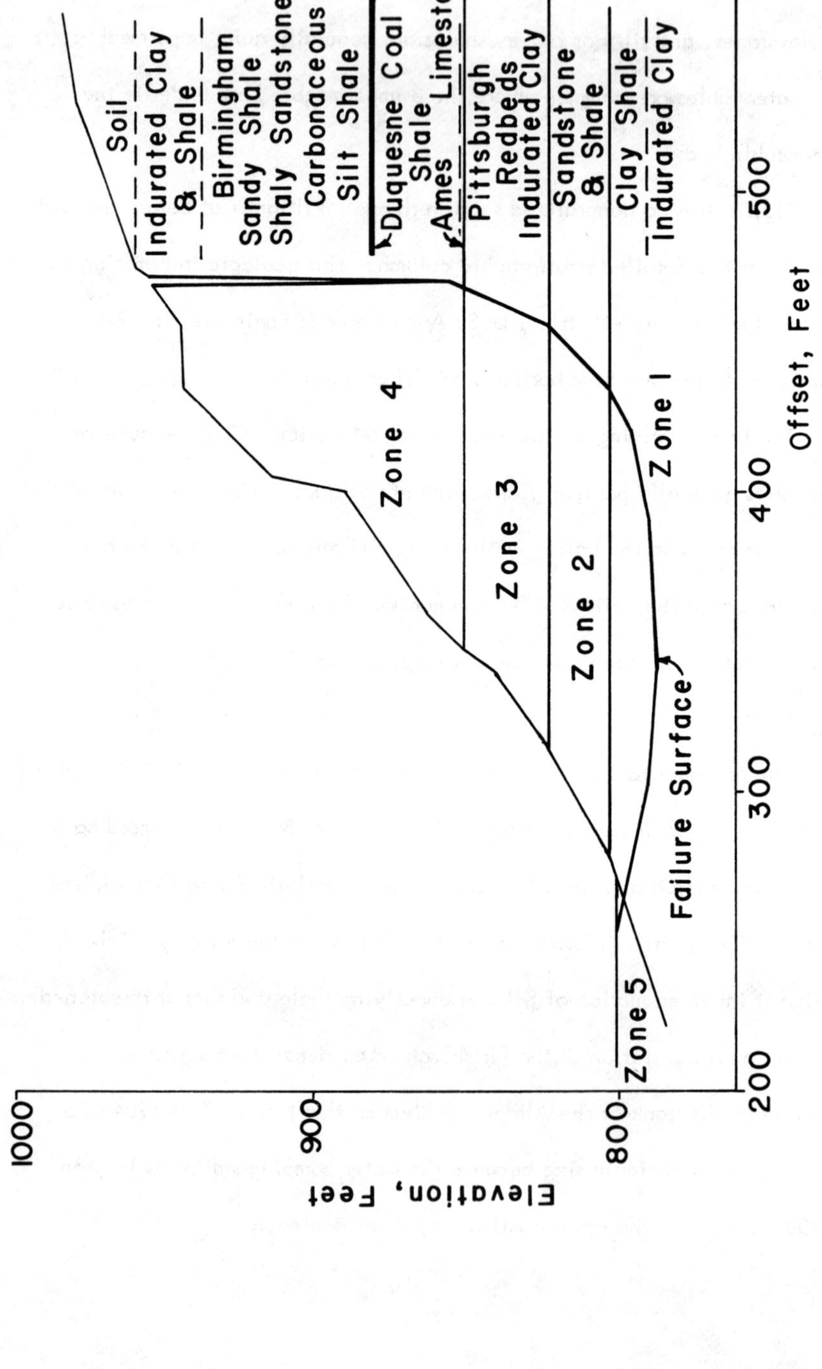

FIG. 3.--GENERALIZED SECTION THROUGH BRILLIANT CUT

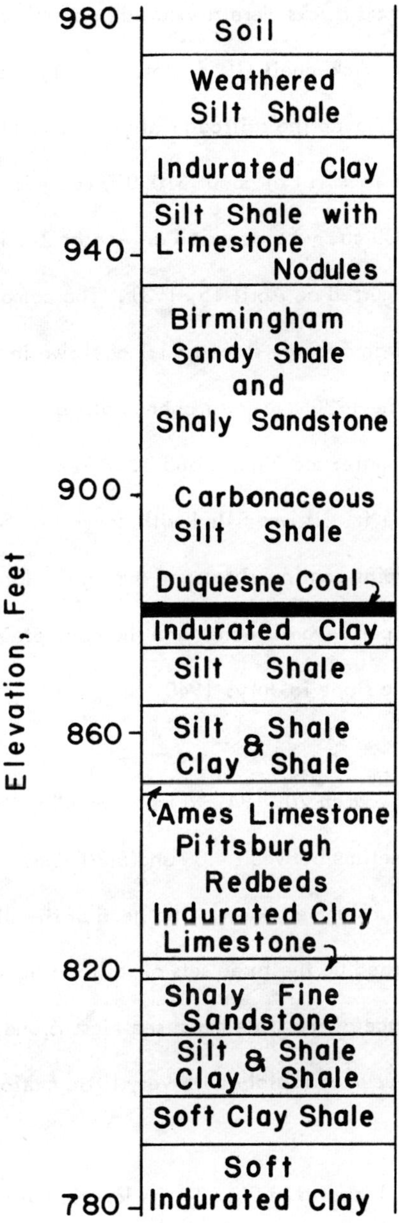

FIG. 4.--STRATIGRAPHIC COLUMN

In 1930, the railroad tracks were relocated to permit construction of Allegheny River Boulevard. Ackenheil (1954) stated that this relocation flattened the track curvature to 11°, moved the railroad right of way 70 ft. further into the hillside, and necessitated a rock cut of about 310,000 cu. yds. During construction of this cut, a slide of 12,000 cu. yds. occurred on August 22, 1930 and another slide of unspecified size occurred on April 15, 1931. The completed cut had a height of about 160 ft. and an average slope of 1:1 as shown in Fig. 3.

Sometime during the 1930's, a joint opened along the crest of the slope, just above the Birmingham Shale; see Figs. 2 and 3. After this joint had gradually opened to a width of about 1 ft., it was filled with concrete. Movement was slow and irregular but the joint continued to open during the late 1930's. By July, 1940, the joint was open several inches and the concrete plug was ineffective. Fig. 5 is a photograph of the slope in July, 1940.

The Slide

On the morning of March 20, 1941, a train rounding the curve at Brilliant was derailed. Sometime between 4:00 and 5:30 a.m., the entire nose of the hill had broken off and slumped down. The head of the slide mass dropped about 15 ft. and the toe pushed up the three sets of railroad tracks. Large tension cracks were visible in the face of the slide mass; see Figs. 6 and 7. The slide movement seemed to have occurred rapidly with very little motion after the initial slump.

The failure surface is shown in Fig. 3. Its location was determined from surface indications and borings made after the slide. The lower part of the failure surface generally followed the contact between the soft clay shale and indurated

FIG. 5.--PROFILE VIEW OF BRILLIANT CUT FROM NORTHEAST; JULY, 1940

FIG. 6.--PROFILE VIEW OF BRILLIANT CUT SLIDE FROM NORTH; MARCH 20, 1941

FIG. 7.--PANORAMA VIEW OF BRILLIANT CUT SLIDE; MARCH 20, 1941

clay at about elevation 790. The upper part of the failure surface met the open vertical joint in the Birmingham Shale. The rear of the failure mass was defined by a line along this joint as shown in Fig. 2.

Precipitation preceding the slide was somewhat above normal. About 1.1 in. of water equivalent in the form of rain and snow had fallen the previous week and a sudden thaw occurred just before the slide. Normally, surface water infiltrating the upper part of the slope would pass down through the vertical joints in the Birmingham Shale and drain laterally to the surface of the cut through the Duquesne Coal (elevation 880). Water in the middle to lower part of the slope would likewise drain laterally along the Ames Limestone (elevation 850) and the un-named limestone below the Pittsburgh Redbeds (elevation 820).

On the day of the slide, ice formations were visible on the slope face at the levels of the Duquesne Coal and the Ames Limestone; see Fig. 6. It is likely that ice plugged essentially all the natural drainage outlets in the slope face and that hydrostatic thrust in the open joint at the rear of the failure mass triggered the slide as suggested by Ackenheil (1954) and Philbrick (1953, 1960). It should be noted that the northerly exposure of the cut slope at Brilliant is such that the slope face is in shadow much of the day. This orientation of the slope face probably contributed to the development and maintenance of icicles and hence to the buildup of water pressure in the slope.

Power shovels began excavating material from the toe of the failure mass on March 21, 1941. As excavation proceeded, the failure mass slumped downward. This rotational motion was slow and continuous until equilibrium was established on May 15, 1941 after about 109,400 cu. yds. of rock had been removed. Fig. 8 is a photograph showing the slope during the final stages of excavation; note the seepage from the Duquesne Coal. The rock mass was benched at mid-height to decrease its rotational thrust and the railroad tracks were rebuilt. A paved ditch was installed behind the crest of the cut slope to intercept surface water and carry it away from the cut.

ANALYSIS OF THE SLIDE

A slope failure can be considered a large scale natural shear test. This in situ test includes all the heterogeneity, anisotropy, and environmental effects that can not be modeled in laboratory or theoretical investigations. If the boundary conditions of a slope failure are known or can be evaluated, it is possible to calculate the average shear strength of the failure surface material at the time of failure. The boundary conditions of the Brilliant Cut slope failure were reasonably well known, so it was decided to calculate the average shear strength of the basal clay shale and indurated clay through which most of the failure surface passed.

This strength calculation required that shear strength values be assumed for the other materials along the failure surface. The joint at the rear of the failure mass was assumed to be open to a depth of 100 ft., i.e., from elevation 955 to elevation 855, and the slope was divided into five material zones as shown in Fig. 3. Zone 1 consists of the basal clay shale and indurated clay through which 70% of the failure surface passes. Zone 2, which consists of the sandstone and shale overlying Zone 1, contains 12% of the failure surface. Zone 3, the Pittsburgh Redbeds indurated clay, also contains 12% of the failure surface. Zone 4 consists of the rocks above the Pittsburgh Redbeds; this zone contains 2% of the

FIG. 8.--PANORAMA VIEW OF BRILLIANT CUT SLIDE AFTER EXCAVATION OF 104,000 CUBIC YARDS OF ROCK DEBRIS; APRIL 25, 1941

failure surface. Zone 5 consists of residual soil, fill, and railroad ballast at the toe of the slope. Only 4% of the failure surface passes through Zone 5.

Since relatively minor percentages of the failure surface pass through Zones 2-5, the influence of the strengths assumed for the materials in these latter zones is not likely to be too great as long as reasonable strengths are assumed. The following effective stress Mohr-Coulomb shear strength parameters were used for the materials in Zones 2-5:

Zone	Cohesion, c', in kips per square ft.	Friction Angle, Ø', in degrees
2	2.0	35
3	1.0	28
4	2.0	35
5	0.0	30

These strength parameters were estimated by the writer from an inspection of the rock and soil materials at the slope site in July, 1968. Progressive opening of the joint at the rear of the failure mass is believed to have weakened the rocks of Zones 2-4 in the vicinity of the failure surface. The strength parameters given above for Zones 2-4 reflect this probable weakened state.

Total unit weights of sedimentary rocks like those at Brilliant range from about 150 to 165 pcf., so a total unit weight of 160 pcf. was used for the rocks in Zones 1-4. A total unit weight of 130 pcf. was used for the soil material in Zone 5.

Groundwater in the slope was assumed to be static and everywhere at the level of the surface of the slope. The open joint at the rear of the failure mass was assumed to be filled with water to elevation 955. These assumptions concerning water levels are consistent with the water conditions inferred for the slope by Philbrick (1953, 1960).

The set of effective stress Mohr-Coulomb shear strength parameters c' and Ø' required in Zone 1 for limiting equilibrium of the failure mass was calculated with the Morgenstern-Price (1965, 1967) method of slope stability analysis. This is a two-dimensional limiting equilibrium method of analysis which treats a failure surface of general shape and requires the failure mass to be in complete static equilibrium. Calculations were performed with the IBM 7090 computer at the University of Pittsburgh Computer Center using a Fortran II version (Hamel, 1968) of the MGSTRN program developed by Bailey (1966).

The calculated values of c' and Ø' are given in Table 1 and values of tan Ø' vs. c' are plotted in Fig. 9. The relationship between tan Ø' and c' is linear from tan Ø' = 0.625 (Ø' = 32°) at c' = 0 to tan Ø' = Ø' = 0 at c' = 5.4 ksf.

TABLE 1. -- STRENGTH PARAMETERS CALCULATED FOR ROCKS OF ZONE 1 AT LIMITING EQUILIBRIUM

Cohesion, c', in kips per square ft. (1)	Friction Angle, Ø', in degrees (2)	Tangent Ø' (3)
0.0	32	0.625
2.3	20	0.364
3.9	10	0.176
5.4	0	0.000

COMPARISON OF CALCULATED AND MEASURED STRENGTH PARAMETERS

No tests for the determination of strength parameters or other physical properties were performed on the rocks from Brilliant Cut at the time of its design or its failure. It is impossible today, from a practical standpoint, to obtain rock

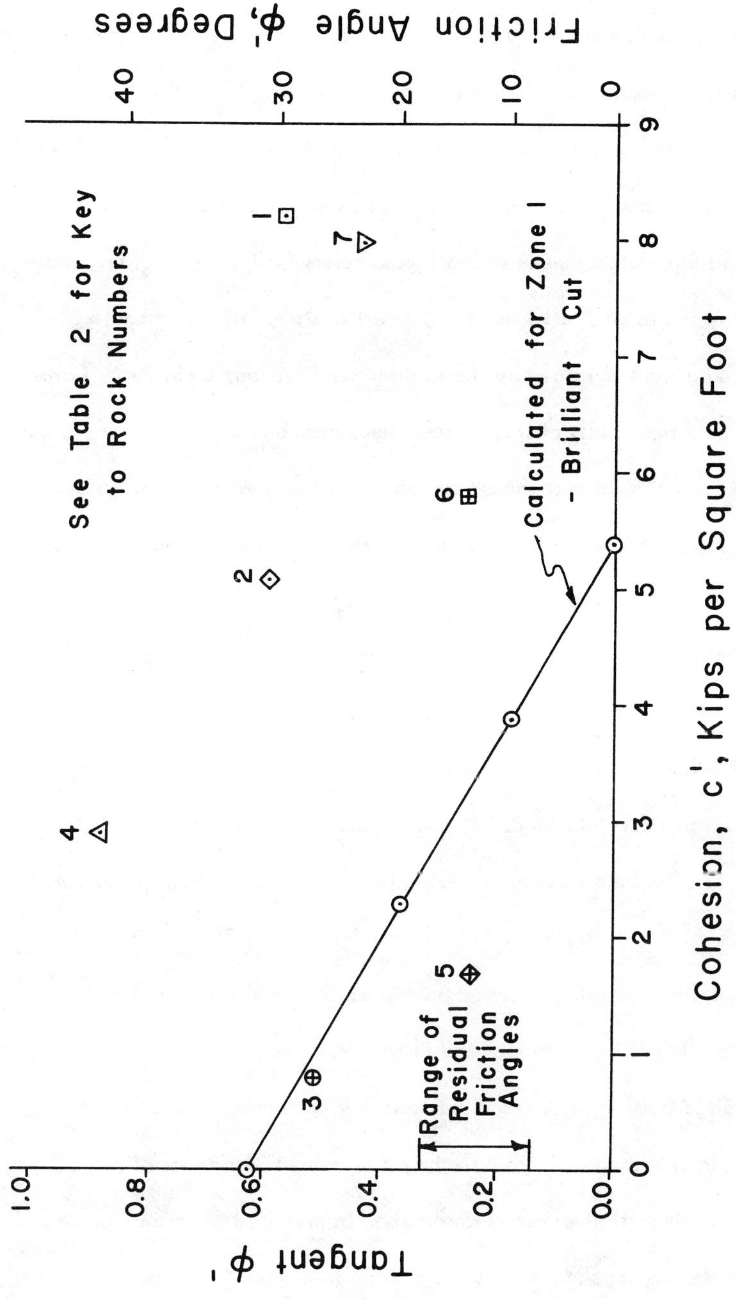

FIG. 9.--CALCULATED AND MEASURED TAN ϕ' VS. c'

samples from Brilliant for testing. Even if samples could be obtained and tested, there is no guarantee that the test results would reflect the in situ strength of these rocks some thirty years ago at the time of the slide.

The strength parameters calculated from the slide can, however, be compared with strength parameters measured for rocks similar to those in Zone 1 at Brilliant. Available data on peak strength parameters for indurated clays, claystones, and clay shales similar to those in Zone 1 are summarized in Table 2. Most of these data were obtained by the Ohio River Division, U.S. Army Corps of Engineers, from the results of field direct shear tests on 12 in. by 12 in. block specimens and laboratory direct shear tests on 6 in. diameter core specimens. Detailed descriptions of most of these rock materials and the tests performed on them were given by Kenty and Meloy (1965) and Mellinger (1966).

In 1969, a series of drained direct shear tests was performed at the University of Pittsburgh on 2 in. by 2 in. specimens of indurated clay from the shear zones of several ancient and recent landslides located near Glenfield, Pennsylvania. The Glenfield landslide site is 12 miles west of Brilliant Cut as shown in Fig. 1. The landslides at the Glenfield site were located in indurated clay (claystone) at the stratigraphic level of the base of the Pittsburgh Redbeds. These landslides and the tests performed on materials from them were described in detail by Hamel (1969) and Hamel and Flint (1969).

A limited number of direct shear tests was performed on undisturbed specimens of soft indurated clay from slightly above the shear zones of the landslides. Peak strength parameters determined from these tests are given in Table 2. Repeated direct shear tests were also performed on these specimens in

TABLE 2. -- SUMMARY OF MEASURED PEAK STRENGTH PARAMETERS

Rock No. (1)	Rock Type and Hardness (2)	Geologic Age and Unit (3)	Location (4)	Cohesion, in kips per square foot: Range (5)	Cohesion, in kips per square foot: Average (6)	Friction Angle, in degrees: Range (7)	Friction Angle, in degrees: Average (8)	Reference (9)
1	Indurated Clay, Soft to Medium Hard	Pennsylvanian, Pittsburgh Redbeds	Pike Island, West Virginia	2.4-11.6	8.2	19.5-42	29.5	Unpublished Data, Ohio River Division, Corps of Engineers
2	Indurated Clay, Soft to Medium Hard	Pennsylvanian, Pittsburgh Redbeds	Pike Island, West Virginia	3.6-6.0	5.1	28-35	30.5	Kenty & Meloy (1965); Mellinger (1966)
3	Indurated Clay, Soft	Pennsylvanian, Pittsburgh Redbeds	Glenfield, Pennsylvania	0.2-1.9	0.8	18-34	27	Hamel (1969); Hamel & Flint (1969)
4	Indurated Clay and Claystone, Soft to Medium Hard	Permian, Creston Redbed	Belleville, Ohio	0.0-5.8	2.9	40-43	41.5	Kenty & Meloy (1965); Mellinger (1966)
5	Clay Shale, Soft to Medium Hard	Devonian, Ohio Shale	Cave Run, Kentucky	0.9-2.6	1.7	-	13.5	Mellinger (1966)
6	Clay Shale, Soft to Medium Hard	Mississippian, Degonia Shale	Cannelton, Indiana	-	5.8	-	14	Kenty & Meloy (1965); Mellinger (1966)
7	Silty Clayey Shale, Soft to Medium Hard	Pennsylvanian	Western Pennsylvania	-	8.0	-	23	Underwood (1967); Ferguson (1968)

order to determine residual strength parameters and a large number of repeated direct shear tests was performed on undisturbed and remolded specimens of clay and indurated clay from the failure surfaces of the landslides. The repeated direct shear test results indicated residual cohesion intercepts of essentially zero and residual friction angles of 8° to 18°. Most of the residual friction angles were in the range of 11° to 16° and a residual friction angle of 14° is considered typical for the indurated clay at the Glenfield site.

The average peak tan ϕ' vs. c' values for each of the seven rock groups of Table 2 are plotted in Fig. 9 along with the range of residual tan ϕ' values measured for the indurated clay from the Glenfield site. It is seen from Fig. 9 that the tan ϕ' vs. c' values calculated for Zone 1 of the Brilliant slide fall well below the average peak tan ϕ' vs. c' values measured for soft to medium hard indurated clay, claystone, and silty clayey shale (Rocks 1, 2, 4 and 7). These calculated tan ϕ' vs. c' values lie in the range of average peak values for soft indurated clay (Rock 3) and soft to medium hard clay shale (Rocks 5 and 6). It is also seen from Fig. 9 that the tan ϕ' vs. c' values calculated for Zone 1 of the Brilliant slide fall well above the range of residual strength tan ϕ' values (with residual c' = 0) measured for indurated clay from the Glenfield site.

The data plotted in Fig. 9 suggest that the average shear strength mobilized in Zone 1 of the Brilliant slide was on the order of the peak strength of the soft clay shale and indurated clay along that part of the failure surface. Progressive failure of the slope related to widening of the open vertical joint at the rear of the failure mass may have reduced the strength of the clay shale and indurated clay somewhat but it seems unlikely that the strains associated with this pre-slide movement were sufficient to reduce the strength to a residual value.

The large strains which occurred in the 1941 slide are, however, believed to have reduced the strength along the failure surface to a true residual value. When the toe of the failure mass was excavated during remedial work following the slide, the mass rotated slowly and continuously on its failure surface. This continuous movement is characteristic of a failure surface along which the strength has been reduced to a true residual value. Such a failure surface can not sustain greater shear stresses than those corresponding to its residual strength. If greater shear stresses are imposed, as in excavating the toe of the slide mass at Brilliant, continuous motion occurs.

CONCLUSIONS

The 1941 slide at Brilliant Cut was triggered by water pressures in the slope. These water pressures existed because natural drainage outlets in the slope face were blocked with ice. The opening of the vertical joint at the rear of the failure mass occurred progressively over a period of several years but the failure itself occurred quite rapidly when water pressures in the slope reached critical values. These critical pressures probably corresponded to a water level at or near the top of the open joint.

The failure surface was largely defined by structural features in the slope. The open vertical joint defined the rear of the failure mass and the lower part of the failure surface followed two weak strata at the base of the cut. The shear strength mobilized along the basal portion of the failure surface was calculated from the slide. This calculated strength was found to be on the order of peak strengths measured by several investigators for rocks similar to those along the basal portion of the failure surface. This peak strength can be characterized approximately by $c' = 0$, $\phi' = 32^\circ$ to $c' = 2.3$ ksf., $\phi' = 20^\circ$ (Fig. 9).

Movement of the failure mass during the slide is believed to have reduced the strength along the failure surface to a true residual value. It is impossible to calculate the magnitude of this residual strength from the information available on the slide. Experience with slides in similar materials (Hamel, 1969; Hamel and Flint, 1969) suggests, however, that the residual strength of the indurated clay at Brilliant Cut can be characterized by $c' = 0$, $\phi' = 11^\circ$ to 16°.

ACKNOWLEDGMENTS

The material in this paper is drawn largely from a dissertation presented by the writer to the University of Pittsburgh as partial fulfillment of the requirements for the degree of Doctor of Philosophy (Hamel, 1969). Analysis of the Brilliant Cut slide was supported by the U.S. Department of the Interior, Bureau of Mines, under Grant No. G0180476 (MIN-11) to the University of Pittsburgh. Laboratory testing of the Glenfield clays was supported by the Pennsylvania Department of Highways and U.S. Department of Transportation, Bureau of Public Roads, under Research Project No. 41472 with the University of Pittsburgh. Unpublished information on the Brilliant slide and unpublished rock shear strength data from the files of the Pittsburgh District, U.S. Army Corps of Engineers, were furnished by H.F. Ferguson, District Geologist. The opinions and conclusions expressed herein are those of the writer and not necessarily those of any of the organizations which contributed information or financial support to this investigation.

REFERENCES

1. Ackenheil, A.C. (1954). "A Soil Mechanics and Engineering Geology Analysis of Landslides in the Area of Pittsburgh, Pennsylvania," Ph.D. Dissertation, University of Pittsburgh.

2. Bailey, W.A. (1966). "Stability Analysis by Limiting Equilibrium," C.E. Thesis, Massachusetts Institute of Technology.

3. Ferguson, H.F. (1967). "Valley Stress Release in the Allegheny Plateau," Bulletin, Association of Engineering Geologists, Vol. IV, No. 1.

4. Ferguson, H.F. (1968). Personal Communications, July, 1968.

5. Hamel, J.V. (1968). "Morgenstern-Price Method of Slope Stability Analysis," Report by Civil Engineering Department, University of Pittsburgh, to U.S. Bureau of Mines, 149 pp.

6. Hamel, J.V. (1969). "Stability of Slopes in Soft, Altered Rocks," Ph.D. Dissertation, University of Pittsburgh.

7. Hamel, J.V. and Flint, N.K. (1969). "Analysis and Design of Highway Cuts in Rock: A Slope Stability Study on Interstate Routes 279 and 79 Near Pittsburgh, Pennsylvania," Report by University of Pittsburgh to Pennsylvania Department of Highways and U.S. Department of Transportation, Bureau of Public Roads, 130 pp.

8. Kenty, J.D. and Meloy, C.R. (1965). "Shear Evaluation of Weak Rock Foundations," Technical Report No. 3-45, Ohio River Division, U.S. Army Engineers, DRAFT COPY, 34 pp., 34 plates, 95 figs.

9. Mellinger, F.M. (1966). "Laboratory and In Situ Tests of Shale," Preprint of Paper presented at ASCE Water Resources Engineering Conference, Denver, Colorado, May, 1966, 31 pp. plus appendices.

10. Morgenstern, N.R. and Price, V.E. (1965). "The Analysis of the Stability of General Slip Surfaces," Geotechnique, Vol. 15, pp. 79-93.

11. Morgenstern, N.R. and Price, V.E. (1967). "A Numerical Method for Solving the Equations of Stability of General Slip Surfaces," The Computer Journal, Vol. 9, pp. 388-393.

12. Philbrick, S.S. (1953). "Design of Deep Rock Cuts in the Conemaugh Formation," Proceedings, Fourth Symposium on Geology as Applied to Highway Engineering, pp. 79-88.

13. Philbrick, S.S. (1960). "Cyclic Sediments and Engineering Geology," Proceedings, Twenty-First International Geological Conference, Pt. 20, pp. 49-63.

14. Scharff, M.R. (1920). "Bigelow Boulevard Slide in Pittsburgh," Engineering News-Record, Vol. 85, No. 23, pp. 1076-1080.

15. Underwood, L.B. (1967). "Classification and Identification of Shales," Journal of the Soil Mechanics and Foundations Division, ASCE, Vol 93, No. SM 6, pp. 97-116.

ROCK STABILITY ANALYSIS AT TWIN BUTTES

by

Ben L. Seegmiller[1]

INTRODUCTION

Stability analysis of the south rock slope of the Twin Buttes Pit, Pima County, Arizona, was carried out by Anaconda's corporate Mining Research Department of Salt Lake City, Utah. This work was requested by Twin Buttes Operations management when mine planning and operational problems significantly increased due to a deteriorating slope condition during the three-month period September through November, 1970. A stability examination was begun in December 1970 and an initial study report was completed in March 1971.

The Twin Buttes Pit is located some 25 miles south of Tucson, Arizona (see Figure 1) in a pyrometasomatic deposit from which some 32,000 tons/day of copper-bearing ores are produced. Mining in the pit is carried out using a conventional pushback system. Alluvial gravels, which form a 350-ft thick blanket over the bedrock, are removed by a scraper-conveyor-wagon system, while ore and waste rock are removed by a shovel-truck-conveyor system. Depth of mining at present has reached approximately 1000 ft in the deepest portion of the pit. Initial pit planning called for 37° slopes plus ramps in the alluvium and 45° slopes plus ramps in the rock.

---

[1] Senior Mining Research Engineer, The Anaconda Company, Salt Lake City, Utah

Fig. 1 Location Map

Purposes of the initial stability study were (1) to gain knowledge as to the mode and probable causes of failure in order to more fully understand why such a failure could occur, and (2) to develop and initiate mining techniques which would result in safer and more economical slope conditions.

Results of the initial study indicated that the slope failure was of the classical wedge type which subsequently became a massive block or slab flow. Failure was attributed to an extremely low strength condition along adversely oriented structures. The structures may not only possess low natural strength but may also suffer a strength loss caused by seismic (blasting) damage or differential rebound from high mining rates at the slope toe.

This paper presents the initial study of the Twin Buttes slope failure and discusses some of the more salient results. A general description of the failure problem is discussed first, followed by a description of the structural and strength-analysis techniques, results and usage. Applicable stabilization techniques and current stability research are then outlined.

GENERAL DESCRIPTION OF FAILURE PROBLEM

Historical Aspects.- Minor stability problems in the arkose which makes up the south slope of the Twin Buttes Pit have existed since early 1969. These problems occurred along two of the most prominent pit structures (the Main Pit and the East-West Faults) which cut through the south slope as shown in Figure 2. Initially, these instability problems amounted to only a nuisance; however, subsequent slope failures became larger as mining continued

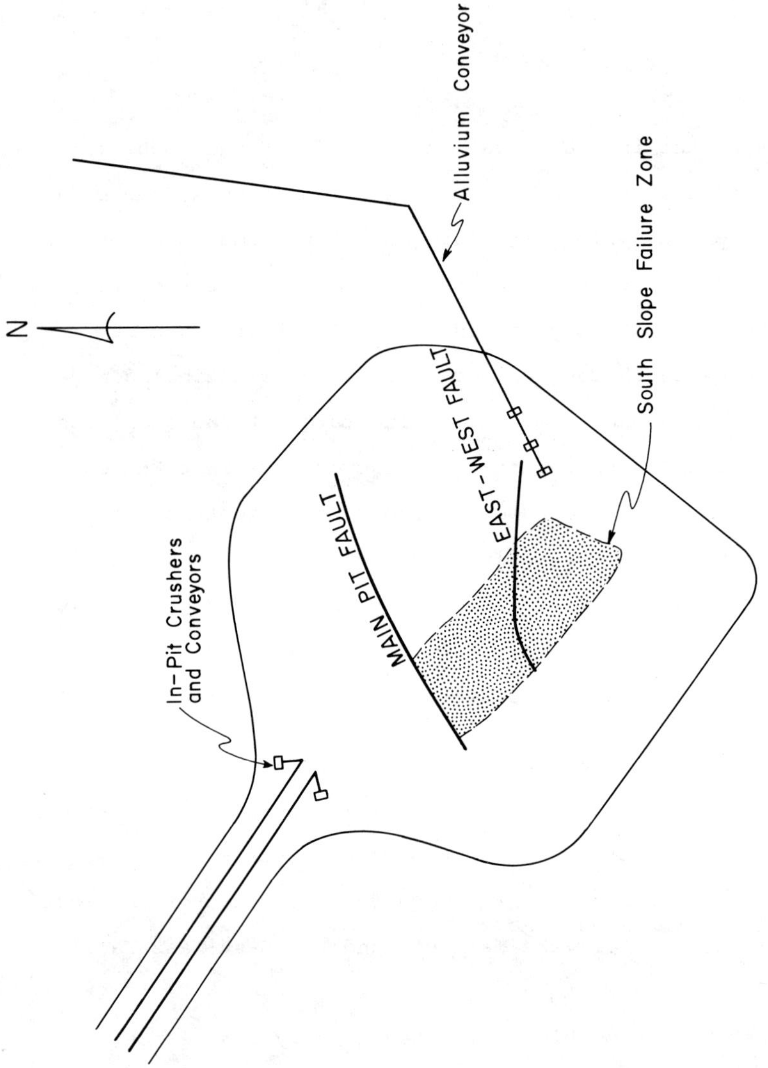

Fig. 2 Sketch of Twin Buttes Pit

during the remainder of 1969 and the first nine months of 1970. By December 1970 sections measuring several hundred feet in length and consisting of two and three slope benches had displaced up to ten feet in some zones. Many new tension cracks were beginning to appear on most benches along the south slope. New areas of displacement and greater magnitudes of movement continued to occur throughout the period from December through March, 1971, when essentially the entire south pit rock slope had failed and was in continuous movement. Crude surface extensometers and survey nets were used to monitor surface displacement and velocity. Cumulative displacement charts were plotted and used to estimate future slope movement. As of July 1, 1971, more than 50 ft of vertical and 60 ft of horizontal displacement had been noted and displacement was continuing.

Mining at or near the slide toe continued throughout the period of slope failure. The material removal rate was as high as one 50-ft bench every three weeks. Such activity aggravated the instability condition and a direct correlation of displacement magnitude and mining was noted. However, this situation could not be avoided because a major zone of ore was being mined at the slide toe. Use of the displacement monitors to forewarn of any possible catastrophic movement was successful in that no injuries or equipment damage resulted.

Failure Observation Details.- The minor failure problems which initially occurred adjacent to the Main Pit and East-West Faults consisted of small zones up to 50 ft in length wich progressively disintegrated into rubble slopes. These zones generally involved only one bench, and never more than two. The larger failure zones which began to appear in late 1970 resembled classic wedges with sliding occurring along the structural intersection line. These wedges would begin on one bench and gradually enlarge to involve two and three benches. By March 1971, numerous wedges had grown and enlarged to the extent that essentially the entire south slope had become a massive block or slab flow. Very minor rotation in isolated zones occurred during failure, indicating the failure zone may have been initially relatively shallow. Detailed examination of the entire failure area later revealed that the failure plane was composed of a number of relatively planar parallel structures striking parallel to the slope face and dipping at approximately 38° into the pit. These basic or lowermost failure planes appeared to limit the present depth of failure and, as of July 1, 1971, visual observation indicated that material below these planes was still stable. Some undercutting of the alluvial gravels has taken place causing ramp maintenance problems in the higher south slope zones. Total slide size including alluvium is estimated to be slightly less than three million tons.

STRUCTURAL ANALYSIS

Accumulation and Handling of Data.- Structural information has been collected by pit geologists since the first rock was exposed in January 1968. These data have been recorded in the field on 50-scale sheets, then transferred onto 100-scale geologic plan maps. By November 1970 some 5,000 structures (faults, joints and veins) had been plotted. Information, including structural type, strike, dip, coordinate location, bench elevation, and rock type, was subsequently digitized and punched into computer cards for processing and analysis.

Lambert-Schmidt Nets.- A modified version of the USBM computer program{1} capable of producing equal area nets, together with sector and dip ring analyses, was used to reduce the structural data. Numerous nets were made for each structural type in each of five different rock types in various zones of the pit. In all cases, variation in structural sets from one part of the pit to another was minor. It was concluded that mapping bias was not a major factor and the structural data would be sufficient for the initial study. A typical net (lower hemisphere projection) for arkose faults is shown in Figure 3. By combining the nets for different rock types with field observations, three structural trends (see Table 1) were determined to be significant from a slope stability standpoint.

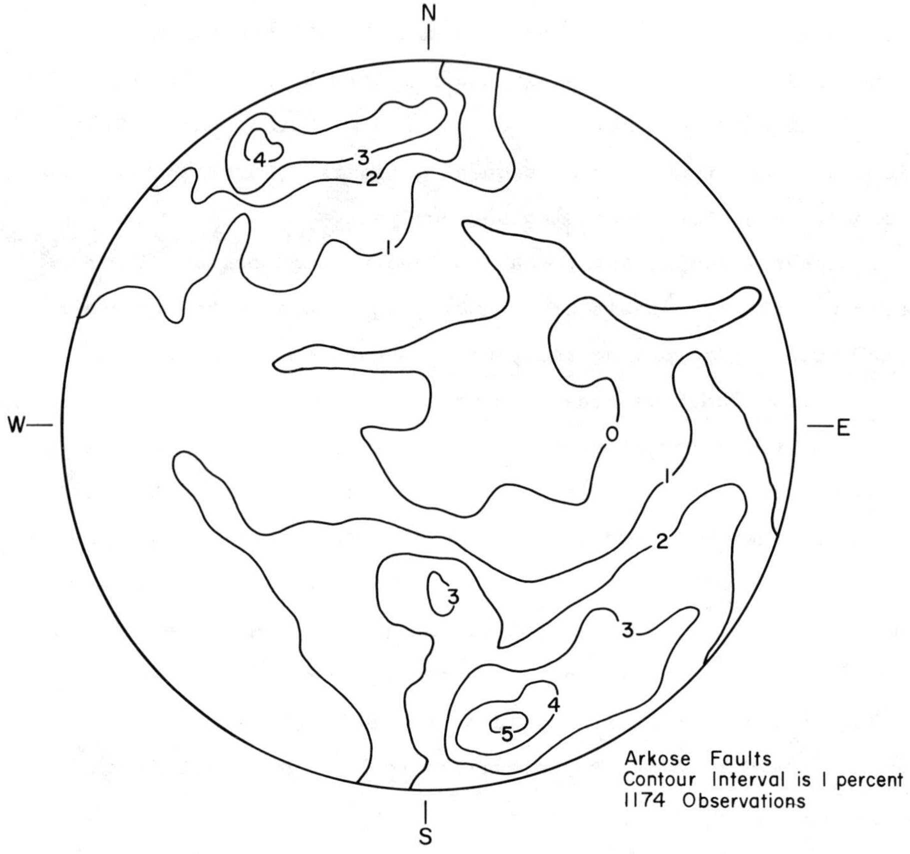

Fig. 3 Typical Equal Area Net
(Lower Hemisphere Projection)

TABLE 1.- IMPORTANT STRUCTURAL TRENDS

Trend or Set	Strike	Dip
A	N 00° to 90° W	30° to 70° E
B	N 00° to 80° E N 80° to 90° E	60° to 90° W 70° to 90° W
C	N 40° to 90° E N 80° to 90° W	70° to 90° E 70° to 90° W

Structural Intersections and Wedge Failure.- Using a Wulff equal angle stereonet, the significant structural sets were combined with the slope geometry of the south pit slope. Shown in Figure 4 is a Wulff stereonet (lower hemisphere) upon which structural sets A and B are shown, as well as various south slope geometries. The areas designated as Set A and Set B structures were formed by denoting all lines of intersection of the planar structures and the lower hemisphere for that particular set. That is, the designated area encompasses all possible lines of intersection of the structure and the lower hemisphere that are possible within the given strike and dip limits specified for a given structural set. Figure 5 was constructed in a manner similar to Figure 4 and shows structural sets A and C, together with south pit slope geometries. The area of common overlap in Figure 4 represents the structural intersections of sets A and B and, similarly, the common overlap in Figure 5 represents the intersections of Sets A and C. In each figure the approximate location of the highest concentration of intersections is indicated. The strike of the south slope (N 55° W), together with hypothetical dips of 10, 20, 30, 40 and 50 degrees, has been drawn on both figures.

Using these figures, one can get an idea of the possible adverse effects that structural intersections will have on a given

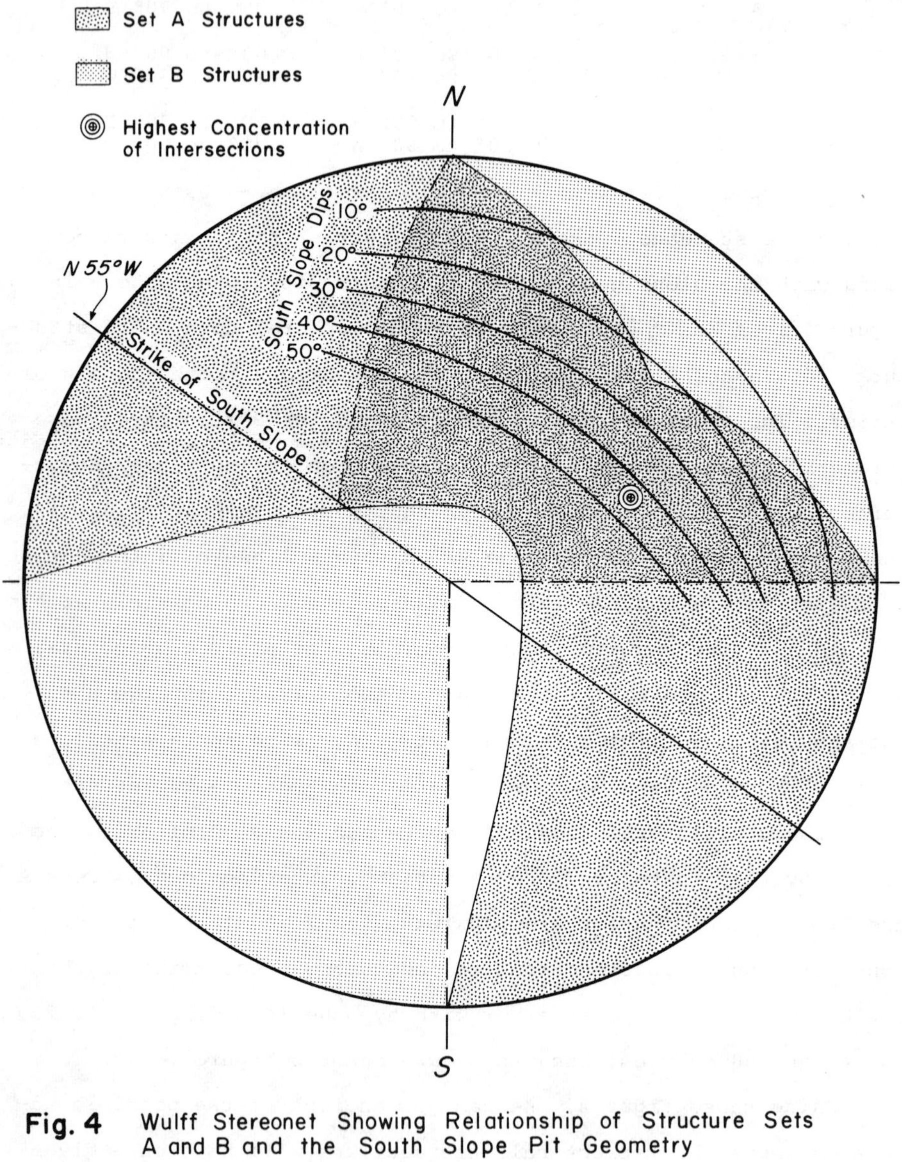

Fig. 4 Wulff Stereonet Showing Relationship of Structure Sets A and B and the South Slope Pit Geometry

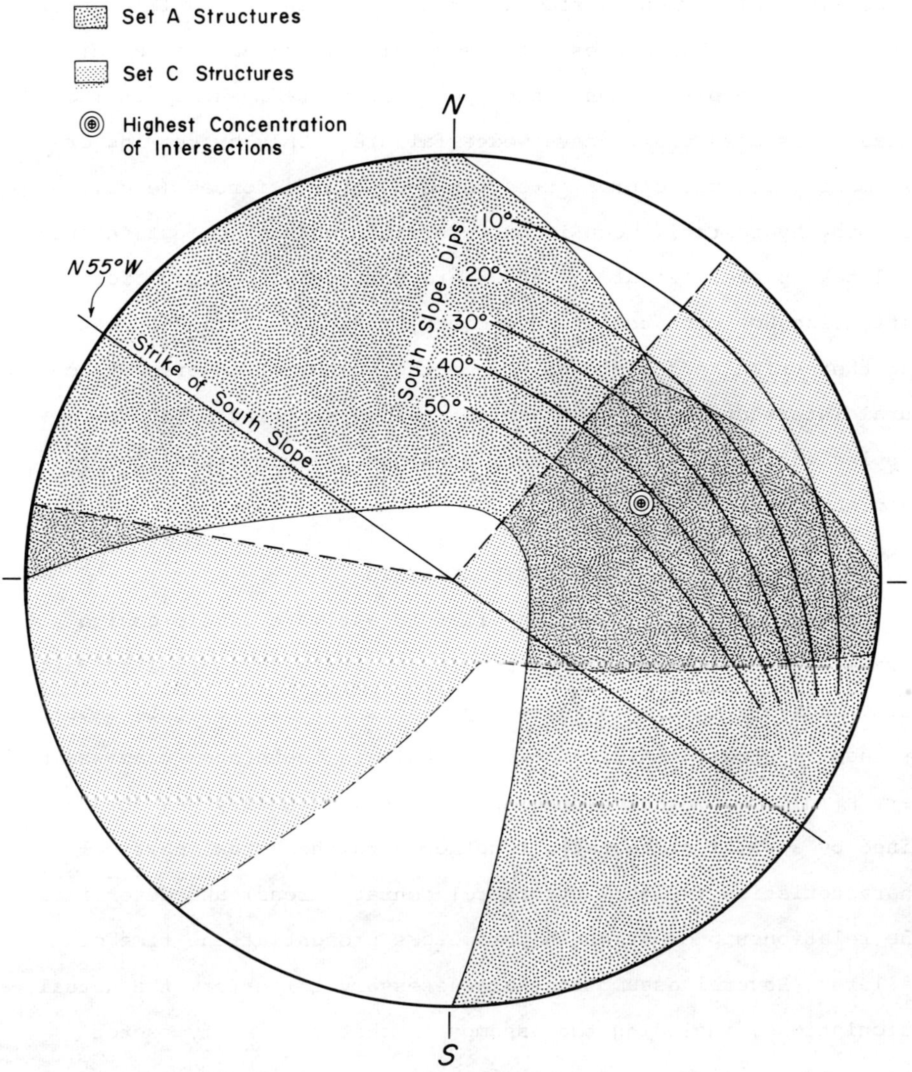

Fig. 5 Wulff Stereonet Showing Relationship of Structure Sets A and C and the South Slope Pit Geometry

slope geometry. It is readily apparent that the original south slope design (strike: N 55° W, dip: 45° E) was a very unfortunate geometric choice when considered in terms of the prevailing structures. Such a choice does not necessarily mean that wedge failures will take place when the slope is excavated; rather it indicates it is possible to have wedge failure. Only by knowing or estimating the structural strengths and outside forces (e.g., seismic, hydrostatic) can it be determined whether or not failure will take place. Figures 4 and 5 are particularly useful for estimating how much the slope geometry must be modified by changing the strike and/or dip to reduce the number of adverse structural intersections. For example, a slope striking N 55° W dipping 45° W would have virtually no adverse structural intersections from Sets A and B.

STRENGTH ANALYSIS

Wedge Back-Analysis.- In order to estimate usable design parameters which could be applied to structurally controlled failures, back-analysis of some twenty wedges was carried out, five[1] of them in great detail. The back-analysis technique used was a very simple method which combines the wedge geometry (as determined by a Wulff stereonet), together with the actual physical characteristics (weight, structural contact areas) and determines the relationship of strength parameters present at the time of failure. Several assumptions are necessary to perform the actual calculations, including the assumption that the safety factor (ratio of resistive to driving forces) is identically equal to

[1] These wedges ranged in total weight from 225 to 1000 tons.

one and that any hydrostatic and seismic forces, if present, are known. Also, the failure criterion[1]--linear (τ_L) or nonlinear (τ_{NL})--must be assumed. Using limiting equilibrium methods and a linear failure criterion (assuming no hydrostatic or seismic forces), the following relationship of cohesion (C) in psf and internal friction (ϕ) was determined to exist at the time of failure

$$C = 428 - 613 \tan \phi \qquad [1]$$

Strength Testing.- Field observations of the failure plane surfaces indicated that failure occurred predominantly along clay-filled[2] structures. The actual shearing on these planes appeared to be in the clays themselves, rather than between the clays and the rock surfaces. For these reasons, a limited series of field and laboratory strength tests was carried out. The field tests involved the use of a portable testing device[3] capable of determining cohesion and internal friction constants along rock joints. The laboratory tests consisted of direct shear tests using four points for the failure envelope determination. Results of both

---

[1] Linear: $\tau_L = (\sigma_n - \mu) \tan \phi + C$

Nonlinear: $\tau_{NL} = B (\sigma_n - \mu)^D$

where τ = shearing stress
σ_n = normal stress
μ = hydrostatic pressure
ϕ = angle of internal friction
C = cohesion
B = friction constant
D = constant

[2] The clay infilling along the structures contains breccia particles which tend to increase the strength of wedges having relatively large structural constant areas.

[3] This device was designed and operated by Royal School of Mines (London) graduate students as part of an overall graduate research study.

field and laboratory tests indicated the value of internal friction was fairly consistent from test to test and that a good estimate for design purposes would be 30°. The cohesion values were much higher than expected, ranging from 1300 psf to 8500 psf.

Determination of Design Equations.- Using the internal friction value of 30° and computing the corresponding cohesion value from equation [1], a cohesion value of approximately 75 psf resulted. This cohesion value is much lower in comparison to the test results than would normally be expected. Lowering the internal friction value to 23° increases the cohesion to only 170 psf, which is still comparatively low in light of the test results. Therefore, it was concluded that a conservative failure criterion for the analyzed wedges could be stated as

$$\tau_L = \sigma_n \tan 30° + 75 \text{ psf} \qquad [2]$$

Use of a nonlinear failure criterion has been suggested by several investigators, notably Pentz,{2} to more closely approximate the actual pit slope conditions. Such a criterion[1] was investigated during the course of initial study and safety factors for the five wedges back-analyzed by linear criterion were estimated. A comparison of these safety factors, together with those

[1] Estimated constant values were $B \cong 0.669$ and $D \cong 0.988$.

computed using a linear criterion ($\phi = 30°$) with cohesion values of 0, 75 and 250 psf are shown in Table 2.

TABLE 2.- COMPARISON OF SAFETY FACTORS FOR FIVE WEDGES

Wedge	Safety Factors			
	Linear			Nonlinear
	C = 0 psf	C = 75 psf	C = 250 psf	
1	0.73	1.08	1.93	0.79
2	0.92	1.07	1.43	0.98
3	0.79	0.89	1.14	0.84
4	0.85	1.00	1.33	0.91
5	0.85	0.96	1.22	0.90

These tabulated values reveal the sensitivity of relatively small changes in cohesion with safety factor. The nonlinear approach is thought to have merit, particularly in cases similar to the present analysis where production blasting may have destroyed or lessened the near surface cohesion. In these cases, the nonlinear usage would be particularly useful to describe both cohesion loss near surface and cohesion retention at depth with one equation. However, the scope of this initial study was so limited that the mathematically simpler linear approach was used for design estimates.

Hydrology Considerations.- Throughout the course of the initial study, lesser emphasis was placed on pore water pressures than on other factors. The reason for this is essentially threefold: First of all, hydrology studies carried out during the past two years indicate the water table in the vicinity of the pit has been effectively lowered several hundred feet below the south slope failure zone. Secondly, observations in and adjacent to the failure area and along the failure plane have not revealed any saturated zones or other evidence of water. Thirdly, the scope of the initial study was limited to the extent that a detailed hydrological study could not have been effectively completed.

STRUCTURAL AND STRENGTH DATA USAGE

Failure Mode.- Field observations of the actual failure planes in the central south slope indicate that while failure began in the classical wedge manner, it proceeded to completion in a slabbing fashion. In other words, failure initially took place on steeply dipping (72°) faults of Set B and medium dipping (38°) structures of Set A in the classical wedge sense. It then continued by slabbing parallel to Set A structures which in this case were striking almost parallel to the strike of the slope or N 56° W and dipping at about 38° E. The faults of Set B acted in effect as freeing devices for the sides of the slabs since they were striking at a very high angle to the slope or N 78° E and dipping at about 72° W. Figure 6 depicts in plan the initial wedge failures and the slab failures which followed.

Structural mapping of portions of the failure plane in the south slope yielded data which support the field observations of the actual failure plane orientations. These data were analyzed and the statistics given in the following table were determined. Both failure planes A_1 and A_2 correspond to structural Set A, the difference between them being that A_1 was mapped at approximately 100 ft up the slope from A_2. Failure plane B corresponds to structural Set B.

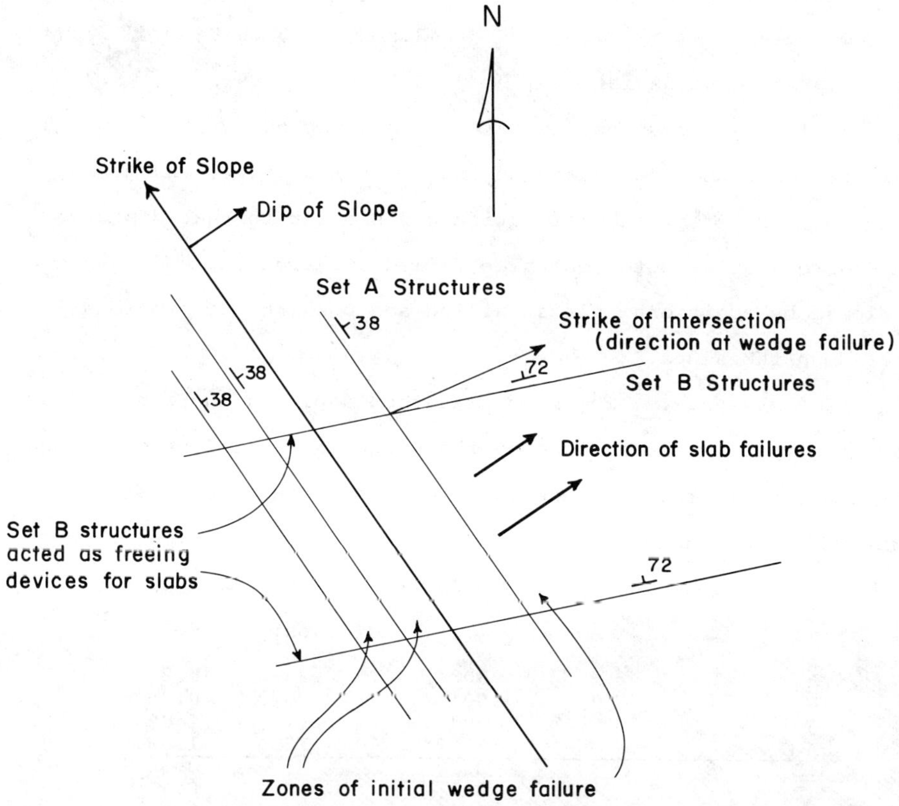

Fig. 6 Plan of Central South Slope Showing Initial Wedge Failures and Subsequent Slab Failures

TABLE 3.- FAILURE PLANE STATISTICS

Failure plane	A_1	A_2	B
Number of data pairs	44	16	32
Strike mean	N 55.8 W	N 57.0 W	N 77.7 E
Strike standard deviation	15.7	7.9	7.0
Dip mean	43.2 E	37.6 E	71.7 W
Dip standard deviation	8.1	3.9	5.7

Failure was observed to take place along one or two benches initially and then to expand to include three, four, and more benches. In effect, the situation was similar to that depicted in Figure 7. The outermost slab failed and broke up and was then followed by additional slabs failing and pushing the previous ones down the slope.

Slab Analysis.- The slab situation depicted in Figure 7 was back-analyzed to get an idea of the strength conditions existing at time of failure. A plane failure mode and a linear failure criterion were assumed. Calculated results are given in Table 4.

TABLE 4.- SAFETY FACTORS FOR VARIOUS SLABS

Slab	ϕ	C, psf	S.F.
A	30°	75	0.82
A	30°	0	0.75
A	30°	291	1.00
A	35.7°	75	1.00
A+B	30°	75	0.80
A+B	30°	0	0.75
A+B	30°	421	1.00
A+B	36.2°	75	1.00
A+B+C	30°	75	0.78
A+B+C	30°	0	0.75
A+B+C	30°	628	1.00
A+B+C	36.7°	75	1.00

One of the most significant aspects indicated in this table is that failure can occur along the large slab (A+B+C) with less

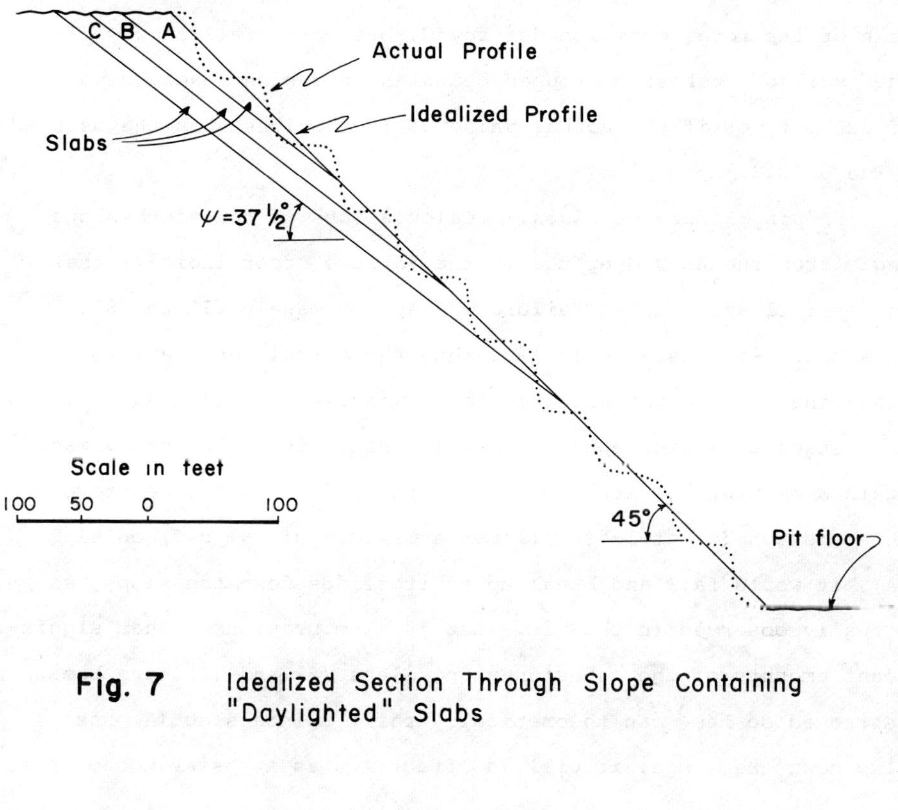

Fig. 7 Idealized Section Through Slope Containing "Daylighted" Slabs

destruction of cohesion than for the small slab (A). In other words, all cohesion in excess of 291 psf must be destroyed on slab (A) for failure to occur, but slab (A+B+C) can have as much as 628 psf remaining and still fail. Because slab (A) had more of its total base area nearer to pit blasting than slab (A+B+C), it could have suffered the necessary cohesion destruction and failed. Being farther from the blasting, slab (A+B+C) would have less of its total cohesion destroyed, but could fail anyway if the available cohesion dropped below an equivalent average value of 628 psf, or if the actual value of ϕ were less than the assumed value of 30°.

Driving Forces on Slabs.- Calculations of the south slope angle from the heaved-up toe to the failure crest indicate that the overall angle after failure was approximately 27° to 28°. This value is considerably less than the actual base failure plane angle which was observed to vary between 35° and 40°. To understand this discrepancy, the slab situation in Figure 7 was again examined. If it were assumed that blasting has destroyed the cohesion (or at least all but a maximum of 291 psf) on slab (A), it would fail and break up as it slides down the slope, as actually observed in the pit. Let it also be assumed that significant amounts of the cohesion along slabs (B) and (C) have been destroyed so they can theoretically fail. Since slab (A) has slid down the slope, it will in effect act as a resistance or toe buttress against slab (B). If slab (B) does fail and slide down the slope along with slab (A), they will both then tend to act as toe buttresses against slab (C). Before any sliding takes place, the safety factor must become less than 1.00 and the driving forces will become greater than the resistive forces. The difference between these forces will be a real force which will tend to push not only the slab itself but any other material in front of it. The force has been termed an excess driving force and some estimates of this potential force were calculated and are summarized in Table 5. These estimates assume an instantaneous buildup of excess force. In reality, the actual forces will never reach these levels because they will be continuously relieved through displacement. However, except for the dynamic side effects of an instantaneous force, the actual force can be assumed to give the same displacement effect occurring over a longer period of time.

TABLE 5.- EXCESS DRIVING FORCES ON SLABS

Slab	ϕ	C, psf	Excess Driving Force per Linear Foot of Slope Width
B	30°	0	41 tons
B	30°	75	29 tons
C	30°	0	113 tons
C	30°	75	95 tons

The driving force calculations, while being only an approximation at best, do serve to indicate that once the cohesion has been destroyed along the structures in a slope, a slab failure along structure dipping 37½° can cause great forces to be developed in the slope toe areas. These forces are large enough to cause toe heaving, not only in broken material, but also in solid in-place rock. Such rock is vulnerable if it has been weakened by blasting or if it contains adversely oriented structures. In addition to toe heaving, low angle shears can be expected to develop in various parts of the broken material. The end result of the excessive forces is an apparent failure angle less than the angle of slab failure, as shown in Figure 8.

SLOPE STABILIZATION AND CURRENT RESEARCH

Stabilization Techniques.- Having completed an initial structural and strength analysis and having gained an insight into the mode and conditions of the south slope failure, the question arose as to what possible approaches could be taken to obtain a more stable slope. Particular concern was directed at mining another pushback in the same south slope area. All other factors being equal, it is believed the present failure was a full-scale demonstration of what could be expected if another pushback were begun. Therefore, if the pit were to be economically expanded to the south, some--if not all--of the factors causing instability would have to be modified. A summary of these factors and their possible causes and remedies are outlined in Table 6.

TABLE 6.- SOUTH SLOPE INSTABILITY FACTORS, POSSIBLE CAUSES AND POSSIBLE REMEDIES

<table>
<tr><th>Factor</th><th>Possible Cause(s)</th><th>Possible Remedies (where applicable)</th></tr>
<tr><td>Adversely oriented structures</td><td>1) Slope too steep
2) Slope oriented adversely</td><td rowspan="3">1) Lessen slope angle
2) Change slope orientation
3) Modify blasting practices
4) Artificial stabilization
5) Dewater pit slopes</td></tr>
<tr><td>Low strength structures</td><td>1) Natural strength is low
2) Differential rebound from high mining rates
3) Seismically induced</td></tr>
<tr><td>Uplift of wedges or slabs</td><td>1) Seismic uplift forces
2) Pore water pressure</td></tr>
</table>

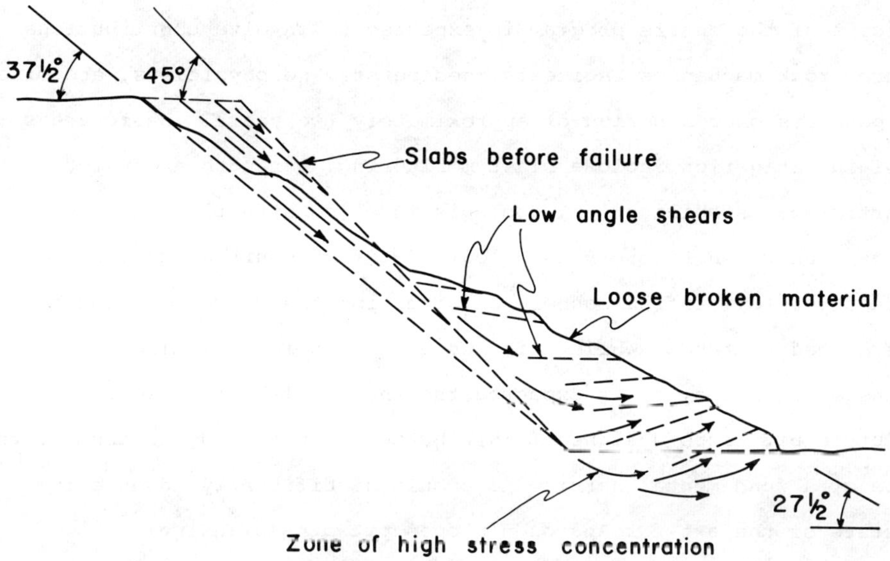

Fig. 8 Section Through Slab Failure on 37 ½° Structures Showing Apparent Failure at 27 ½°

Current Stability Research.- Following the initial stability study of the south slope failure, the corporate Mining Research Department and Twin Buttes Operations began a full-scale stability research effort. The research has been divided into four phases: Phases I and III are data-gathering stages, while Phases II and IV are essentially for analysis and design. Each phase is contingent on the results of previous work and is flexible enough to permit expansion or contraction of any particular facet. The scope of the entire program is expected to involve contributions from rock mechanics engineers, geologists, geophysicists, and mine operators over a period of approximately two years. Basic areas of investigation include blast monitoring, strength analysis, artificial stabilization, geologic model construction, and hydrology. Shown in Figure 9 is a logic-decision chart which indicates in a very simplified manner the direction the current research is expected to take. While this research is primarily directed at the south slope, it is expected the end results will aid in all future pit slope designs at Twin Buttes. This study is thought to be of a fundamental nature and should significantly advance the state of the art for Anaconda's open pit mine planning.

ACKNOWLEDGMENTS

The author is indebted to the Twin Buttes operating and geological personnel for their assistance in the initial study. Particular thanks are extended to W. G. McCurry, Project Geologist, for his help in the collection of structural data.

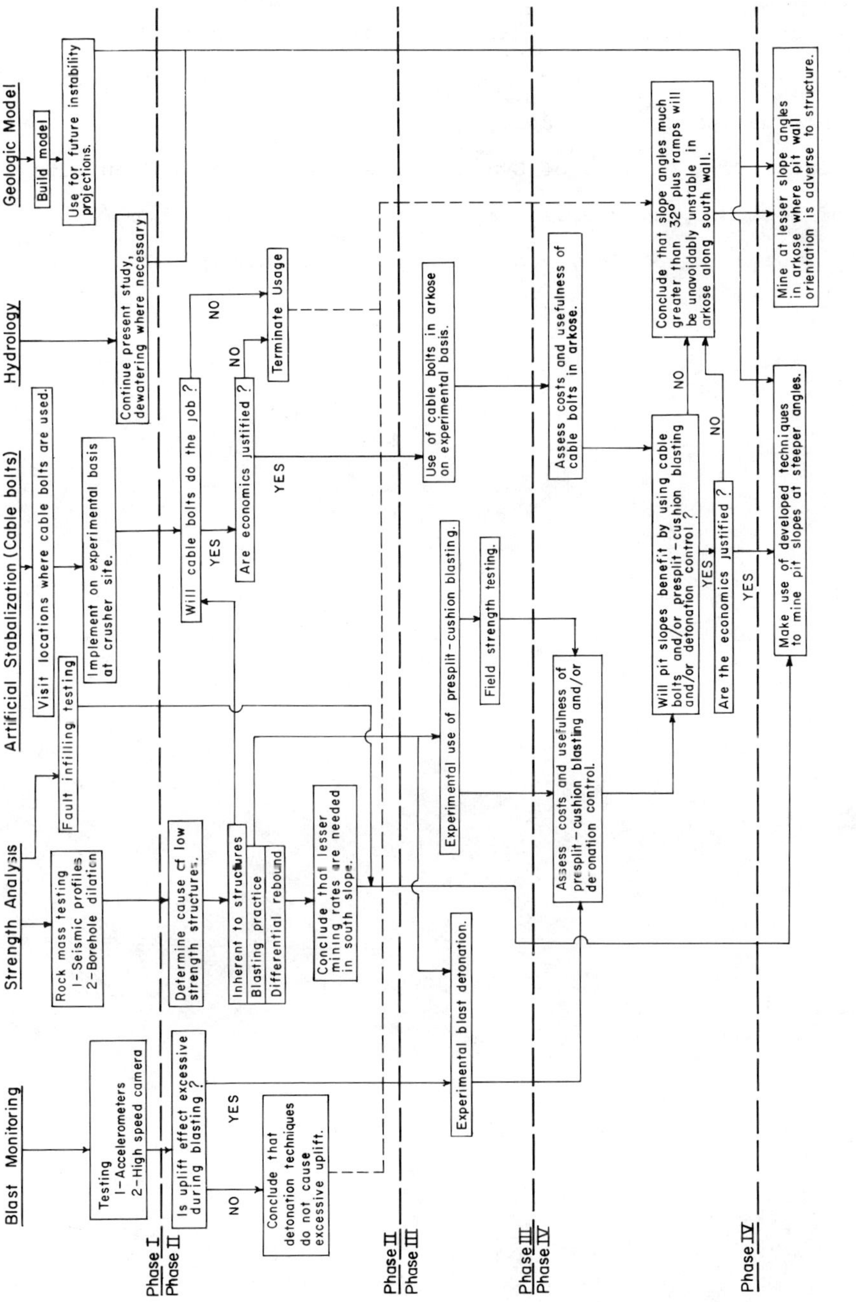

Fig. 9 Logic - Decision Chart: Current Plan for Twin Buttes Stability Research

REFERENCES

1. Jeran, P. S. and J. R. Mashey, *A Computer Program for the Stereographic Analysis of Coal Fractures and Cleats*, USBM Information Circular 8454, March 1970

2. Pentz, D., "Methods of Evaluation and Analysis of Stability of Rock Slopes," *Proceedings of the Symposium on Stability for Open Pit Mining*, Vancouver, B. C., November 1970

METHODS OF MONITORING OPEN PIT SLOPES

by

Bruce A. Kennedy[1]

SUMMARY

There are two basic reasons for monitoring open pit slopes: First, as an insurance that a given slope is safe and exhibiting no movement or small movements within acceptable design limits, and, secondly, as an indicator and prediction tool for slopes that are moving and exhibiting signs of failure. Present monitoring methods often leave much to be desired; however, well-known techniques such as standard surveying methods are sometimes adequate. This paper summarizes some of the monitoring systems which have been tried by The Anaconda Company in its open pit mines.

INTRODUCTION

The Anaconda Company has been actively involved in slope stability studies in its open pit mines since 1966. Slope stability studies have been made or are in progress at Butte, Montana; Yerington, Nevada; Twin Buttes, Arizona; Inspiration, Arizona; Cananea, Mexico; Chuquicamata, Chile; and Exotica, Chile. During this work the need to measure the physical behavior of slopes was required many times. A large number of methods of measuring ground movement have been tried and others are being developed for the future. This paper details some of the monitoring systems tried by The Anaconda Company and summarizes some of the advantages and disadvantages of each.

[1] Senior Mining Research Engineer, The Anaconda Company, Salt Lake City, Utah, U.S.A.

THE NEED TO MONITOR OPEN PIT SLOPES

As man expands his understanding of the behavior of rock slopes, his need to measure the physical movements of those slopes increases. From the practical engineer's viewpoint, there are two basic reasons for slope monitoring:

1. To ensure that the slope is safe and is exhibiting no movement or small movements within acceptable design limits; or
2. To record excessive slope displacements associated with total or partial failure. Recording of displacement data can be used to predict the behavior of a failure. This would enable the mine operators to take steps to minimize production and equipment losses and danger to personnel.

The systems of monitoring available to the engineer today are often crude and frequently leave much to be desired as regards the accuracy and reliability of the measurements. The saving grace is that they at least give an indication of the movement and its magnitude. The one mistake that should be avoided is that of collecting data without first determining the usefulness of that data.

The slope stability engineer should answer the following questions before designing his monitoring system: First, what is the reason for monitoring the slope? The above noted reasons for monitoring slopes can vary because of the uniqueness of the individual slopes. Secondly, what area is to be monitored? If the slope is not in a failure condition and is being monitored purely as an insurance for protection of critical installations, etc., the area to be monitored could be extensive. However, if a specific stability problem exists or is suspected, the area to be monitored can be very precisely delineated. Thirdly, the location and type of monitoring systems should be determined. Ideally, monitoring is required in three dimensions on and in the slope. The crest of the slope,[1] the face of the slope,[2] and at depth within the slope should all be monitored, if possible. Fourthly, what accuracy and precision (repeatability) of measurements will be required? The accuracy and precision required are related directly to the problem, although at the present time no rule of thumb has been developed for determining these values. The values to be used should be decided at the start of monitoring and only changed if absolutely necessary. As the accuracy and precision required increase, so does the cost of monitoring.

The forms of monitoring discussed in this paper measure either linear or angular displacement. Generally, the measured values are reduced to a resultant movement with calculated magnitudes and directions. In some cases, data can be reduced using an X, Y and Z coordinate system with magnitudes and resultant direction of the change stated as changes in coordinate values. Plots of change in

---

[1] Crest of the slope is defined as the line of intersection of the excavated slope and the natural topography.

[2] Slope face is defined as the rock surface exposed by the excavation.

displacement vs. time and variation of velocity vs. time serve as indicators of the stability of the mass. The cyclic occurrence of increases or decreases in activity are also a sign of the stability. If a mass is cyclically accelerating and decelerating yet the cycles are occurring closer together as time progresses, the mass may be becoming more unstable.

Monitoring techniques can be divided into two basic categories: 1) Techniques for determining mass movement of a large volume of material and 2) detail measuring systems for monitoring the movement of individual structures, joints, fractures, tension cracks, etc. Systems for monitoring pore water pressures, water levels, and other hydrological parameters associated with slope stability, will not be discussed in this paper.

MASS MEASUREMENT

Survey Nets

The Control Net

All open pits have a basic control survey grid consisting of a baseline and a number of permanent survey stations set up on stable locations some distance from the perimeter of the pit. A check baseline is usually incorporated into the grid to permit closure checks and correction factor calculations. The survey of these primary points should be made with a first order theodolite, direct reading to one second of arc. The angles should be turned by doubling twice and repeating this three times. Temperature, humidity and barometric pressure readings should be taken while the readings are being made and a refractive correction calculated and applied to the measurements. Chaining should be done with an Invar or steel tape held in tension. Temperature and catenary corrections should be calculated and applied to the measurements. If a distance measuring device such as a Geodimeter, Laser ranger, Tellurometer, Mekometer, etc., is used, a cross check by standard chaining should be made. The accuracy of such a survey should be one in 25,000.

The primary survey should now be carried to a series of secondary points around the pit perimeter, also with a first order survey. As above, a direct reading one-second theodolite should be used. The secondary survey points should be permanent, solidly built monuments. An accuracy of one in 25,000 should also be the aim on this survey. A check should be run from the primary points to the secondary points at least once every three months, and more frequently if movement is noticed. The survey should next be carried down into the pit to the tertiary points. These will be monuments set up on the benches and the slope crest in areas of suspected instability. A second order survey can be used to tie the secondary points to the tertiary. In this survey, angles should be turned by doubling twice and repeating this twice. An accuracy of one in 10,000 should be obtained. All surveying in the tertiary network should be done with reference to the secondary points. Surveying with the tertiary network should be second or third order looking for an accuracy of one in 5,000. The end points of any line of bench or profile monuments should be tied into the secondary points each time the tertiary network is surveyed. The tertiary network can consist of three possible pin arrays--profiles, bench lines, or random array. (Figure 1)

Profiles

Profiles are lines of survey stakes set on the crest of the slope at right angles to the direction of strike of the slope. The survey stake at the end of the line farthest away from the crest is tied into the secondary survey points. Two methods of surveying the profile can be used: 1) The transit is set up on the point of the profile farthest from the crest. Angles are turned to each of the other stakes in the profile with the secondary point being used as a backsight. A level is then run with the transit set up remaining in the same place, transferring the elevation through from the secondary point to the profile points. The distance from the transit set up to each profile point is then taped. X, Y and Z coordinates can now be calculated for each profile point. Comparing these to the values obtained on the previous survey will give the ΔX, ΔY and ΔZ. The horizontal movement ΔH and vertical movement ΔV, and the three dimensional direction of movement can next be calculated. In method 2) a normal traverse is made down the profile line moving the transit from point to point until all the points are surveyed in. Once again, ΔX, ΔY, ΔZ, ΔH and ΔV and the direction of movement can be calculated.

Method 1) is recommended over 2), providing the profile line is of reasonable length, because of the necessity for only one transit setup and the consequent reduction of introduction of error.

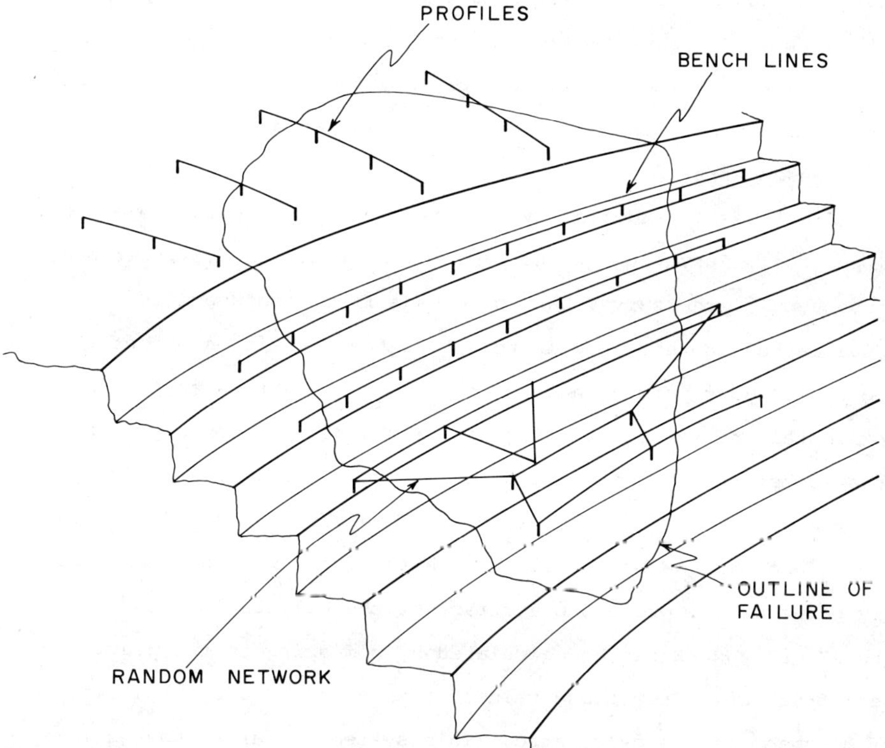

SURVEY SYSTEMS FOR MONITORING FAILURES

FIGURE I

Bench Lines

These are lines of stakes set out along the length of a bench across a suspected or known slide area. The point at one end of this line should be placed in what would be considered a relatively stable area. This point is surveyed accurately from the seconardy points and the survey of the line is carried out by either method 1) or 2) mentioned above. Once again method 1) is recommended because of the necessity for only one setup of the transit and the minimum amount of work on the slide area.

Random Array

Sometimes a random array of survey points is placed over the slope face. Surveying on the slope face itself is difficult and, consequently, the survey has to be made from remote points. The usual method in this case is to set up the transit on one of the known secondary points on the pit perimeter and turn angles, both horizontal and vertical, using one of the other secondary points as a backsight. The transit is then moved to another secondary point and the procedure repeated. The values on the point on the slope face can now be calculated. A third setup on another secondary point will improve the accuracy of the calculation. Once again the ΔX, ΔY and ΔZ can be calculated by comparing to the previous set of values. From these values, the ΔV and ΔH and the direction of movement can be calculated. This system of survey has extreme limitations of accuracy due to the distance of the unknown point from the triangulation baseline.

Recording, Reduction and Interpretation of Data

Records should be kept of all data noted in the field books and all calculations of the ΔX, ΔY and ΔZ and consequent calculations of the change in horizontal and vertical positions, ΔH and ΔV.

Where the old coordinates of a point are X_1, Y_1 and Z_1 and the new coordinates are X_2, Y_2 and Z_2

$$\Delta H = \sqrt{(\Delta x)^2 + (\Delta Y)^2} \qquad \text{where } \Delta X = X_2 - X_1,\ \Delta Y = Y_2 - Y_1,\ \Delta Z = Z_2 - Z_1$$

$$\Delta V = \Delta Z$$

bearing of the movement $\qquad \alpha = \tan^{-1} \frac{\Delta X}{\Delta Y}$

dip of the movement $\qquad \psi = \tan^{-1} \frac{\Delta Z}{\Delta H}$

Plots should be kept of the cumulative displacement vs. time (Figure 2) for both ΔH and ΔV and for the resultant movement ΔR where $\Delta R = \sqrt{(\Delta Z)^2 + (\Delta H)^2}$. Experience has shown that a displacement vs. time chart scale of one vertical unit = 3 ft and one horizontal unit = one month is the most suitable for these plots.

A large-scale plan, with the original locations of the pins plotted on it, should be kept up-to-date by plotting the ΔH and the ΔV for each pin as horizontal and vertical vectors from that point (Figure 3). The resultant ΔR is also plotted to scale from the point, both in its true magnitude and dip (dotted line) and at the actual bearing of the movement from the true north, α (solid line). ψ, the dip of the direction of movement, is the angle formed by the ΔR (dotted line) and ΔH. The next values of ΔH, ΔV and ΔR calculated for this point are plotted using the end point of the previous ΔR (solid line) as an origin.

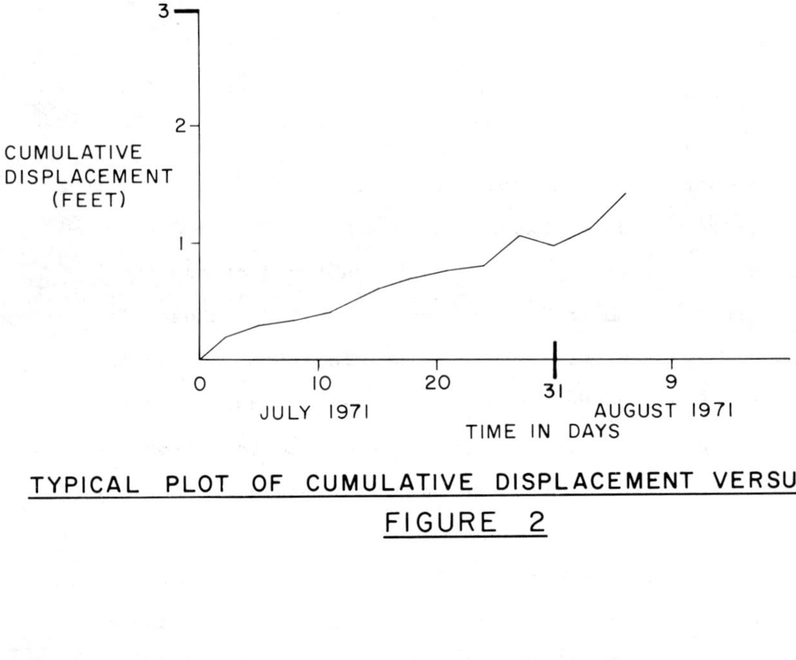

TYPICAL PLOT OF CUMULATIVE DISPLACEMENT VERSUS TIME

FIGURE 2

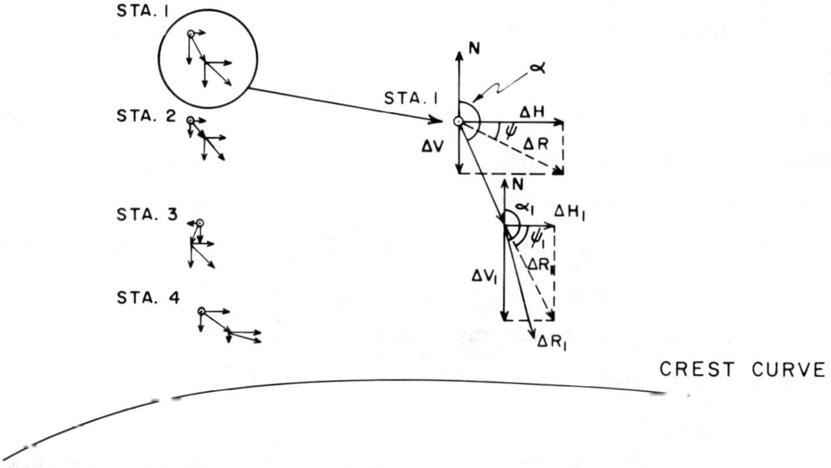

METHOD OF PLOTTING VERTICAL, HORIZONTAL AND RESULTANT MOVEMENT ON PLAN MAP OF PROFILE LINES

FIGURE 3

These charts are used to keep track of the velocity and acceleration of the movement, as well as direction. As the displacement vs. time curve tends toward a slope of infinity (vertical), catastrophic failure approaches. Isograms or contours of velocity (Figure 4) can be drawn at various times during the progress of the slide to give a useful guide to the behavior of the sliding mass. Points of equal velocity are connected together, giving a picture of the moving mass, its size, and direction of movement. This can be done on both the crest and slope face.

Advantages of the survey net for monitoring are that the system is simple to operate, techniques are well understood, interpretation of the results is easy, a large area can be covered in a short time, and it can be fairly inexpensive to operate. Disadvantages to the system are that it is not continuous in recording of data, a certain amount of manpower is required to collect data, initial capital outlay for equipment can be large although generally the equipment is already available, and the movements have to be reasonably large before they are picked up.

Seismic Systems

Monitoring of microseismic transmission from rock was first used in the U.S.A. and South Africa to predict the location and occurrence of rock bursts in deep level mines. Wide band, relatively high frequency equipment was used in this work. High frequency monitoring, however, is fairly difficult to interpret and requires very sophisticated and expensive equipment for recording and data reduction. In 1968 an attempt was made to monitor low

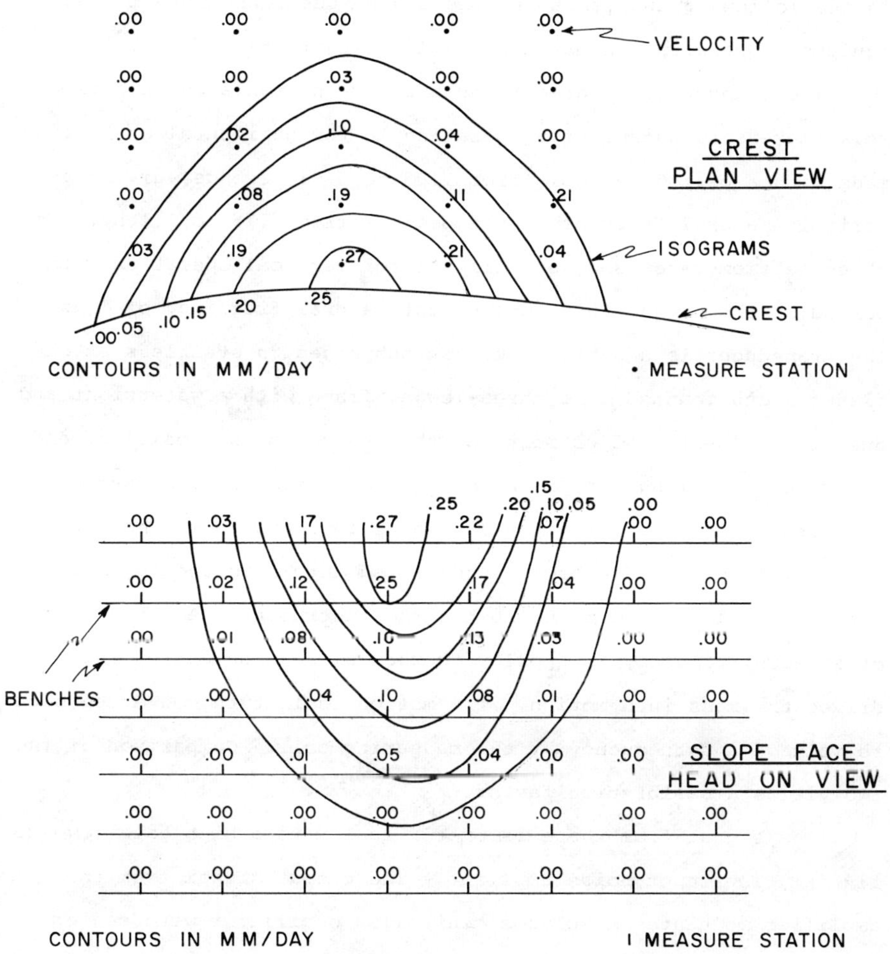

EXAMPLE OF VELOCITY ISOGRAMS

FIGURE 4

frequency seismic activity in an unstable slope at the Chuquicamata Mine, Chile. A low frequency seismic system with a frequency range of from 0.5 to 12.5 Hz was installed. The equipment was modified in the following two years and the system described here is the equipment operating in March of 1971. (Figure 5)

The transducers, three in number, are portable short-period seismometers adjustable for operation in the horizontal or vertical mode with a period variable from 1.33 to 0.91 seconds, giving a variable natural frequency (undamped) of from 0.75 to 1.1 Hz. The three seismometers used were all set for vertical operation with a period of one second and a consequent natural frequency of 1 Hz. The transducer is a 5-kilogram mass suspended on stainless steel flexures and mounted in a three-legged frame with a watertight and dust-tight cover. Adjustment for the period, mass, position, and instrument leveling are provided, as are a mass lock, a mass position indicator, and an instrument level indicator. Any movement of the mass on the flexures is converted into a voltage by a coil. A second coil is used to calibrate the instrument. A known voltage of a known frequency is applied to the calibration coil. This drives the mass into motion. The motion is in turn converted back to voltage and frequency by the measuring coil. Comparison of the two yields a calibration factor.

The signal from the seismometer is fed to a high gain (70 to 112 dB), low input noise amplifer. The output signal from the amplifier modulates a voltage controlled oscillator with center frequencies of 1300 Hz, 1700 Hz, and 2300 Hz for Channels 1, 2 and 3, respectively. This FM signal is fed into a radio transmitter

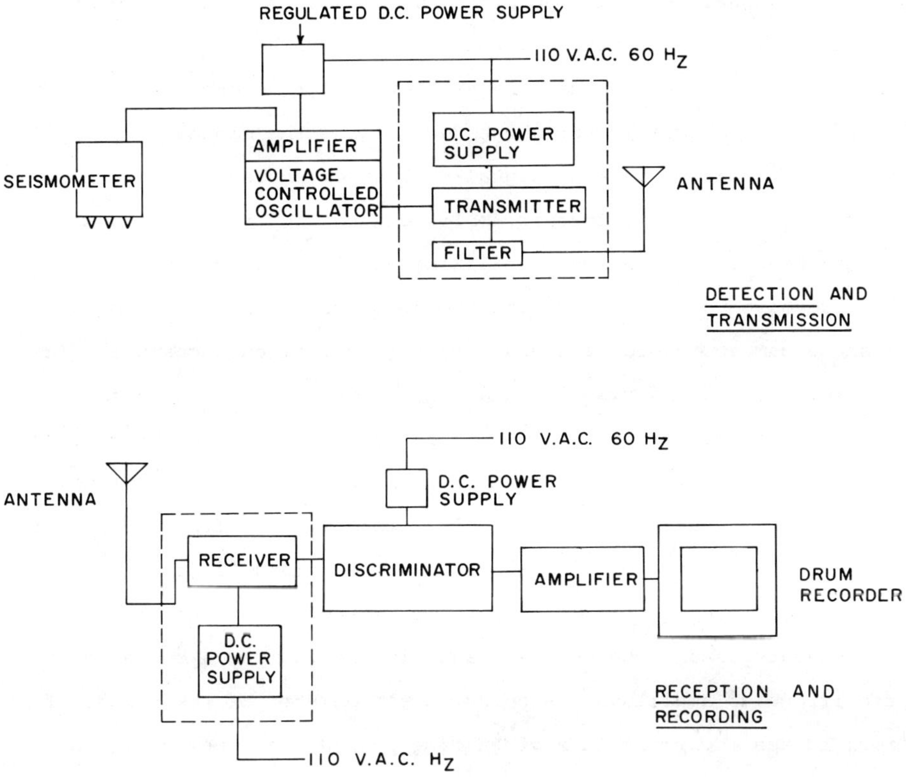

DYNAMIC SEISMIC MONITORING SYSTEM

FIGURE 5

and modulates a carrier frequency of 72.00 MHz, 72.40 MHz, and 76.00 MHz for Channels 1, 2 and 3, respectively. The transmitted signal, 1/2 watt, is passed through a matching impedance filter and then broadcast by a heavy duty directional station antenna.

The signal is picked up by an identical antenna in line of sight with the transmitting antenna. The receiver demodulates the FM on FM signal and removes the carrier wave frequency. The signal is sent to a discriminator which demodulates the signal further, producing the same initial signal as that generated by the seismometer. This signal is further amplified and then recorded on a three-channel drum recorder utilizing heat sensitive paper. The drum revolves once per hour and the pens are automatically moved along a cam to record 12 hours of activity from each channel. The records are scanned visually and the various signals identified. With experience, it is possible to recognize with ease the cultural noise generated in the pit by drills, trucks, shovels, locomotives, etc.(1) Earthquakes are also easily identifiable with a very definite wave form envelope. The rock noises emanating from an unstable zone are identifiable by their characteristic envelope shape, a frequency usually between six and nine Hertz, and a characteristic irregular amplitude. Records should be kept of the number of rock noises occurring in a given time period, say twelve or twenty-four hours. A record should also be kept of the frequency, duration, and equivalent earth motion of each seismic occurrence. As the instability of the rock mass increases so the number of rock noises released in a given time period should increase. Plotting the accumulative total number of rock noises against time may give a useful prediction chart similar in shape to those obtained from displacement plotted against time.

The use of seismic systems for open pit monitoring has been reported by others.[2,3,4,5] However, at the present time the use of seismic monitoring of unstable slopes in open pits is in its infancy and a good deal of development work is required to make it a reliable and inexpensive system. A certain amount of training is necessary for the technicians who maintain the equipment and reduce the data. The advantages of the system are that it is exceptionally sensitive and is a continuous recording method.

Other Systems of Mass Measurement

A number of other systems of mass measurement have considerable promise but have yet to be tried by Anaconda. These are 1) terrestial stereophotography and 2) aerial stereophotography. Neither of these systems has the accuracy of standard survey techniques but are sometimes adequate for determining changes in location of survey points.

DETAIL MEASUREMENT

Detail measurement is usually necessary where a slope or an area of a slope is in a critical or failure condition. Detail measurement is also used when monitoring a slope at depth in the rock because present-day techniques do not permit mass measurements at depth within a slope. The first four methods of measurement discussed are surface detail measurement systems and early warning devices; the last three are borehole measurement systems for depth monitoring.

Tension Crack Stakes

This is a very simple system of measuring the movement of a tension crack. Two steel stakes, generally made of reinforcing bar or drill steel pointed on one end and cut and ground flat at the other, are driven into the ground on either side of the tension crack. Holes can be dug for the stakes and the stakes concreted in if necessary. The stakes can be allowed to stick out of the ground for two or more feet, or they can be cemented in to protrude approximately one inch above ground level. The latter is preferable as it lessens the chance of damage to or movement of the stake by vehicles, etc. Taping with an Invar or steel tape between reference marks on the two stakes will give a resultant displacement measurement. A level can be carried from one stake to the other to give the true vertical change, and from this and the resultant measurement the true horizontal change can be calculated.

A further refinement of this system is to form an approximately equisided quadrilateral by placing two stakes on either side of the tension crack. Measurements are made with an Invar or steel tape across the tension crack along the quadrilateral sides and diagonally across the corner stakes. This system allows computation of a vertical and horizontal movement and the resultant direction of that movement.

In both cases, at least one stake is referenced by standard survey techniques to a known stable point at frequent intervals to correct for any massive movement not registered by the stakes. Plots of cumulative displacement and velocity vs. time should be maintained. This system is extremely cheap, easy to install, and easy to monitor. The movements have to be reasonably large to be significant.

Joint Extensometer (Figure 6)

This is a fairly crude instrument for measuring gross movement on joints or fractures on a slope face or crest. The instrument has two major drawbacks: 1) There is little or no resolution of the exact direction of movement, and 2) the structures on which these instruments are located have to be selected with care as there will be a very large number of structures in any given failure zone and to instrument them all would be difficult, if not impossible.

The instrument consists of two lengths of 3 in. by 1/4 in. strap steel bent into Z shapes and fastened to the rock on each side of the joint of fracture with expansion shell bolts approximately two feet in length. An epoxy cement is also forced into the hole for added strength. A dial gauge inserted in the countersunk holes measures any displacement between the two straps. (Figure 6-A)

An alternative method is to use an LVDT and a strip chart recorder. In this system one of the measuring legs of the straps is twisted at right angles to the rest of the strap (see Figure 6-B) and drilled to take an electrical conduit clamp. The LVDT body is clamped to the strap by this conduit clamp and the core is clamped to the measuring leg of the other strap. Care should be taken in clamping the LVDT as it is subject to damage from crushing.

The data obtained from these instruments are in the form of displacement and should be plotted on a standard cumulative dis-

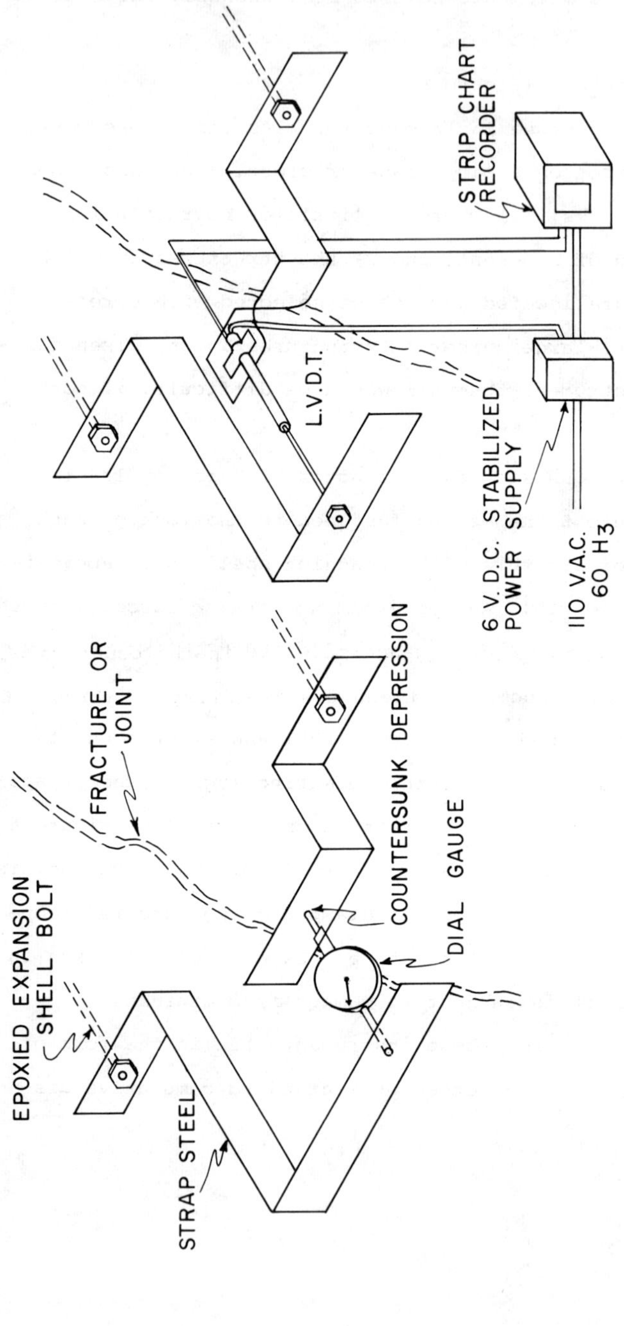

JOINT EXTENSOMETERS

FIGURE 6

placement vs. time charts. Further reduction of the data can be made to produce velocity and acceleration vs. time graphs. These instruments are very cheap and easy to use; however, the location has to be very carefully determined and because of the specific placement of the instruments care should be taken in extrapolating data.

Sliding Wire Contacts

The sliding wire contact system was designed as a very cheap early warning system for use in a slide area where the movement of individual large blocks of rock had to be monitored continuously for excessive movement. (See Figure 7-A.) A stake is driven into the unstable block and another into the stable area. Two separated copper contacts are set up on a spring and wire system between the two stakes. A spring is attached to the stake in the unstable block and a steel wire connected to it and run to a point approximately four feet short of the stake on the stable ground. An eighteen-inch length of heavy gauge insulated copper wire is taken and two inches at either end are stripped of their insulation. A loop is made with one of the bare ends and bent at right angles to the insulated section of the wire. The other bare end is left straight but is also bent at right angles to the insulated section of the wire. The steel wire is attached to the insulated copper wire close to the loop and another steel wire is attached to the other end of the insulated wire and run to the stake on the stable side. Another steel wire is run from the stake on the unstable side and attached to another insulated wire, stripped and bent in the same fashion as the other but threaded over the other with the

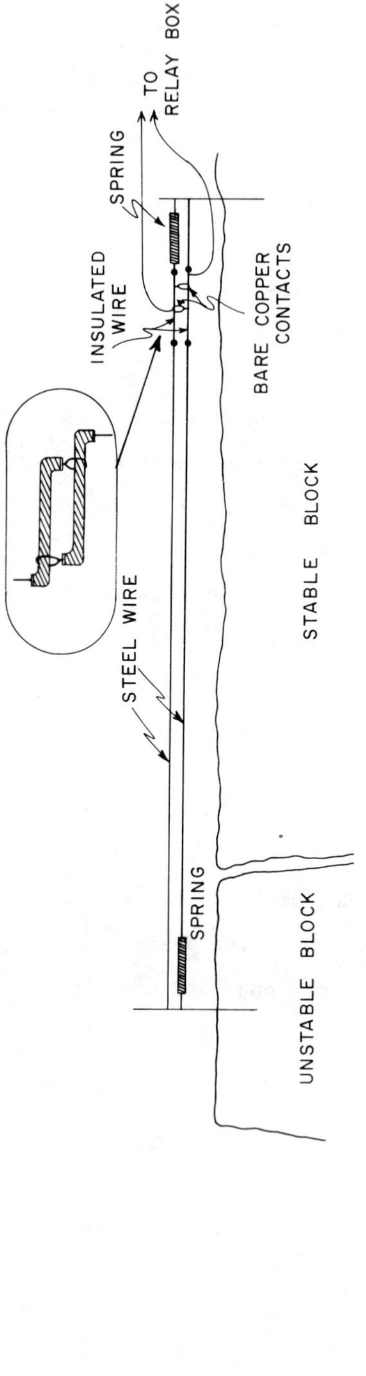

ORIGINAL SLIDING WIRE CONTACT SYSTEM

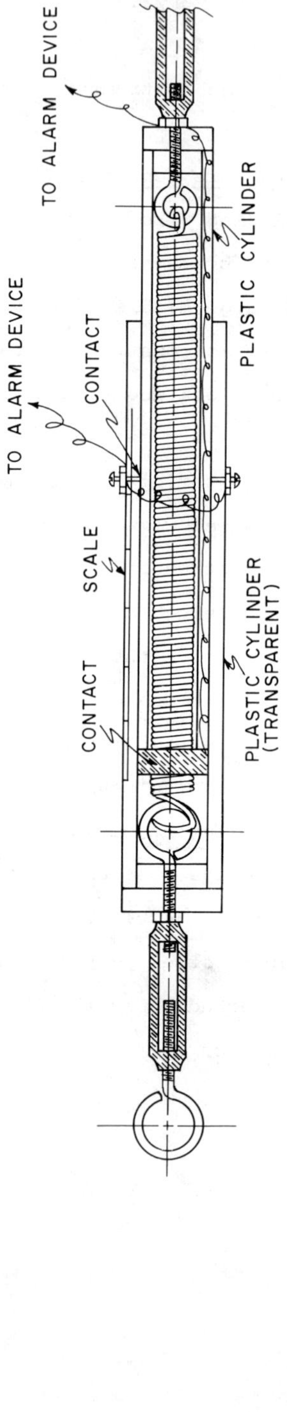

MOVEMENT WARNING DEVICE AND EXTENSOMETER

FIGURE 7

loop closer to the stake on the stable side. The loop is attached to a spring by a short length of wire and the spring is in turn attached to the stake on the stable side.

A remote warning relay box was designed and built by Anaconda Geophysical Department for use with the sliding wire contacts. As many as 24 sliding wire contacts can be connected to one relay box. When contact is made on any of the circuits, a relay is tripped and a warning light or siren is switched on. The spacing between the contacts can be set to a known distance at preset intervals and if the distance is exceeded in the time interval the warning will sound and a visual inspection of the area can be made to determine what action should be taken.

Two more sophisticated versions, one utilizing a compression spring and one a tension spring, have been designed by Anaconda MRD. These instruments double as linear extensometers. (Figure 7-B) Sliding wire contacts of both types are cheap, simple instruments and have been very successfully used in slide situations.

Linear Extensometers

A linear extensometer is a more sophisticated version of the sliding wire contact. A large number of designs have been tried and the most successful appears to be a yoke design. (See Figure 8.) In this system a stake is fixed in the unstable ground and a wire is led from it to the stable zone attached to a yoke made from two pieces of strap steel approximately four inches long and one inch wide. The two pieces of strap steel are attached by two pieces of wire fixed to each end to form the yoke. The other end of the yoke is attached to a spring and then to a stake set in the stable

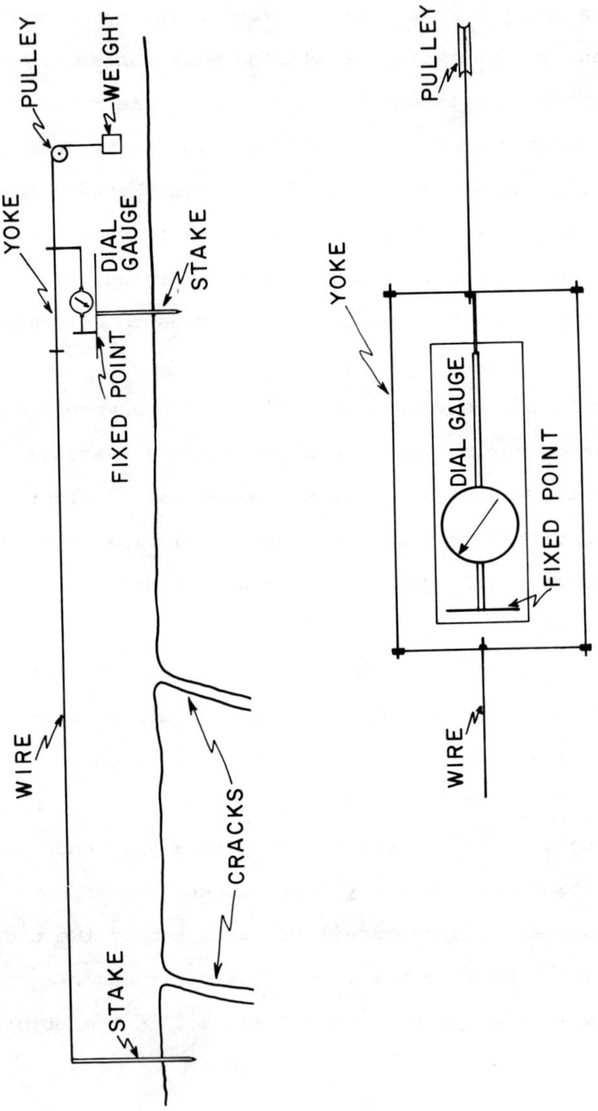

SIMPLIFIED DESIGN OF LINEAR EXTENSOMETER

FIGURE 8

ground. Alternatively, the wire from the yoke is led over a pulley and attached to a heavy weight for constant tension.

Movement between the stakes can be measured with a dial gauge or LVDT by measuring between one side of the yoke and a fixed point on the stable side. These instruments can be built at a reasonably low cost and are easy to read and maintain.

MULTIPLE POINT BOREHOLE EXTENSOMETERS (MPBX)

A multiple position borehole extensometer is an instrument for measuring movement at depth in a borehole by means of a series of anchors set rigidly in the hole and individually connected to a measuring device at the collar of the hole. Two types of MPBX's have been used by Anaconda MRD.

Rod Type Multiple Position Borehole Extensometers

Six position, rod type, borehole extensometers[6] have been used by Anaconda for instrumenting in-pit crusher foundations. These instruments are designed for use in a 2¼-in. dia, 48-ft long borehole. The hole can be drilled by diamond, rotary, or rotary percussive drilling equipment, and should be reasonably straight and clean. High impact plastic anchors filled with epoxy are used. A ¼-in. dia aluminum measuring rod connects the anchor to the collar of the hole.

The first anchor is inserted in the collar of the hole and slowly pushed into position down the hole with the setting rods (usually five- or ten-ft lengths of ½-in. pipe). Lengths of ¼-in. aluminum measuring rod are added to the anchor as it is pushed into position. The anchor is set by pulling back on the setting rods, hence, upsetting the epoxy around the plastic anchor. A marker rod is attached to the measuring rods at the collar to identify the anchor. The setting rods are then unscrewed from the anchor and withdrawn from the hole.

The next anchor is then prepared and inserted in the collar of the borehole after threading the marking rod from the deepest anchor through the anchor. The anchor is then positioned and set. The sequence is repeated, threading the marker rods through the anchors until all six anchors are installed.

At this time there are six marker rods of different lengths protruding from the collar of the borehole. The measuring head, which is made in the same fashion as the anchors, is now installed. The measuring head has six holes drilled through its perimeter along the long axis to take the measuring rods from the anchors. The head is epoxied into the collar of the hole and allowed to set. The marker rods are removed and brass measuring rods are screwed into the aluminum measuring rods in the head. A dial gauge with a machined adaptor collar fits snugly in the holes, allowing the distance to be measured between the top of the brass measuring rod and the face of the measuring head. A protective waterproof cap is screwed onto the face of the measuring head when readings are not being taken. The sensitivity of the instrument is claimed to be .001 in. and replaceable extension points for the dial gauge give the instrument a range of three inches. If set correctly, the instrument is reasonably blast proof.

It is estimated that this instrument could be used in holes up to 70 ft deep, providing enough anchors were installed to reduce sag of the rods. This type of extensometer is easy to install and is recommended for site instrumentation (crusher sites, shafts, etc.) but is not recommended for general slope stability use as it can only be used for shallow holes. It is relatively inexpensive and does not suffer from the calibration and wire twist problems encountered with wire extensometers.

Wire Type Multiple Position Borehole Extensometers (Figure 9)

Long multiple position borehole extensometers[7] of the wire type have been installed at Chuquicamata, Chile; Cananea, Mexico; and Inspiration, Arizona. These instruments enable displacements to be measured at far greater depths in boreholes than the rod type instruments. The instrument works on the principle that movement of the anchors at depth is transmitted to a measuring device at the collar of the borehole by means of stainless steel wires in tension. Three types of anchors have been used, namely:

1) Groutable -- a cast aluminum anchor in the form of a tube with a ridged outside surface.
2) Spring Loaded Watertight -- an anchor, also built on a tube principle, with a triggering mechanism which, when fired by pulling on the measuring wire, fires three anchoring tines into the wall of the borehole.
3) Flat wedge anchor -- a flat anchor which acts on a wedge principle and can be pushed into the hole with a setting rod but cannot be pulled out due to the wedging action.

The instrument strings for the groutable anchor and the spring-loaded watertight anchor are made up in the same fashion; however, the flat wedge anchors are set individually. The flat wedge anchors are recommended for installations up to 200 ft in

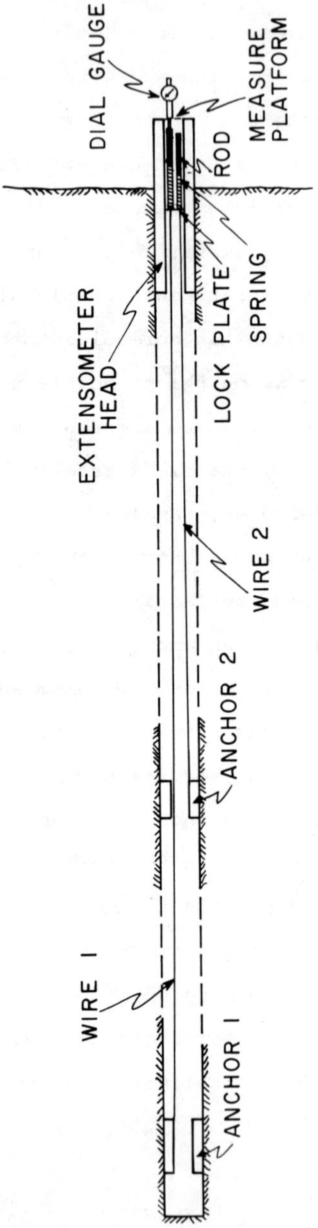

SIMPLIFIED LAYOUT—MULTIPLE POSITION BOREHOLE EXTENSOMETER

FIGURE 9

length. Attempts have been made to install these anchors to depths of 1000 ft, but wire twist, abrasion and kinking have caused numerous problems. The stainless steel wire is attached to the flat wedge anchor and the anchor pushed to its correct position in the borehole with setting rods and set. The next anchor is then threaded over the wire already in the hole and pushed into position. This is repeated for all other flat wedge anchors.

The groutable and spring-loaded anchors are made up into a string using 1.25-in. dia PVC tube as a spacer. The wires from the anchors run inside the PVC tube and through the anchors higher in the hole than themselves. The entire tube is installed in the borehole and the anchors fired into the wall of the hole in the case of the spring-loaded anchors, or stage grouted into the hole in the case of the groutable anchors.

A measuring head is threaded over the wires and installed in the collar of the hole. Each wire is connected to a measuring element in the head. The measuring element can be a steel rod running in a track, pulling against a tension spring or pushing against a compression spring, or it can be an electronic sensor such as an LVDT, linear potentiometer or a strain gauge. In the case of the mechanical measuring elements with the springs, a dial gauge is used to measure the distance between the top of the rod and a measuring platform rigidly fixed in the head. The wires in the hole should be under fairly high tension.

Problems exist with the wire type extensometers in that calibration of the instrument is difficult. Allowances have to be made for wire stretch, friction between wires, bird-caging of the wires, friction against the borehole wall, temperature, etc. Anaconda has developed methods to calibrate these instruments in place in the ground, but even this does not solve the entire problem. These instruments can be read to the nearest 1/1000th of an inch. However, owing to the inherent design problems of this system, the deeper the installation is the larger the changes in the readings should be before they are considered significant. The borehole extensometer is one of the few methods available today to give indications of movement at depth within a borehole.

Very careful analysis of the advisability of borehole extensometer installation should be made before deciding to install this type of instrumentation. The instrument and borehole costs can be quite large and often alternative forms of instrumentation should be considered.

Shear Strip (Figure 10)

The shear strip is a device for locating subsurface shear failure. It consists of a continuous length of 3/4-in. wide printed circuit board with parallel conductors down each side. The conductors are connected by a 10 K ohm resistor every foot.[1] The entire strip is wrapped and sealed in a waterproofing material. Waterproof, well insulated, two-conductor cable is connected to both ends of the shear strip. The shear strip is then installed in the borehole through the suspected zone and grouted into position, leaving the two signal cables protruding out of the hole.

---

[1] Spacing of the resistors can be varied dependent upon the accuracy of the location of the shear that is required.

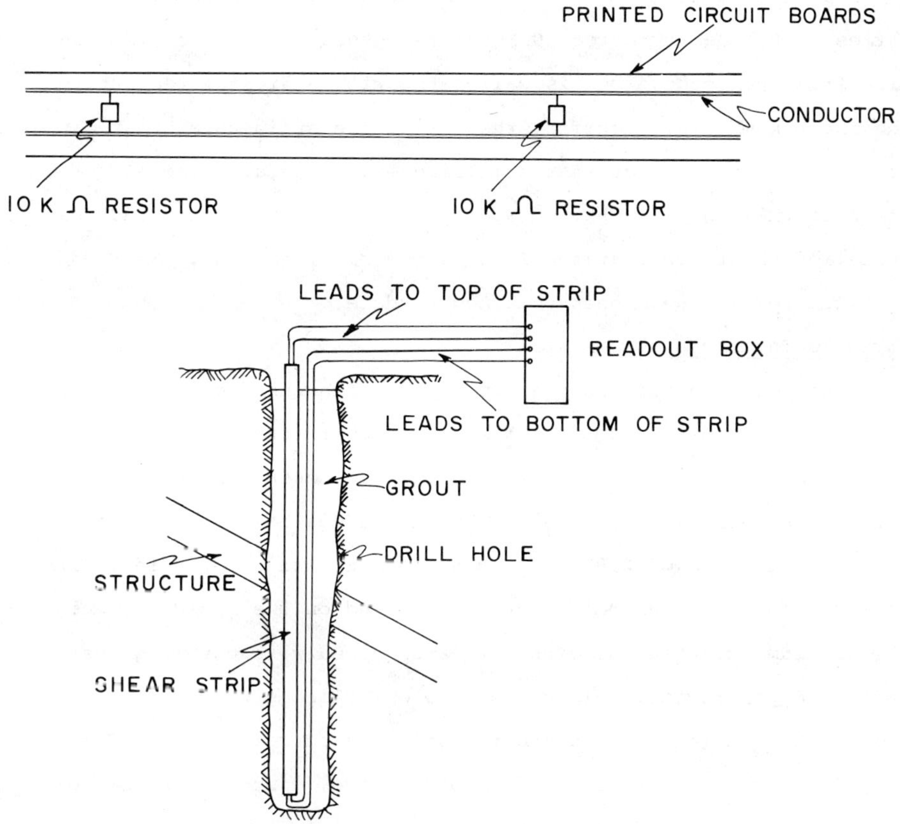

SHEAR STRIP

FIGURE 10

The ends of the signal cable are connected into a Wheatstone Bridge circuit with the shear strip acting as one side of the bridge and the balancing potentiometer as the other. If a break occurs in the shear strip, the circuit is broken and forms two separate resistance circuits. The resistance of the circuits varies as to the number of 10 K ohm resistors in parallel in each circuit. The readout box is calibrated with a digital readout so the exact number of feet from the end of the strip to a break can be read directly. About three millimeters of shear movement across the hole will shear the strip and break the circuit. Anaconda has installed shear strip at the Twin Buttes Mine, Arizona. No specific results have been obtained to date.

Borehole Inclinometer

The borehole inclinometer is a precision instrument designed to measure the displacements of rock and soil surrounding a borehole in a slope. A special grooved plastic casing in five-ft lengths is installed in a 4½-in. or larger dia hole. This casing has an inside dia of 2.87 in. and an outside dia of 3.34 in., with four equispaced longitudinal grooves milled on the inside to guide the instrument. Lengths of casing are joined with couplings using epoxy and pop rivets. Plastic borehole inclinometer casing has been installed in four churn drill holes which were drilled for sampling purposes at the Berkeley pit, Butte, Montana. A borehole inclinometer was used to log the holes. Initially, problems occurred in grouting the casing in the churn drill holes. These problems were caused by grout entering the casing either through ruptures caused by the hydraulic pressure of the fluid grout or by seepage of grout through the couplings. One installation had the casing couplings reinforced with fiberglass-resin to combat this problem, and pea gravel (1/8 in. to 3/8 in.) was used as backfill in preference to grout. This installation was completed with no problems or delays.

A 15-ft high headframe was constructed in Butte to lower the plastic casing into the churn drill hole. A considerable savings in costs was achieved. The headframe is made of 2½-in. steel pipe and is bolted together for ease of transportation. A hand-operated winch was used to lower the casing on 3/16-in. airplane cable.

The borehole inclinometer instrument consists of a segment of precision wound resistance element mounted on a lucite block and enclosed in a watertight aluminum cylinder 2-3/8 in. in dia and 30 in. long. Attached to the resistance element is a relay which, when activated by an electrical current, pulls a conductor into contact with the resistance and forms one half of a conventional bridge circuit. The other half of the bridge, including switches and batteries, is enclosed in a control case at the ground surface. The instrument and control box are connected by a heavy duty cable with yellow bands at one-ft intervals and red bands at five-ft intervals. The instrument has three guide wheels at each end, equispaced around the instrument (120° apart). Two wheels are fixed and ride on the inside surface of the casing. The third wheel is spring mounted and rides in one of the casing grooves.

The instrument is designed so the inclination of the plane defined by the spring-mounted guide wheels is proportional to the reading on the control box when the circuit is in balance. The instrument is lowered down the grooved plastic casing, the grooves controlling the orientation of the instrument in a predetermined direction. Inclination readings are taken at frequent intervals of depth and are subsequently converted to displacements. Consecutive readings at the same depth intervals, at periodic intervals of time are used to determine rate of ground movement at that point. The borehole inclinometer used has a sensitivity of one part in 1000 and can detect a tilt as slight as three minutes of arc. This corresponds to a lateral displacement of one inch in 100 ft of depth. The angular range of the instrument is ±12°.

Borehole inclinometer holes are monitored monthly, or as time permits. Field data are reduced on a 9100A Hewlett-Packard desk-top computer with printer output. The computer output is in the form of displacements and cumulative displacements (N-S and E-W) corrected for the rotation of the pipe (which can occur during installation).

A lengthy test period at the Berkeley pit in Butte has shown that minimal useful information can be obtained from borehole inclinometers in very deep holes. The data showed only general trends of movement in a large rock mass. Some inconsistencies in the data were noted and not satisfactorily explained. Cost of the data obtained was reasonable with installation costs of about $5.00 per ft and logging costs of $0.80 per ft.

CONCLUSIONS

Present-day slope monitoring systems still leave a lot to be desired as far as convenience, capability and cost are concerned. Possibly the most effective monitoring system available is the survey net and consequently considerable effort is being devoted to looking for fairly accurate, automated, low-cost methods of surveying. Anaconda's six years of experience with many forms of open pit slope monitoring equipment have resulted in the following conclusions:

1. In most cases where failure is in progress, the simplest, crudest and cheapest forms of monitoring devices are often quite sufficient for prediction purposes.
2. In critical areas, such as crushers, main access routes to the mine, etc., where monitoring is being carried out for protection of personnel and equipment, more sophisticated instruments, such as multiple point borehole extensometers and seismic systems will be used to give a far earlier warning of any movement.

REFERENCES

1. Kennedy, B. A. and K. E. Niermeyer "Slope Monitoring Systems Used in the Prediction of a Major Slope Failure at the Chuquicamata Mine, Chile" South African Institute of Mining and Metallurgy, Johannesburg, September 1970

2. Broadbent, C. D. and C. W. Armstrong "Design and Application of Microseismic Devices" 5th Canadian Symposium on Rock Mechanics, Toronto, December 1968

3. Paulsen, J. C., R. B. Kisler and L. L. Thomas "Slope Stability Monitoring at Boron" *Mining Congress Journal, Vol 53,* September 1967, pp 28-32

4. Stateham, R. M. and J. Vanderpoor "Microseismic and Displacement Investigations in an Unstable Slope" RI 7470, Bureau of Mines Report of Investigations, January 1971

5. Wisecarver, D. W., R. H. Merrill and R. M. Stateham "The Microseismic Technique Applied to Slope Stability" AIME, February 1969

6. Anon, "Borehole Rod Extensometer, Instruction Sheet," Soiltest, Inc.

7. Dutro, H. B. and D. R. Stewart "Multiple Position Borehole Extensometers, Instruction Manual" Terrametrics, Inc., August 1968

RHEOLOGIC ASPECTS OF ROCK SLOPE FAILURES

Broadbent, C. D.[1] and Ko, K. C.[2]

Introduction

Fundamental field-scale slope stability research by the Kennecott Copper Corporation, a $3-million effort done in part with the USBM at the Kimbley Pit in Nevada, has been directed primarily at developing slope design techniques to optimize mine slope angles and minimize failure potential. The Company has not overlooked, however, the probability of future failures and has directed a realistic portion of its effort, probably 6,000 man-hours so far, to failure engineering. The premise is that only the unknown becomes a costly surprise, whereas thorough knowledge of a failure condition allows rational planning and a minimum ultimate cost and risk to the operation.

Literature on slope stability contains abundant technology related to the stability analysis, but references to the kinetics of failures are rare. Many investigators have postulated the mechanics or modes of failure initiation to further their stability studies, but the operating mechanisms of active, moving failures are not known to the degree that prediction of failure behavior based on known fundamentals is possible.

---

[1]Mining Engineer
[2]Associate Mining Engineer,
Kennecott Copper Corporation, Metal Mining Division
Engineering Department, Salt Lake City, Utah

The Kennecott approach on failure research has been, in part, to study the displacement vs. time data, the related mining activity, and the failure geometry. This has been done on a formal basis for seven years and each year there has been an attempt to increase the relative number, frequency, and accuracy of data measured at any failure site. The work has produced several useful observations regarding failures and a rheologic model that has enabled accurate predictions of failure movements. The model is a postulation of sliding mechanics with a very acceptable data fit. It is not suggested, however, that the model is the true model of sliding, only that it is generally equivalent and can be useful for analyses.

General Discussion of Slope Failures

The first opportunity accorded the present Company mining research group to study failure phenomenon occurred in 1965. At that time a failure situation of substantial proportions occurred at a western mine. The mine personnel and Company staff engineers cooperated in a study of the problem in depth. By 1966, the failure displacement trends had been established. Fig. 1 shows the data from a typical gage station from the time the failure condition was recognized in 1964 until the time of collapse. Note that both horizontal and vertical components of displacement are apparently of simple logarithmic form and that the only interruption of this pattern occurred during the stripping period. As displacements became

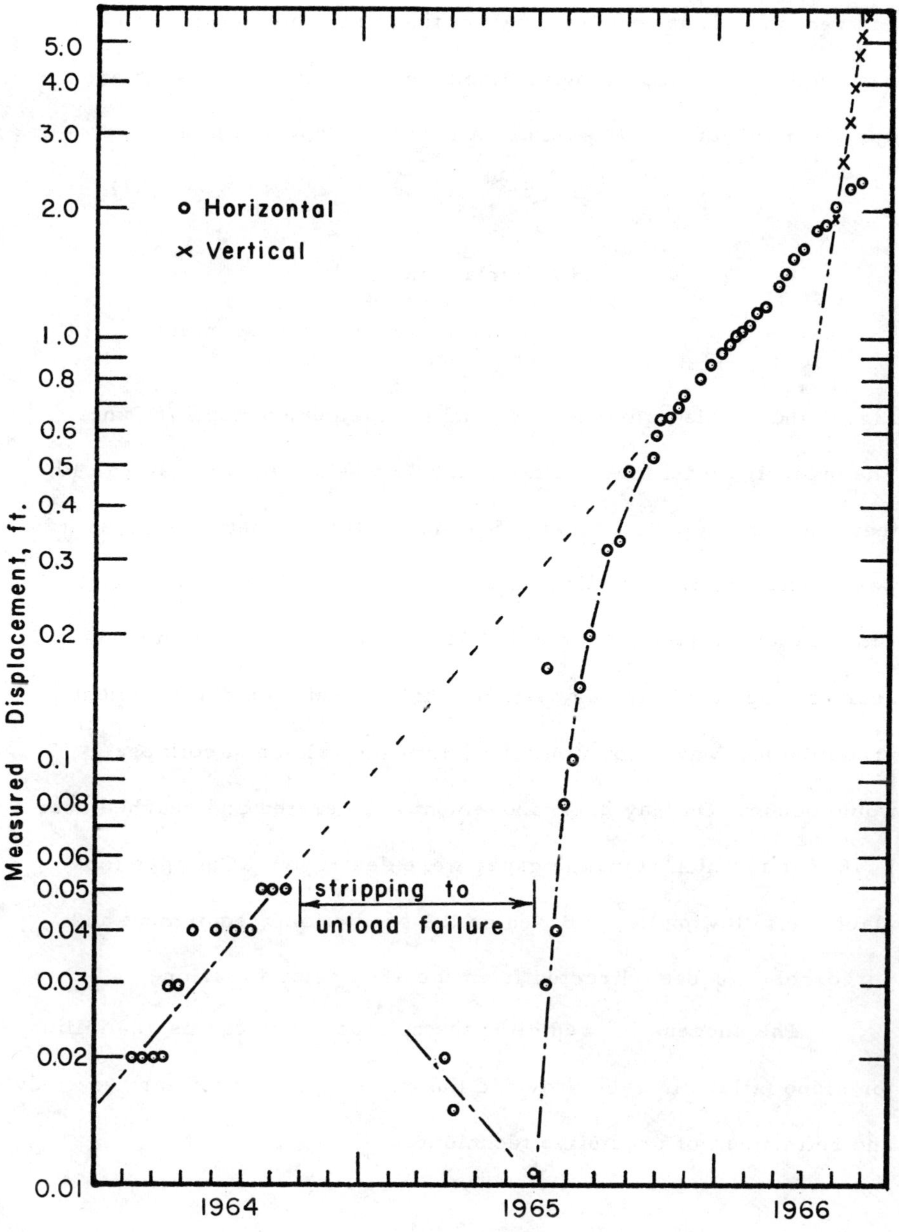

Fig. 1. Displacement Data at Gage 7.

larger and the potential impact of the failure on mine operations was recognized, it became apparent that a reliable predictive curve was necessary. Accordingly, displacement data from March and April were further evaluated and a curve fitted of the form:

$$S = Pe^{kt} \qquad (1)$$

Where
S = displacement, ft
t = time, days
e = basis for the natural logarithm
P, k = constants

Fig. 2 shows this equation evaluated for the period April to June. The projected total movement on May 15 was 5 feet, on May 25, 10 feet, and on June 10, 50 feet. Because of the geometry involved it was considered impossible that a movement of 50 feet could exist without total collapse. Somewhat arbitrarily, but based on engineering judgement, it was assumed that 10 feet, the displacement predicted for May 25th, would be a limiting value and collapse would occur. On May 21st, movement accelerated and reached nearly 10 feet at which point the gages were destroyed. Collapse took place the following day. Based on the predictions, equipment and personnel had been directed from the area prior to failure.

The success in predicting the collapse time of this 12-million ton slope failure in 1966 provided the encouragement for further study and refinement of predictive techniques.

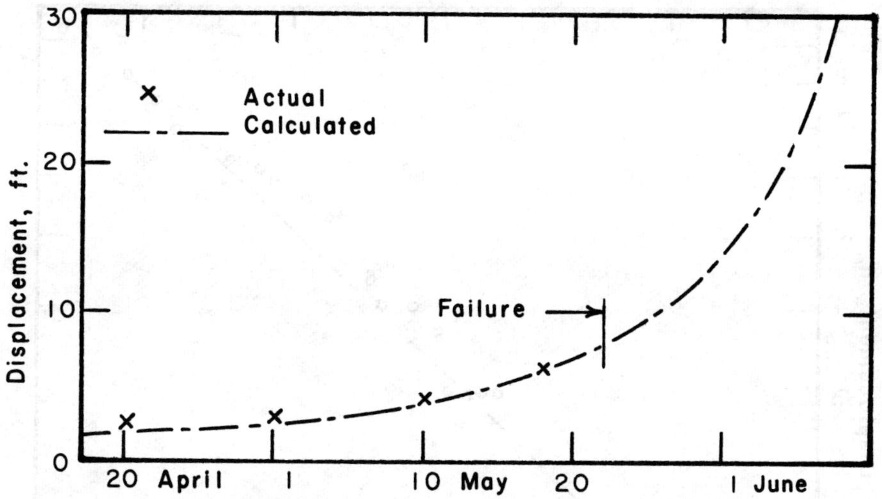

Fig. 2. Comparison of Actual Vs. Projected Displacement.

During the next several years, other smaller failures were surveyed using more precise techniques. The data were evaluated and perserved. Although similar patterns were identified, (Fig. 3) there were a disturbing number of exceptions in which the displacement function was apparently one of exponential decay, such as shown in Fig. 4, rather than acceleration. Because of the frequency of decay-type curves, a model for this type of movement was postulated.

<u>Basis of Model</u>

The relative movement of a sliding block with respect to failure plane was simulated by the rheological model shown in Fig. 5. Each characteristic unit of the block is composed of springs, dashpots, and a friction block in series and parallel combinations.

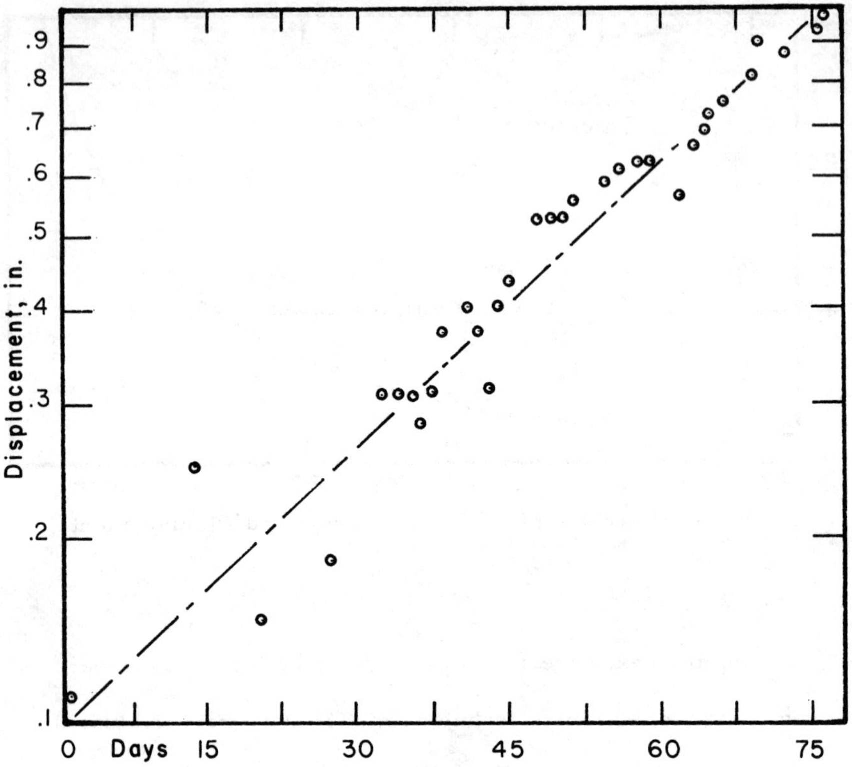

Fig. 3. Typical Accelerating Movement.

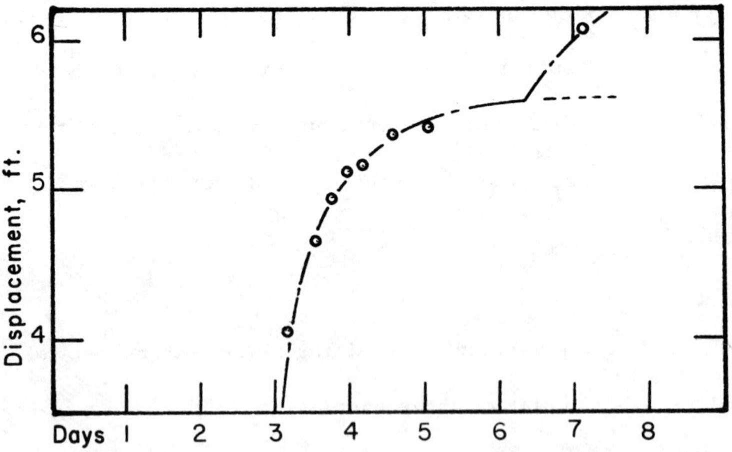

Fig. 4. Typical Decay-type Movement.

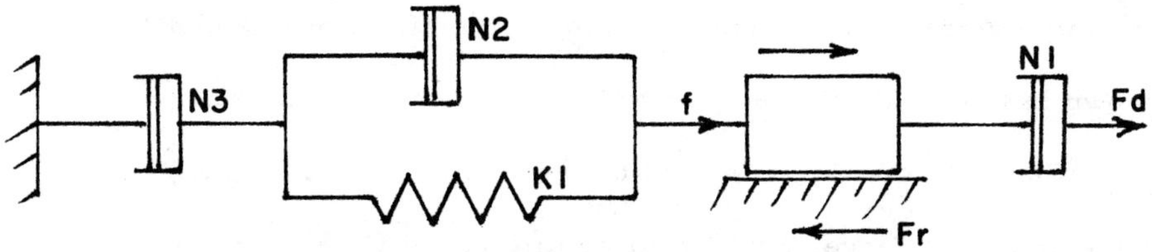

Fig. 5. The Displacement Model.

The spring and dashpot N2 resume the original displacement characteristics after each movement has reached the equilibrium point, because of the relaxation of dashpot N3. The dashpot N3 is assumed to be many times stiffer than N2 so that its effect is not shown in the displacement during the movement of the block.

The friction block represents the actual frictional characteristics of the sliding plane. The arrangment of the components is such that the force acting on the block will not be transmitted to the elements behind it until the load overcomes the frictional resistance of the failure surface.

The properties of each element are dependent upon the properties of rock, rock mass and the sliding plane. The dashpot N1 is related to the inherent creep characteristics of the rock mass (the block as a whole). Its action begins with excavation. The displacement of this element is shown in the general deformation of the block without any apparent slippage along the potential failure plane. The rate of creep is small for intact rocks and can be neglected for a stable slope. For an unstable slope, the effect of this constant is shown in the small displacement at the beginning of the failure.

When the force, Fd, acting on the friction block is greater than the resisting force, Fr, the force difference f = Fd - Fr will act on the combined element of a spring, K1, and a dashpot, N2. The combined element is characterized by a non-linear time dependent displacement. The maximum displacement, S, is controlled by K1 and the initial slope of the displacement curve (velocity) is inversely proportional to the value of N2.

Once the element reaches the equilibrium point, the spring relaxes at that position and the force within the element becomes zero. Thus, when a new force, f, acts on this element, the properties of the element will be similar to that at the original position. The force, f, can be generated by a change in friction angle due to water, by a sudden external force due to removal of the confinement, or by the stress waves due to blasting.

Because the N1 is very stiff compared to that of the combined element, its effect on the overall displacement may be neglected for practical purposes. Therefore, the problem is essentially reduced to studying the behavior of the combined element which is recognized as the Kelvin or Voigt model. This model has a limit in the maximum total displacement it can withstand. When the limit is reached the element is destroyed, leaving the resistance block and N1 alone to act against the force f. The displacement curves of the combined element operated on by a constant force are shown in Fig. 6.

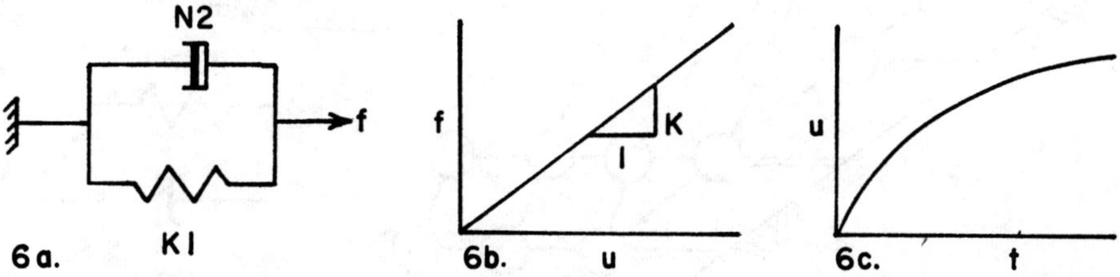

Fig. 6. Displacement Characteristics of the Combined Element.

The block mass itself has some non-elastic properties and undergoes a non-linear deformation. The construction of the model is shown in Fig. 7. The block mass is divided into a number of unit masses, each unit being dependent on the rock type and the size of the block. The unit masses are interconnected by strings to the neighboring units. The strings represent the intraunit behavior characteristics of the mass. If the mass is intact, continuous, and elastic, a string represents the spring and the spring constant K2 is the same as the stiffness used in the finite element technique. Depending on the slope, there are many other ways of representing the property of the connecting string. The use of such a model is more realistic (theoretically) than the use of a rigid block model but requires additional information on the rock and extra measurement analysis. Because measurements of the slope movement are rarely, if ever, monitored on the failure plane, such an analysis would be necessary to infer from measured surface displacements the actual movement on the failure plane.

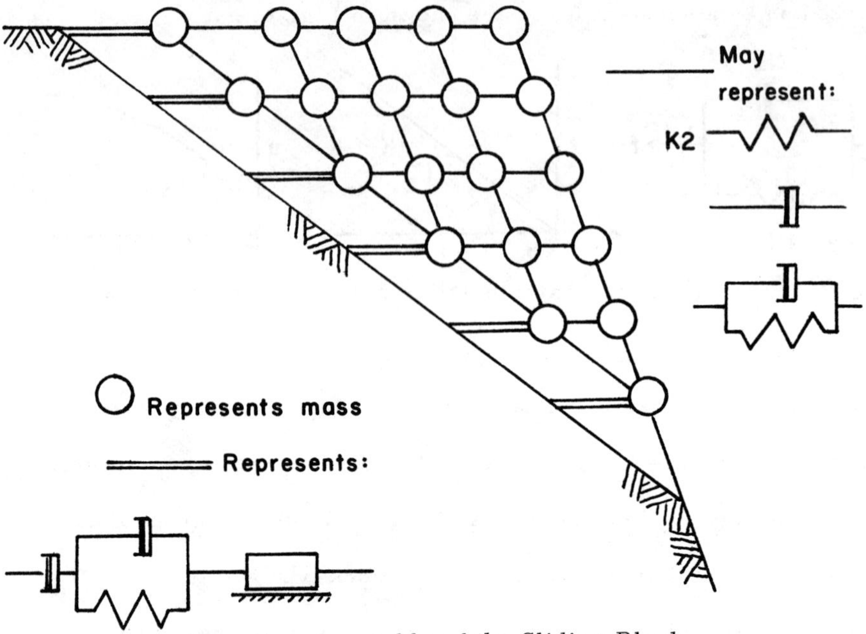

Fig. 7. Assembly of the Sliding Block.

For completeness of the model, the elements representing the failure plane would be added to each unit block in contact with the plane (Fig. 7). In a more rigorous analysis, the properties and constants for each element and block unit might be different; but for the practical purpose, they may be considered to be uniform. The equations for the elements and for the combined unit are:

Spring Element

$$u = Kf + u_o \tag{2}$$

where u = Displacement
u_o = Initial displacement
K = Stiffness constant
f = Positive force difference

Dashpot Element

$$u = Nt + u_o \tag{3}$$

where N = Viscosity coefficient
t = Time

Friction Element

$$u = 1/2\, at^2 + u_o \tag{4}$$

where a = Acceleration

Combined Element

$$u = \frac{f}{K}\left(1 - e^{\frac{-Kt}{N}}\right) + u_o \tag{5}$$

where e = Basis for the natural logarithm

Application of the Model for Failure Analysis

Observations indicate that most, if not all failures displace in multiple cycles and that the characteristic of any cycle is largely dependent on the previous cycle. It is, in fact, from the first identified cycle of an actual failure that the amplitude (f/K) and velocity characteristics (K/N) are determined, as the constants (K) and (N) cannot themselves be evaluated. The following hypothetical example is considered in terms of the described model. It is assumed that:

(a) the sliding block is rigid (i.e., mass units as in Fig. 7 are tied by infinitely-rigid springs),

(b) the characteristics of the failure surface do not change

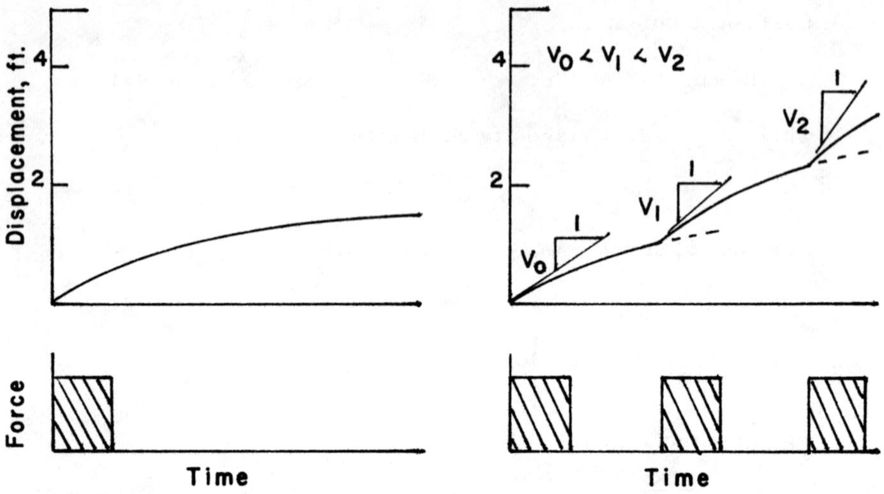

8a. Single Force 8b. Multiple Forces

Fig. 8. Resultant Displacements Due to Different Types of Force Application.

with time or displacement and are defined by the following constants:

f/K (amplitude) = 1.5 ft.

K/N (velocity factor) - 0.075

(c) the force difference (f) is an instantaneous function applied at the evaluation of t and is always the same.

Fig. 8 illustrates the displacement-time functions evaluated using the above values. Fig. 8a shows the result of a single application of force, and Fig. 8b shows the effect of multiple force applications. Because the total displacement is a sum of displacements for each cycle, the initial velocity of each succeeding cycle must, according to the model, be greater than the last, although the long-range trend could either accelerate or decelerate depending on the period and magnitude of each burst of activity (force). The application of force on a continuous basis would cause infinitely short cycles and long-term displacement acceleration. The elimination of force at any point would allow decay and ultimate arrest of movement.

The most nebulous aspect of the model is the force factor. It could, and probably does, include:

(a) any unbalance of gravity load and resisting force potential on the failure surface,

(b) a seismic load, as from blasting, and

(c) a temporary load of elastic readjustments related to mining progress.

It could be deduced that if there is a net disturbing imbalance, as (a) above, the failure would be irreversible and ultimate collapse would be expected. If, on the other hand, there were no imbalance (i.e., the f is derived from temporary elastic loads and seismic loads), the displacements might be stopped if activity were terminated.

The sensitivity of the model to the parameters K, N, coefficient ratio, and to cycle frequency has been evaluated over parameter ranges found in actual failures. The results are shown in Fig. 9. The curves are drawn by connecting the initial point of each displacement cycle and show, therefore, the long-range trend.

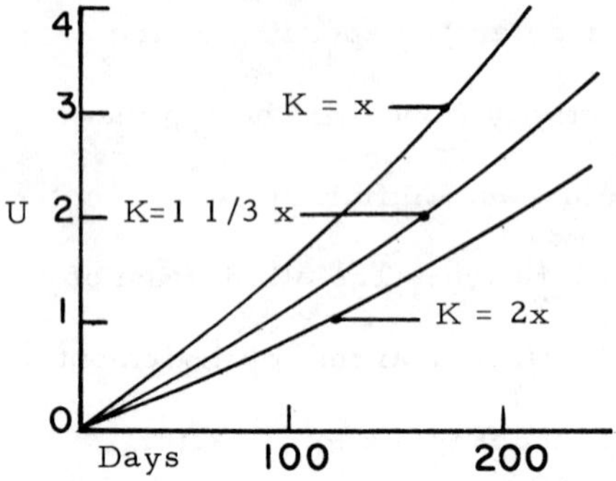

9a. Effect of K Values

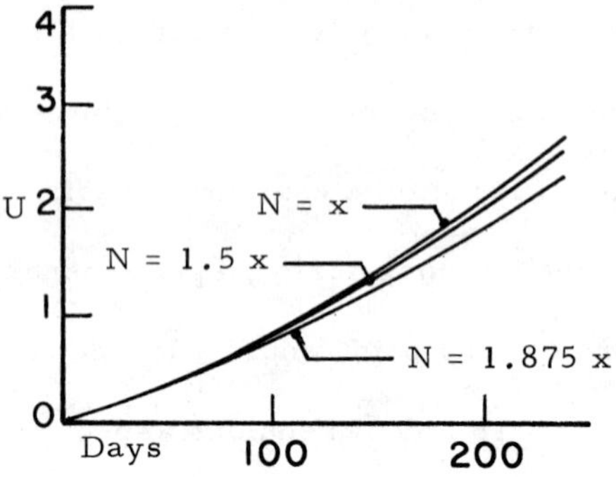

9b. Effect of N Values

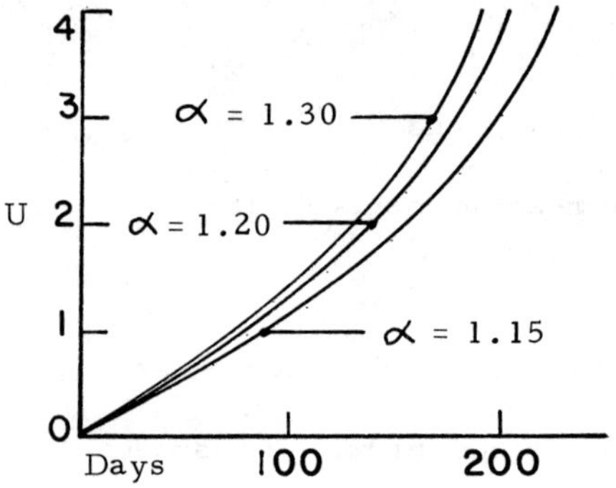

9c. Effect of Coefficient Ratios

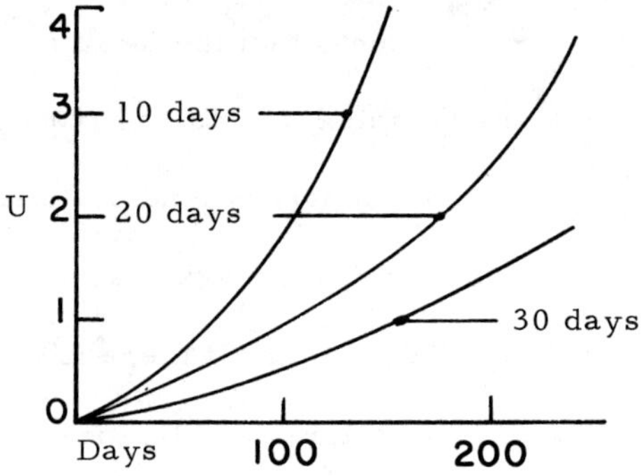

9d. Effect of Cycle Frequency

Fig. 9. Functional Effects of Failure Parameters to the Total Displacement (U) of the Failure.

The coefficient ratio is introduced because experience shows that contrary to (b) (p. 11) the surface characteristics of the failure apparently do change with time and/or displacement. Furthermore, the data indicate that the product of coefficients at any one location is relatively constant over several cycles. That is, if f/K = A and K/N = B, then $A_1 B_1 \approx A_i B_i \approx$ constant. It has also been noticed that the coefficients, both A and B, vary in a regular manner (i.e.: $\alpha = A_2/A_1 \approx A_3/A_2 \ldots$). These two observations have been used to establish the predictive curves on several occasions. From Fig. 9a - d it may be anticipated that the order of importance of parameters is (1) cycle frequency, (2) coefficient ratios, (3) K values, and (4) N values. In any case, the displacement is, of course, directly proportional to the non-defined f value.

Comparison of Model and Actual Failure Data

Available data (Ref. 1, 2, 3) from past slope failures at both Company and non-company operations have been evaluated in terms of the model. Figures 10 to 13 show some typical examples of measured displacements and computed curves. Note both the tendency to constant coefficient products and ratios and the exceptions as given in Figs. 10 and 11. Particular attention is directed to the Figs. 12 and 13 which are linear scale plots of the data presented in Figs. 1 and 3 respectively.

Figs. 1 and 3 show the trend curves evaluated in 1966, whereas Figs. 12 and 13 show the cyclical aspects of these failures in

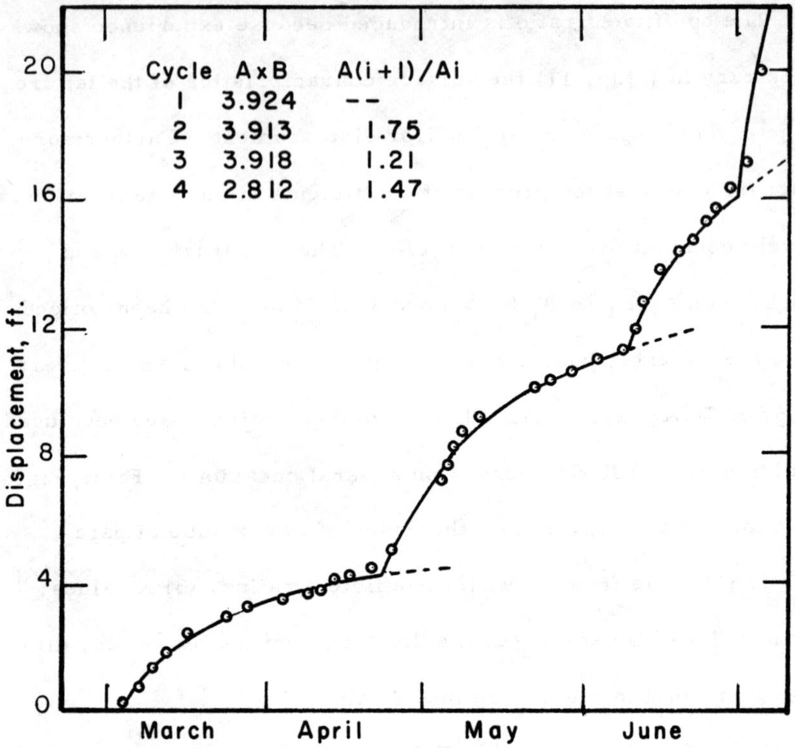

Fig. 10. Corner Slide Measured Displacement (300, 000 tons).

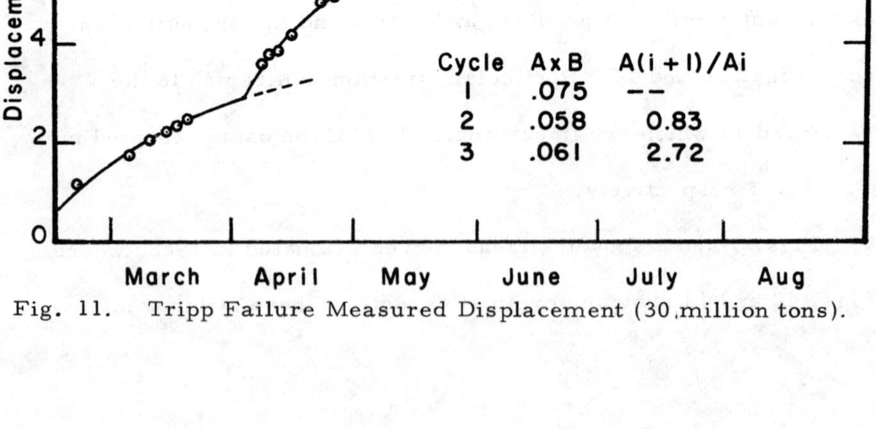

Fig. 11. Tripp Failure Measured Displacement (30,million tons).

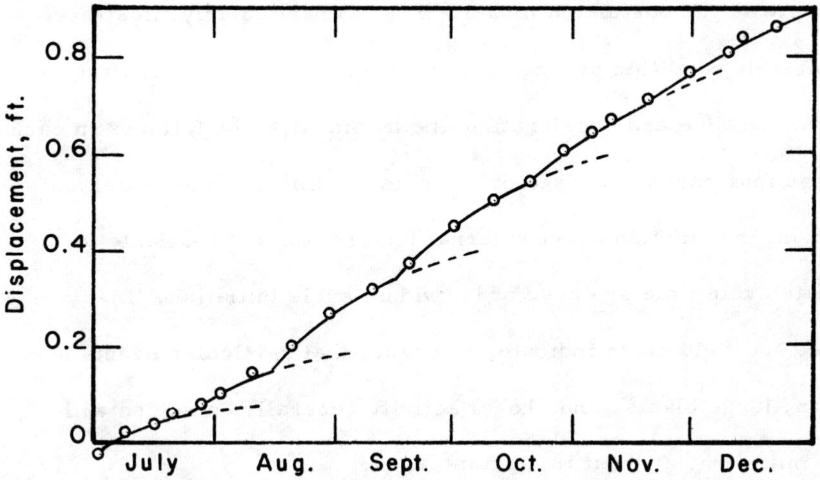

Fig. 12. Normal Plot of Displacement at Gage 7 (see Fig. 1).

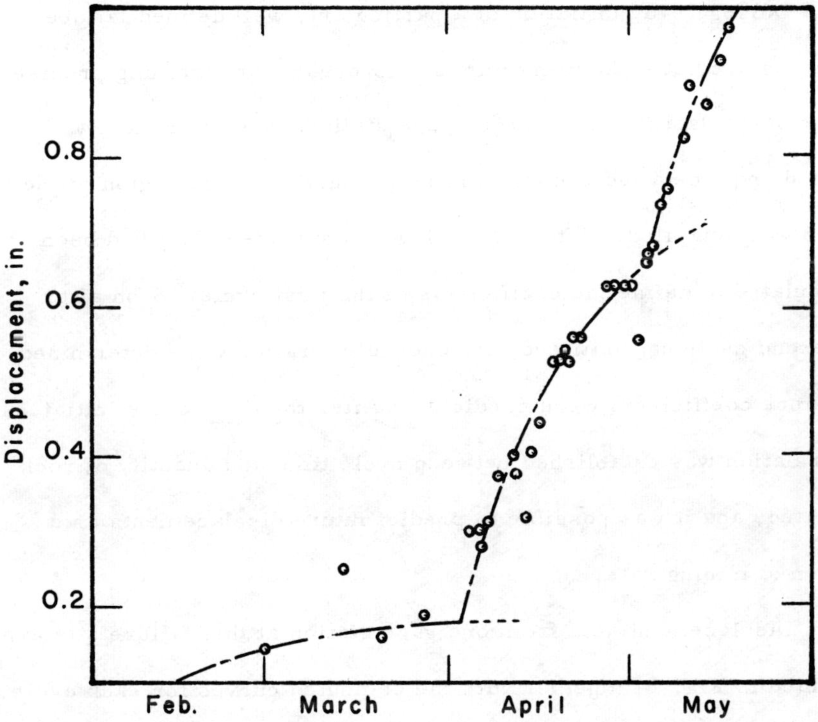

Fig. 13. Normal Plot of Displacement Data of Fig. 3.

keeping with the postulated model. Fig. 13 particularly illustrates the possible condition of a continuous force application, resulting in short cycle time and accelerating displacement. The failures in each of these four cases occurred prior to development of the model and sufficient information about external factors was not available to evaluate cycle time or cause of individual cycle initiation. Diary entries and field notes indicate, however, that particular events such as rain, large blasts, and shovel activity generally coincided with cycle initiation, at least in certain cases.

Use of the Model for Evaluating the 1860 Failure

During 1970, initiation of a particularly well defined failure was recognized at a Company mine. A program for obtaining precise displacement data from numerous gage stations was immediately initiated and a detailed record of mine production in the region of the failure was compiled. After several weeks sufficient data had been accumulated to define the coefficients of the first cycle. Soon after the second cycle had initiated, the coefficient ratios were determined and future coefficients were predicted. After the third cycle initiated, a correlation was established between cycle time and quantity of rock excavated, and it was possible to predict future displacements as a function of mining rate.

Displacement data from one gage station at this failure site are presented in Fig. 14 together with the computed curves for each cycle. Curves through January 20 were evaluated from measured data and

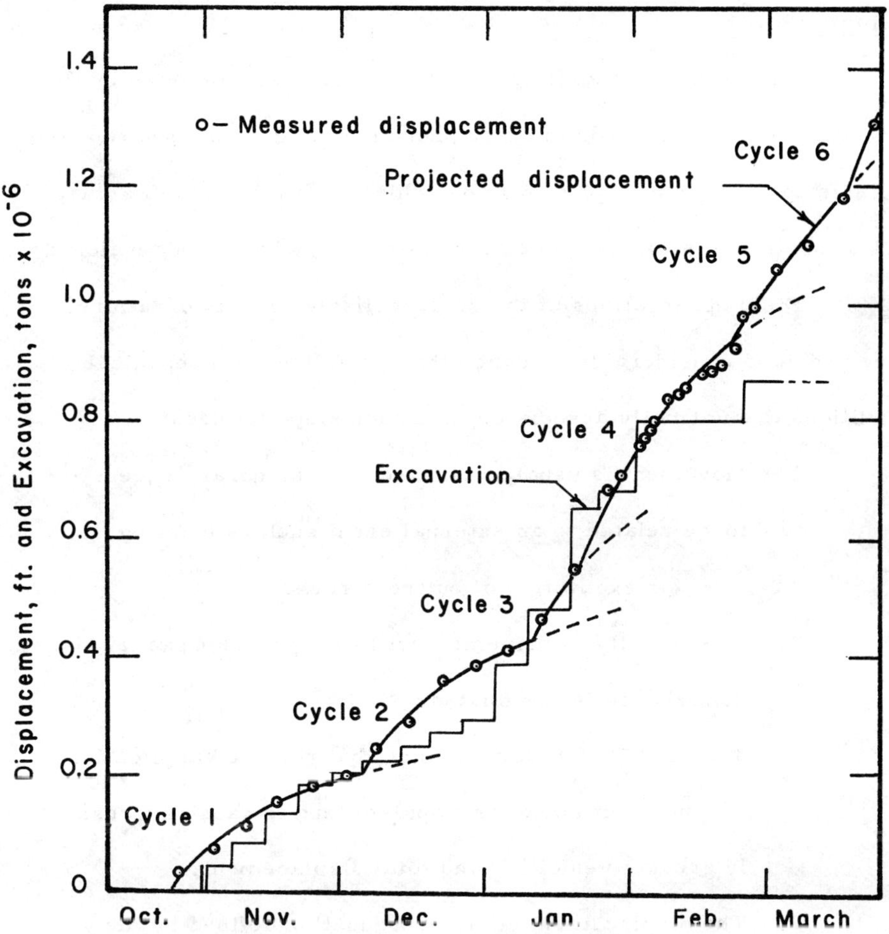

Fig. 14. Comparison of Displacement and Quantity Excavated.

those after January 20 were projections. Also presented on the figure are the quantities excavated each week from the immediate region of the failure.

Recent data obtained from this site have fit very closely the extended curves. The mining rate was reduced during February and curtailed during March. Displacement rates decreased accordingly.

Summary

The need for predicting long-term failures is dictated by the potential costs and hazards of these phenomena. Because positive and reliable analyses cannot presently be achieved on the basis of actual and theoretical mechanics, an empirical rheology has been developed.

Field observations of the surface displacements of failures have provided considerable evidence that the following mechanical conditions are normally associated with rock slope failures:

1. Movement is usually initiated by a temporary excess force related to an external event such as blasting, rain, or excavation of buttress rock.
2. The velocity of movement will decay if the external disturbance is eliminated.
3. The inertia of failure is such that repeated applications of a uniform force (or impulse) tend to cause successively greater velocities and total displacement.
4. The displacement curve is a function defined by the characteristics of the failure surface and a usually unknown force function.

Experience with the common Voigt or Kelvin model as described in this paper, indicates that it is a practical and reasonably accurate model of failure movement cycles. Furthermore, the failure sliding block, although recognized as having both elastic and viscous properties which affect the relationship between measured displacements and concealed failure surface displacements, can be considered as a rigid block for the practical analysis.

This empirical theory has been successfully used to formulate positive action and control of active slope failures. Use of the model as a generally applicable tool for failure analysis appears to be justified and may ultimately permit more reliable control of unstable slope areas.

REFERENCES

1. Dimock, R. R.; Slope Failure - A Continuing Problem, Paper presented at Reno meeting on Slope Stability, 1970.

2. Kennedy, B. A., et al; A Case Study of Slope Stability at the Chuquicamata Mine, AIME Preprint 70-AM-81, 1970.

3. Kimura, E. T. and Rennie, C. C.; The Craigmont Open Pit Slide, Paper presented at annual meeting of Geological Association of Canada, Vancouver, B. C., April 1968.

ALTERNATIVE INTERPRETATION OF THE VAIONT SLIDE

K.Y. Lo, C.F. Lee, and P. Gélinas *

*Respectively, Associate Professor and Graduate Students, Faculty of Engineering Science, The University of Western Ontario, London, Ontario, Canada.

On October 9, 1963, a major disaster resulting from a rock slide shook the entire world, and the engineering profession in particular. A mass of rock at the upstream side of the Vaiont Dam, on the south bank, involving a volume of approximately 300 million cubic yards, plunged into the Piave River in Italy. A wave, 300 feet above the crest of the dam, was generated and earth tremors created by the slide were recorded as far away as Brussels. The flood swept both up and down stream and close to 3,000 lives were lost. The thin arch dam itself withstood the tremendous force and remained intact but the large volume of the slide debris in the valley prevented further use of the dam for the generation of hydro power up to the present.

A description of the slide was given by Kiersch (1964). The history, observation of movements, and geology of the rock formations were reported by Muller (1964). Post-mortem analyses of the static equilibrium of the slide were carried out by Mencl (1966), Nonveiller (1967), Kenney (1967) and also discussed by Skempton (1966). The dynamics of the slide was considered by Jager (1968). A comprehensive investigation of the geological details of the rock formation and the slide surface was reported by Broili (1967). A re-evaluation of these investigations was undertaken by Muller (1968), who considered that the limiting equilibrium analysis of the slide had not been satisfactorily resolved.

A review of the previous works shows that:

(a) The analyses performed by Mencl (1966), Kenney (1967), and Nonveiller (1967) indicated that the friction angle ϕ' required for limiting equilibrium is around 20°, taking $c' = 0$. Subsequent analyses by Muller (1968), using more accurately defined slip surfaces from subsequent site investigation, indicated ϕ' to be 21°, using the same sets of assumptions as the other authors. This low value of the angle of shearing resistance can only be attributed to the presence of clay seams of considerable extent, almost entirely covering the slip surface. The recent detailed investigation by Broili (1967) convincingly proved that such a large area of clay layers is non-existant. The "clay films" in members C1 and C2 (classification according to Broili, 1967) which embodied most of the slip surface are described as "very rare". For these "clay films" the percentage of clay minerals is no more than 16% on the average.

(b) Detailed analyses of the slide were confined mainly to the western part of the slide. The stability of the eastern part was referred to only casually. Nonveiller (1967) stated that ϕ' for equilibrium was 30°. However, other conditions being equal, the stability condition of the eastern part of the slide is, in fact, more critical on account of the geometry of the slope alone.

(c) In all these analyses, the ground water conditions in the slope were taken as those corresponding to the reservoir level, or only very slightly higher (Nonveiller, 1967). The relative permeability of the rock formations have not been considered in detail in a region where the annual rainfall is substantial (about 60 in. rainfall per year before the slide).

Because of the great importance of the cause and consequence of the slide, and in view of the inconsistency of the analyses discussed above, a more complete analysis of the static equilibrium conditions of both the eastern and western part of the slide is performed in this paper using Janbu's method of general slip surface (Janbu, 1957). The essential features of the geology of the rock formations are discussed, leading to a revised concept of the ground water condition which, in turn, results in a more realistic value of the residual strength of the rock mass.

GEOLOGICAL SETTING

Rock Succession

Detailed geological work had been performed before the slide but it was mainly near the abutments of the dam itself and only three exploratory holes were bored in the westernmost part of the slide. The geological section in the reservoir was known mainly from these boreholes, of which only two actually did go through the potential slip surface, and the outcrops of the rocks along the Vaiont gorge. Eighteen more boreholes were ordered after the slide in the mass of slumped material and the geology of the outcroping slip surface was studied in detail, as reported by Broili (1967). A plan of the slide is shown in Fig. 1.

Four formations totalling 1000 m. were recognized in the immediate vicinity of the slide. Formation A, composed of limestone, outcrops mainly in the downstream part of the dam and was not involved in the slide. Formation B, maximum thickness of 350 m., is composed of oolitic limestone, sometimes affected by dolomitization, laid in beds about one meter thick. This formation is referred to as the Dogger formation and acted as the seat above which most of the slide occurred. The beds are rigid and not very susceptible to deformation. Formation C, with a thickness varying from 230 m. to 350 m., was the one affected by the slide. The lithology is mainly limestone bearing nodules or beds of chert, and occasionally thin intercalations of marly limestone. Due to its importance, this formation has been subdivided into eight members of varying thicknesses. The strata have a thickness ranging from 5 to 20 cm. in members 1, 3 and 7, and from 30 to 100 cm. in other members. Finally, formation D, in the eastern part, which is 300 m. thick, was not involved in the slide.

The Problem of Interbeds

Various opinions have been expressed about the influence of clay in the interbeds and even within the beds. Within the beds some authors report limey marls and marls intercalated with limestone. The material, however, was proved to contain no more than 25% to 30% clay minerals, the rest being mainly calcite. The formation does not include any "clay layers" or "clayey rocks" in regard to their mechanical properties.

As for the interbeds, Broili (1967) reports very thin films of pelitic material in the bedding joints (from 1 to 3 mm.). They are very sparse in formation B and very rare at the base of formation C (members 1, 2, and 3). Otherwise, the joints are empty and have a rough surface due to the growth of stylolites perpendicular to the surface of the joints. Even if these films have been referred to by the term "clayey films", mainly upon visual examination, the content in clay minerals averages 16%, chiefly composed of montmorillonite as reported by X-ray diffraction techniques. However, these films have been consolidated under such high pressures that they are very hard and do not dissolve or become plastic when soaked in water for a long time. Therefore, those pelitic films are not thought to affect the stability for the following reasons:

(1) because of their petrographical and mechanical properties;

(2) because they are very thin compared to the roughness pattern of the bedding joints, and

(3) because they are very sparsely and scarcely distributed throughout the rock mass.

As for Nonveiller's "lumps of bentonitic clay", (Nonveiller, 1967) they probably come from weathered zones of the upper part of the slide and are derived from members 1 and 2. Since the material was in a loose state, it looks more like a residual soil from which the carbonates have been leached and there are good indications, from the boreholes, that the plastic properties did not exist under pressures of 250 m. of rock.

Structure and Attitude of the Beds

The Vaiont Valley is situated in the core of a syncline with an axis running East-West. The rock formations have a very marked chair-like form. At the back of the chair, the beds have an E-W strike and dip toward the North at 45° to 50°. At the seat of the chair, the strike is N-S and the dip toward the East varies from 18° near the dam, to horizontal in the central part of the slide, and again 20° towards the East in the eastern part of the slide. The transition from back to seat is rather abrupt. The best picture is given by the attitude of formation B which is massive and less deformable. Formation C, due to its greater deformability, was slightly deformed by folds and faults during tectonic activities.

The most important structural element seems to be the stratification. The frequency and character of bedding joints become especially important when one considers that nearly 60% of the slip surface coincides with bedding joints. In the lower members of formation C (1 and 3) the spacing of the bedding joints is very small (5 to 15 cm.) compared to the other members (20 to 100 cm.). The surfaces of the joints are rough, often undulated and buckled, and often show a mortise effect due to stylolites growing in members 1, 2, 3, 5, 6 and 8.

Among other structural features, there are many sets of major and minor joints that intersect the rock mass: the principal set strikes N-S and is vertical; it marks the eastern and western boundaries of the slide. A second set, perpendicular to the first, accounts for the fractures parallel to the gorge observed before the slide. Thrust faults also developed having an E-W strike and dipping southward at 45°. They deform the base of formation C and the top of formation B and were developed during tectonic processes forming the "chair". Other minor sets of joints were also recognized and indicate that the whole mass had undergone a high degree of fracturing.

Geology Along the Slip Surface

The shape and depth of the slip surface is very important in the stability analysis. For this reason, all boreholes reported by Broili (1967) did intersect the failure surface. For purposes of analysis, the slide area has been divided in three zones. In the western zone, the slip is sensibly N-S in direction and the chair-like form is very accentuated. The seat is 500-700 m. wide and the back is inclined at 35-40°. The transition is sharp with a short radius of curvature. The central part is similar in attitude but there is a depression in the seat probably due to tectonic irregularities. In the eastern part, the chair-like form is less evident. The seat extends only 200-300 m. and the radius of curvature between the seat and the back is large.

Along the slip surface, the failure occurred in member 1 and at the interface of members 1 and 2, in the western zone. In the eastern-central zone, the slip was situated mainly in members 3 and 4, and partially in member 2. In the eastern zone, the slip surface cuts across most members of formation C, since the surface assumes a more circular shape. Therefore, in the western and central parts most of the deformation took place in those members where the thickness of the beds is the smallest.

Another interesting point is the relationship between the attitude of the slip surface and the attitude of the bedding joints as shown in Figure 1. In this figure, J indicates zones where the bedding joints and the slip surface have the same attitude; K, where the strike is the same but the dip of the slip surface is more inclined than the beds; L, where the strike is the same but the dip of the slip surface is less inclined than the beds; M, where the strike and dip are discordant. In western and central zones, as well as the upper part of the eastern zone, the slip surface follows the bedding. The areas where the bedding coincides with the slip surface represent about 55-60% of the whole surface.

In the boreholes, above the slip surface, Broili (1967) reports a zone of mylonite or broken rock material of several meters. The composition is mainly sand, silt and clay. The minus two micron fraction contains about 35% clay minerals, the rest being chert and carbonates. Such a zone of mylonite was never reported in the exploratory boreholes of 1960. This observation suggests that the failure surface was not formed at that time.

GROUNDWATER CONDITIONS

Before the Vaiont disaster, the groundwater conditions were assessed mainly through observations from four "piezometers" and from parameters estimated from geological information (Muller, 1964). These "piezometers" could not have indicated the correct water pressures within the potential sliding surface since:

(1) from the description (Muller, 1964, p.168-169), the "piezometers" appeared to be only observation boreholes without proper sealing and casing;

(2) the "piezometers" are in the central and eastern zones of the slide, and the elevations of their tips vary from El. 620 m. to 670 m., well above members C1 and C2, while the tops of formation B are El. 520-550 m. in these vicinities;

(3) they are within or at the edge of the mass which moved in November 1960 (Muller, 1964) and are therefore within the highly fractured mass. It is therefore reasonable to expect that their water level varied fairly closely with that of the reservoir level, as was observed. The permeability of the rock mass near the surface was estimated to be of the order of 10^{-2} to 10^{-3} cm./sec. (Muller, 1968). The permeability of the rock before the 1960 movement would be orders of magnitude less.

The possibility of artesian pressure existing on the plane of failure has been mentioned by Kiersch (1964), Mencl (1966), Muller (1968) and by the Court Experts. Muller (1968) based his argument on the possibility of a prehistoric slide, the mylonite zone which would act as an impervious layer. However, without resorting to such an assumption, there are some other factors that can lead to the same conclusion. First, Muller (1968) and Mencl (1966) point out that the layers of Malm and Upper Cretaceous (formation C) have a low permeability at depth. The values for the oolitic limestone are appreciably higher due to secondary porosity developed during the process of dolomitization of the oolitic limestone. This fact is supported by the observation of very active water circulation in a number of boreholes drilled in the upper part of the oolitic formation (Broili, 1967; Muller, 1968). "Zones of decayed and porous rock becomes friable, and evidence of water circulation is further provided by the yellowish-reddish (limonitic) colour of the rock." (Muller, 1968, p.14). Therefore, rainfall infiltrating the upper reaches of formation B can circulate quite easily within that formation.

It may also be noted that the porosity of oolitic limestone lies in the range of 13%-22% while that of compact limestone is of the order of 1%-2% (Castany, 1966).

In order to develop artesian pressures, an impervious layer is required. In the western and central part of the slide, large portions of the failure surface lie at the interface of members 1 and 2. The rock material in members C-1 and C-2 is practically impervious, the permeability being of the order of 10^{-10} to 10^{-7} cm./sec. (Serafim, 1968). Therefore, the field permeability is derived mainly from joints and fissures in the rock.

Consider first the frequency of the horizontal bedding joints of the two members. From the detailed investigation reported by Broili (1967), the thickness of the interbeds is 15 cm. on the average. The total thickness of the member is 35 m. Hence, the number of bedding joints is about 230. On the other hand, the spacing of bedding joints is about 1 m. in member C-2, whose thickness is 20 m. There are, therefore, approximately 20 bedding joints in C-2. Hence, the frequency of bedding joints of the two members differs by an order of magnitude.

Vertical joints and fissures also provide preferential paths for water flow. Previous work, summarized by Price (1966), showed that the frequency of vertical joints in sedimentary rocks, including limestones, is inversely proportional to the thickness of the strata. It may, therefore, be expected that the frequency of vertical joints in member C-2 is 10 times lower than that of member C-1. From the above considerations of geological details, it is reasonable to expect that member C-2 would form a relatively impervious layer compared to member C-1. Thus, a geological setting is provided for the existence of artesian pressures on the potential sliding surface.

ROCK STRENGTH

The total movement of the rock mass recorded before the slide is about 250 cm. This large deformation indicates clearly that the strength of the rock along the sliding surface must have passed the peak strength and reached its residual strength given by the angle of friction ϕ'_r. No data on the residual strength of the limestones at Vaiont has yet been published. There is, however, sufficient information in the literature to make a reasonable estimate of the residual angle of friction.

The published data on the angle ϕ'_r are summarized in Table 1. It must be noted that ϕ'_r varies not only with rock type, but also with the normal pressure acting on the failure zone.

The slip surface at Vaiont was on the average some 200 metres below the original ground surface and this corresponds to an overburden pressure of the order of 600 psi (40 kg/cm^2). Referring to Table 1, one would probably regard the value given in Krsmanovic and Popovic (1966) as most relevant to the case at Vaiont, in view of the nature of the rock, the testing conditions, and the range of stresses applied. Starting with this value of $\phi'_r = 30°$ and allowing for the possible presence of a small amount of clayey material, one would probably estimate a friction angle of the order of 28° to be operative along the slip surface at Vaiont. This value agrees closely with that obtained by Mencl (1967) on a limestone formation similar to those involved in the slide at Vaiont. It is also significant to note that none of the ϕ_r' reported has a value as low as 20°, as required in the previous analyses.

STABILITY ANALYSIS

Based on the foregoing geological considerations, the stability of the western slope has been re-analysed for the two ground water conditions: (a) artesian pressure, with hydraulic gradient inclined as shown in Fig. 2; (b) at same water level as that in the reservoir at time of failure, namely, at El. 700 metres.

Of the sections given by Broili (1967) in the western zone of the slide, sections B-B and A-A are very similar. Section B-B was first chosen for analysis. As the slip surface is of a "chair-shaped" nature, it would be appropriate to use Janbu's method of non-circular surface adopted in a computer program. The slide mass was divided into two wedges or rigid bodies: (1) an upper, (2) a lower, passive wedge on its "seat" portion. The vertical plane separating these two wedges was taken as that passing through the point of minimum displacement along the slope surface. The latter can be located by plotting out the displacement distribution along the slope surface (Fig. 2), based on the measurements reported in Muller (1964, 1968). This point was found to lie in the sharp bend of the chair-like surface, where energy consumption by interior resistance during the slide was believed to be considerable.

In the present analysis, a value of $\phi_r' = 28°$ has been assigned to the angle of friction along the "back" of the slip surface for reasons previously discussed. The inter-wedge forces and angle of friction required for limiting equilibrium of the lower wedge were then computed for each of the two ground water conditions. The angle of friction required for the artesian pressure case was found to be 25°, and that for the reservoir-water-level case 13°. It follows that while it is difficult to interpret the rock slide in terms of the reservoir water level alone, a consistent and reasonable strength of the rock mass can be obtained if one allows for the presence of an "artesian" pressure condition. As pointed out in the foregoing geological considerations, this condition should not be ruled out in view of the relative permeabilities of the strata adjacent to the slip surface and of the altitude of Mount Toc.

To lend further support to this concept, the stability of the eastern slope was also analysed. Section F-F was thought to be most critical (Figure 3), and Janbu's method again was employed to determine the average angle of friction required for limiting equilibrium of the rock mass on slope. In this case, as the slip surface passes well above the oolitic limestone of Formation B, the artesian pressure would have no effect on the slip surface. Consequently, the slope has been analysed with a ground water level at El. 700 m. (i.e. water level in the reservoir at time of failure). The required angle of friction was found to be between 27° and 28°. For the eastern slope of the Vaiont slide, the lower portion of the slip surface cuts across the bedding planes. However, it should be noted that under high normal stresses, the angle of friction approaches a unique value regardless of whether shearing is performed along or oblique to the beddings. The reason why the slip surface had not been formed at a greater depth and parallel to the bedding planes is to be found in the confining effect of the rock mass at the toe and below the gorge. At the toe of the eastern slope, the gorge was relatively shallow, so that the aquifer (oolitic formation B) passed well below its bottom. The possibility of having a slip surface formed adjacent to formation B is thus somewhat reduced. Nevertheless, the condition of stability associated with a potential slip surface parallel to the strata and meeting the bottom of the gorge (Figure 3) was also investigated. As this potential slip surface is rather close to the oolitic formation, artesian pressures should be taken into account. This artesian head would probably have dissipated to a considerable extent in the vicinity of the fault near the gorge, so that it could probably be approximated by the hydraulic gradient shown in Figure 3. Using Janbu's method and taking an angle of friction of 28°, the factor of safety against sliding along this potential slip surface was computed to be 1.06. The potential slip surface herein analyzed was thus less critical than the one along which the slide actually occurred (for which F.S. = 1.0).

In order to test further this concept of artesian pressures in the other sections, stability analysis was also performed on Section E-E, located on the eastern slope and close to the central portion of the sliding mass. In this case, the slip surface is close to the oolitic formation B for its upper part, and is of the order of 120 metres above B in the vicinity of the bend. If the foregoing concept holds, artesian pressure would have influenced the stability condition of this section significantly. Again, this artesian head would probably have dissipated considerably in the vicinity of the fault near the gorge, so that it could probably be approximated by the hydraulic gradient shown in Figure 4. The factor of safety against sliding along the slip surface, with the angle of friction being taken as 28° as before was found to be 1.0, again suggesting that the slope was in a state of quasi-equilibrium before the catastrophic slide.

It thus follows that with this concept of artesian pressure in the oolitic limestone of formation B, the Vaiont slide can be interpreted in terms of a consistent and reasonable value for the strength of rock mass.

SUMMARY AND CONCLUSIONS

A study of the structural details of the rock formations of the Vaiont slide suggests the probable existence of artesian pressures in the oolitic limestone (Formation B). The area has an annual rainfall of about 150 cm. with concentration of precipitation in October and November. The necessary conditions for the artesian case were therefore satisfied, and its existence is supported by the observation of water condition in borings carried out after the slide (Broili, 1967).

The detailed geological investigation reported by Broili (1967) showed that there was no significant amount of clay seams in the bedding planes, or within the interbeds of the rock units. This finding refuted the assumption of a continuous clay layer covering the entire slip surface made in the previous stability analyses. Even this extreme assumption could hardly account for the instability of the eastern zone of the slide, where the sliding surface cuts across the beds.

A review of the residual strength of limestones in the literature indicated that the most probable ϕ_r' , even taking into account a minor extent of clay layers, is about 28°. No value was reported to be as low as 20°, which was required by the previous analyses.

Using the value of $\phi_r' = 28°$, together with the assumption of artesian pressures acting on the slip surface, sections in both the western and eastern portion of the slide have been analysed using Janbu's method of general slip surface. The results are summarized in Table 2.

It may be seen from Table 2 that the stability condition, not only of the western zone, but also that of the eastern zone of the slide, may be consistently interpreted. The reason why the eastern zone of the slide failed across the strata is firstly due to the external geometry of the slope and the Vaiont gorge, and secondly, the actual slip surface is more critical as borne out by the analyses.

It is recognized that additional information on the Vaiont slide may exist but has not been publicly released yet, and that some of the factors considered in this paper might have been appreciated. Furthermore, some of the information utilized in the analyses herein, were only available after the slide occurred. The purpose of this investigation is solely to indicate that with a self-consistent set of assumptions, the entire slide may be alternatively interpreted.

ACKNOWLEDGMENT

The investigation performed was supported by the National Research Council of Canada under Research Grant No. A-7745.

REFERENCES

Broili, L. (1967). "New Knowledge on the Geomorphology of the Vaiont Slide Slip Surface", Felsm. u. Ing. Geol. vol. v/l, 1967.

Castany, G. (1966). "Traité pratique des eaux sousterraines", Paris, Dunod.

Giuseppe, B. (1970). "The Shear Strength of Some Rocks by Laboratory Tests", Proc. 2nd Int. Cong. Rock Mech., Belgrade, 1970, vol.2, p.3-24.

Handin, J., and R.V. Hager, Jr. (1957). "Experimental Deformation of Sedimentary Rocks under Confining Pressure: Tests at Room Temperature on Dry Samples", Bull. Am. Assoc. Petrol. Geol., vol.41, no.1, pp.1-50.

Jaegar, J.C. (1969). "Behaviour of Closely Jointed Rock", Proc. 11th Symp. on Rock Mech., Berkeley, 1969, pp.57-68.

Jäger, C. (1968). "New Considerations of the Vaiont Slide - The Dynamics of the Slide", Rock Mech. and Eng. Geology, vol.VI/4, pp.243-247.

Janbu, N. (1957). "Earth Pressure and Bearing Capacity by Generalized Procedure of Slices", Proc. 4th Int. Conf. on Soil Mech. & Found. Eng., vol.2, pp.207-212.

Kenney, T.C. (1967). "Stability of the Vaiont-Valley Slope", Felsm. u. Ing. Geol. vol.v/l, 1967.

Kiersch, G.A. (1964). "Vaiont Reservoir Disaster", Civil Engineering, March 1964, pp.32-40.

Krsmanovic, D. and M. Popovic (1966). "Large Scale Field Tests of the Shear Strength of Limestone", Proc. 1st Int. Cong. Rock Mech., Lisbon, 1966, vol.1, pp.773-9.

Krsmanovic, D. (1967). "Initial and Residual Shear Strength of Hard Rock", Geotechnique, June 1967, pp.145-160.

Lane, K.S. (1966). "Stability of Reservoir Slopes", Proc. 8th Symp. Rock Mech., Univ. of Minnesota, 1966, pp.321-336.

Locher, H.G. and U.G. Rieder (1970). "Shear Tests on Layered Jurassic Limestone", Proc. 2nd Int. Cong. Rock Mech., Belgrade, 1970, vol.3, p.3-1.

Lundborg, N. (1966). "Triaxial Shear Strength of Some Swedish Rocks and Ores", Proc. 1st Int. Cong. Rock Mech., Lisbon, 1966, vol.1, pp.251-255.

Mencl, V. (1966). "Mechanics of Landslides with Non-circular Surfaces with Special Reference to the Vaiont Slide", Geotechnique, vol.16, pp.329-337.

Muller, L. (1964). "The Rock Slide in the Vaiont Valley", Felsm. u. Ing. Geol., vol.II/3-4, 1964.

Muller, L. (1968). "New Considerations on the Vaiont Slide", Felsm. u. Ing. Geol., vol.6, pp.1-91, 1968.

Nonveiller, E. (1967). "Shear Strength of Bedded and Jointed Rock as Determined from the Zalesina and Vaiont Slides", Proc. Geot. Conf., Oslo, 1967.

Nonveiller, E., (1967). "Discussion on Mencl's paper, (1966)", Geot., vol.17, no.2, pp.170-171.

Patton, F.D. (1966). "Multiple Modes of Shear Failure in Rock and Related Materials", Ph.D. Thesis, Univ. of Illinois, Urbana, 1966.

Price, N.J. (1966). "Fault and Joint Development in Brittle and Semi-Brittle Rock", London, Pergamon.

Serafim, L. (1968). "Influence of Intersticial Water on the Behaviour of Rock Masses", pp.55-97 in Stagg, K.G. and Zienkiewicz, O.C. (ed.), Rock Mechanics in Engineering Practice, New York, John Wiley & Sons.

Skempton, A.W. (1966). "Bedding-Plane Slip, Residual Strength and the Vaiont Landslide", Corr., Geotechnique 16, pp.82-84, 1966.

SOURCE OF INFORMATION	DESCRIPTION	SPECIMEN SIZE	ϕ_r	NORMAL STRESSES	REMARK
(1) Handin & Hager (1956)	Fusselmarr limestone (Texas) Marianna limestone (Florida) Devonian limestone (Texas) Wolfcamp limestone (New Mex.)	1/2"D 1"L	31° 32° 33° 32°-35°	>3000 psi	Ductile failure due to very high confining pressure
(2) Krsmanovic & Popovic (1966)	Fissured Grancarevo limestone with thin fill layers (Yugoslavia)	Field test, shearing surface 2.8m.x1.8m.	30°	Constant ϕ_r=30° for σ_n>200 psi.	Confining pressure 70-200 psi
(3) Patton (1966)	"Hard" limestone	2.95x1.75x2.0 in.	33°-37°	<300 psi	Lab. direct shear
(4) Lundborg (1966)	Granboforsen limestone (Sweden)	15 mm. D 50 mm. L	50°	<70 psi	ϕ' decreases for higher σ_n
(5) Krsmanovic (1967)	"Hard" limestone	40x20x20 cm.	56°	<100 psi	"
(6) Mencl (1967)	Senonian limestone	/	27°	/	Sheared across beddings
(7) Locher & Rieder (1970)	Jurassic limestone (Switzerland)	Shearing surface 200 cm^2	31°-41°	<100 psi	/
(8) Guiseppe (1970)	Genova limestone (Italy)	40x40x20 cm.	40°	<125 psi	/
(9) Lane (1966)	Stockton dolomite (jointed)	/	30°	0-2400 psi	/
(10) Jaegar (1969)	Mt. Isa dolomite & shales (closely jointed)	6 in. D 12 in. L	27°-35° (mean 31°)	<1000 psi	/

TABLE 1. ANGLE OF FRICTION (ϕ_r) FOR LIMESTONE AND RELATED ROCKS

SECTION	HYDROSTATIC CONDITION ON SLIP SURFACE	FACTOR OF SAFETY	REMARKS
A-A	Artesian	0.88	Western
B-B	Artesian	1.03	Western
E-E	Artesian	1.00	Eastern
F-F	Reservoir Level	0.98	Eastern
F-F	Artesian on Deeper Surface	1.06	Eastern

TABLE 2. SUMMARY OF STABILITY ANALYSES, $\phi_r' = 28°$

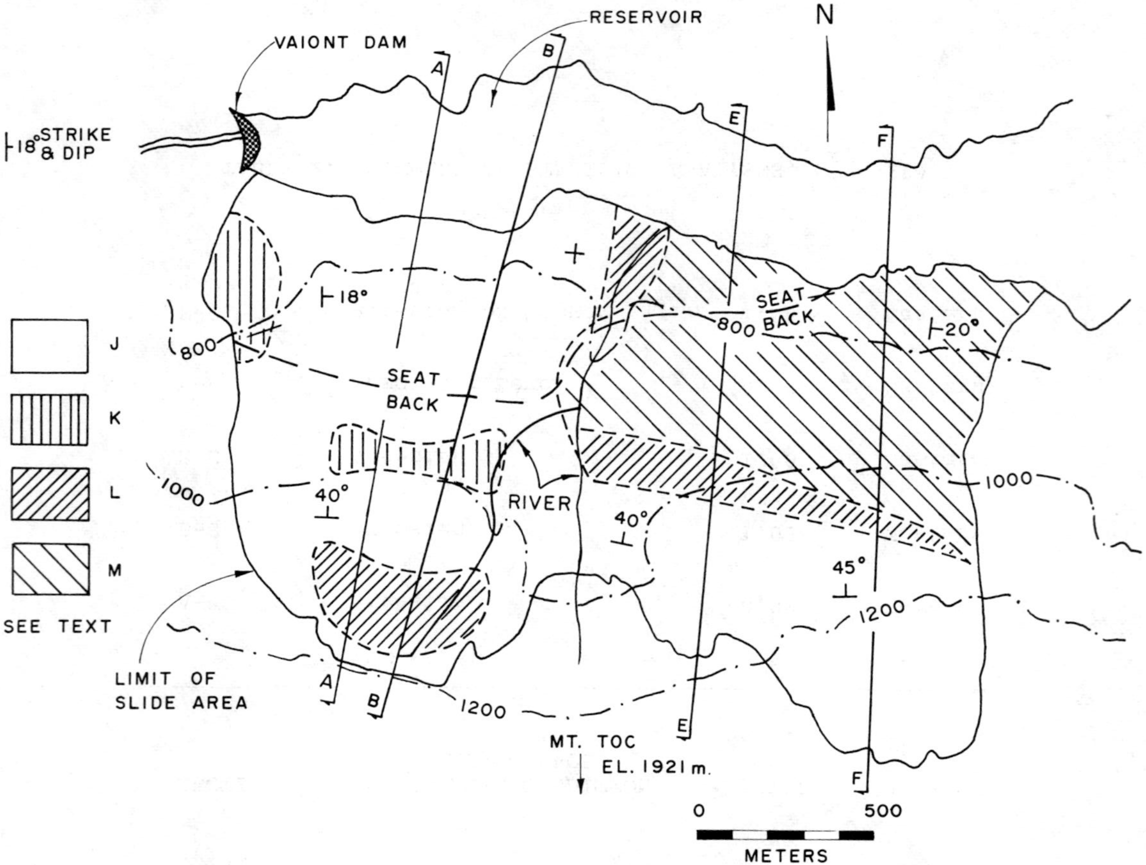

1. Plan of slide, showing attitudes of rock formations and slip surface.

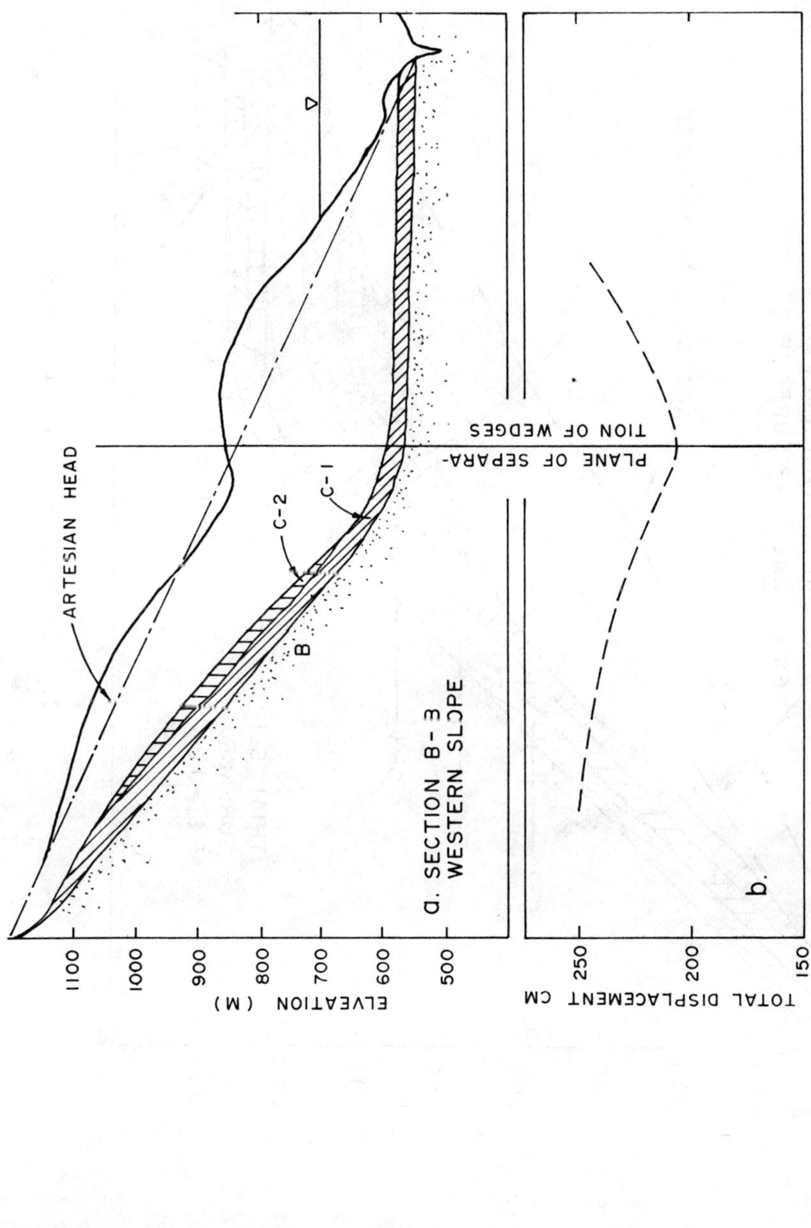

2. (a) Section B-B, western slope

(b) Total displacement of surface of slope before slide.

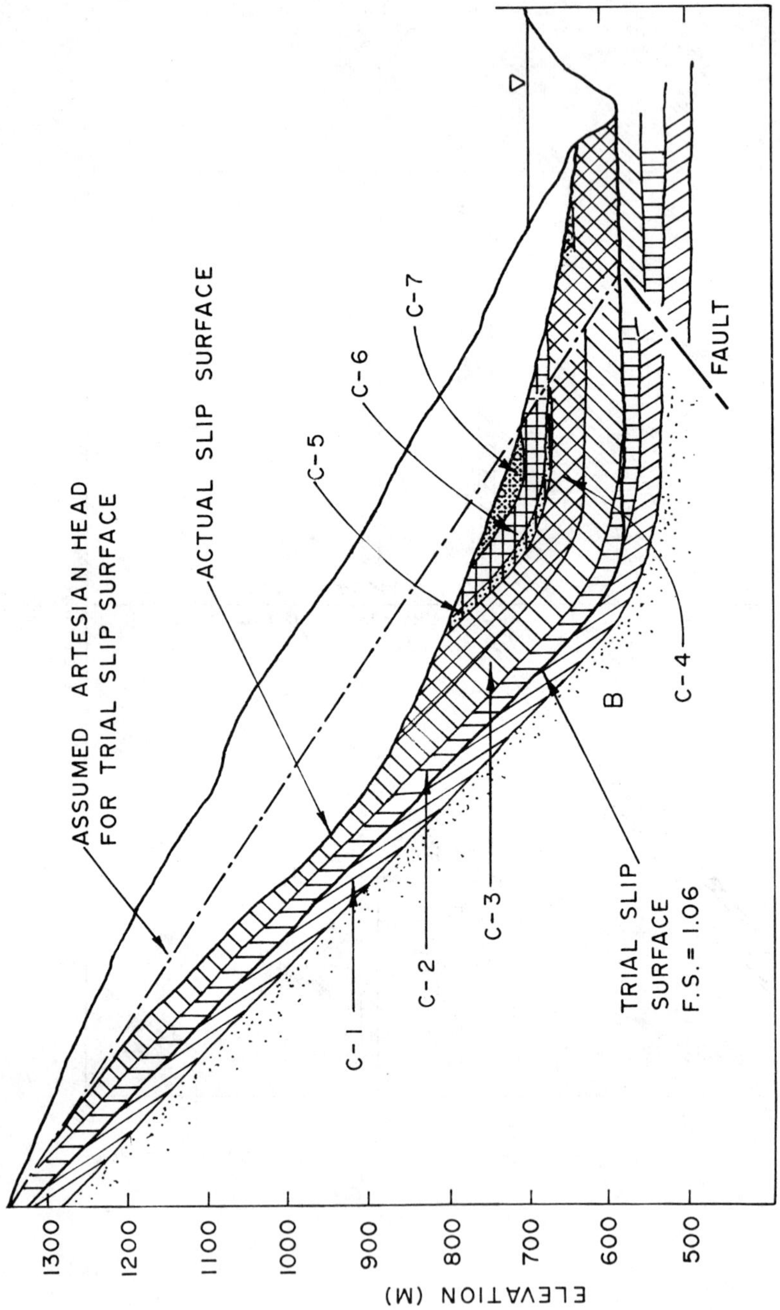

3. Section F-F, eastern slope

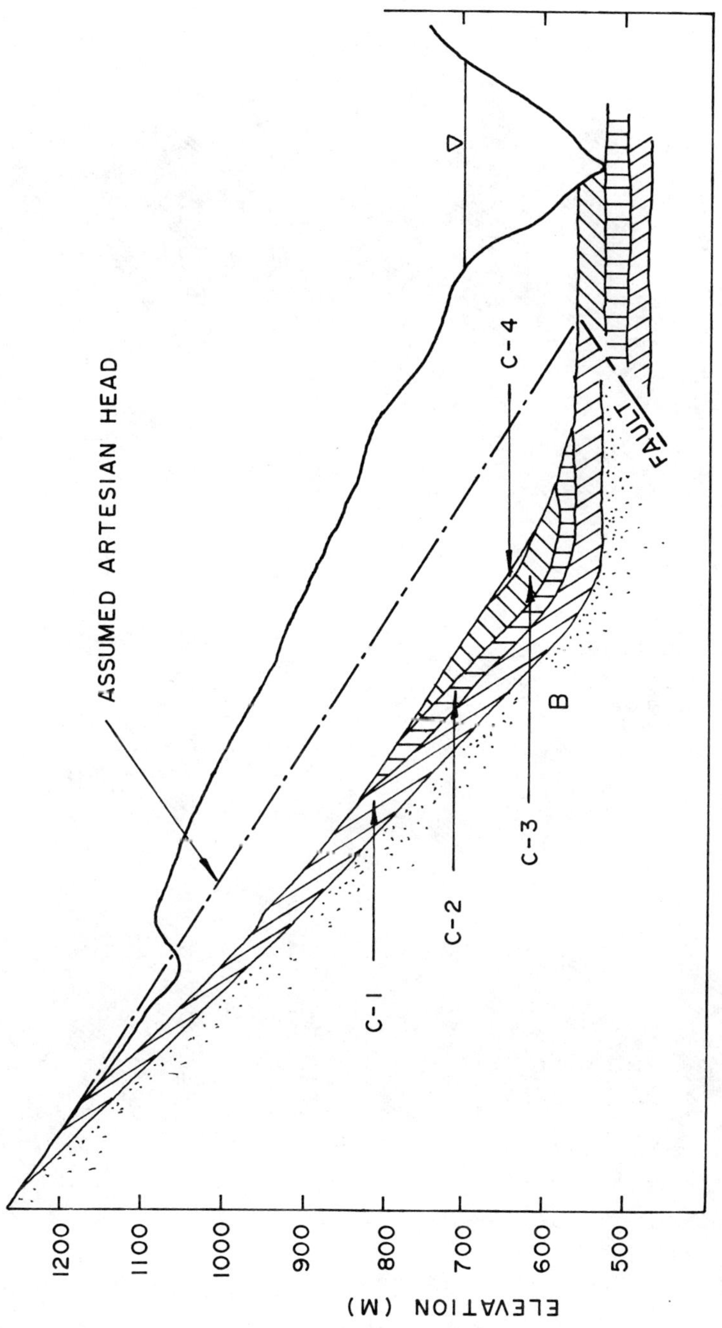

4. Section E-E, eastern-central slope

SLOPE STABILITY IN TROPICALLY WEATHERED DIORITE

by

Andrew H. Merritt**

INTRODUCTION

Tropical weathering of crystalline rock produces a wide range of materials which include clayey residual soil, sand, spheroidal boulders and soft rock with severe chemical alteration on joint surfaces. The entire typical profile is rarely found at Alto Anchicaya. Since the degree of development of the weathering profile in diorite is quite variable, slope properties change remarkably in a limited area. The design engineer, therefore, must be aware of the specific materials present and must anticipate their unique engineering behavior.

In this paper several examples of weathered diorite slopes are presented. The variable nature of the weathering products, the performance of the slopes and the design and construction requirements are discussed.

** Senior Engineering Geologist - Acres International Limited
Niagara Falls, Ontario

GENERAL DESCRIPTION OF THE PROJECT

The Alto Anchicaya Power Project is located in the western cordillera of the Andes mountains in Colombia. The power plant complex, 340,000 kilowatts, consists of a 140-meter high concrete-faced rock fill dam, a 9-kilometer power tunnel, underground powerhouse and transformer gallery and other appurtinent structures and tunnels. A 16-kilometer access road, connecting a pre-existing road with the dam site, was begun in 1969 and completed in June 1971 through difficult jungle terrain.

GEOLOGY

Schists, hornfels, diorite and amphibolite are exposed in the access road cuts. The schists have been deformed into isoclinal folds and occasional shear zones are visible. The hornfels was formed by the metamorphism of the schists near the contact with the diorite intrusions. Diorite is the exclusive rock type in the underground powerhouse and occasional dykes of diorite have been exposed along the access road. The amphibolite, also associated with diorite intrusions , is rarely exposed and is of little consequence to the engineering works.

The diorite has been slightly metamorphosed, but is essentially fine grained with no prominent planes of foliation. The mineralogy from hand specimen analysis is nearly identical throughout the site; plagioclase feldspar, amphibole and biotite mica. At least four sets of joints have been mapped in the diorite and some shearing exists at the contacts with the schist.

FACTORS AFFECTING DIORITE SLOPES

The typical profile of weathering in crystalline rocks in tropical climates has essentially five principal components: (a) organic sandy soils; (b) sandy textured decomposed rock with the original discontinuities still preserved (saprolite); (c) spheroidally weathered boulders, generally surrounded by sandy material; (d) jointed, deeply weathered rock with sand or clay along the major surfaces; (e) fresh rock.

The typical results of diamond exploration drilling in diorite at Alto Anchicaya indicate the following characteristics in the top 15 meters: core recoveries less than 30 percent, RQD's less than 10 percent; weathered core fragments which can be broken by hand; pieces of fresh core between weathered soft material; high drilling water losses and irregular drilling rates. The description of the recovered core and drilling performance is consistent with the typical profile given above.

The variability of the weathering products makes a prediction of the engineering significance of the geologic conditions difficult. However, when any portion of the weathering profile predominates over another, the slope will have unique properties. For example, if the surficial zone consists of deeply weathered rock with preserved interlocking mineral fabric, the slopes in this material can be cut steep, whereas if the sand and boulder zone is thick, say 5 meters, then the slopes must be laid back to between 30 degrees to 35 degrees before they are stable. In this case, it is important to note that in the Anchicaya type of terrain, where the natural slopes are generally steeper than 35 degrees,the size of a design cut would have to be large, if it is at all possible to excavate.

The variable nature of the weathered rock profile can be divided into four main categories as observed at Alto Anchicaya:

Type A. Well developed profile, thick sand and spheroidal boulder zone. Inclination of critical joint angle is less than 40 degrees and the natural slope angles are relatively flat. The critical joint is defined as the joint set parallel to the slope.

Type B. Thin sandy surficial deposits grading into open joints and fresh rock. Inclination of the critical joint angle is greater than 40 degrees.

Type C. Deep weathering along joints with a gradual transition into fresh rock. Thin surficial organic sandy soil.

Type D. Predominance of a clay-rich surficial zone (saprolite) with a rapid transition to fresh rock.

<u>TYPE A</u> The complete typical weathering profile has been developed in this area (Fig 1). This is the result of two factors; a uniform rock type with a relatively low drainage gradient providing thorough infiltration of rain water, and the angle of the critical joint planes being flat enough so that the weathering products remain on the slope beneath a cover of vegetation. The complete soil-rock profile includes 1 meter of organic sandy soil, 5 meters of sandy residual soil-saprolite and at least 4 meters of weathered open jointed rock with spheroidal boulders.

The dip of the critical joints varies from 30 degrees to 40 degrees and the weathering products, soil and boulders, are generally stable on the joint surfaces. The original topographic surface in this area also varies from 35 degrees to 40 degrees.

Excavation of the access road into this area precipitated numerous landslides and the slopes eventually stabilized when the weathering products worked back to an angle of 30 degrees to 35 degrees, or when the slope failed to one of the continuous joint planes which was undercut by the road.

<u>TYPE B</u> One section of the access road in diorite (Fig 2) has joints which dip steeply into the cut. There is a thin zone (0 - 2 meters) of sandy surficial soil underlain by a

transition zone of up to 5 meters of open jointed rock and then into fresh diorite with tight joints.

Throughout this area, the angle of dip of the critical joints varies from 50 degrees to 70 degrees (Fig 3) and the natural slope angle corresponds to this dip. The weathering products such as sand, spheroidal boulders and rock blocks are not stable on such a slope and consequently are continuously failing. The natural process of large scale slope failure occurs when weathering reduces the shearing resistance on the joint planes below the point at which the rock blocks are stable at this angle, and the river undercuts the toe of the slope.

The combination of the road cuts parallel to the strike of the major joints and the cutting of natural drainage systems at right angles to the strike have produced sections of rock with no end restraint and these behave in a typical two-dimensional failure mode. Excavation in these areas reveal large open joints, the rock near the surface being close to the point of failure. These conditions are sensitive to blasting during road excavation and many of the dangerous landslides occur then. The volume of rock remaining must be assessed for its stability and any potentially hazardous zones removed. Additional excavation and hand cleaning is preferable to remedial rock support because of the large areas involved and the relatively low labor costs in this area.

Fig 1 Well Developed Weathering Profile: Landslide Limited by the Depth to the Visible Joint Surface

Fig 2 Natural Slopes Developed Parallel to Major Joint Set.

Fig 3 Critical Joint Set Dipping Between 50 degrees to 70 degrees. Note thin Soil Cover and Open Joints

<u>TYPE C</u> The topographic relationships in the portal area of the access tunnel to the powerhouse (Fig 4) are more complex than those described in Type A. There is a geologic contact with massive hornfels about 100 meters above the portal and this influences the morphology of the terrain.

The natural slopes in the diorite are 38 degrees to 40 degrees and in the hornfels they are nearly vertical. The soil and bouldery sections of the typical weathering profile are not well developed here which is the result of the high gradient of the natural run-off. These small creeks have cut deeply into the diorite producing steep side slopes and leading to the natural removal of the unstable weathering products. The remaining material consists of blocks of rock bounded by nearly vertical joints (Fig 5). These joints may have up to 15 centimeters of clay and sand. The depth of significant weathering is at least 20 meters in this zone.

At this site, the portal cut was slotted and the slopes were rock bolted as the excavation advanced. The major joints were nearly vertical and slope angles were designed accordingly. Rock cuts for permanent structures in this type of terrain should be kept as small as possible. On a local scale,the attitude of the jointing systems may be predictable. However, as the size of a rock cut increases so does the chance of encountering an unfavorably oriented joint or shear zone.

Fig 4 Access Tunnel Portal. Geologic Contact is above the Tops of the Small Ridges

Fig 5 Weathered Open-Jointed Diorite With No Surficial Soil Cover. First Stage of Pilot Tunnel Visible

<u>TYPE D</u> The tunnel portal (Fig 6) was excavated through approximately 6 meters of deeply weathered material with occasional small boulders of weathered rock. This zone overlies fresh diorite and no sandy transition zone is present. The absence of a loose sandy zone may be explained by the occurrence of a shear zone above the portal (which can be seen inside the tunnel). This zone originally contained a high percentage of clay and, therefore the weathering products would not have a sandy texture. The clay undoubtedly acts as a binder to aid in preserving the original interlocking mineral fabric. The close spacing of the fractures, 5-10 centimeters, also explains why no large spheroidal boulders were developed.

The clayey impermeable nature of the surficial material also acts as a barrier to groundwater flow as the first 10 meters of this tunnel are drier than the other tunnels in diorite.

The upper slopes in the residual soil were cut at 1:1, and the lower slopes in rock were cut vertically. Prior to tunnelling and installation of the ribs, approximately ten rock bolts were placed through key rock blocks. The steep face in the soil material has kept the exposure to heavy rain at a minimum and prevented excessive erosion.

The joints in the upper portion of the exposed rock are slightly open (Fig 6). However, the major joint sets strike parallel to the tunnel axis and thus present no problems in the portal excavation. Although the joints

Fig 6 Portal Cut in Clayey Weathered Diorite with the Original Interlocking Mineral Fabric Preserved (Saprolite). Shear Zone Dips into the Cut Above the Portal

are unfavorably oriented with respect to the tunnel, with proper rock bolting and careful blasting techniques the tunnel itself was excavated without difficulty.

CONCLUSIONS

The engineering behavior of the slopes in tropically weathered diorite at Alto Anchicaya is a function of the natural weathering products of rock which are present on the original slopes. These weathering products depend upon the general morphology of the terrain and the angle of the major joint sets parallel to the slopes.

Therefore, the design of slopes for permanent structures will vary depending on whether the slope is governed by: (a) soil-like conditions whose principal component is either completely weathered rock or sand; (b) weathered rock whose stability depends upon the angle of inclination of the major joint systems; (c) the combination of both.

If the material consists of a deeply weathered rock with clay and an interlocking relict mineral fabric, the slopes can be cut steeply,i.e. at least 1:1. Where the natural material is slightly weathered rock with open joints, limited size slopes may be cut essentially vertical, 1H:6V, if thejoints are vertical or are dipping into the hillside. Field observations also indicate that when the slope is composed of sand and weathered boulders, the natural apparent stable angle after a succession of landslides is about 32 degrees. Therefore the cuts should be designed as such unless an artificial retaining system is desired. These slopes are liable to ravel during heavy rains, however,and remedial treatment such as drainage and slope covering would be required for a permanent structure.

When the critical joint sets are steeply inclined, 40 degrees to 70 degrees, and will be undercut by the excavation, then extensive artificial support is necessary. If the rock is weathered and the joints are filled with sand or clay, standard techniques such as rock bolting may require specialized drilling equipment to prevent the hole from collapsing prior to inserting the bolt.

It is apparent that an effective slope stabilization system in this type of terrain can be extremely costly. In the case of road excavation, it is undoubtedly cheaper to initially cut as steeply as possible and clean off the landslides as they occur. However, for all permanent structures, cuts must be stable and all excavations must be designed and reinforced accordingly. Underground construction should be considered where possible and standard engineering designs altered in an effort to reduce the size of the surface cuts and to make the structure fit the rock conditions and natural terrain.

ACKNOWLEDGEMENTS

The writer wishes to acknowledge the Corporacion Autonoma Regional del Cauca for permission to write this paper, to Dr. Bayardo Materon for his reading and suggestions in technical content and to J. Arana for photographic services.

EXCAVATION QUALITY CONTROL FOR THE NCMC EXPANSION

By John D. Smart[1], A.M. ASCE,
and Ronald L. Friestad[2], A.M. ASCE

INTRODUCTION

The NORAD Cheyenne Mountain Complex (NCMC) near Colorado Springs, Colorado is currently being expanded. Previous excavation at the NCMC has been described by L. B. Underwood and C. J. Distefano (1)[3]. The new chambers and tunnels will total approximately 50,000 cubic yards of excavation in granite. The new excavation at NORAD will house a new diesel power plant and associated cooling towers and air handling equipment. The Expansion Project was designed for the U. S. Air Force by the Omaha District, Corps of Engineers.

This paper presents a discussion of the design of the new chambers, discusses specific quality control requirements of the plans and specifications, and summarizes construction progress to date (August 1971).

EXCAVATION DESIGN

Early design efforts were directed towards achieving a layout of the required openings that would provide as much distance as

---

[1]Structural Engineer, Omaha District, U. S. Army Corps of Engineers

[2]Structural Engineer, Omaha District, U. S. Army Corps of Engineers

[3]Numerals in Parenthesis Refer to Corresponding Items in Appendix-References

possible between openings and satisfy all mechanical, electrical and structural requirements. Close cooperation between electrical, mechanical and structural engineers and geologists was required during the design period to ensure that the requirements of each discipline were satisfied. The design engineers responsible for preparation of the excavation contract plans and specifications were fortunate in that valuable experience gained during previous excavation at NORAD was available, often as first-hand knowledge. Prior knowledge of the structural geology in the highly jointed and often sheared granite was adequate for planning purposes and tentative chamber locations were selected and intensive geological investigations were begun in these areas.

The model shown in Figure 1 depicts the marginally pleasing compromise between rock engineering, mechanical engineering and economics which was developed for chamber locations. Close proximity to existing chambers was undesirable to the excavation designers but highly desirable to the mechanical engineers. To ensure that the chambers could be successfully excavated and to provide information for design, three core borings were made into the tentative areas and an exploratory drift was mined for the full length of the power plant chamber. Based on the information obtained in these investigations the design was finalized. Minor adjustments were made in configuration to avoid adverse rock conditions, but in general opening sizes and shapes were dictated by mechanical engineering requirements.

The darker shaded chambers shown in Figure 1 are portions of the initial complex. The tunnels and chambers shown in white are those being mined for the expansion project. The larger chamber to the right is the power plant chamber and has a span of 67 feet, a height of 50 feet, and a length of 172 feet. The other large chamber on the lower level is the cooling tower chamber having a span of 38 feet, a height of 56 feet, and a length of 185 feet. The small chamber on the upper level has a span of 20 feet. The other major excavations are the 18 foot diameter air tunnel and the 19 foot diameter exhaust tunnel. In Figure 2 it can be seen that the openings are at three different elevations. This made possible maximum use of the available rock mass.

PLANS AND SPECIFICATIONS

After final chamber sizes and locations were selected, the remaining rock engineering requirements were quite clear; namely to prepare plans and specifications that spell out excavation and reinforcement procedures that would result in stable rock mass structures after the Contractor had completed his work. The engineering efforts and inspection efforts necessary to prepare and enforce the contract documents were labeled as excavation quality control.

The basic philosophy adhered to in preparing the specifications for excavation and reinforcement was to be specific on procedures expected to be successful by the designers. It may be argued that such an approach stifles the ingenuity of mining contractors. However, when the disparity between time allowed for a Contractor to bid a job and time available for design was considered, it was decided that the burden of determining critical procedures should

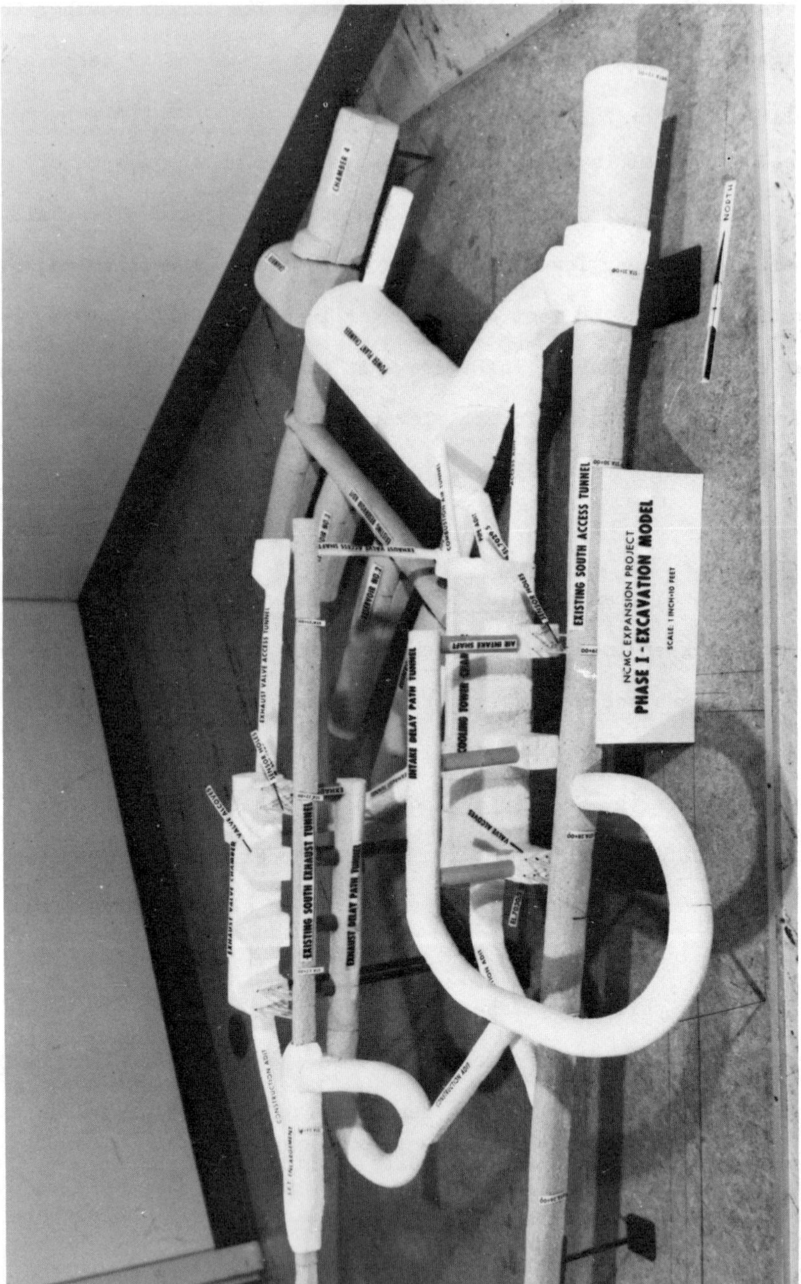

FIGURE 1. EXCAVATION MODEL

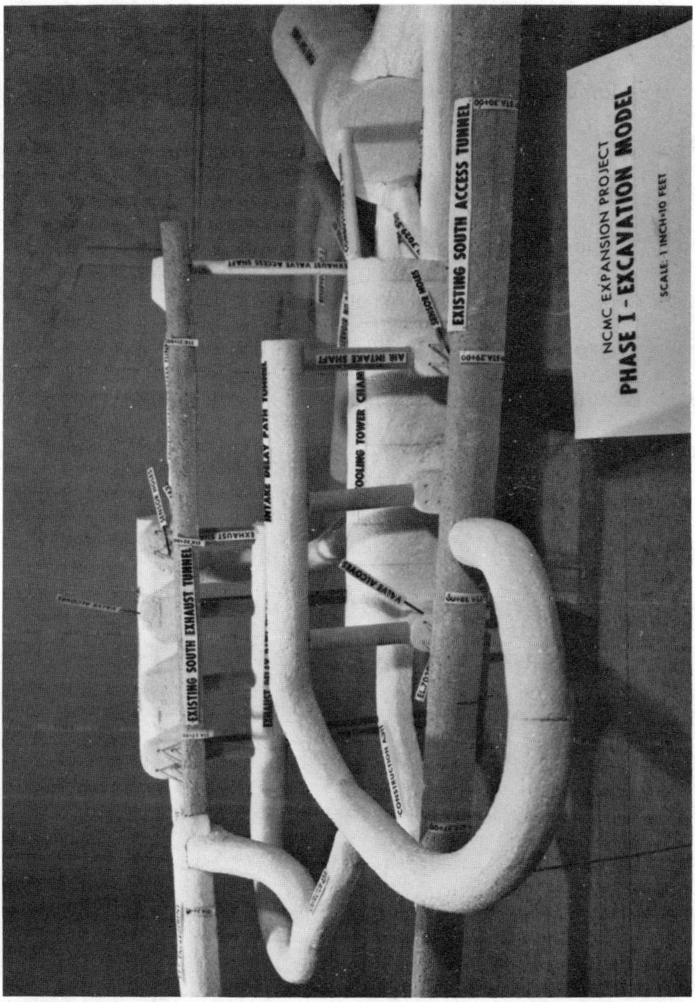

FIGURE 2. EXCAVATION MODEL

rest squarely on the shoulders of those responsible for preparing the plans and specifications. Therefore, in three areas considered critical, namely; smoothwall blasting procedures, reinforcement procedures, and special treatment of intersections, specific procedures were required.

Major requirements of the smoothwall blasting specification are as follows:

1. Hole spacings of 12 to 30 inches.
2. Burden to spacing ratio of 1.2 to 1.5.
3. Alignment of perimeter holes held to within 6 inches of intended location.
4. Loading of 0.1 to 0.25 lbs/ft of low to medium velocity explosives.
5. Perimeter holes fired simultaneously.
6. Trim cut fired separately after the pilot advance or main cut is removed (2).

Each of the first four restrictions is conventional practice for smoothwall blasting. Simultaneous, instantaneous and separate firing of the trim holes was required for three reasons. First it adds emphasis to the control required on the perimeter. Secondly it achieves the technical advantage of precise simultaneous firing of the trim holes that is not achievable with the scatter that is inherent with regular tunnel delays. The third advantage is that this blasting sequence affords the opportunity for the added control achieved by using recessed rock anchors (prebolting) in weak rock.

The feasibility of rock reinforcement in the rock encountered was established during the mining of the existing chambers. The primary reinforcement chosen for the expansion project was the hollow-core 1-inch diameter deformed rock bolt. Direct pull tensioning was required by the specifications. A second reinforcement element was selected for prebolting in certain areas. This is an untensioned "perfo" sleeve type rock anchor shown in Figure 3. Field and laboratory tests of reinforcement hardware and procedures were conducted prior to advertising the job. The Contractor was also required to perform test installations prior to excavation work to verify his procedures.

The excavation reinforcement sequence required by the contract documents where prebolting was not required is shown in the upper portion of Figure 4. Round lengths were limited to 10 feet. The main excavation or pilot advance is carried approximately one and one half round lengths ahead of the trim blasting. As discussed earlier, the trim holes were to be fired simultaneously with zero delay caps. Rock bolts were to be installed and tensioned to within 2 to 10 feet of the trim blasting. It was required that not closer than 10 feet or further than 30 feet from the trim blasting the bolts be retensioned and grouted.

The crowns of the power plant and cooling tower chambers are intersected by clay filled shears. To avoid potential problems with these areas, a pattern of recessed perfo type anchors was specified in addition to the tensioned bolt pattern. These are installed as shown in the lower portion of Figure 4. They are recessed through the burden of the trim cut to the required excavation line and carried as close as possible to the main face.

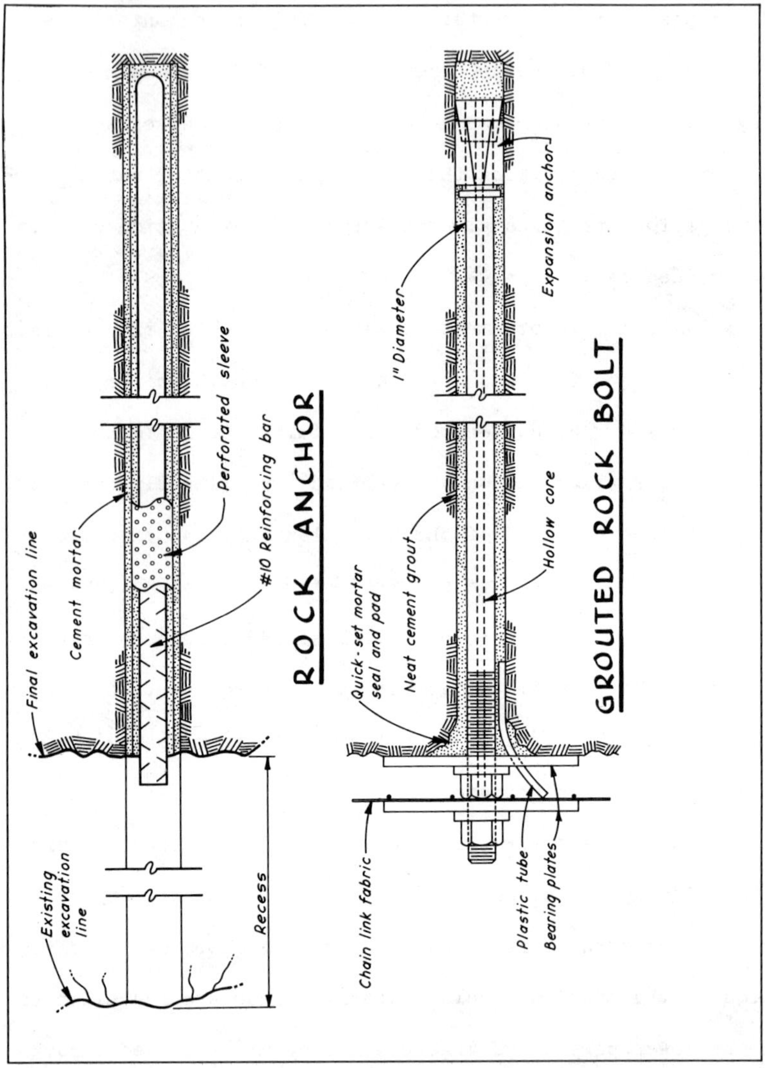

FIGURE 3. ROCK ANCHOR AND ROCK BOLT DETAILS

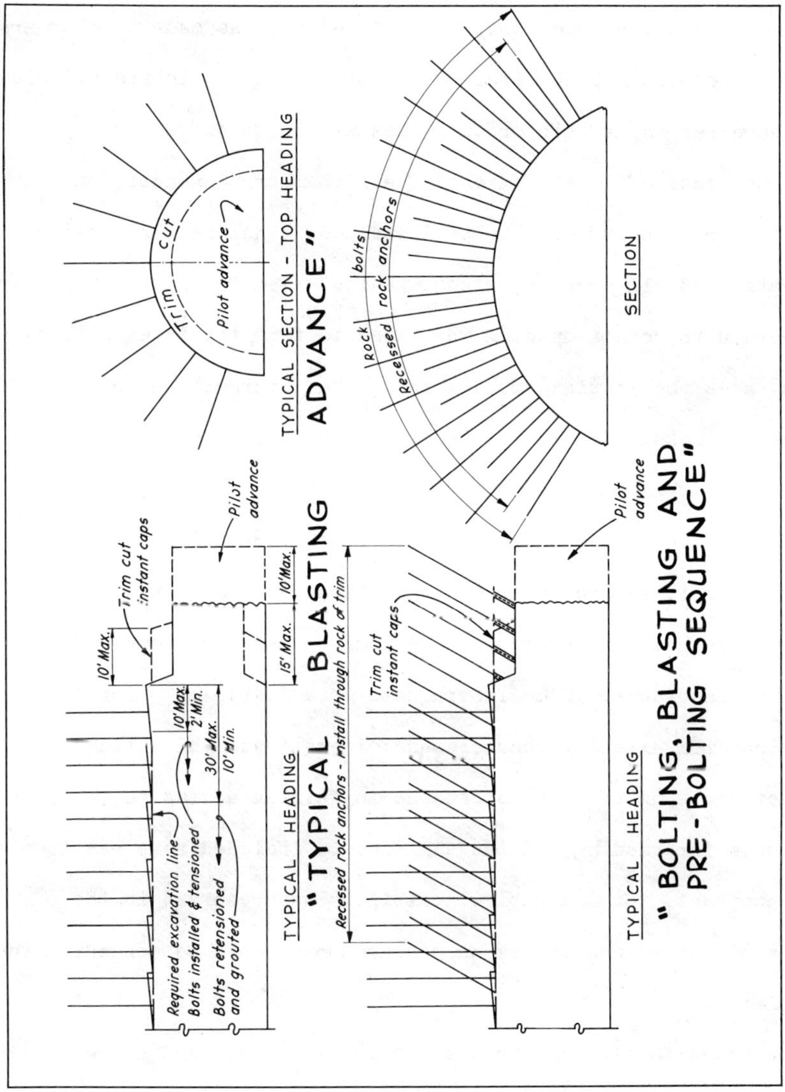

FIGURE 4. BLASTING AND PRE-BOLTING DETAILS

Three basic means of handling localized zones of rock not stabilized by the reinforcement shown on the drawings were included as pay items in the contract documents. These were additional bolts, additional prebolting and shotcrete. An effort was made to discourage the use of temporary stabilization measures such as timbers or safety bolts where permanent reinforcement was at all feasible.

In the case of intersections of all chambers, tunnels, and shafts, all rock bolts were to be installed, tensioned, retensioned, and grouted and all anchors were to be installed in the first opening mined before the other opening was mined to form the intersection. In some cases the initial round forming the intersection was limited to 5 feet or less. In all intersections an extra ring of bolts on 2 foot centers was included on the plans.

In addition to the means provided by the specifications for quality control, instrumentation consisting of single and multiple position bore hole extensometers and rock bolt deformeters was planned as quality assurance. Some instruments were installed prior to excavation from existing chambers and others installed during construction. Contractors assistance in gaining access to the instruments for reading and installation is included as a pay item in the contract. Considerable flexibility was reserved in the location of the extemsometers so measurements could be concentrated in problem areas.

Continued involvement in the job by geologists and design engineers provides guidance in the use of extra reinforcement, proper inspection, and interpretation of instrumentation data.

The instrumentation man and field geologist share the responsibility for the reading of instrumentation.

CONSTRUCTION PROGRESS

With excavation well underway at this time an assessment of the experience gained while using the specified procedures can be reported. The excavation accomplished thus far includes: (1) all major excavation on the upper level except raise boring the four 8 ft. diameter shafts, (2) more than 80 per cent of the circular air tunnels, (3) the crown area and a portion of the first bench in the cooling tower chamber, and (4) approximately 75 per cent of the crown area in the power plant chamber.

Rock reinforcement installation has been accomplished in the sequence specified with few problems. Prebolting has been exceptionally successful. Continued checks of mortar consistency are required to keep the mix stiff enough. With the use of the hydraulically positioned booms on self-propelled jumbos, 16 foot anchors are often installed in 4 minutes or less. Two major changes have been made in the rock bolting procedures. The Contractor is now grouting the bolts immediately after the first tensioning thus eliminating the need for retensioning. The second change is that in some areas torquing of the 1 inch rock bolts to 350 foot pounds has been substituted for direct pull hydraulic tensioning to 24 kips. In the two large chambers where direct pull tensioning was retained, the bolts are torqued to 350 foot pounds following the direct pull tensioning. This was the Contractor's preference over taking more care in snugging the nut prior to release of the direct pull device.

For the bolts where direct pull tensioning is not now required, the Contractor has agreed to a lower unit price on the bolts. There seems to be some psychological preference on the part of the miners for using brute force on a torque wrench rather than a hydraulic tensioning device. However, the hydraulic tensioning has proven satisfactory and feasible and provides a positive check of anchor capacity for every bolt. The question of relative costs of torquing versus pulling has not been resolved by this case history.

The use of the delayed trim smoothwall blasting procedure has been very successful. The only difficulty has been with trim hole alignment. The hydraulically positioned drills operated from the self-propelled jumbo require experienced operators with a good sense of direction. In some areas poor alignment of trim holes has resulted in over excavation. The flexibility of these jumbos for a multi-heading operation seems to justify their use, though trim hole alignment must be watched closely.

The response of the Contractor to requests for additional reinforcement and shotcrete has been immediate. At this time none of the instrumentation indicates instability. Inward movement of the rock surface has been approximately 0.1 inch or less in the crown areas. The only indicated area of possible concern is the intersection of the power plant chamber with the 32 foot wide access adit. Blast reports for each round, furnished by the Contractor, make better interpretation of the instrumentation data possible.

Figures 5 through 12 give an indication of the results to date and serve as a summary of results achieved through quality control on the project.

FIGURE 5. AIR INTAKE TUNNEL

This is a view back down the spiral portion of the tunnel. The 1 on 5 slope and continuous curve made trim hole alignment particularly difficult in this area.

FIGURE 6. CONSTRUCTION ADIT

This adit leads to the top heading of the cooling tower chamber. The rock surfaces are quite good here but some overdrilling can be seen in the upper right.

FIGURE 7. CONSTRUCTION ADIT

This photo illustrates a low angle intersection which was designed for muck removal. Careful treatment of such intersections has resulted in sound pillars. Note the shotcrete in the crown area to the right. This was applied prior to turning under on the left.

FIGURE 8. AIR INTAKE TUNNEL

This photo illustrates a straight section of this tunnel. The trim blasting was required on the entire periphery of the circular tunnel.

FIGURE 9. AIR INTAKE TUNNEL

In this photo the burden of the trim cut may be seen behind the main heading. As is the usual case, the bolting is carried much closer than the required 10 feet behind the trim blasting.

FIGURE 10. COOLING TOWER CHAMBER

This photo illustrates the crown of the cooling tower chamber. Three headings were used here. The plates on the bolts in the center heading were occasionally undermined when blasting the side headings. However, since the bolts were fully grouted their effectiveness was virtually undiminished. The rock in the left rib of this chamber is some of the poorest quality on the job and some shotcrete was used in this area.

FIGURE 11. COOLING TOWER CHAMBER

This photo shows the mobile jumbo drilling rock bolt holes.

FIGURE 12. EXHAUST VALVE CHAMBER

This photo illustrates well the results achieved through quality control. Note the drill casts, the bolt grouting operation and rings of perimeter bolts being installed prior to turning under both in the left crown and the right rib.

CONCLUSIONS

Excavation procedures and rock reinforcement procedures were specified clearly and in detail in the contract documents for the NCMC Expansion Project. The quality control inherent in the contract documents and supplemented by continuous field inspection is considered to be responsible for the excellent results achieved thus far. Instrumentation and continued observations by the field geologist have played invaluable and necessary roles in the excavation quality control.

APPENDIX - REFERENCES

1. UNDERWOOD, L. B., and DISTEFANO, C. J., "Development of a Rock Bolt System for Permanent Support at NORAD," Transactions of the Society of Mining Engineers, March 1967.

2. REED, J. J., "New Concepts of Rock Engineering," The Mines Magazine, March 1964, pp. 18-22.

STABILITY OF DISCONTINUOUS ROCK MASSES UNDER POLYAXIAL STRESS SYSTEMS

by

Peter B. Attewell[1] and John P. Woodman[2]

SUMMARY

The paper describes the concepts behind the computer processing and plotting of density data referred to the arbitrary orientations of planar discontinuities in rock masses. Orientation zones of potential instability based on a Coulomb-Navier-Mohr criterion of failure can be delimited on equal area projections for a preliminary assessment of the discontinuity stability status. Further processing utilizes the discontinuity information in the evaluation of a maximum and limiting shear strain energy criterion of instability. It is shown that potential failure, on the basis of this latter criterion, can be quite strongly dependent on both the value of the intermediate principal stress and on the orientation density distribution of the discontinuities.

---

[1]Reader in Engineering Geology

[2]Research Assistant in Engineering Geology

Both of University of Durham, Durham City, England

INTRODUCTION

It is generally accepted that the mechanical stability of rock masses is controlled more directly by the strength characteristics of large-scale structural discontinuities than by the intact strength of solid elements. There are many ways of investigating the stability status of such a rock mass as a discontinuum or even of assessing the potential instability of a single bounded unit within that mass; that is, the instability with respect to sliding and/or rotation. In certain situations, an empirical or semi-empirical two-dimensional treatment allied to rather simple stress concepts may suggest quite valuable criteria upon which combative measures against failure can be based. On the other hand, it may sometimes be necessary to study the problem three-dimensionally if the influence, for example, of a multiplicity of joint sets upon mass stability is to be realistically appraised. A basic requirement involves the geological mapping of discontinuities in terms of their spatial orientations, an estimation of principal stress amplitudes and directions so that the stresses can be resolved normal to and parallel to the discontinuities, and an assessment of the shear strength parameters with respect to each discontinuity. If all this information is available then it is possible to analyse for stability on the basis of standard Coulomb-Navier-Mohr theory (see Jaeger and Cook, 1968) provided that data acquired with respect to global space is transformed to principal stress space through a knowledge of the directions in which the principal stresses act. In fact such analyses can be performed manually, if albeit laboriously, using graph paper, compasses and a stereographic net, but since there is inevitably a range of uncertainty attached to each input parameter the magnitude of the work load becomes discouragingly heavy. Fortunately, it is possible to program the analysis and, as a preliminary exercise to a more detailed <u>in situ</u> engineering investigation, to explore the sensitivities of the input parameters with negligible effort. It is to a description of the program function and its application in terms of failure criteria that this paper is devoted. The early analysis has the added advantage of being useful also to the structural geologist who, faced with the tedium of contouring numerous poles to feature planes for the evaluation of concentration densities, would readily accept an automatic print-out of his data analysis. For this phase of the work reference has been made to Mellis (1942), Flinn (1958), Spencer and Clabaugh (1967), Lam (1969) but, in the event, the present approach proves to be somewhat different from that of Spencer and Clabaugh. The purpose of the present work is three-fold and can be summarized as follows:

a) to plot equal area projections of fabrics of any planar feature by entering directly the polar coordinates of the poles to those features determined by field measurement;

b) to superimpose on these projections and delimit those areas in space which represent potentially unstable regimes with respect to shear failure along joint or other discontinuities, the instability being based on a Coulomb-Navier criterion of failure;

c) to investigate quantitatively another criterion for mass instability based on a total limiting shear strain energy generated by incremental movements along planar discontinuities, the energy being a characteristic of the particular rock material and the program accommodating any arbitrary, non-isotropic distribution of discontinuities.

Thus, although these three aims may be regarded as being distinct, they are at the same time complementary, and since the program logic treats them as sequential operations on the data it is also useful to discuss them in this way.

FABRIC DENSITY REPRESENTATION

To be able to specify accurately the distribution of elements that composes a fabric, the total number of elements N' should be observed. For any finite number of samples $N < N'$ the probability distribution $P(\theta, \psi)$ for this subset of the population is a sum of Dirac delta functions. If θ and ψ are the polar coordinates (respectively latitude and azimuth) of the discontinuity normals, then

$$P(\theta, \psi) = \frac{1}{N} \sum_{i=1}^{N} \delta(\theta_i, \psi_i) \ldots\ldots\ldots\ldots\ldots\ldots (1)$$

where for integration over part of the sphere of projection:

$$\int_{\mathcal{S}} \delta(\theta_i, \psi_i) \sin\theta \, d\theta \, d\psi = 1 \text{ if the range of integration } \mathcal{S} \text{ includes } (\theta_i, \psi_i) \ldots\ldots (2)$$
$$= 0 \text{ otherwise.}$$

P obviously satisfies the normalization condition:

$$\int_{\text{Sphere}} P(\theta, \psi) \sin\theta \, d\theta \, d\psi = 1 \ldots\ldots\ldots\ldots\ldots\ldots (3)$$

where the integration is carried out over the whole sphere of projection.

In the situation where the N samples represent the whole population and are not merely part of it, equation (1) gives the correct probability density, but in the general case it is obvious that in order to obtain a better representation of the total population, it is necessary to smooth out the delta functions in some way. This can be achieved by allocating to each data point a small circular area on the sphere of projection centred on the projection of that point. If $\underline{\underline{a}}$ is the area chosen, then it is a simple matter to calculate the locus $C(\hat{\underline{r}})$ of the circle that bounds this area, vis:

$$\hat{\underline{r}}\cdot\hat{\underline{r}}_i = \cos\theta_a \quad \ldots\ldots\ldots\ldots\ldots\ldots\ldots\ldots(4)$$

where $\hat{\underline{r}}_i$ is the unit radius vector of the projected data point and θ_a is given by:

$$a = 2\pi\,(1 - \cos\theta_a) \quad \ldots\ldots\ldots\ldots\ldots\ldots\ldots\ldots(5)$$

We now define a function S for each data point such that

$$\begin{aligned} S(\hat{\underline{r}},\hat{\underline{r}}_i) &= w \text{ if } \hat{\underline{r}}\cdot\hat{\underline{r}}_i \geq \cos\theta_a \quad \ldots\ldots\ldots\ldots\ldots\ldots\ldots\ldots(6)\\ &= 0 \text{ if } \hat{\underline{r}}\cdot\hat{\underline{r}}_i < \cos\theta_a \end{aligned}$$

Therefore,

$$\int_{\mathcal{S}} S\, d\mathcal{S} \begin{cases} = wa & \text{if } \mathcal{S} \text{ wholly includes } \underline{\underline{a}} \\ = wfa & (0<f<1) \text{ if } \mathcal{S} \text{ partly includes } \underline{\underline{a}} \ldots\ldots(7) \\ = 0 & \text{if } \mathcal{S} \text{ excludes } \underline{\underline{a}} \end{cases}$$

if the $\frac{\delta i}{N}$ in equation (1) are now replaced by S_i, then

$$P(\theta,\psi) = \sum_{i=1}^{N} S(\theta,\psi;\theta_i,\psi_i) \ldots\ldots\ldots\ldots(8)$$

Therefore,

$$\int_{\text{Sphere}} P(\theta,\psi)\ \sin\theta\, d\theta\, d\psi = \sum_{i=1}^{N} \int S(\theta,\psi;\theta_i,\psi_i) \sin\theta\ d\theta\, d\psi = wNa \ldots\ldots\ldots\ldots\ldots\ldots(9)$$

It can be seen that there are two ways in which $\int P d\mathcal{S}$ can

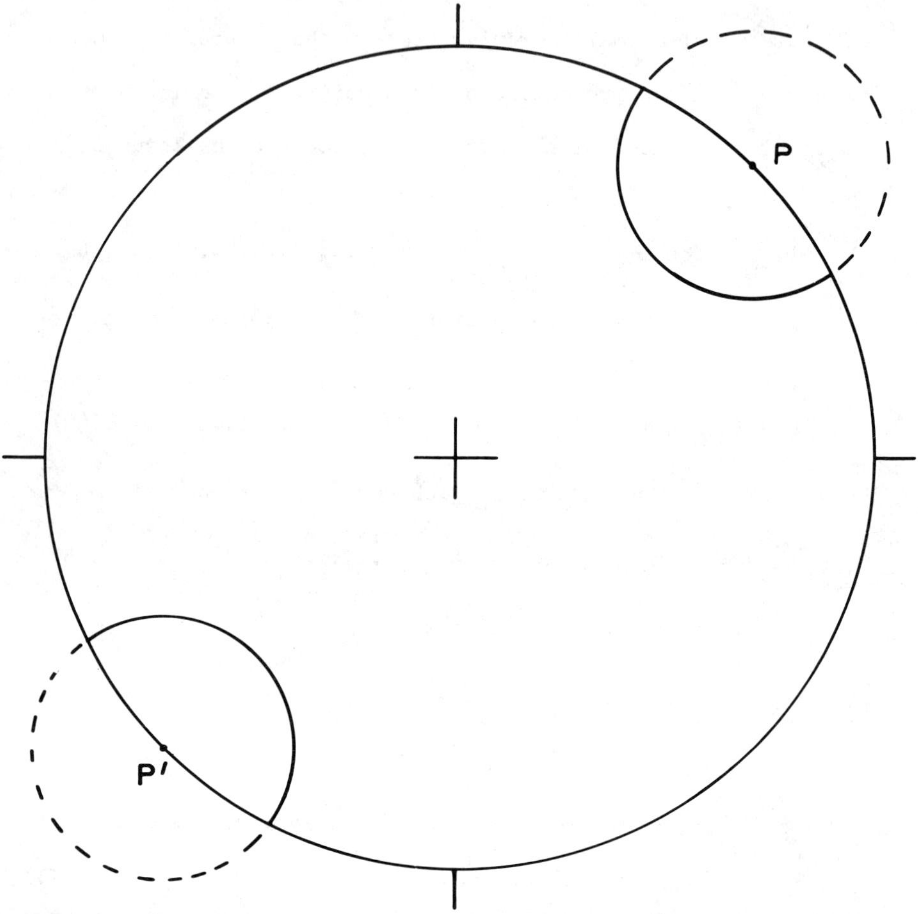

Figure 1. Conventional inversion of a pole P, at the boundary of the projection, to P' using a 10% area. Area plotted within major circle is only 9.0%.

be normalized:

(i) $a = \frac{1}{N}$ with $w = 1$

or, alternatively,

(ii) if $\underline{\underline{a}}$ is fixed, then S can be weighted by the factor $\frac{1}{Na}$.

$\underline{\underline{a}}$ is usually fixed by convention at 0.01 (1 percent area).

The choice of (i) or (ii) can have a considerable bearing on the final representation of the data for small N. In general, area normalization (i) produces better results for a nearly uniform distribution (projecting near-spherical fabric symmetry) than for a highly concentrated distribution and vice versa.

Quite clearly, $C(\underline{\hat{r}})$ does not project on to an equal area net as a circle, although its area $\underline{\underline{a}}$ is preserved. Consequently, the conventional method of plotting circles on the equal area projection itself - which is adopted as a hand operation and which also formed the basis of Spencer and Clabaugh's computer program for plotting fabric diagrams - leads to asymmetrical smoothing of data away from the centre of the projection (see Mellis, 1942; Flinn, 1958).

There is another consideration, not taken properly into account by this conventional method. Figure 1 shows how a data point P, which falls on the peripheral boundary of the projected hemisphere, is conventionally represented. P', the inversion of P through the centre of the sphere of projection, is calculated and a similar area is re-centred on this new point. In the simple case considered, P' is on the circle diametrically opposite to P. It can be seen that the sum of the two areas within the major circle is not equal to $\underline{\underline{a}}$, except at the limit

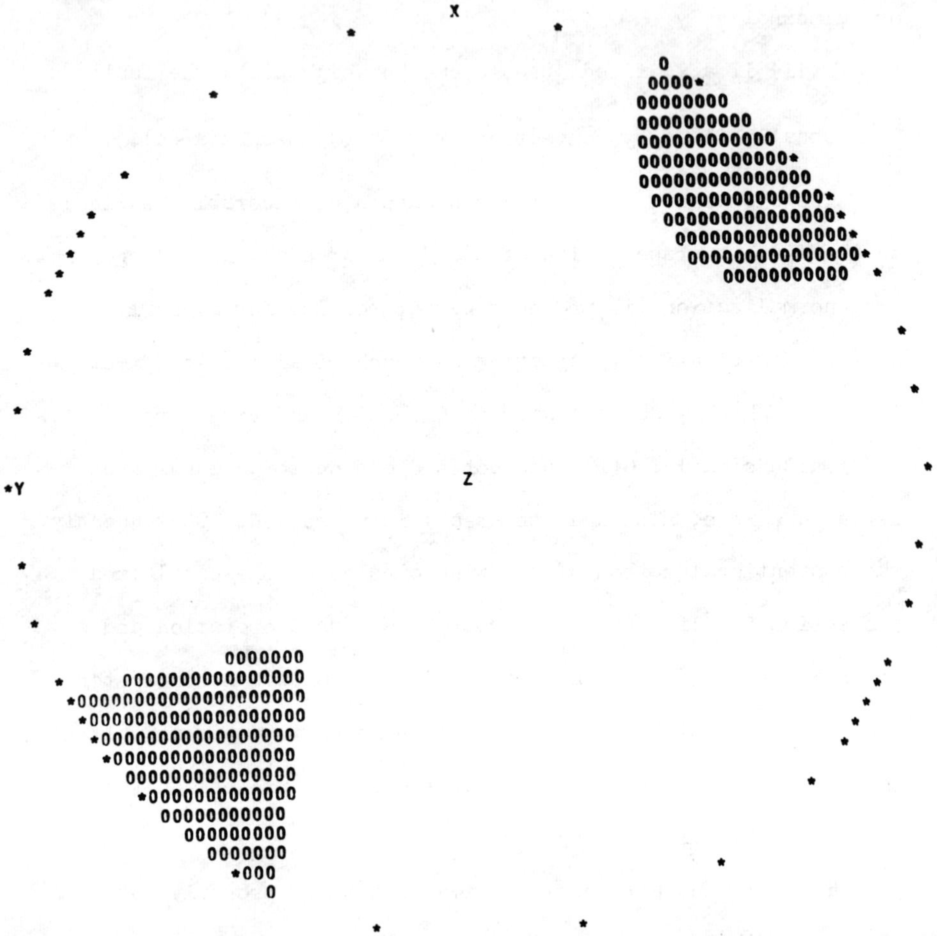

Figure 2. Modification of the conventional method, which does satisfy the normalization condition for probability density by correctly inverting the area that falls outside the major circle at P in Figure 1.

when $\underline{\underline{a}}$ is very small, and the underweighting consequent to this approach is considerable when area normalization ($a = \frac{1}{N}$) is used for small N. The correct technique is to calculate the inversion of the small circle centred on P, as in Figure 2, by which means no area is lost. But since in any case this involves appreciably more computation it was considered desirable to adopt the more rigorous approach discussed earlier.

Calculations have been performed using a matrix, the elements of which matched - on a one-to-one basis - the desired output format. The matrix element nearest to the data point was used to initiate a systematic outwardly-directed search for elements which satisfied the condition.

$$\underline{\hat{r}} \cdot \underline{\hat{r}}_i \geqslant \cos \theta_a \qquad \text{.........................(10)}$$

(see equation (4))

For each such element, the weighting constant $\underline{\underline{w}}$ was added to the value already held in that location. When no more elements could be found to satisfy equation (10), the search was terminated and the next data point examined. This search procedure has the advantage of requiring only a small fraction of the matrix to be checked for each point. When all of the S_i functions have been superimposed on the matrix, a corresponding output matrix holding character information can be generated and printed out (see Figure 3).

The PL1 program written for these and for the subsequent purposes to be described provides for data rotation to any new set of areas (such as, for example, from global axes to principal stress/strain ones) prior to plotting and the equal area projections can either be upper hemisphere, lower hemisphere or whole sphere projections. The latter option also includes inversion of zenith and nadir if required.

STABILITY REGIMES BASED ON THE COULOMB-NAVIER CRITERION

On the basis of this theory, shear movement takes place along a discontinuity - and therefore localized sliding leading to general instability develops - when the shear stress along it exceeds a limiting value given by the well-known functional relationship:

$$|\tau| = \mu\,(\sigma - u) + c \qquad \text{.........................(11)}$$

where μ and c are the shear strength parameters referred to the discontinuity (note that if, for example, there is a lower strength clay infilling a shear zone, μ, c for that material must be determined and entered*) and $(\sigma - u)$** is the resolved pressure (active normal pressure less water pressure) normal to a discontinuity.

Taking σ_1, σ_2, σ_3 $(\sigma_1 > \sigma_2 > \sigma_3)$ as the principal stresses directed along orthogonal axes x,y,z, and 1,m,n to be the direction cosines of the normals to the discontinuities with respect to the x,y,z axes, σ' and τ can then be calculated for every element on the projection through the standard equations:

$$\sigma = 1^2\sigma_1 + m^2\sigma_2 + n^2\sigma_3 \qquad \text{...................(12)}$$

$$\tau^2 = 1^2\sigma_1^2 + m^2\sigma_2^2 + n^2\sigma_3^2 - \sigma^2 \qquad \text{...................(13)}$$

---

* Quite apart from the need to estimate the <u>insitu</u> drainage facility for a cohesive infilling material, the question of whether to input peak or residual shear strength parameters also arises.

** $(\sigma - u) = \sigma'$, the effective stress.

```
DOLERITF
HEMISPHERE UPPER PROJECTION        20.00CM CIRCLE
N:   359   W:  1.000
1.00 TIMES NORMALISED FRACTIONAL AREA WINDOW USED
```

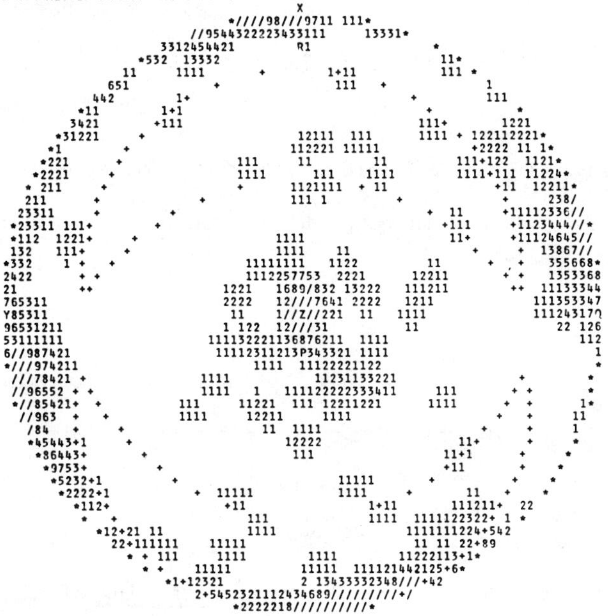

```
SIGMA1:     1.0000 ( -0.173648  0.000000  0.984808)
SIGMA2:     0.2000 ( -0.000000 -1.000000  0.000000)
SIGMA3:     0.1000 (  0.984808  0.000000  0.173648)
CFT OF FRICTION:     0.7600    COHESION:     0.0500
STRAIN ENERGY A:     G.4320    I:            9.4864
```

I	J	P
IC(-23,	-8)	=16.00
IC(-23,	-7)	=19.00
IC(-23,	-6)	=22.00
IC(-23,	-5)	=24.00
IC(-23,	-4)	=19.00
IC(-23,	-3)	=19.00
IC(-23,	-2)	=18.00
IC(-23,	-1)	=17.00
IC(-23,	0)	=13.00
IC(-23,	1)	=10.00
IC(-22,	-14)	=14.00
IC(-22,	-13)	=16.00
IC(-22,	-12)	=15.00
IC(-22,	-11)	=13.00
IC(-22,	-10)	=11.00
IC(-22,	-9)	=14.00
IC(-22,	-8)	=12.00
IC(-22,	-7)	=12.00
IC(-22,	-6)	=13.00
IC(-22,	-5)	=13.00
IC(-22,	-4)	=12.00
IC(-21,	-15)	=10.00
IC(-21,	-14)	=11.00
IC(-21,	-13)	=10.00
IC(-9,	36)	=11.00
IC(-8,	36)	=14.00
IC(-8,	37)	=15.00
IC(-7,	36)	=13.00
IC(-7,	37)	=24.00
IC(-6,	37)	=14.00
IC(-6,	38)	=14.00
IC(-5,	36)	=12.00
IC(-5,	37)	=17.00
IC(-5,	38)	=19.00
IC(-4,	36)	=12.00
IC(-4,	37)	=13.00
IC(-4,	38)	=15.00
IC(-3,	37)	=12.00
IC(-3,	38)	=11.00
IC(-1,	-1)	=17.00
IC(-1,	0)	=16.00
IC(-1,	1)	=16.00
IC(0,	-2)	=14.00
IC(0,	-1)	=14.00
IC(0,	0)	=13.00
IC(0,	1)	=14.00
IC(0,	2)	=12.00
IC(0,	39)	=11.00
IC(1,	-1)	=19.00
IC(1,	0)	=16.00
IC(1,	1)	=15.00
IC(2,	-1)	=11.00
IC(5,	-38)	=17.00
IC(5,	-37)	=11.00
IC(6,	-38)	=16.00
IC(6,	-37)	=11.00
IC(7,	-37)	=16.00
IC(7,	-36)	=10.00
IC(8,	-37)	=14.00
IC(8,	-36)	=13.00
IC(9,	-36)	=10.00
IC(22,	13)	=10.00
IC(22,	14)	=13.00
IC(23,	0)	=11.00
IC(23,	1)	=12.00
IC(23,	2)	=13.00
IC(23,	5)	=11.00
IC(23,	6)	=11.00
IC(23,	7)	=14.00
IC(23,	8)	=13.00

Figure 3. Equal area projection of P, probability density for poles to joint planes in a dolerite rock derived from 359 samples. The probability of a pole falling within a particular fractional area ΔA of the projection is given by the mean value of P over that area times ΔA. X(north), Y (west) and Z (vertical) are global axes and the characters P, Q, R denote σ_1, σ_2, σ_3 axes respectively. Direction cosines 1,m,n of the principal stress axes are bracketed after the stress amplitudes. Potential unstable space is delimited by + characters. Probability density values greater than 9 are shown by the / character and an overflow table with referenced locations is printed out.

As shown in Figure 3, the locus of $|\tau| = \mu\sigma' + c$ is printed on the projection and the data then examined visually to determine the extent of the density distribution falling within the unstable regime.

It might be argued that there is a case for retaining and printing the raw input data with respect to discrete pole orientations and this facility is accommodated in the processing.* The existence of a single pole within the fail area of the projection would then imply shear failure not only for that particular discontinuity but probably for the rock mass in general. Situations of single block instability, such as might arise in connexion with rock slopes, could be appropriately analysed on this basis, but in the general context of mass instability and <u>sampling</u> of discontinuity orientations, a statistical picture of orientation density distributions must be presented. Tolerable and intolerable distribution levels on overlap into the unstable regime then become a matter of engineering and geological judgement, but the decision-making process can be eased by making use of the rapid computational facility to increment the variables σ_1, σ_2, σ_3, their polar coordinates, c and μ and so assess their individual sensitivities with respect to the stability picture. Such a procedure is justified since the amplitudes of these variables will not be known with any degree of precision.

The concept of a stability regime in principal stress space embodies the notion of discontinuity movement over a <u>range</u> of orientations determined by the amplitudes of the measured variables.

---

*Note that this can equally well be achieved by condensing the $\underline{\underline{a_i}}$

Unfortunately, the Coulomb-Navier-Mohr criterion tells us nothing about the relationship of this movement to the stresses resolved normal and parallel to its direction, but it might be felt that any mass instability is a function more of an integrated movement over a range of discontinuity orientations than of limiting τ, σ'. This concept must be examined in terms of the critical strain energy acquired by the rock mass.

STRAIN ENERGY FAILURE CRITERION

It is assumed* that failure occurs in a stressed material when the total strain energy generated by sliding between opposing surfaces of closed discontinuities exceeds some characteristic value.

If, as before, μ is the coefficient of friction between the opposing surfaces of a discontinuity, sliding takes place when

$$|\tau| - \mu\sigma' > 0 \qquad \text{.........................(14)}$$

$(|\tau| - \mu\sigma')$ is defined as the shear stress available for the promotion of sliding, τ'. When $\tau' > 0$ for a particular discontinuity, sliding occurs to induce a corresponding shear strain energy:

$$W_d = sl^3 \tau'^2 \qquad \text{.........................(15)}$$

where $\underline{\underline{s}}$ is a function of the material, the shape of the discontinuity, and its position and orientation relative to other discontinuities, and $\underline{\underline{l}}$ is a dimension of the discontinuity. Hence,

---

*Wiebols and Cook (1968) used this criterion to examine intrinsic rock failure due to shear movement along Griffith cracks having a random orientation density distribution. In this present paper, the criterion is applied to 'large cracks' having a non-random distribution compatible with the usual field situation of joint sets.

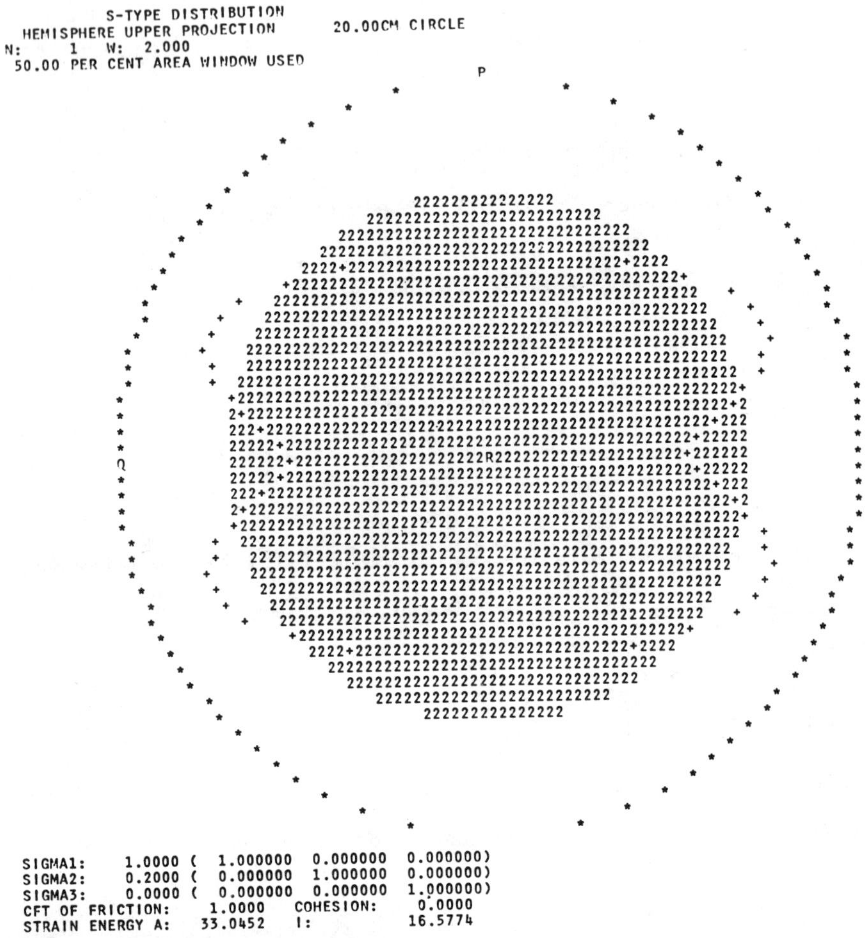

Figure 4. Equal area projection for one of S_3 series (a = 0.5, w = 2) showing unstable area between σ_3 axis and $|\tau| = \mu\sigma'$ curve. Symbols P, Q, R denote respectively σ_1, σ_2, σ_3 axes.

the total shear strain energy due to movement along N' surfaces is:

$$W = N' \int_{\tau' > 0} Psl^3 \ \tau'^2 \ d\mathcal{S} \ ds \ dl \qquad \ldots\ldots\ldots\ldots\ldots\ldots(16)$$

For N' discontinuities uniformly distributed in size and spatially, but not necessarily directionally, equation (16) becomes:

$$W = \bar{k} N' \int_{\tau' > 0} P \tau'^2 \ d\mathcal{S} \qquad \ldots\ldots\ldots\ldots\ldots\ldots(17)$$

where $\bar{k}$ is some mean factor.

Substitution for P from equation (1) leads to:

$$W = \bar{k} \frac{N'}{N} \sum_{\substack{i=1 \\ \tau'_i > 0}}^{N} \tau'^2_i \qquad \ldots\ldots\ldots\ldots\ldots\ldots(18)$$

where $\tau'_i = \tau'$ with respect to the i th data point.

Substitution from equation (8) leads to:

$$W = \bar{k} N' \sum_{i=1}^{N} \int_{\tau' > 0} S \tau'^2 \ d\mathcal{S} \qquad \ldots\ldots\ldots\ldots\ldots\ldots(19)$$

$$= w\bar{k} N' \sum_{i=1}^{N} \int_{\substack{a_i \\ \tau' > 0}} \tau'^2 \ d\mathcal{S} \qquad \ldots\ldots\ldots\ldots\ldots\ldots(20)$$

(the integral being taken over the area a_i representing the ith point).

W is computed according to equation (20). For comparison purposes, W is also computed for an equivalent isotropic distribution (a probability density of unity):

$$W = \bar{k} N' \int_{\tau' > 0} \tau'^2 \ d\mathcal{S} \qquad \ldots\ldots\ldots\ldots\ldots\ldots(21)$$

The effect on W of the detailed form of P can be demonstrated by choosing just two particular series of S-type distributions:

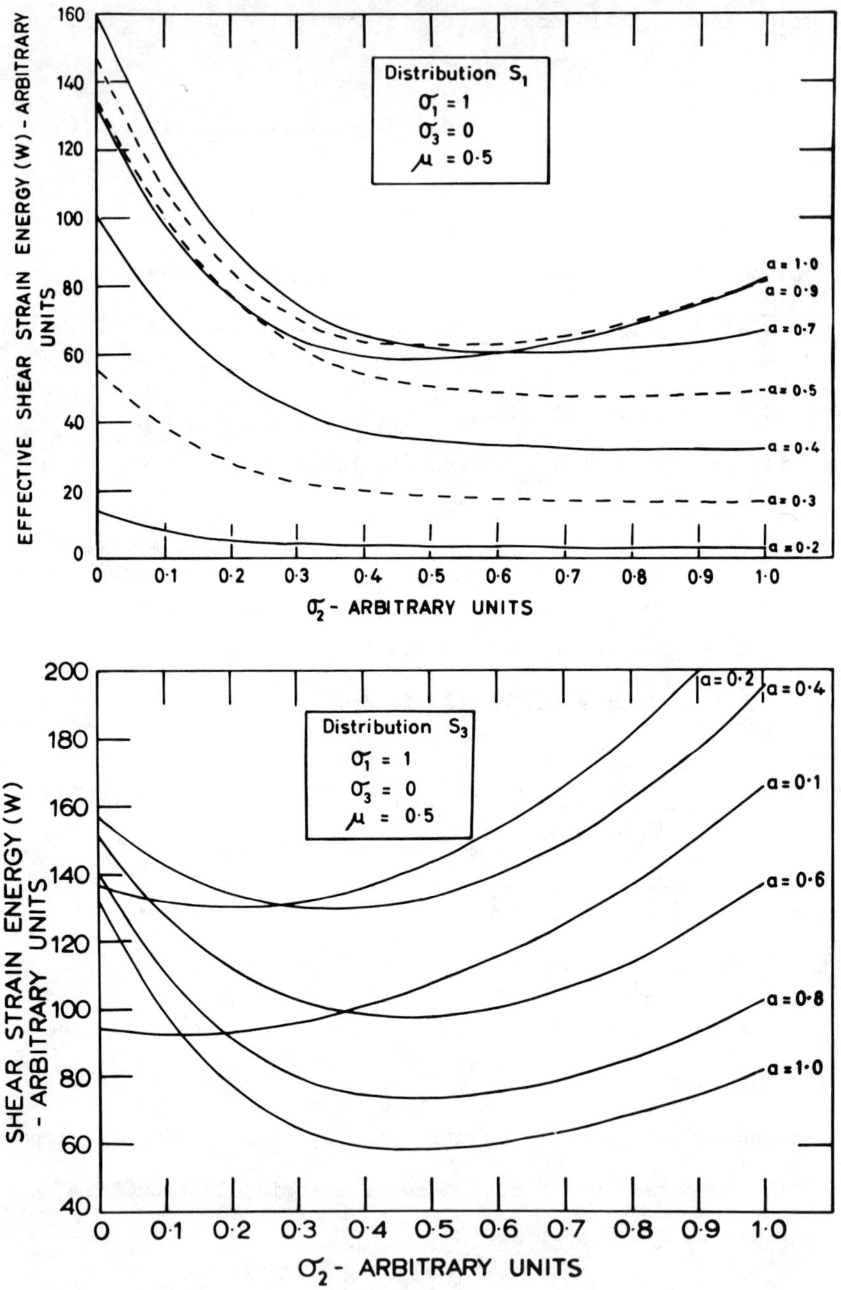

Figure 5. Dependence of shear strain energy on the intermediate principal stress for both S_1 and S_3 - type distributions of varying anisotropy when $\sigma_1 = 1$, $\sigma_3 = 0$, $\mu = 0.5$.

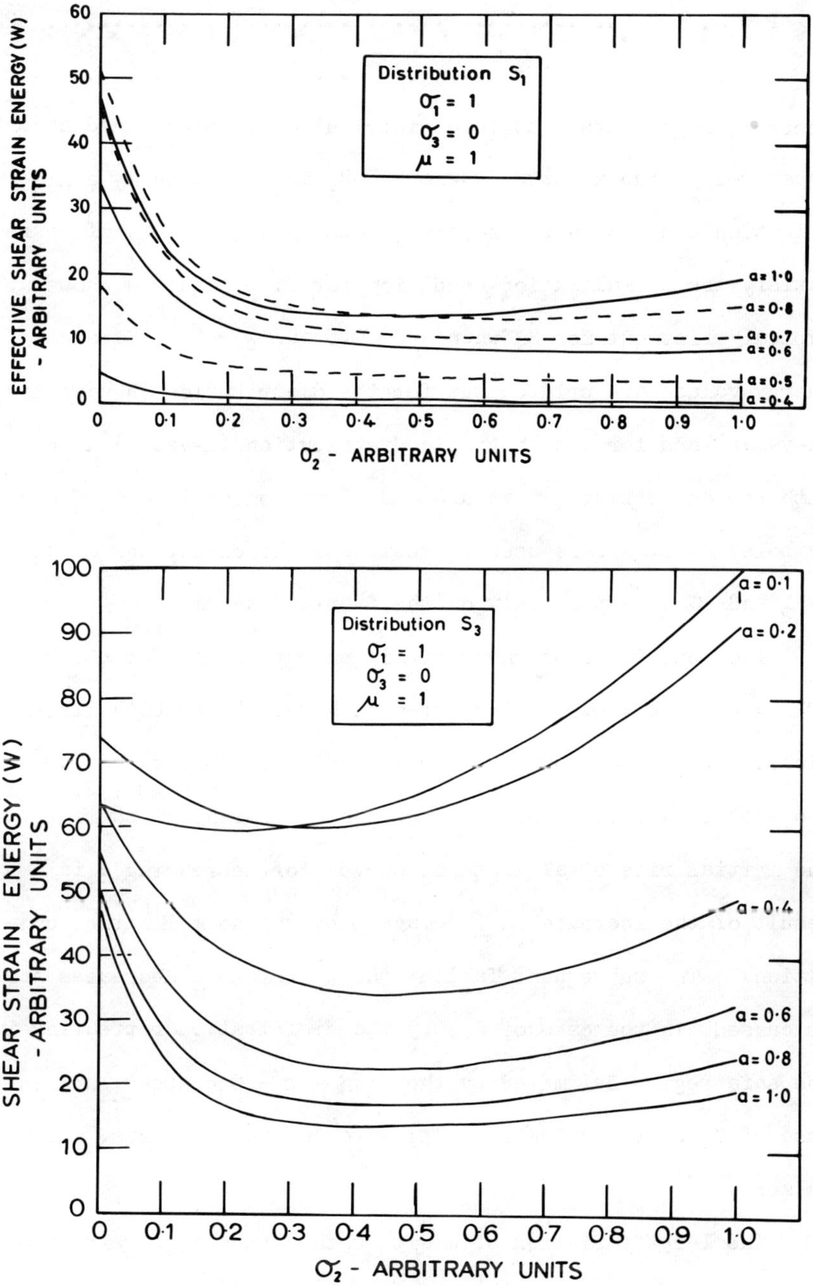

Figure 6. Dependence of shear strain energy on the intermediate principal stress for both S_1 and S_3 - type distributions of varying anisotropy when $\sigma_1 = 1$, $\sigma_3 = 0$, $\mu = 1.0$.

$$P = S_1 \quad \text{.......................................} \quad (22)$$

and $$P = S_3 \quad \text{.......................................} \quad (23)$$

where S_1 represents a uniform distribution of area $\underline{\underline{a}}$ and density $\underline{\underline{w}}$ centred on the x- or σ_1 -axis and S_3 is equivalently a distribution centred on the z- or σ_3 -axis. S_1 and S_3, of course, satisfy the normalization condition for P. Figure 4 shows a typical print-out for S_3 when $\underline{\underline{a}} = 0.5$ and $\underline{\underline{w}} = 2$. Strain energy values are printed out for the distribution in question (A-value) and for the isotropic distribution (I-value) along with the appropriate σ values and direction cosines. The symbols P, Q, R are used to designate the directions of σ_1, σ_2 and σ_3 respectively on the fabric diagram.

The dependence of shear strain energy on σ_2 for $\sigma_1 = 1$, $\sigma_3 = 0$ and $\mu = 0.5$ is indicated in Figure 5 for both distributions. Figure 6 shows the corresponding cases for $\mu = 1.0$. Two matters arising from these graphs are worthy of comment. The initial rise of all sets of curves for decreasing $\underline{\underline{a}}$ is a result of the increase in $\underline{\underline{w}}$ demanded by the normalization condition. The subsequent fall of the curves as $\underline{\underline{a}}$ decreases further is caused, in the case of S_1, by the distribution retreating into the safe region delimited by the $|\tau| - \mu\sigma' = 0$ contour, and in the case of S_3 by the value of $|\tau| - \mu\sigma'$ on the σ_3-axis, which is zero.

The large variation of W with both $\underline{\underline{a}}$ (the areal extent of the distribution) and σ_2 carries considerable implications for, if it is supposed that the maximum strain energy the material can withstand is an intrinsic property of that material, and is represented on graphs such as Figs. 5 and 6 by a horizontal line of appropriate height, then the mass strength properties must be highly dependent both on the type of discontinuity distribution and on the relative values of the principal stresses.

APPENDIX 1. - REFERENCES

Flinn, D., "On tests of significance of preferred orientation in three-dimensional fabric diagrams", Jour. Geol. Vol.66, 1958, pp.526-539.

Jaeger, J.C. and Cook, N.G.W., Fundamentals of Rock Mechanics, Methuen, London, 1968, 513p.

Lam, P.W.H., "Computer method for plotting beta diagrams", Am. Jour. Sci., Vol.267, 1969, pp. 1114-1117.

Mellis, O., "Gefügediagramme in stereographischer Projecktion", Zeitschr. Min. Pet. Mitt., Vol. 53, 1942, pp.330-353.

Spencer, A.B. and Clabaugh, P.S., "Computer program for fabric diagrams", Am. Jour. Sci., Vol.265, 1967, pp. 166-172.

Weibols, G.A. and Cook, N.G.W., "An energy criterion for the strength of rock in polyaxial compression", Int. J. Rock Mech. Min. Sci., Vol.5, 1968, pp.529-549.

STRAIN AND TILT OBSERVATIONS NEAR AN ACTIVE FAULT

by Charles G. Bufe[1]

INTRODUCTION

The Stone Canyon Observatory of NOAA's Earthquake Mechanism Laboratory is located in the San Andreas Fault Zone southeast of San Francisco (see inset, Figure 1). The active fault trace shown in the figure is the site of slow fault-creep at the rate of about a centimeter per year. This trace is straddled by extensometers (creep meters) and arrays of downhole geophones and electrodes. Section P-P' crosses the fault and the local topographic trend nearly at right angles. There is no vertical exaggeration in the section. The active fault trace is located in the hillside sag. The strain and tiltmeter site is in fractured Tertiary sediments, 100 meters southwest of the fault and at the highest point of section P-P'.

INSTRUMENTATION

The strainmeter consists of three 30-meter extensometers, oriented north-south, east-west, and northwest-southeast. The length reference consists of a system of 3-meter quartz rods suspended end to end and coupled by folded aluminum sleeves to achieve thermal compensation. One end rod is set in a concrete pier. The other end rod projects into an "end house" or vault where a displacement transducer is used to measure ground expansion or contraction.

[1]Geophysicist, Earthquake Mechanism Laboratory, National Oceanic and Atmospheric Administration, San Francisco, Calif.

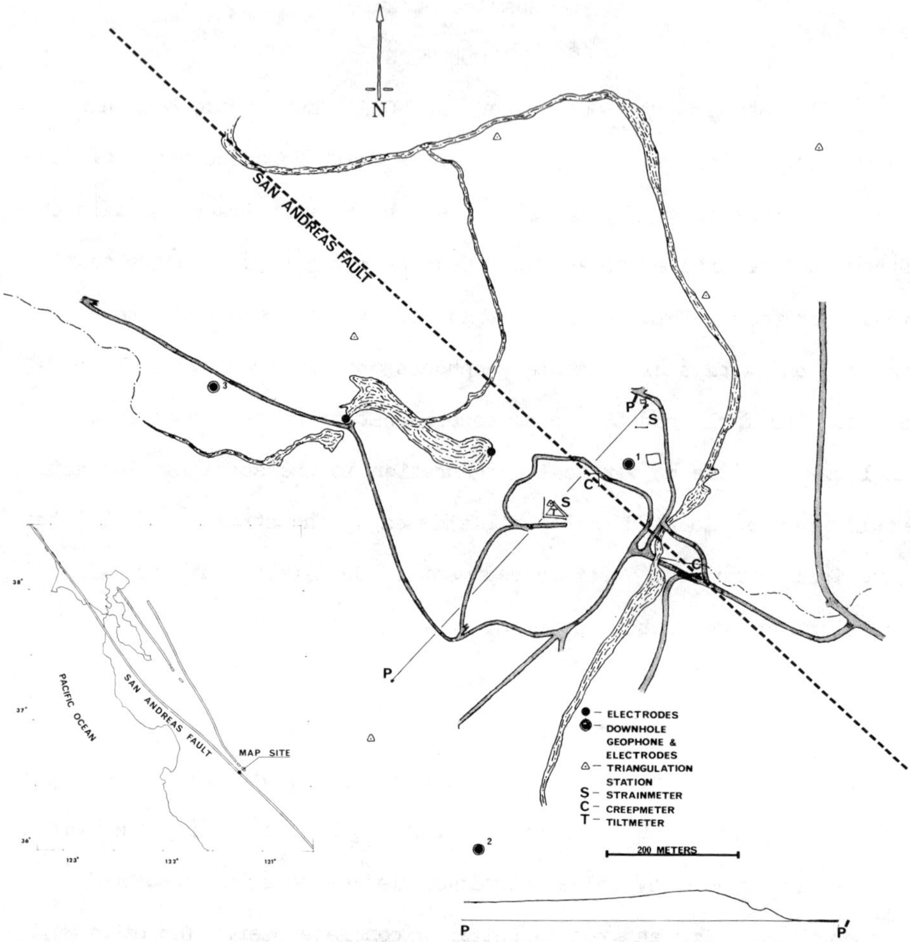

FIGURE 1. LOCATION AND LAYOUT OF STONE CANYON OBSERVATORY

The tiltmeter consists of an interconnected 20-meter triangular array of pools of mercury set in piers attached to invar rods driven to resistance. Weekly micrometer readings of elevation changes relative to the mercury level are taken and used to compute tilt vectors. Tilt and strainmeter occupy essentially the same site and are buried 2 to 3 meters beneath the surface.

OBSERVATIONS

Because of insufficient transducer range and a strain rate much higher than expected, the strain data at Stone Canyon were intermittent until mid 1970. The data were sufficient, however, to indicate that large seasonal areal strain changes were occurring. The strain data appeared to approximate a sine curve which was about 90° (or 3 months) out of phase with air temperature observations. The areal strain data of Figure 2 for the period March 1970 through March 1971, however, show an interesting departure from the sine curve pattern. The episode of rapid areal contraction beginning at the end of November was apparently triggered by a 3-inch deluge in a single day (November 30) at the beginning of the rainy season.

The 3 strain-change ellipses of Figure 2 represent deformation of the circle shown at the left of the figure. Apparent ellipticity was produced by introducing a deformational exaggeration of 10^4. Size of ellipse was adjusted for a 10^4 exaggeration in areal strain. The ellipses trend northwesterly, roughly parallel to the fault and to the local topographic trend.

Tilt observations, made weekly at Stone Canyon since 1967, are presented in vector form in Figure 3. Arrow heads are placed at the

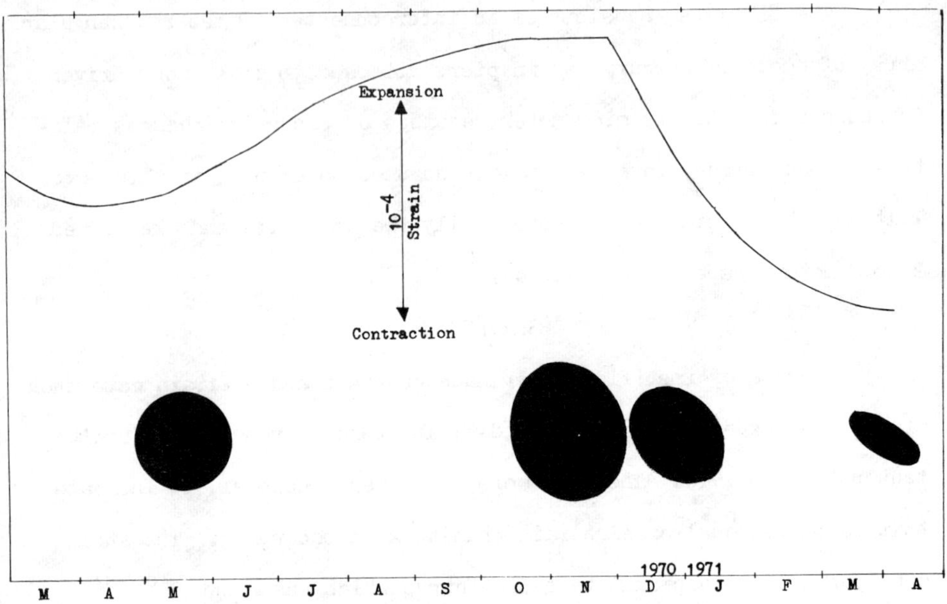

FIGURE 2. AREAL STRAIN AT STONE CANYON, MARCH 1970 TO APRIL 1971

beginning of each month and numbered accordingly. Net tilt at any time may be determined by drawing a vector from the point of origin (3-67). Two patterns are evident upon inspection of Figure 3:

(1) Maximum tilt rates occur during the summer months; little tilt occurs during the winter.

(2) While secular tilts are large, there seems to be a tendency to return to a previous datum.

This tendency is most obvious in the large tilt "event" which occurred at the beginning of August, 1970, reversing a tilt trend of several months duration. This rapid tilt happened during a period of high tectonic activity in the region around Stone Canyon. Rapid fault creep was observed northwest of the site in late July

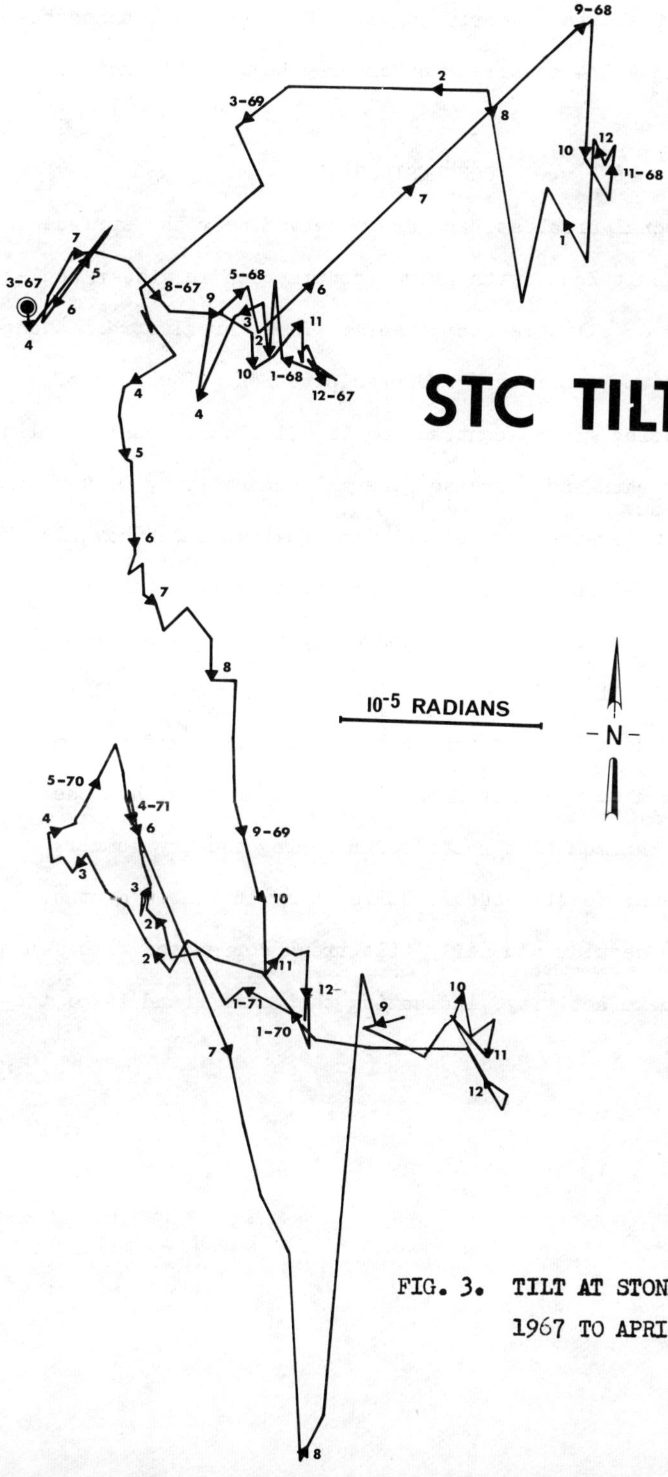

FIG. 3. TILT AT STONE CANYON, MARCH 1967 TO APRIL 1971

and southwest of the site in early August. On August 4th an earthquake of magnitude 4 3/4 occurred in Monterey Bay, 75 kilometers west of Stone Canyon.

CONCLUSIONS

Large seasonal areal strains are observed near the surface in the San Andreas Fault Zone, with ground contraction in winter and expansion in summer on all three components. These areal strain changes appear to be tied to temperature and precipitation. The role of rainfall in producing ground contraction is not clear. Rapid cooling of the surface by rain and increased thermal conductivity of wet soil may be factors. Water-pressure effects on fault surfaces may also be a consideration. Sliding within the strainmeter does not seem to be a factor, since the strains are apparently reversible, although not necessarily elastic. The orientation of strain change ellipses appears to relate to the local topographic, geologic, and tectonic regimes. Sliding also does not appear to be occuring within the tiltmeter, since maximum tilt rates occur during the dry summer months and little or no tilt occurs during the wet winter months. A reversal of the secular southerly tilt trend correlates with fault creep and earthquake activity, indicating that this trend is of tectonic origin.

HIGH PRESSURE EXPERIMENTS ON SOIL AND ROCK

By Donald J. Murphy Jr.[1]

INTRODUCTION

A number of recent experiments, relative to the behavior of soil and rock at octahedral normal stress levels as high as 80,000 psi have indicated that behavioral characteristics of the two materials can be distinctly similar. Further, it appears that the strength and compressibility of soil, as well as the brittle or ductile nature of rock depend not only on stress level, but also on mineralogical composition.

PURPOSE

The primary purpose of this investigation was to study the influence of mineralogical composition and stress level on the behavior of monomineral granular assemblies and of the parent rocks from which they were obtained. Initial laboratory testing led to the evolution of a secondary goal, examination of the role of petrofabric and microstructure in rock deformation.

---

[1]Associate, Joseph S. Ward & Associates, Consulting Engineers, Caldwell, New Jersey, formerly Graduate Fellow, Duke University, Durham, North Carolina.

EXPERIMENTAL PROGRAM

The experimental phase of the study consisted of approximately sixty tests including isotropic and triaxial compression tests on soil and rock and unaxial compression tests on rock and on soil samples that became coherent during isotropic compression. The isotropic compression tests were performed in order to isolate the effects of particle degradation during shear from the effects of crushing during consolidation. In addition, they were used to obtain preconsolidated samples for the study of possible relationships between degree of induration and preconsolidation pressure. The triaxial compression tests formed the basis of comparison between soil and rock behavior. The uniaxial tests served as indices of coherence developed by soil samples during isotropic compression and allowed for evaluation of "cohesion" intercepts of rock samples. In addition, the uniaxial and triaxial tests on rock were used to illustrate the influence of stress level on the nature of rock deformation.

MATERIALS

A fourfold criterion was used in selecting materials for the study. Primarily, representation of a broad range of properties, particularly crushing strength or hardness, was sought. Availability, specifically in a size convenient for both coring and grinding, was considered.

Parent rocks of essentially pure mineralogical composition were preferred. In addition, practical considerations concerning the extent of testing required, dictated selection of a limited number of different materials. The group of minerals, quartz (hardness M=7), feldspar (M=6), calcite (M=3) and chlorite (M=1) appeared to satisfy the criterion. These were obtained in bulk form from the sources indicated in Table 1. Samples of these minerals were prepared either by means of a bench mounted core barrel or by means of laboratory jaw crushers and ball mill grinders.

TEST EQUIPMENT

Two different equipment systems were used during this investigation. The first, a 5000 psi capacity system, is shown in Figure 1 and has been described in detail by Tai (1)[2]. The second, a 100,000 psi system is shown in Figures 2, 3, and 4. Details pertinent to this equipment, as well as specifics concerning sample preparation and testing techniques are given elsewhere (2).

TYPICAL TEST RESULTS

Isotropic and Triaxial Compression Tests on Soil - Data from drained triaxial tests conducted on dry, 1.40 inch diameter by 3.00 inch high soil samples indicate that

---

[2]Numerals in parentheses refer to corresponding items in the Appendix - References.

TABLE 1. MATERIAL SOURCES

Mineral	Source
Quartz (milky)	Grafton, New Hampshire
Feldspar (microcline)	Parry Sound, Ontario
Calcite	Cherokee County, Kansas
Chlorite	Ipsheming, Michigan

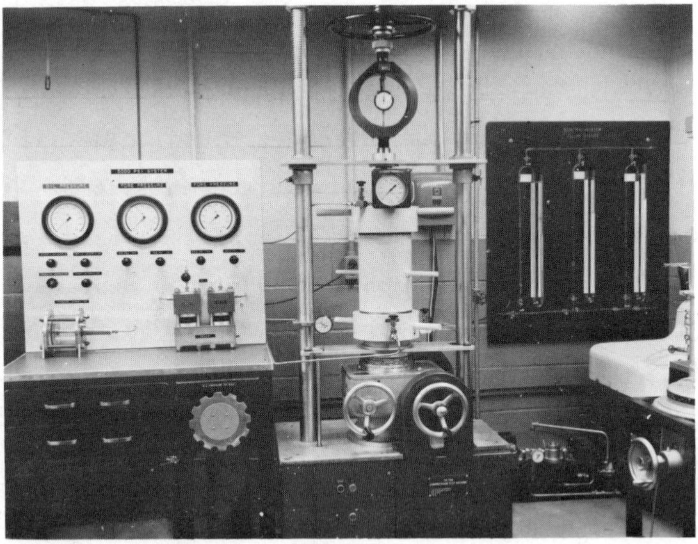

FIGURE 1 5000 PSI SYSTEM

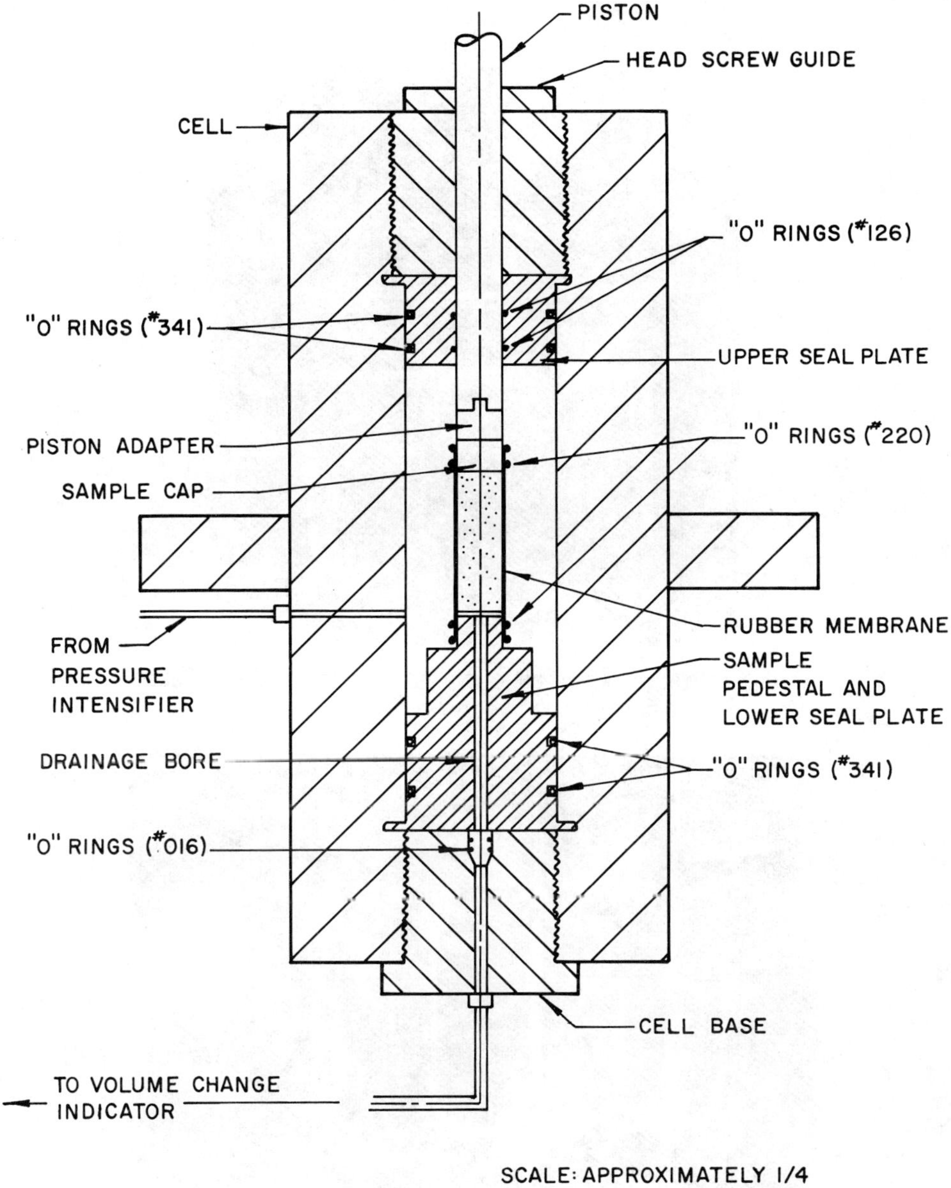

FIGURE 2 SECTION VIEW - 100,000 PSI CELL

FIGURE 3 100,000 PSI CELL AND PRESSURIZATION EQUIPMENT

FIGURE 4 400,000 LB. PRESS FOR 100,000 PSI SYSTEM

a more than casual relationship exists between mineralogical hardness and mechanical behavior. Specifically, the strength and volumetric strain occurring during shear appear to increase and the true volumetric strain occurring during consolidation appears to decrease with increasing hardness, as shown in Figures 5, 6, and 7. Tabor (3) has shown that a unit increase in Mohs hardness corresponds to a 60 percent increase in indentation hardness. Since both crushing strength and indentation hardness are functions of molecular structure, it appeared that compressibility at high stresses (being due essentially to particle degradation) should be related to hardness. Figure 8 shows this relationship. True volumetric strain is plotted against Mohs hardness such that, in arithmetic scale, each hardness abscissa is 60 percent greater than the one preceeding it. As shown, the relationship is linear for both true volumetric strain occurring during consolidation and that occurring during shear. A similar plot, but on logarithmic scale, relating principal stress ratio at failure to Mohs hardness is shown in Figure 9. Again, the relationship is linear.

Triaxial Compression Tests on Rock - Typical results from 45,000 psi tests conducted on 1.20 inch diameter, 2.80 inch high rock cores are shown in Figure 10. Based on a comparison of this data and that presented in Figure 5 for tests on soil, it appears that for all samples tested the initial tangent modulus of deformation was, as expected, higher for rock than for soil. In terms of principal stress

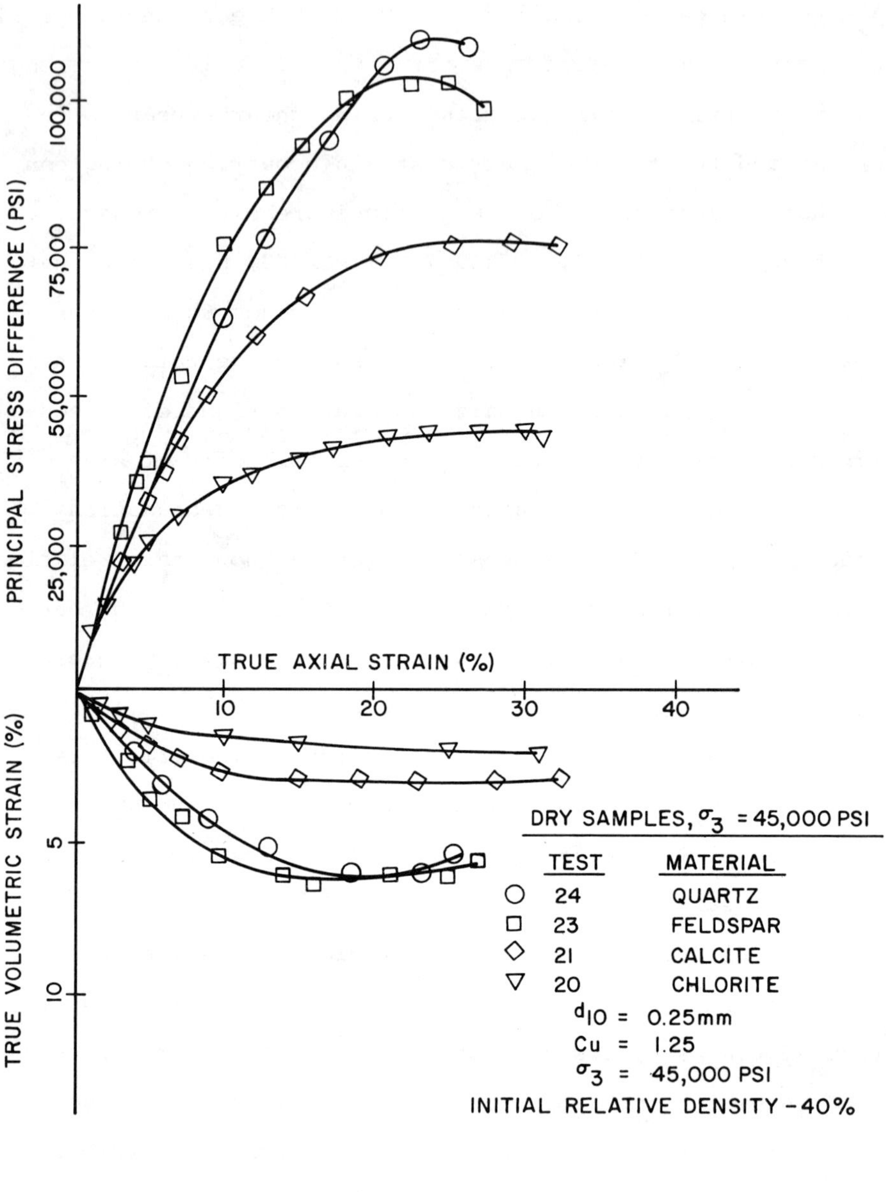

FIGURE 5 TRIAXIAL COMPRESSION TESTS ON SOIL

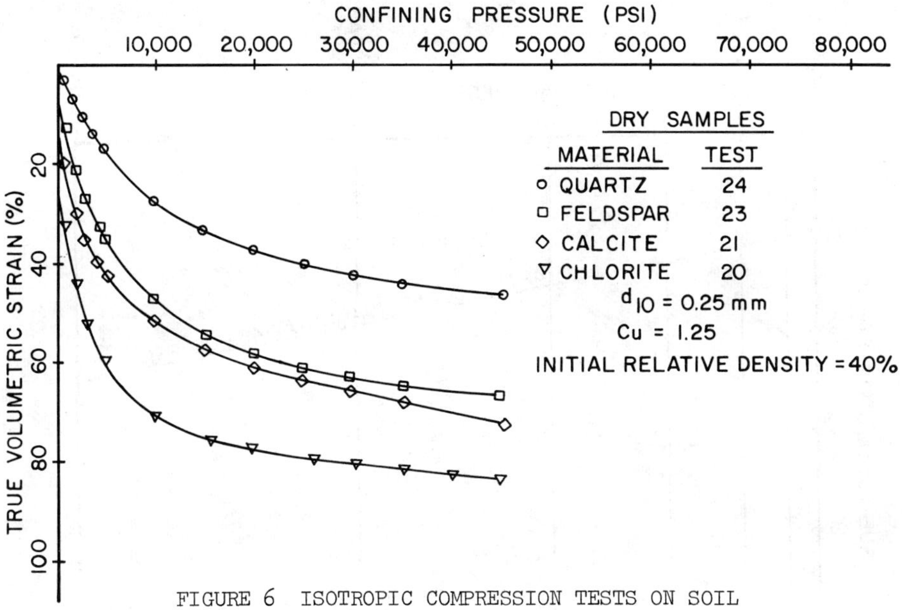

FIGURE 6 ISOTROPIC COMPRESSION TESTS ON SOIL

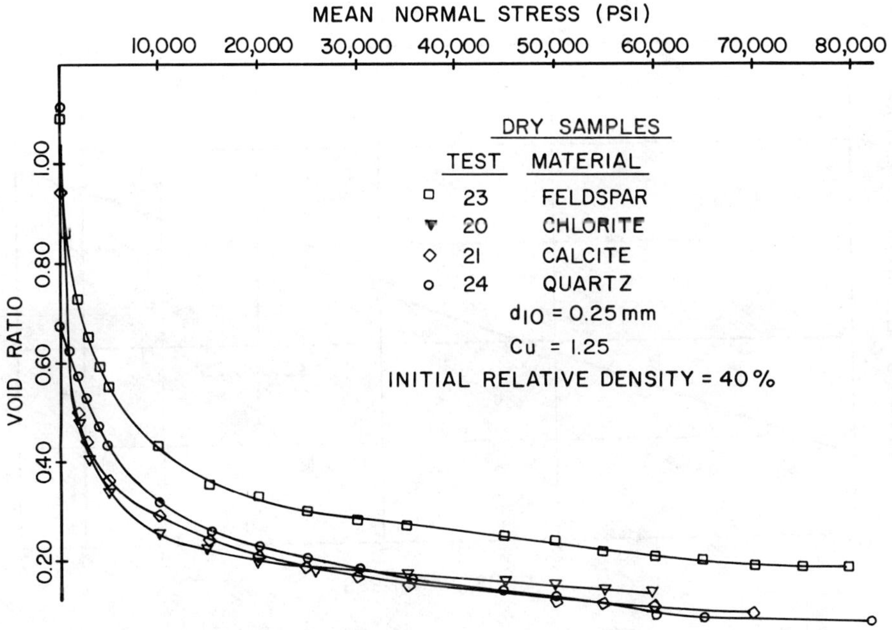

FIGURE 7 VOID RATIO VS. MEAN NORMAL STRESS

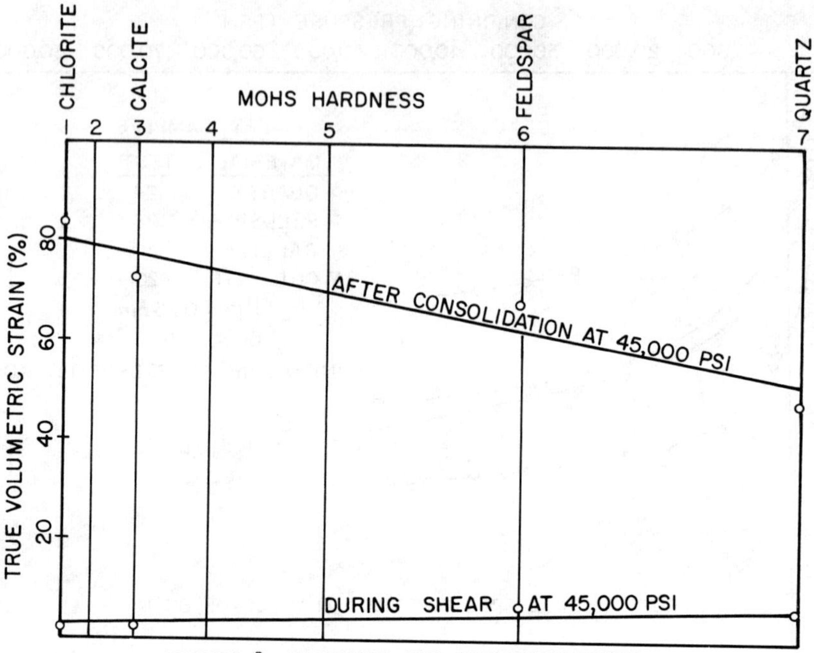

FIGURE 8 HARDNESS AND COMPRESSIBILITY

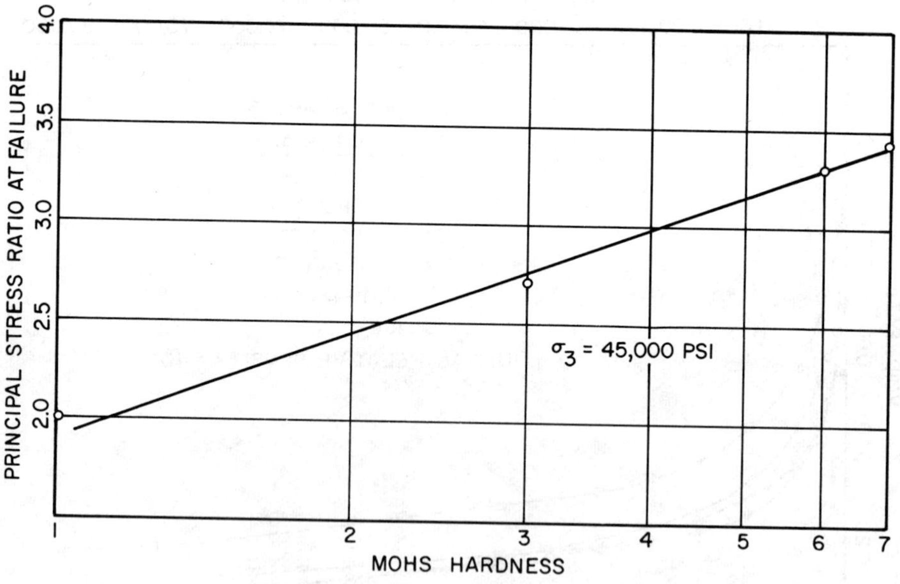

FIGURE 9 STRENGTH AND HARDNESS

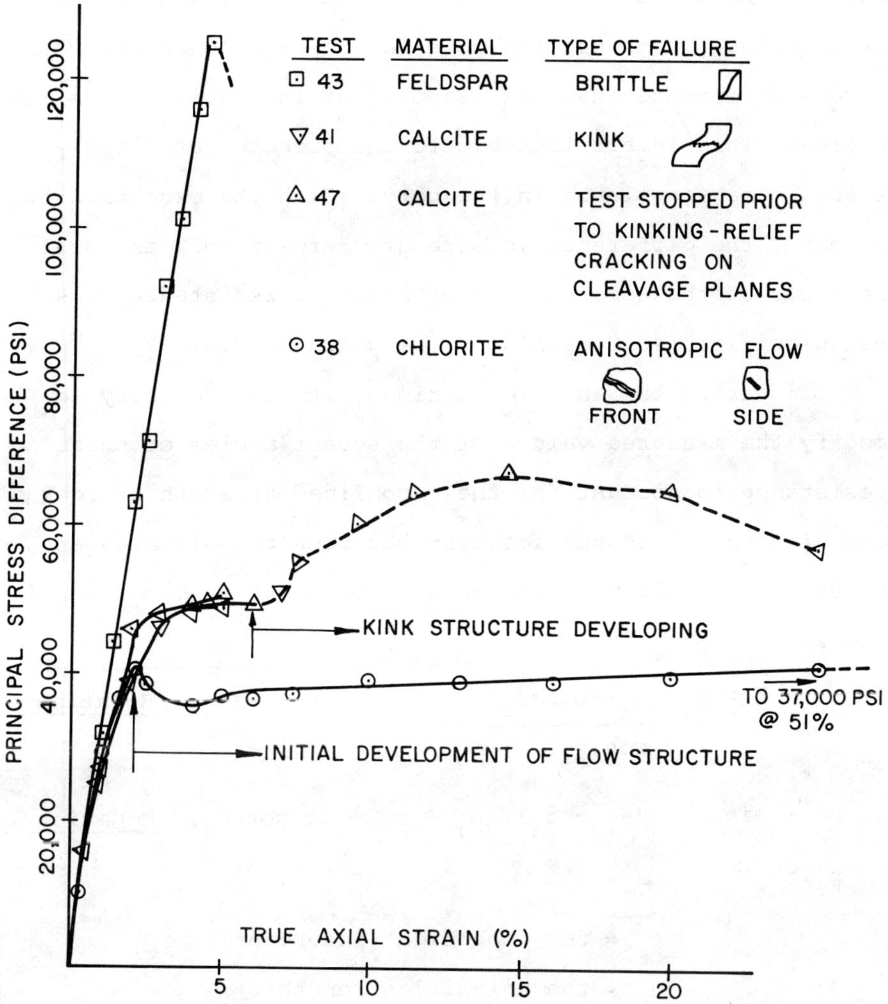

FIGURE 10 TRIAXIAL COMPRESSION TESTS ON ROCK

difference at failure, the difference in strength between rock and soil seems to decrease with decreasing hardness. Interestingly, the calcite core failed at a lower principal stress difference than the assembly of initially sand sized grains. This is attributable to the presence of closely spaced cleavage planes in the core. With the exception of calcite, the difference in strength between rock and soil, in terms of the secant angle of shearing resistance, was not significant. Comparisons are given in Table 2.

In making the above comparison, it was necessary to modify the measured values of the secant angles of shearing resistance to account for the unconfined strength of rock and of granular assemblies that had been consolidated at 45,000 psi. This was accomplished by means of Equations 1 and 2.

$$c = \frac{q_u \, (1 - \sin \phi_s)}{2 \cos \phi_s} \qquad \text{Equation 1}$$

$$\sin \phi = (\sigma_1 - \sigma_3)/(\sigma_1 + \sigma_3 + 2c \cot \phi_s) \qquad \text{Equation 2}$$

where

c = the "cohesion" intercept

q_u = the uniaxial strength

ϕ_s = the measured secant angle of shearing resistance

ϕ = the "corrected" angle, based on the assumption of a straight line envelope from the uniaxial to the triaxial Mohr circle

TABLE 2. COMPARISON OF BEHAVIOR: ROCK AND SOIL

Mineral	State	Initial Tangent Modulus (lbs.per sq.in.)	Angle of Shear Resistance Measured	Modified
Feldspar	Rock	3.1×10^6	35.5	31.9
	Soil	0.8×10^6	32.2	32.2
Calcite	Rock	1.6×10^6	21.3*	21.0*
	Soil	0.6×10^6	27.0	26.5
Chlorite	Rock	2.9×10^6	18.0	16.8
	Soil	0.7×10^6	19.0	17.9

*Both triaxial and uniaxial samples were influenced by the presence of cleavage planes.

σ_1 = the major principal stress at failure

σ_3 = the minor principal stress at failure

The Role of Petrofabric, Microstructure and Hardness - A number of unexpected but significant phenomena were observed during the triaxial tests on rock. Of particular interest were the extremely brittle behavior of feldspar and ductile behavior of calcite and chlorite. The former was indicated by an explosive energy release at failure, the latter by formation of kink bands in calcite and flow structure in chlorite. In addition, barrelling (circular cross sections becoming elliptical) was noted in chlorite. Patterson and Weiss (4) have indicated that such behavior is common in rocks with naturally anisotropic elastic "constants." The broad range in observed ductility is evident in Figures 11, 12, and 13. The influence of stress level on the nature of rock deformation is hinted at by the brittle uniaxial fracture and ductile triaxial failure shown for chlorite in Figure 13. Though additional data is required, it appears that brittle-ductile transition may be related not only to temperature and stress level but also mineralogical hardness.

Uniaxial Tests on Rock and Soil - As indicated earlier, these tests were performed not so much for the purpose of obtaining direct data, but rather to obtain information needed to modify the results of the triaxial compression tests. However, some significant phenomena were observed during this testing. Of particular interest, were the

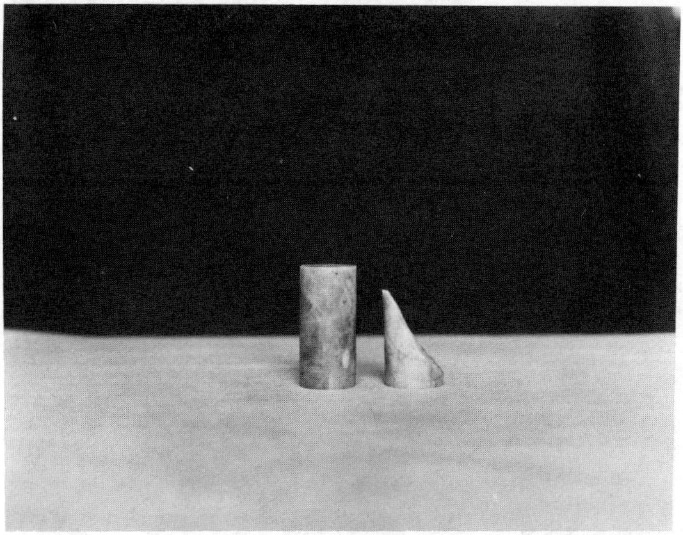

FIGURE 11 FELDSPAR AFTER TRIAXIAL FAILURE

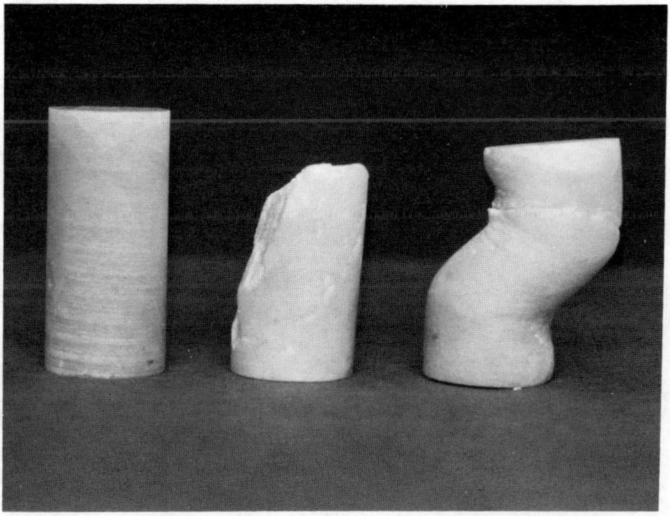

FIGURE 12 CALCITE: ORIGINAL CORE (LEFT), AT ONSET OF KINKING (CENTER), AFTER KINKING (RIGHT)

FIGURE 13 CHLORITE: ORIGINAL CORE (LEFT), AFTER UNIAXIAL FAILURE (CENTER), AFTER TRIAXIAL FAILURE (RIGHT)

observations that calcite cores were weaker than preconsolidated (to 45,000 psi) assemblies of calcite grains and that assemblies of chlorite grains were essentially transformed into rock during isotropic compression. Details concerning the behavior of chlorite are given below.

Chlorite: Rock to Soil to Rock - Chlorite was obtained as bulk samples of schist. These were cored or ground to medium sand size and then tested. During isotropic compression tests on the granular assemblies, individual sand size grains broke down to individual clay size platelets. These then regrouped and formed what would be considered by most definitions (5, 6) rock. This is evident in the slickensided appearance of the sample shown in Figure 14 and in the electron micrographs shown in Figures 15, 16, and 17. Interestingly, it appears that strength development during "laboratory metamorphism" is a function of void ratio attained during isotropic compression. As shown in Figure 18, the relationship between void ratio and the logarithm of unconfined strength is linear. This information can be used to estimate the preconsolidation pressure of the parent chloritic schist. This was done by assuming that isotropic compression of granular assemblies of chlorite would (at some stress level) result in a void ratio close to zero (the void ratio of the parent rock). Extrapolation of the data shown in Figure 7, led to the conclusion that this stress would be about 245,000 psi. This result, along with the

FIGURE 14 SLICKENSIDED CHLORITE SURFACE

FIGURE 15 SCANNING ELECTRON MICROGRAPH - SAND SIZED CHLORITE

FIGURE 16 SCANNING ELECTRON MICROGRAPH - CHLORITE AFTER ISOTROPIC COMPRESSION AT 4500 PSI

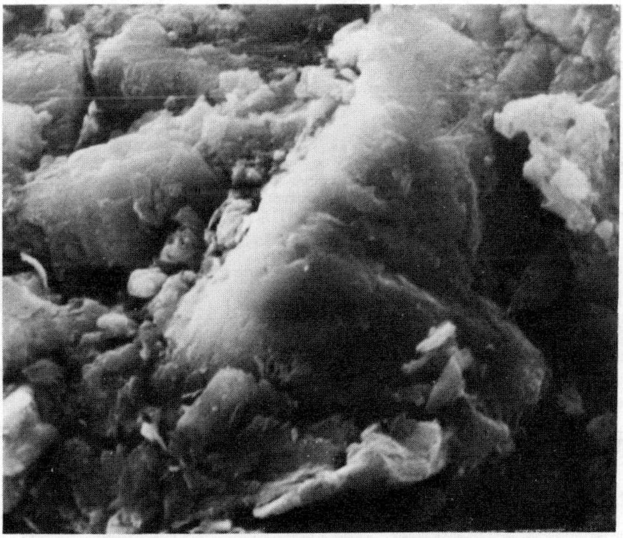

FIGURE 17 SCANNING ELECTRON MICROGRAPH - CHLORITE AFTER SHEAR AT 45,000 PSI

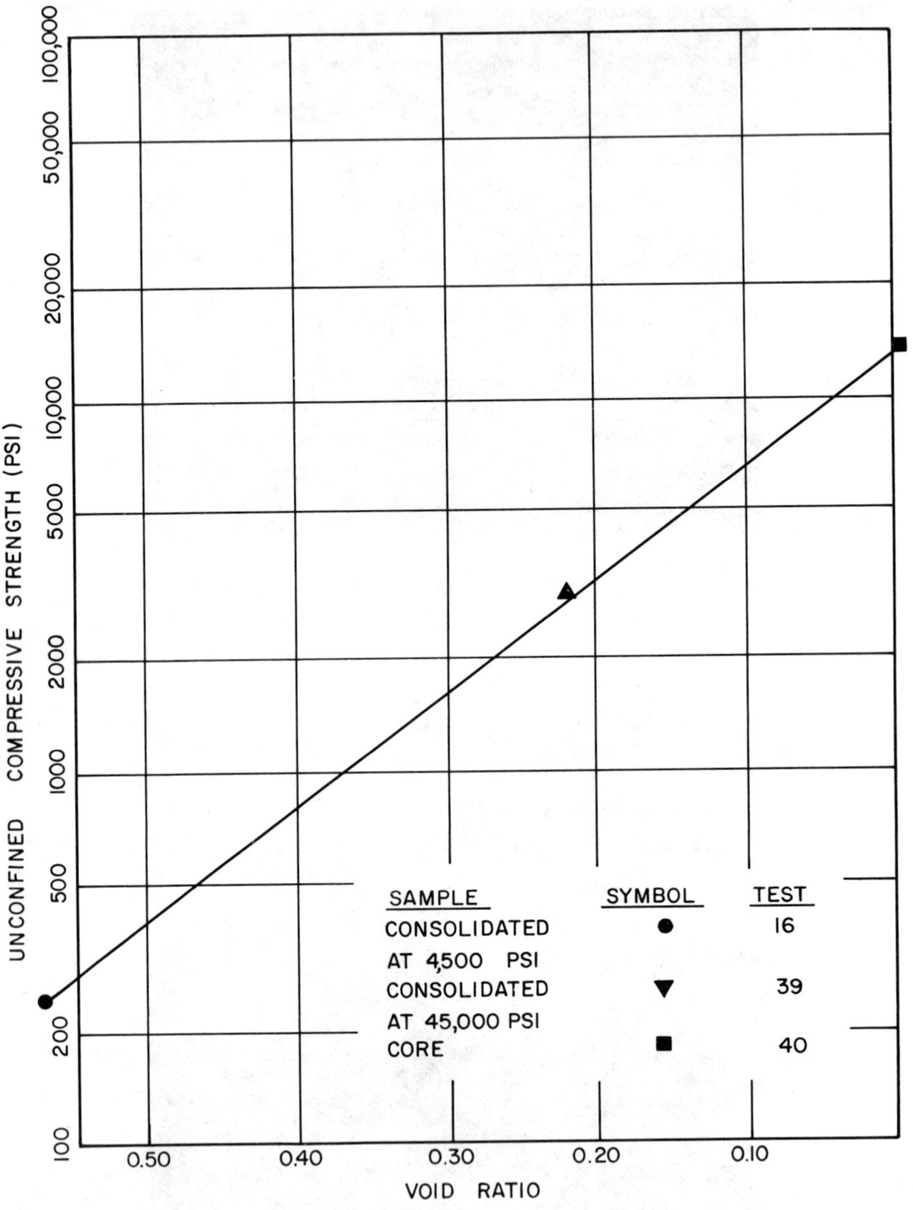

FIGURE 18 UNCONFINED STRENGTH VS. VOID RATIO (CHLORITE)

results of uniaxial tests conducted on rock cores and preconsolidated granular assemblies was plotted as shown in Figure 19. The linearity of this plot, actually demanded by that of Figure 18, is indicative of the validity of the 245,000 psi estimate. The similarity between a linear q_u vs. p_c plot and the Hvorslev concept of true cohesion is significant. It implies that true physico-chemical bonding developed during isotropic compression.

CONCLUSIONS

The results of this investigation indicate that, at high stress levels, strong inter-relationships can exist between shear strength, compressibility and mineralogical hardness. It appears that at some stress level, the magnitude of which is related to hardness or molecular structure, it is possible to evaluate the behavior of rock by means of tests conducted on sand with identical parentage. From a practical point of view, such a procedure obviates the need for testing very large samples to evaluate the strength and deformation characteristics of rock fill. This is especially true for soft minerals since at high stress levels rock and gravel size particles would crush to sand size or smaller during isotropic compression. In addition, it seems that it may be possible to evaluate the preconsolidation pressure of soft to medium hard sedimentary and metamorphic rocks by means of relatively simple laboratory techniques.

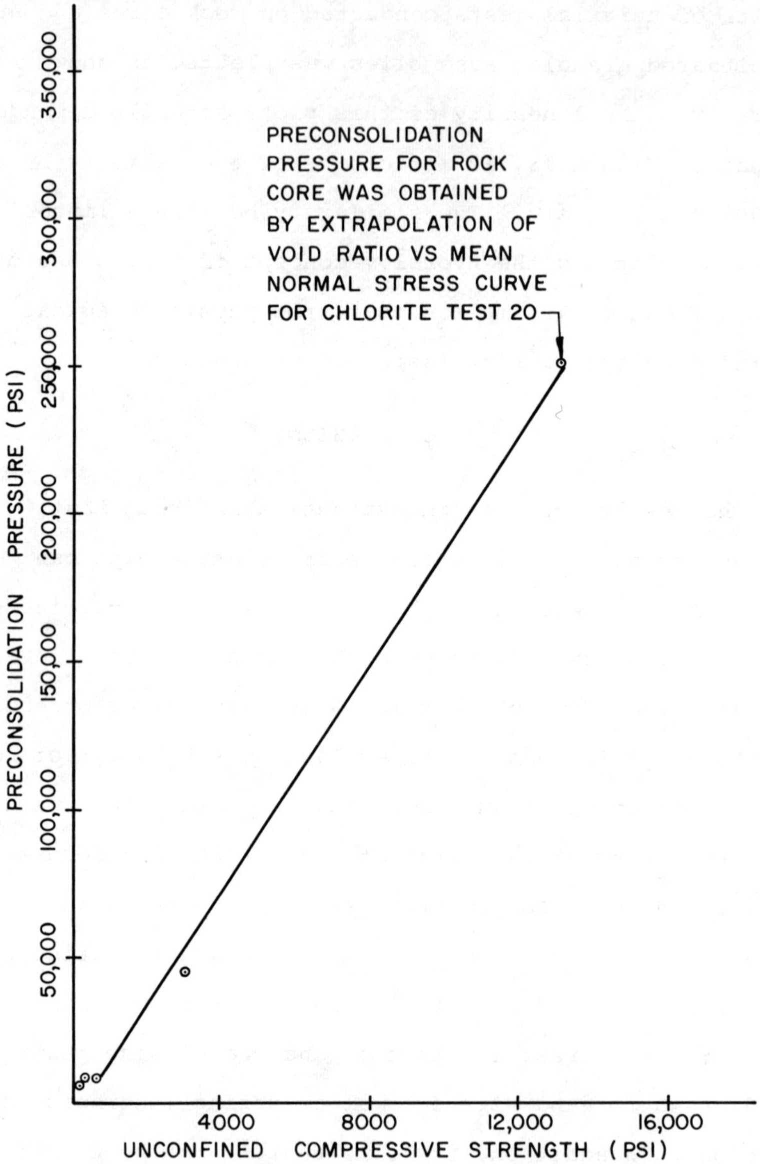

FIGURE 19 PRECONSOLIDATION PRESSURE VS. UNCONFINED STRENGTH (CHLORITE)

ACKNOWLEDGEMENT

The work reported on herein was conducted at Duke University as partial fulfillment of the authors requirements for the Doctor of Philosophy degree. The author is indebted to his major advisor, Dr. A. S. Vesic', Chairman of the Civil Engineering Department at Duke, for the guidance and encouragement he offered during the course of the study. In addition, the author wishes to thank his minor advisor, Dr. G. Wayne Clough and his colleagues Dr. Tein-lie Tai and Mr. D. Michael Hollaway, for their advice and constructive criticism.

APPENDIX I - REFERENCES

1. Tai, T., Strength and Deformation Characteristics of Cohesionless Materials at High Pressures, Ph.D. thesis, Duke University, 1970.

2. Murphy Jr., D. J., Soils and Rocks: Composition, Confining Level and Strength, Ph.D. thesis, Duke University, 1970.

3. Tabor, D., "Mohs Hardness Scale--A Physical Interpretation," Proceedings, Physical Society of London, Vol. 67, Section B, 1954, pp. 249-257.

4. Patterson, M. S. and Weiss, L. E., "Experimental Deformation and Folding in Phyllite," Bulletin of the Geological Society of America, Vol. 77, 1966, pp. 343-374.

5. Nasciemento, U., "Identification of Petrification in Soils," Proceedings, 6th International Conferences on Soil Mechanics and Foundation Engineering, Montreal, 1965, Vol. 1, pp. 80-81.

6. Leggett, R. F., Geology and Engineering, McGraw-Hill, New York, 1962.

APPENDIX II - NOTATION

The following symbols are used in this paper:

c = "cohesion" intercept

q_u = uniaxial strength

ϕ_s = secant angle of shearing resistance

ϕ = modified angle of shearing resistance

σ_1 = major principal stress at failure

σ_3 = minor principal stress at failure

"TL" A New Approach to Stress Distribution in Rocks

Gabrielle Douglas*

The phenomenon of thermoluminescence -TL- has been known for at least three centuries (1) but it is only in recent years that it is coming to be understood, due mainly to advances in solid state physics. The following can be considered as a general definition: when some crystalline materials are heated to temperatures somewhat below that at which incandescence is observed, small amounts of light are emitted. The maximum emission may occur at one or more temperatures and if plotted as emission versus temperature the resulting graph is referred to as "glow curve."

The explanation, based on crystal field theory, which accounts for the phenomenon can be very briefly summarized in the following way: various imperfections of the crystal lattice distort the electron clouds, and in a narrow temperature range, the electrons which have been forced by the distortions into higher energy states can return a lower energy state. In so doing there is a release of energy quanta in the form of light which is equivalent to the difference between the two energy states. In the same material, the phenomenon may occur at several temperatures, in which case the material is said to have several "glow peaks." These glow peaks are due to different energy levels which in turn are due to the different types

* Research Scientist, Mineral Physics Laboratory, Dept. Geology, Loyola College, Montreal, Canada.

of imperfections of the lattice which gave rise to the initial distortions of the electron clouds. These elevated energy level states are usually referred to in thermoluminescence literature as "traps" or "electron traps," although the latter expression is misleading. Thermoluminescence has been found to be a very useful tool for the investigation of defects in crystalline materials, in particular such geological materials as quartz and feldspar-bearing rocks, limestone, marble and in a number of less common minerals such as flourite, halite (rock salt) and zircon. Since all these minerals are thermoluminescent at convenient temperatures, and there are few geological formations which contain none of them, it would appear that the use of thermoluminescence would be almost mandatory in geological solid-state investigations. The best general reference on the subject is the proceedings of a NATO Advanced Research Institute (7).

Historically, thermoluminescence was first believed to be due to radiation damage in the crystal lattice of minerals and artifical crystals, since ionizing radiation was found to give an increase in the thermoluminescence signal (9). For a time it was assumed that only radiation damage affected the thermoluminescence characteristic of a material. Somewhat later, anomalous results were observed and it was found that very small amounts of impurity elements were responsible for large differences in the amount and type of thermoluminescence emitted under similar circumstances by materials previously considered to be identical. The effects of these "trace elements" was investigated for a number of years, partially by physicists and chemists who were concerned with improvements in radiation dosimeters, and partially by geologists who hoped to be able to use the thermoluminescence effect either for stratigraphic correlation or as a prospecting tool (6). The presence of trace amounts of impurity elements, either in interstitial positions in the crystal lattice or in isomorphous substitution will necessarily cause some distortion of the crystal lattice, even if this is no more than a deformation of the electron clouds.

At a still later date, additional anomalous conditions became apparent. It was first noted that crushing and/or grinding of samples sometimes introduced some errors in results (3). To a considerable extent this can be overcome by keeping the conditions of crushing as uniform as possible. In this connection it is obvious that it is difficult if not impossible to avoid all disruption or distortion of the crystal lattices during crushing but it can be kept at low levels, and it has been demonstrated that rapid crushing will cause a minimum development of new dislocations. Fine grinding on the other hand tends to cause a considerable amount of lattice distortion and deformation of the fragments. On a larger scale, geological field studies have demonstrated that the pattern of thermoluminescence which might have been expected as the result of the distribution of "trace elements" and natural radioactivity in rocks is rather considerably altered in the vicinity of dikes, faults, underground nuclear explosions and craters due to meteorite impact. It is only in the last few years that serious attempts have been made to come to an understanding of this pressure effect through geological field investigations and controlled pressure experiments in the laboratory. (2,4,8).

Strain due to stresses imposed on a rock mass in the range between elastic deformation and true plastic flow will propagate dislocation in the crystals of the rock. This will increase the amount of atomic disorder and in turn will be reflected in the thermoluminescence glow curves. In rocks where faulting has occurred (slow rate of loading) the thermoluminescence is increased, while in rocks obtained from the near proximity of meteorite craters and underground nuclear explosions (high rate of loading) thermoluminescence is decreased.

Experiments in the laboratory to reproduce these effects have been carried out on small cylindrical samples of drill core, but due to the small size of the sample it has been frequently found that only an average thermoluminescence could be obtained, often with an annoying spread in values. To satisfactorily examine the distribution of the thermoluminescence response to loading, so that the results could be mapped in a similar manner to large scale field studies, required the use of large rock samples from which many small samples could be taken after straining. Large cylindrical rock samples were difficult to obtain and cylindrical castings of an "organic limestone conglomerate" (better known to us all under the familiar name of concrete) were substituted. In the concrete, the cement is non-thermoluminescent and does not affect the results. The silica sand is only moderately thermoluminescent, and in a special experiment it was found that under the test conditions used it was not appreciably affected by strain. The predominent signal was that of 1/4" size limestone aggregate. The limestone is the most ductile material in the "conglomerate" and in consequence was the most seriously deformed during the compression of the sample.

A set of six concrete test cylinders (6" diam x 12" long) were supplied by the Francon Company of Montreal. All six cylinders were prepared from the same mix and were theoretically identical. They contained 1/4" Trenton limestone aggregate from the Francon quarry and sand from the Ottawa River. Three of the cylinders were subjected to the standard unconfined compression test, with additional measurements to determine their elastic moduli. In each case the compression testing was halted at the first indication of failure. The remaining three cylinders were retained in an "as cured" condition.

All six cylinders were sampled by taking 1" diamond drill core samples at 2" intervals along each of three lines spaced 120° apart around the cylinders for a total of 15 regularly spaced samples per cylinder. The samples were crushed in a jaw crusher and sieved to yield a uniform size range of -60 +100 mesh. These samples were then tested in a modified version of a Madison Research thermoluminescence reader and "glow curves" obtained for each. The thermoluminescence reader has a heating rate which heats the sample from room temperature to 400°C in two minutes, with the thermoluminescence emission being recorded through a photomultiplier with an S-11 type response.

The data obtained was plotted so as to give a contour map of the distribution of thermoluminescence on the unrolled surface of the cylinder. Examples of contour maps of strained and an unstrained cylinder are shown in Figure 1. Comparison of the two shows that the unconfined compression test has produced a much greater range of thermoluminescence as shown by the larger number of contours. In addition, there is a considerable similarity of the contour pattern to the "butterfly" in lucite strain patterns. Careful inspection will show that a thermoluminescence "high" developed on one side of the cylinder, which is balanced by a "low" on the opposite side. This would appear to indicate that the "cone of failure" was not centered in the cylinder and that the cylinder may have been under tension on one side during the unconfined compression test (presumably due to non-parallelism of the end faces and/or anisotropic behavior of the material).

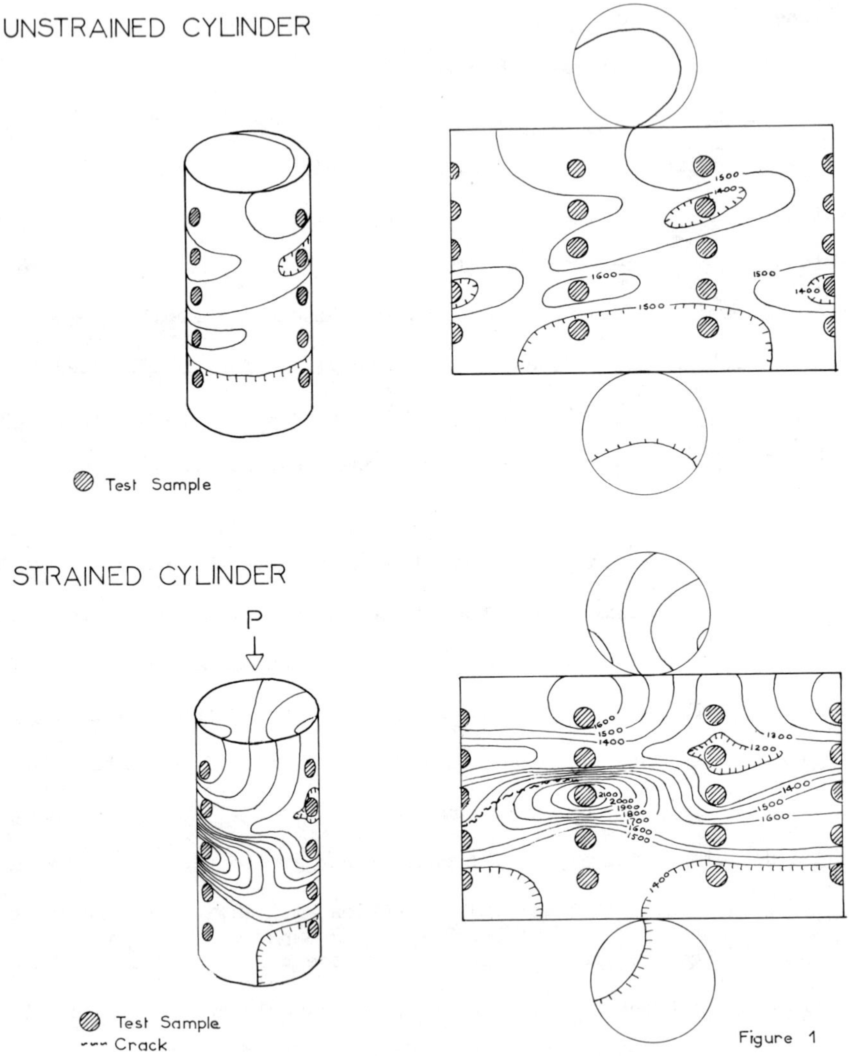
UNSTRAINED CYLINDER
1500
1400
1600
1500
1400
1500
Test Sample
STRAINED CYLINDER
P
1600
1500
1400
1200
1200
2100
2000
1900
1800
1700
1600
1500
1400
1500
1600
1400
Test Sample
Crack

Figure 1

Figure 2 is a plot of the standard deviation from the mean of the thermoluminescence readings for each of five horizontal planes through the strained and unstrained cylinders. The standard deviation from the mean is considered to be proportional to the differential strain. From Figure 2 it can be seen that at the loaded end of the cylinders the standard deviation from the mean is fairly high and remarkably constant from one strained cylinder to another (end effect). Below the loaded end of the cylinder the lower values of standard deviation suggest conditions of more uniform strain. At the center, (where the zone of failure is indicated on the surface of the cylinder by a surface crack in Figure 1) there is a marked increase in the standard deviation from the mean which differs from cylinder to cylinder. Below the center, the standard deviation from the mean decreases indicating less differential strain, and the bottom appears to be similar to the conditions in the unstrained cylinders (suggesting that the bottom of the cylinders have not been affected by load).

In Table I, a comparison is made between the average thermoluminescence, standard deviation from the mean, elastic moduli (Ec) and compressive strength (psi). There is an indication of a correlation between the elastic moduli and the standard deviation, which suggests that with greater strength, a greater differential strain is sustained before failure.

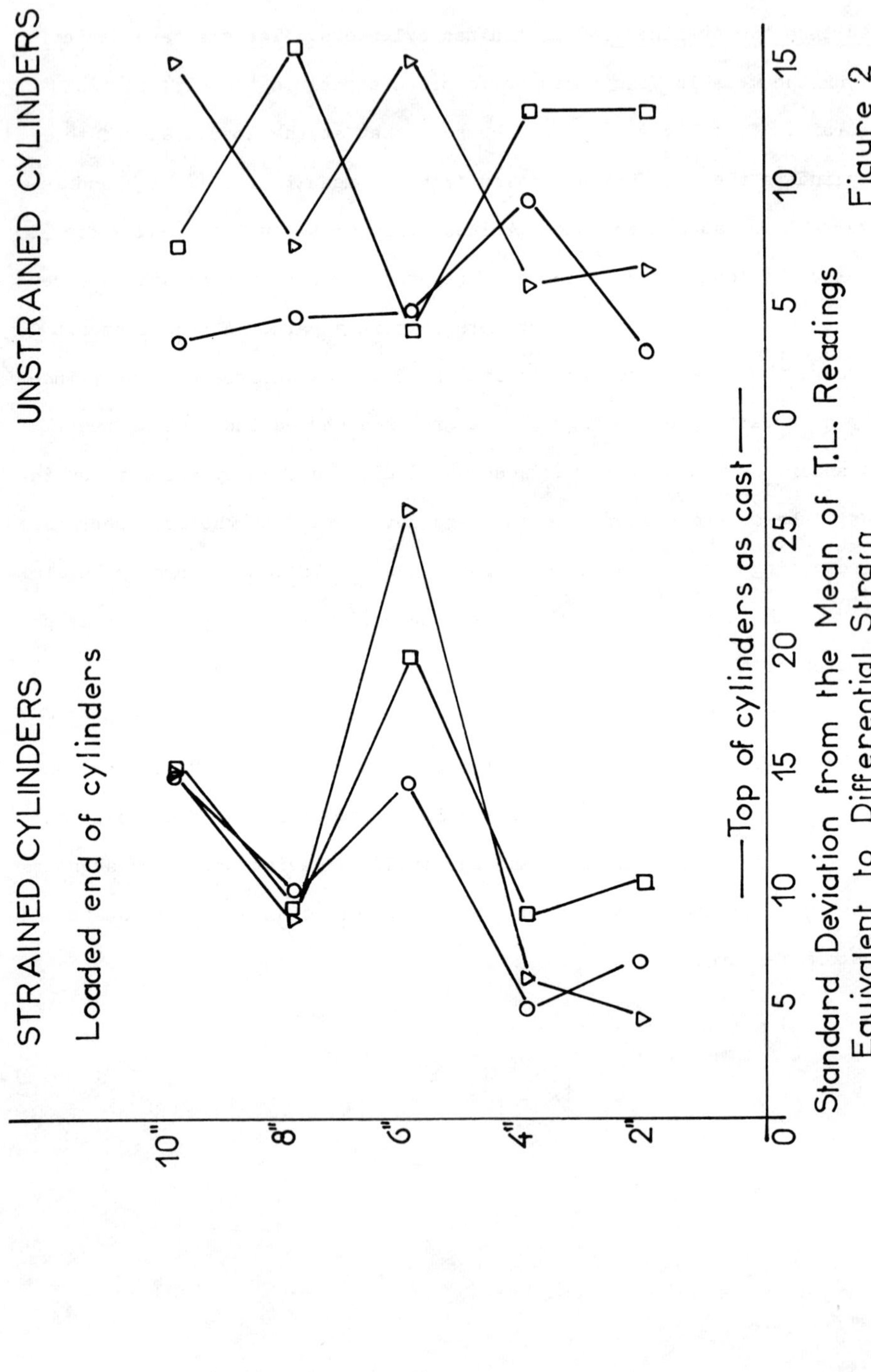
STRAINED CYLINDERS
Loaded end of cylinders
UNSTRAINED CYLINDERS
10"
8"
6"
4"
2"
0
0
5
10
15
20
25
0
5
10
15
—Top of cylinders as cast—
Standard Deviation from the Mean of T.L. Readings
Equivalent to Differential Strain

Figure 2

TABLE I

Cylinder	Average TL	Standard Deviation	Ec	Compressive Strength psi
B21	1477.1	16.8	5.42	6,982
B22	1410.1	14.1	5.49	7,266
B23	1498.3	16.4	5.31	6,928

In a recent survey of rock mechanics techniques, Hardy (5) has commented that thermoluminescence presents considerable potential as a means of measuring the degree of permanent strain present in rocks. The experimental results described above suggest that with more closely spaced samples it should be possible to obtain a detailed three-dimensional picture of the stress distribution under load in both concrete test cylinders and large rock mechanics test samples. As an extension of this, "deviation" plots appear to provide a means of predicting failure in both concrete structures and rock.

APPENDIX - REFERENCES

1. Boyle, R. "Experiments and Considerations upon Colours with observations on a Diamond that Shines in the Dark." Proceedings of the Royal Society, (London), 1663.

2. d'Albissin, M. "Les Traces de la Deformation dan les Roches Calcaires." Revue Geographie physique et Geologie dynamique, (Paris), 2nd series, Vol. 5, 1963, pp. 121-140.

3. Debenedetti, A. "On the Chemical Activation of Thermoluminescence in Calcite." Nuovo Cimento, (Rome), Vol. 7, No. 2, 1958, pp. 25-254.

4. Douglas, G., Morency, M., and McDougall, D. J. "Detection of Strain in Rocks using an Intrinsic 'Semi-insulator' characteristic of some Minerals." Modern Geology, Vol. 1, 1970, pp. 211-217.

5. Hardy, H. Reginald. "Rock Mechanics - A Kit of Tools for the Structural Geologist," Proceedings - Seminar on Appalachian Structure: Mont Chateau Lodge, West Virginia, March 3-5, 1971. Sponsored by West Virginia University and West Virginia Geological and Economic Survey.

6. Medlin, W. L. "Comments on Thermoluminescent Properties of Calcite." Journal Chemical Physics, Vol. 30, 1959, pp. 451-458.

7. Thermoluminescence of Geological Materials. D. J. McDougall, ed. Academic Press (London), 1968, 678 pages.

8. Morency, M. "Determination of Strain Effects in Crystalline Rocks using Thermoluminescence." (abstract) Transactions, American Geophysical Union, Vol. 51, No. 11, 1970, pp. 830-831.

9. Urbach, F. "Zur Lumineszenz der Alkalihalogenide." Sitzungsberichte der Akademie der Wissenschaffen Wien, Mathematisch-Naturwissenschaftliche Klasse. (Vienna), Vol. 139, 1930, pp. 353, 363, 482.

DEVELOPMENT AND POTENTIALS OF MOIRÉ METHOD OF STRAIN ANALYSIS FOR ROCK TESTING

By K. Thirumalai[1] and R. L. Fischer[2]

ABSTRACT

Created by an optical interference of two closely spaced arrays of lines or grids, the moiré effect is a phenomenon offering a convenient means of detecting large field inplane displacements for strain analysis. This pioneering study carried out by the Bureau of Mines develops an experimental technique for using the moiré method in making remote full field strain distribution measurements for rock fragmentation study. The paper briefly explains the basic principles of the moiré method and its significance to measure strain distribution during rock testing. Medium and high intensity (up to 1,000 lines/inch) moiré grids were formed on test rock surfaces by vacuum depositing a thin layer of grid material suited to elevated temperature and environmental testing of rocks. The fringes were obtained by projecting the image of the deformed specimen grid onto a transparent glass etched master grid with the help of a precisely aligned high resolution optical system. The study indicates the potential of the method to examine full field biaxial strain distribution up to fracture and fragmentation of rock and coal materials.

---

[1] Supervisory mining engineer
[2] Research physicist
Both authors are with Twin Cities Mining Research Center, Bureau of Mines, U.S. Department of the Interior, Twin Cities Minnesota 55111.

INTRODUCTION

New and improved methods of rock fragmentation are necessary to meet the ever increasing mineral and excavation requirements and the emerging environmental demands. An understanding of rock reaction to different forms of incident energy is the key for a successful formulation and evaluation of different innovative rock breaking methods by using thermal, electrical, hydraulic and other energy processes. Rocks are complex solids with heterogeneous mineral aggregates which are characteristically anisotropic. A complete understanding of rock behavior to elevated temperature and adverse environmental conditions of the fragmentation processes is hard to reach without supporting experimental investigations (11,12).

Measuring displacement and strain is of fundamental importance in determining the mechanical response of rocks to a given energy system for fragmentation. The present techniques used for experimental strain analysis during rock testing are limited to obtaining integrated values of displacements over limited small areas or lengths by strain gauges, or by electrical, induction, and capacitor type transducing systems. These methods give average surface dilations over the length of the contact zone and are highly dependent on the associated transducer systems and the environmental conditions during rock testing. Since the nonuniformity of rock structure induces a nonuniform stress distribution, the study of rock fragmentation processes demands the development of new techniques for measuring large field strain distribution at different points of the surface planes of specimens during testing. Such a method should also facilitate measurements of rock response to thermal shock and other high energy incidences and environmental conditions. The deformation measurements under adverse environmental conditions are best made using optical methods for remote sensing. The photoelastic model technique employing the principles of birefringence to analyze stress distribution of complex structures and the photoelastic coating methods have only limited application to rock testing. The photoelastic model technique is restricted to measurements only in the elastic range of material behavior, free from the material inhomogeneities. The model materials and the photoelastic coating tend to respond nonlinearly to large deformations and are unsuitable for making tests under conditions of elevated temperature and adverse environment.

Full field strain measurements, however, can be made remotely b applying the principles of an optical interference. The moiré effect is an optical phenomenon created by a geometric (mechanical) interference of light passing through two closely spaced arrays of lines. If the two arrays of lines are not identical, fringes are formed by mechanical interference. Though this phenomenon was first described by Lord Raleigh in 1874 (8), only in the past decade have scientists and engineers become interested in understanding and applying the moiré phenomenon of displacement and strain measurement to material testing (1,2,10). The fringes produced by the moiré effect provide a potential means of precisely determining large field displacement and strain distribution in both the static and dynamic testing of rock materials for all magnitudes of deformation and enables an analysis of strain distribution in elastic, inelastic, and fracture ranges during testing of rocks.

This paper describes a pioneering work carried out by the Twin Cities Mining Research Center, U.S. Bureau of Mines, to apply the moiré method for testing of rocks and coal. New techniques were formulated to deposit grids on rock surface using high temperature materials to withstand adverse environmental conditions during rock testing. The application of the moiré method to analyzing large field strain distributions of rocks is examined and its potential of the method for determining non-uniform strain distributions during testing of rocks is shown qualitatively with the help of some experimental results in Salem limestone and bituminous coal.

THE PRINCIPLES OF THE MOIRÉ METHOD

The moiré method offers a simple and fundamental means of making direct large field determination of strains by noncontact optical techniques. Moiré fringes are created by geometric interference induced during deformation of an array of lines or dots fixed to a test specimen (specimen grid) when viewed through an initially identical reference grating (master grid). With opaque specimens, reflected light techniques can be used to produce the mechanical interference. The image of the specimen grid can be projected without distortion onto a master grid to form fringes through a properly aligned precision optical system placed at a distance from the specimen test system.

THE MECHANISM OF FRINGE FORMATION

Figure 1 gives a simple illustration of the fringe formation. Consider the light as transmitted through two similar and parallel line arrays, A and B. Assume a distortion is induced by external forces in array A. Array A will now possess a slightly different line spacing (pitch) from array B (Fig. 1). Let the pitches of arrays A and B be p and p' respectively. The transmitted light intensity through the arrays A and B will give the series of light steps shown in figure 1 with their widths depending on the amount of distortion or deformation of array A. The line of average light intensity produce light bands at

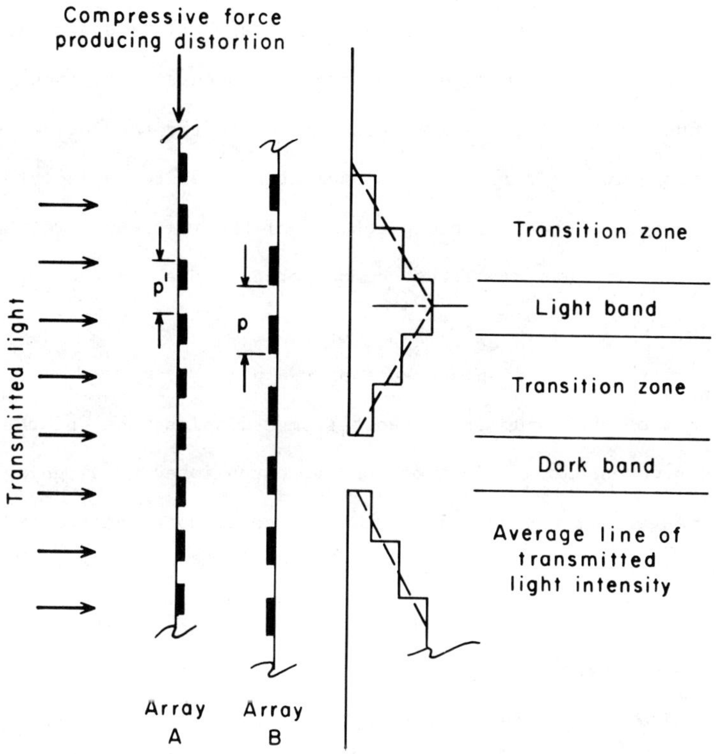

FIG. 1. - FORMATION OF MOIRÉ FRINGES

positions of maximum light intensity and dark bands or moiré fringes at positions of minimum light intensity. The moiré fringe is thus formed when array A is distorted to produce displacement equal to the pitch of array B. These fringes constitute the loci of points of equal displacement in a direction perpendicular to the array of the reference lines.

USE OF MOIRÉ FRINGES FOR FULL FIELD DISPLACEMENT AND STRAIN MEASUREMENT OF ROCKS

The moiré fringes can be considered as isothetic lines representing the locus of points of equal displacement components in a given direction. The continuous moiré fringes are thus similar to the topological contours. The difference in the value of displacement between two adjacent fringes will be a constant equal to the pitch, p, of the reference grating. The relative displacements along the common perpendicular to the reference grating will be:

$$\text{displacements } u, v = n \cdot p \tag{1}$$

where n is the order (number of bands from 0 displacement) of the fringes. The u and v components of deformation are established by aligning the array of master grids parallel to the x, y coordinates of the grid placement.

Since the moiré phenomenon is related to the displacement of a point, a determination of the derivatives of the displacement obtained from the moiré pattern helps compute the strains induced in the plane independent of its magnitude. For a uniform large strain field, the general strain displacement relations from geometric considerations in Lagrangian description reduce to (6):

$$\epsilon_x = [1 + 2\,\partial u/\partial x + (\partial u/\partial x)^2 + (\partial v/\partial x)^2]^{1/2} - 1 \tag{2}$$

$$\epsilon_y = [1 + 2\,\partial v/\partial y + (\partial v/\partial y)^2 + (\partial u/\partial y)^2]^{1/2} - 1 \tag{3}$$

$$\gamma_{xy} = \arcsin \frac{\partial u/\partial y + \partial v/\partial x + (\partial u/\partial x)(\partial u/\partial y) + (\partial v/\partial x)(\partial v/\partial y)}{(1 + \epsilon_x)(1 + \epsilon_y)} \tag{4}$$

where ϵ_x, ϵ_y, and γ_{xy} are direct, normal, and shear strains respectively in x, y coordinates. The strains induced during experimental tests of rocks are relatively small and are associated with little or no rotation during testing. For rock testing of this study the products and powers of the derivatives of displacements shown above (equations 2 - 4) can be ignored. The direct and shear strains for practical rock testing therefore reduce to:

$$\epsilon_x = \partial u/\partial x,\ \epsilon_y = \partial v/\partial y \text{ and } \gamma_{xy} = \partial u/\partial y + \partial v/\partial x \qquad (5)$$

Figure 2 illustrates how the displacement gradients are obtained from the moiré fringes obtained in a direction perpendicular to the grids of the reference array. The difference between any two adjacent displacement components, u, u_{n+1}, u_{n+2}, belonging to the corresponding fringes, n and n+1 or n+1 and n+2, is p. The limit of the incremental values of displacement components at a point A is $\Delta u/\Delta x = p/\Delta x = \partial u/\partial x$. The displacement gradients, $\partial u/\partial y$, can similarly be obtained by interpolating along the y axis. A similar set of v displacement gradients can be derived from fringes obtained by positioning the reference grid parallel to the x coordinate. A complete inplane strain distribution can thus be obtained at all critical locations from the full field inplane displacement of the moiré fringes.

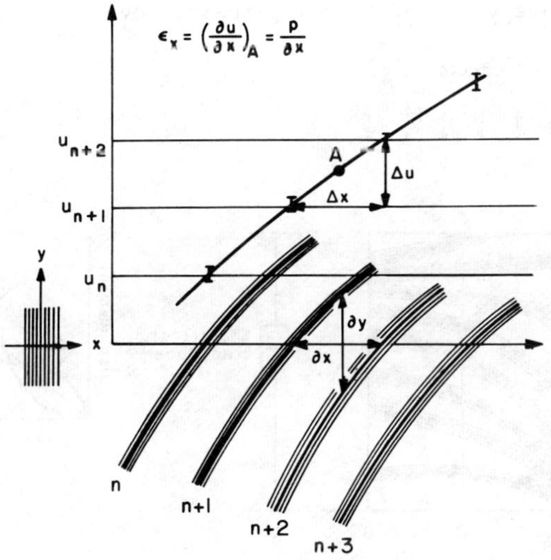

FIG. 2. - DETERMINATION OF DISPLACEMENT GRADIENTS FROM MOIRÉ FRINGES

SIGNIFICANCE OF STRAIN DISTRIBUTION MEASUREMENTS BY THE MOIRÉ METHOD FOR ROCK TESTING

The moiré fringes display isothetic lines of equal displacement. These large field displacement contours help determine the strain distribution for the surface plane of the test specimen and identify the zones of high strain concentrations during rock testing.

Figure 3 shows a simple example of the significance of full field inplane displacement distribution measurements using the moiré method. Let us consider the longitudinal field (v field) of the moiré fringes over a small segment A B B' A' of a rock specimen subjected to compressive stresses at the boundary as shown in figure 3. The structural variations of the rock material induce nonuniform strain distributions. Assuming zero displacement at the boundary A'B', the successive isothetic fringes will have an increasing order from A'B'. Let us consider, for comparative purposes, an electrical or mechanical transducer is placed between the points C and C' to measure the specimen deformation. The integrated average value of the displacement over the length CC', given

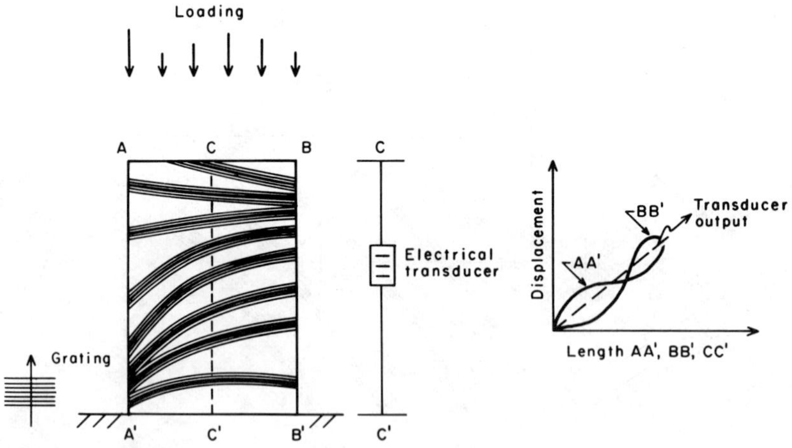

FIG. 3. - THE TREND OF DISPLACEMENT MEASUREMENTS IN ROCK TESTING USING THE MOIRÉ METHOD AND OTHER TRANSDUCER TECHNIQUES

by the transducer output is represented by the linear dotted gradient line (Fig. 3). The corresponding displacement-strain distribution at sections AA' and BB' at different points along the longitudinal plane indicated by the isothetic moiré fringes is also projected in the figure. Since the mechanism of fringe formation is independent of the mode and the magnitude of specimen deformation, the local variations of the displacement and strain concentrations and subsequent fracture and failure of rock specimens can be identified by the moiré fringes. The lateral strain distribution of the specimen can similarly be obtained by orienting the reference grids to measure the u field distribution. The moiré isothetics thus permit a complete determination of the inplane strain distribution of all critical sections during rock testing.

DEVELOPMENT OF THE MOIRÉ METHOD FOR LABORATORY ROCK TESTING

FORMATION OF GRIDS ON ROCK SURFACES FOR THE STUDY

Forming suitable grids on a rock surface compatible to the elevated temperature and environmental conditions of the fragmentation study is the key to a successful application of the moiré method for rock testing. The conventional methods of forming grids by using a process of photo-etching are unsuitable for rock surfaces and for environmental testing of rocks. Although a machine ruled grid surface of coarse to medium frequency lines (up to 400 lines/inch) has been frequently used in testing metals, this method cannot be applied to rocks since machine ruling on brittle rock surface produces micro- and macrostructural failures. Transferable grids of different types have been recently made available for testing materials by the moiré method. A very thin deposit of etched metal grid is carried on a backing plate and the pattern is transferred to the specimen surface in a manner similar to that used in strain gage bonding. The bonding materials are normally low modulus cements having thin layer viscoelastic properties. Although suitable for testing of low strength and high deformation brittle materials, these bonding materials are unsuitable for hard rock testing under adverse conditions of temperature and environmental conditions.

A method of making a suitable grid pattern directly on the rock surface without intermediate bonding agents is necessary if the moiré method is to be successfully applied to hard rock fragmentation study. The grid material must withstand elevated temperatures and environmental effects during rock testing. A thin layer of high temperature material coating can be deposited on the surfaces of many solids using vacuum evaporation or by using cathodic sputtering (4,5). The process mechanism of vacuum evaporation is associated with evaporation at reduced pressures and the phenomenon of cathodic sputtering is associated with electrode (cathode) disintegration and deposition caused by ionized gas molecules being charged under reduced pressure. Although the process of vacuum deposition requires lower pressures (0.5 micron Hg) than cathodic sputtering (10 to 100 microns Hg) vacuum evaporation rapidly deposits a sufficiently thick film to produce good fringe contrasts on a rock surface.

Thermally stable compounds, including metals, evaporate rapidly at reduced pressures if the temperatures are raised to a point where the vapor pressure reaches a value in excess of 10 microns Hg. Most metals evaporate from the liquid phase under reduced pressure. The development of modern high speed oil diffusion pumps has simplified the process of creating the vacuum conditions for the evaporation. A wide range of high temperature metals and compounds (metal halides and sulphides) are suitable for thin film deposition by vacuum evaporation on rock surface. High pumping speeds with minimum backstreaming of the vacuum system are necessary to ensure evacuation during degassing and vaporization of the moisture in porous rock materials.

METHOD OF MEDIUM TO FINE FREQUENCY GRID DEPOSITION ON ROCK SURFACE BY VACUUM EVAPORATION

A high quality nickel screen mask with a 25 percent open area square dot pattern was custom made for depositing medium (200 lines/inch) and high intensity grids (1,000 lines/inch) on a rock surface. The mask was fabricated by the process of electroforming. This process is used in the industry for producing direct view storage tubes and other special purpose screens for the electronic industry. By a photochemical process a photographic negative of the screen was produced and then etched and electrodeposited with nickel to achieve the wire mesh of required dimensions and thickness. To form the grids on the rock surface, the mask was placed in close contact over the surface area of the test rock specimen and vacuum deposited with high temperature metals. The mask was then stripped off the rock surface after the grid deposition.

Figure 4 shows the equipment layout used for the grid deposition. To achieve a uniform and controlled deposition, the evaporant was volatilized from a heated tungsten basket filament whose temperature was controlled during the evaporation. The mechanical interceptor between the filament and the rock surface helped to provide a precise control of the duration of deposition and thus minimized the radiation heating and possible buckling of the mask during deposition. The same experimental procedure was applied in depositing both the medium intensity (200 lines/inch) and high intensity (1,000 lines/inch) grids on the rock surfaces studied.

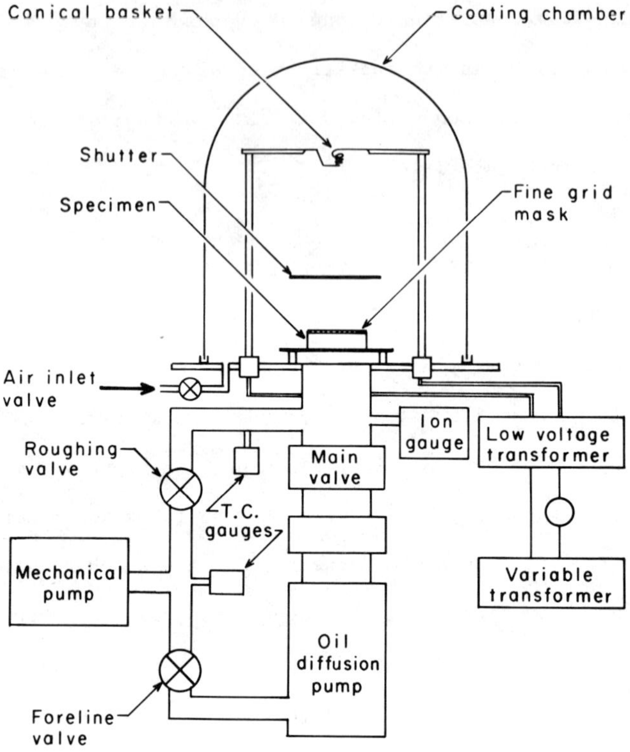

FIG. 4. - EQUIPMENT LAYOUT FOR GRID DEPOSITION ON ROCK SURFACE

This laboratory examined several high temperature materials for their suitability in depositing medium and fine intensity grids up to 1,000 lines/inch on rock surfaces ranging in color from light to dark gray. The coating materials are described in the table in order of their ability to give good fringe contrasts for applying the moiré method for rock testing.

TABLE 1. - MATERIALS FOR GRID DEPOSITION ON ROCK SURFACES

Material	Melting Point	Vaporization Temperature (at 10 microns Hg)
Aluminum	660° C	990° C
Gold	1,063° C	1,465° C
Nickel	1,455° C	1,510° C
Chromium	1,900° C	1,205° C

All the materials listed in the table were deposited on the rock surface with the tungsten basket filament shown in Fig. 4. The surfaces of the rock specimens were fine ground and cleaned before the deposition. Because of low temperatures of vaporization, high reflective properties and good fringe contrast, aluminum was selected as the grid material on the rock surfaces in the experimental study of this paper.

Salem limestone and bituminous coal were selected for the study to examine the application of the moiré method for rock and coal testing. Salem limestone consists of 98 percent cryptocrystalline carbonates with many dark gray grains of calcite in a light gray to white matrix with numerous disseminated porous cavities. Figure 5 shows the general interöolitic structure of the limestone and the topography of a fine-ground rock surface prepared for grid deposition at x 500. The photomicrographs were obtained using a scanning electron microscope. The magnification of continuous grid deposition on the limestone surface is also shown in Fig. 5. Note the undamaged and continuous grid filling in the pores and in the structural discontinuities of the rock surface. The figures indicate that despite the rough, porous, and interöolite aggregate structure of the limestone, the process of vacuum deposition gives a good continuous grid deposition on rock surfaces.

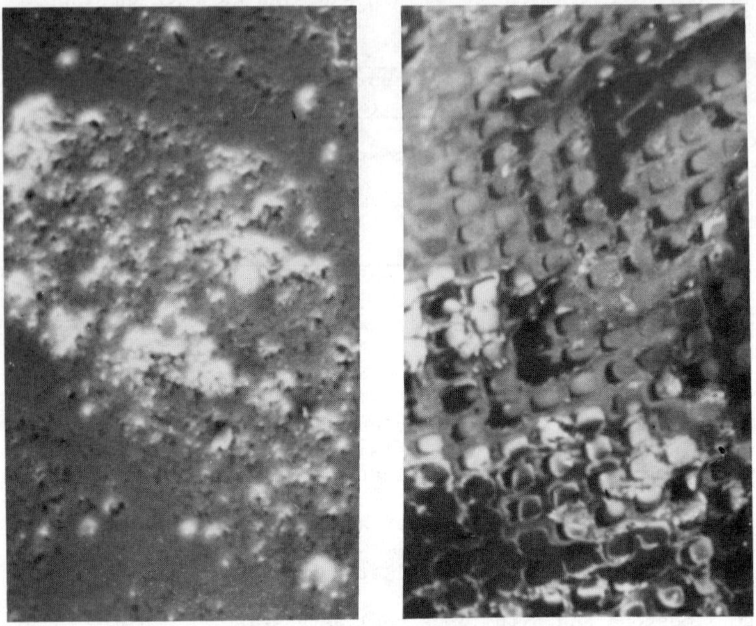

FIG. 5. - SCANNING ELECTRON MICROSCOPE PHOTOGRAPHS OF LIMESTONE TEST ROCK SURFACE AND ALUMINUM GRID DEPOSITION

LAYOUT OF THE OPTICAL TEST SYSTEM FOR A NONCONTACT FULL FIELD DISPLACEMENT - STRAIN MEASUREMENT

Figure 6 shows the features of the precision optical projection system used for the study. The image moiré technique was adopted to eliminate the need for physical contact between the specimen and the reference grid. The image of the specimen grid was projected on a ground glass plate with etched master grids. This arrangement allowed a pitch mismatching of the specimen grid image to increase the fringes for the tests. The specimen was placed in a 10,000-lb screw type mechanical

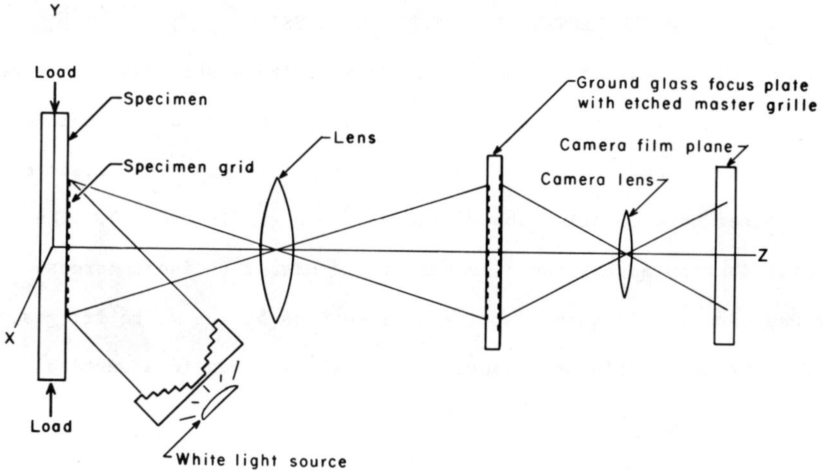

FIG. 6. - OPTICAL LAYOUT FOR MOIRÉ FRINGE MEASUREMENT

loading frame equipped with a strain gage dynamometer which accurately measured the load during testing. During the tests, the pitch was mismatched optically in either the positive or the negative direction depending on whether the specimen is loaded by compression or by tension. The optical interference of the specimen grid image, superimposed on the reference grid, produced the moiré fringes in the ground glass plate. The fringes were photographically recorded during testing. The optical system was arranged to rotate 90 degrees to record the fringes of both the inplane components of displacements. A variable power multihead light source was incorporated to illuminate the specimen grid to achieve sufficient fringe contrast for the study. The system permits a noncontact measurement of biaxial strain distribution by rotating the optical system by 90 degrees.

EXPERIMENTAL RESULTS AND DISCUSSION

The optical alignment of the experimental moiré system and the large field illumination for the strain distribution measurements were initially tested with 500 lines/inch nickel transfer grids fixed on the surface of a 2-in diameter hard elastic rubber disc specimen (India rubber). The strain distribution across the diameter of the disk for line compressive loading was measured in both coordinate directions by the moiré fringes. Figure 7 shows an example of measurements of the fringes in a vertical

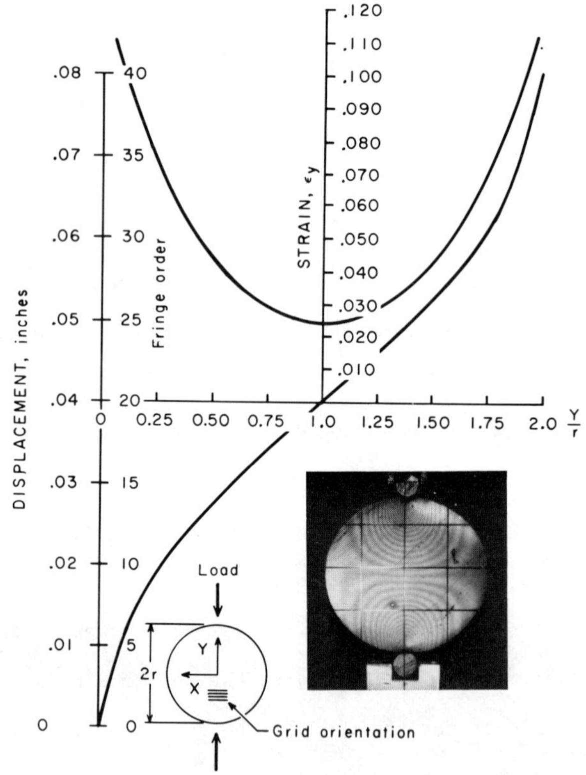

FIG. 7. - STRAIN DISTRIBUTION IN A CIRCULAR DISC OF ELASTIC MATERIAL DURING COMPRESSIVE LOADING USING THE MOIRÉ TECHNIQUE

field and the corresponding displacement-strain distribution across the diameter of the disk. Good fringe contrast was obtained to determine large field strain distributions by the noncontact methods this study describes.

A study of large field displacement-strain distribution is of particular interest to analyze fragmentation processes of heterogeneous rock and coal materials. The application of the moiré method to determine large field displacement-strain distribution up to fracture and fragmentation of a large specimen (2x2x4 in) of bituminous coal consisting of 12 percent moisture, 25 percent volatile combustible matter, and 40 percent fixed carbon was examined using the experimental system. The specimen was prepared with a 500 lines/inch nickel transfer grid on its surface (2x4 in) and tested in compression. The bedding and cleavage planes of the coal specimen were placed parallel to the direction of the compressive loading. The structural variations of the bituminous coal induce a nonuniform distribution of strains during the compressive loading. An initial mismatch of approximately 8 fringes/inch was placed in both fields of fringe alignments before loading the sample. The initial mismatch induces fringe additions in the longitudinal field and fringe deletions in the lateral field of the moiré isothetics during loading.

The progressive pattern of the full field displacement isothetics of the specimen surface in the lateral field (u field) of the specimen placement for stress levels at 0, 375, 625, and 1,000 psi are shown in Fig. 8. The displacement distribution in the lateral field (Fig. 8) during loading is nonuniform. The dilation of the fringe grouping indicates a high strain concentration in a localized zone at the right

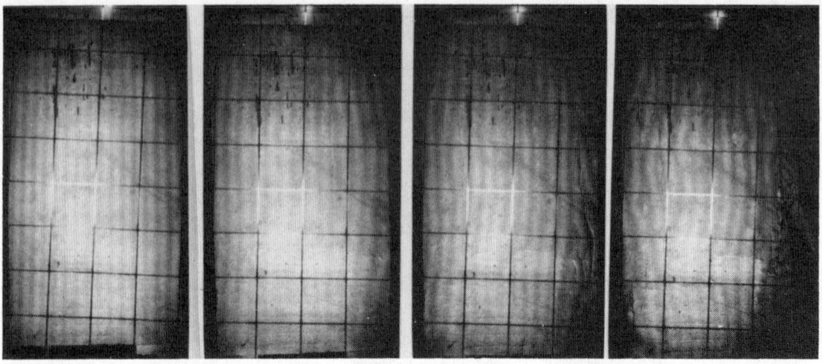

FIG. 8. - LATERAL FIELD (u FIELD) MOIRÉ ISOTHETICS DURING COMPRESSIVE LOADING UP TO FRACTURE AND FRAGMENTATION OF BITUMINOUS COAL

middle section during loading to 375 psi. The progressive loading of the specimen to 625 psi initiates cracks in this localized zone of fringe dilation, followed by fracture and fragmentation of this section by increasing the stress to 1,000 psi. These lateral field moiré isothetics help determine the strain distribution at all points across the width of the sample placement. Figure 9 shows an example of the results of the analysis on strain distribution for sections A and B above the middle of the specimen for compression at 375 psi. The strain distribution along both the sections A and B are nonuniform and the section B is subjected to higher strain levels than section A during the compression. The strains have peak values in section B at about 1/4 in from the right specimen edge. During progressive loading to 625 psi, fractures initiate in this zone of high strain concentration.

Figure 10 shows an example of the longitudinal field moiré isothetics recorded for identical positions indicated in Fig. 8. These displacement isothetics help determine the strain distribution along the

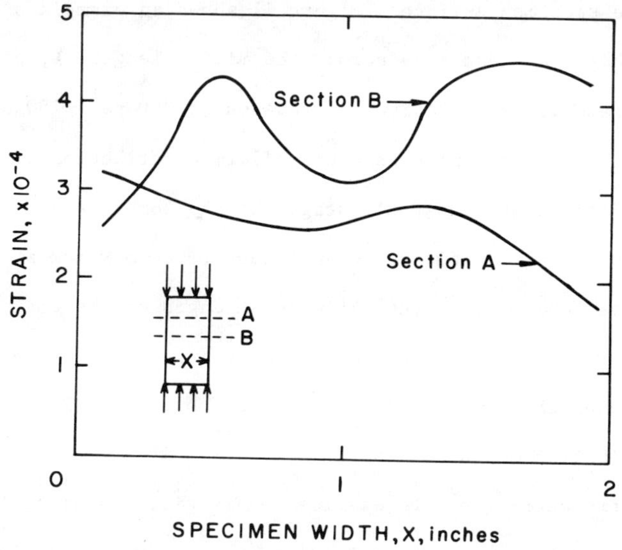

FIG. 9. - AN EXAMPLE OF LATERAL STRAIN DISTRIBUTION DURING COMPRESSION OF A COAL SPECIMEN (σ = 325 PSI)

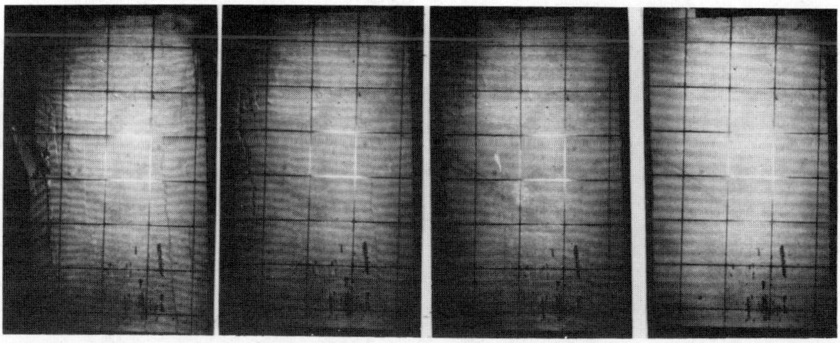

FIG. 10. - LONGITUDINAL FIELD (v FIELD) MOIRÉ ISOTHETICS DURING COMPRESSIVE LOADING UP TO FRACTURE AND FRAGMENTATION OF BITUMINOUS COAL

length of the specimen surface. Figure 11 shows an example of the results of the strain analysis across the middle length, L, of the specimen surface for progressive compression at 250 psi, 750 psi, and 1,375 psi. The results indicate a nonuniform distribution of strain across the specimen length at all stages of the compressive loading. The variations in strain distribution appear to be maximum at intermediate stress levels below the ultimate strength of the coal specimen. Despite the fracture of the specimen side seen in Fig. 10, the unfractured portion of the specimen display fringe continuity for displacement-strain distribution measurements.

Tests were carried out to examine the feasibility of applying the moiré method for rock material. Because of the porous granular texture,

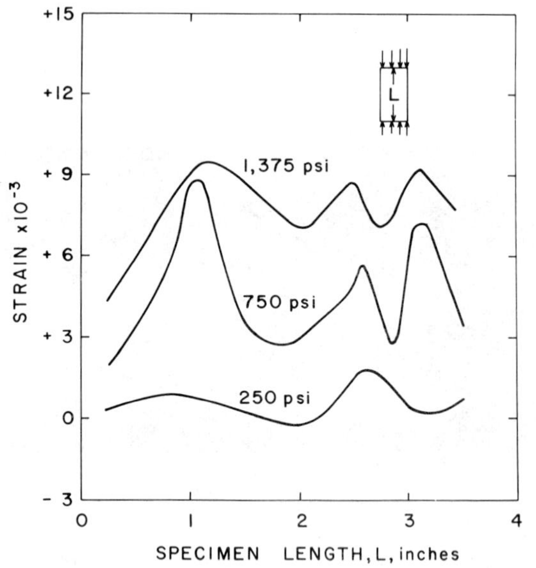

FIG. 11. - LONGITUDINAL STRAIN DISTRIBUTION DURING PROGRESSIVE COMPRESSION OF A COAL SPECIMEN

the Salem limestone test rock was selected in order to determine the applicability of the moiré method for rock studies. Medium and high intensity grids up to 1,000 lines/inch were successfully deposited on the test rock surface by the vacuum evaporation techniques described in this study. Despite the porous structure of the limestone, the grid was uniformly deposited as shown by the scanning electron micrographs in Fig. 5, and a good fringe contrast was obtained for the measurement. Figure 12 shows an example of the lateral (u field) and the longitudinal (v field) moiré isothetics in a compressive loaded specimen of Salem limestone (1x1x2 in) with vacuum deposited aluminum grids of 200 lines/inch. The displacement isothetics over the full surface area of the specimen is shown in the figure. In order to increase the sensitivity of the measurements, fine grids (1,000 lines/inch) were successfully

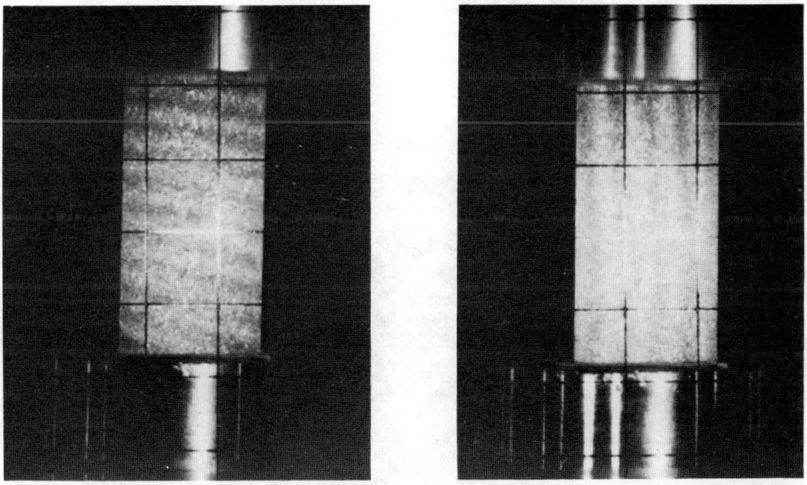

FIG. 12. - THE MOIRÉ ISOTHETICS DURING COMPRESSION TESTING OF SALEM LIMESTONE

deposited on rock specimens. The versatility of the method for use in all types of rock testing is further demonstrated by the fringes obtained during three point bending of a limestone specimen (½x½x4 in) with 1,000 lines/inch aluminum grids (Fig. 13). The deflection isothetics seen in Fig. 13 is obtained for a low bending stress of 8 psi. Despite the low stress levels, a dilation of the fringes are observed. A full field displacement-strain distribution can be obtained by analyzing these fringes.

The resolution of the moiré method for rocks is highly dependent on the line density of the grids and on the sensitivity of the optical system used in making fringe measurements. Though fine grids up to 1,000 lines/inch have been used successfully on porous limestone in this study, increasing the density of the grids beyond about 1,000 lines/inch

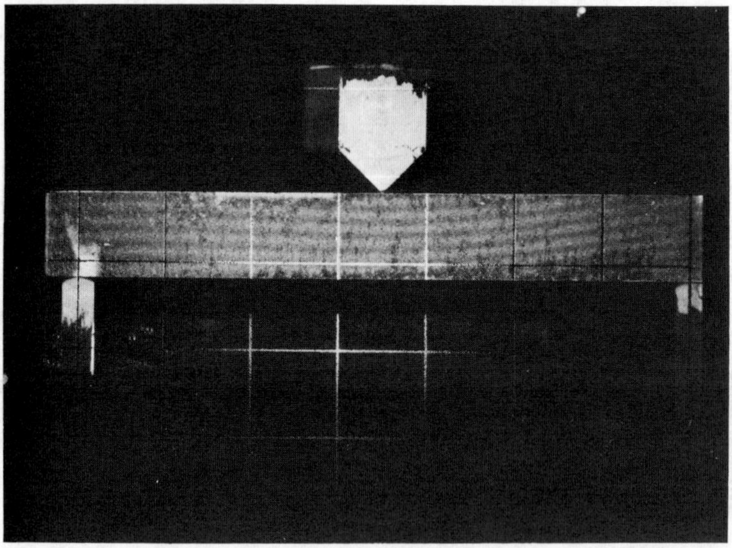

FIG. 13. - MOIRÉ FRINGES DURING THREE POINT BENDING OF SALEM LIMESTONE

increases the problems of fringe disorder caused by the structural and surface features of the rock and reduces fringe sharpening and fringe contrast. The resolution specific grid frequencies give can be increased by improving the optical techniques for making fringe measurements. High resolution optical fringe interpolations can be obtained by the microdensitometer methods Sciammarella suggests (9) or by the fringe multiplication methods with diffraction grating Guild proposes (3). A fringe multiplication system adapted from Post (7) is now being devised at this laboratory to improve the sensitivity of the fringe measurements for rock studies.

CONCLUSIONS

This study shows the feasibility of applying the moiré method for rock and coal studies. The method helps determine the full field inplane strain distribution up to fracture and fragmentation point of these materials. The noncontact moiré method has the potential to examine novel and advanced rock fragmentation processes which involve rock testing at elevated temperature and under adverse environmental conditions.

ACKNOWLEDGEMENTS

The authors are indebted to Robert P. Lund, physical science technician, Twin Cities Mining Research Center, for his support in the experimental study.

APPENDIX I. - REFERENCES

1. Dantu, P., "Extension of the Moiré Method to Thermal Problems," Experimental Mechanics, Vol. 4, No. 3, Mar., 1964, pp. 64-69.

2. Durelli, A. J., and V. J. Parks, Moiré Analysis of Strain, Prentice-Hall, Inc., 1967, 399 pp.

3. Guild, J., The Interference System of Crossed Diffraction Grating, Oxford Clareden Press, 1956, 320 pp.

4. Holland, L., Vacuum Deposition of Thin Films, Chapman Hall, 1966, pp. 1-33, 169-198, 320-355.

5. Maissel, L. I., and R. Glang, Handbook of Thin Film Technology, McGraw Hill, 1970, pp. 1.3-7.1

6. Parks, V. J., and A. J. Durelli, "Various Forms of the Strain-Displacement Relations Applied to Experimental Strain Analysis," Experimental Mechanics, Vol. 4, No. 2, Feb., 1964, pp. 37-47.

7. Post, D., "Moiré Fringe Multiplication with a Non-Symmetrical Doubly Blazed Reference Grating," Applied Optics, Vol. 10, No. 4, 1971, pp. 901-906.

8. Raleigh, Lord, "On the Manufacture and Theory of Diffraction Gratings," Philosophical Magazine, Vol. 47, No. 81, 1874, p. 193.

9. Sciammarella, C. A. "Basic Optical Lane in the Interpretations of Moiré Patterns Applied to the Analysis of Strains," Experimental Mechanics, Vol. 5, No. 5, May, 1965, pp. 154-160.

10. Theocaris, P. S., Moiré Fringes in Strain Analysis, Pergamon Press, 1969, 426 pp.

11. Thirumalai, K., "The Process of Thermal Spalling Behavior in Rocks - An Exploratory Study," Ch. 34 in Rock Mechanics - Theory and Practice,

Proceedings of 11th Symposium on Rock Mechanics, ed. by W. H. Somerton, AIME, New York, 1970, pp. 705-727.

12. Thirumalai, K. and S. G. Demou, "Effect of Reduced Pressure on Thermal Expansion Behavior of Rocks and Its Significance to Thermal Fragmentation," Journal of Applied Physics, Vol. 41, No. 13, Dec., 1970, pp. 5147-5151.

FAULTING ACROSS DISCONTINUITIES IN ANISOTROPIC ROCK

By Fred A. Donath[1]

INTRODUCTION

Most rocks encountered by engineering projects in the upper parts of the earth's crust are characterized by some type of planar structure--such as bedding, cleavage, or jointing--and thus are distinctly anisotropic. Previous experimental studies (1,2,3,9,10) have demonstrated that the presence in rocks of simple planar anisotropy can have a pronounced effect on both the strength and the geometry of deformation. Not only do faults tend to parallel well-developed planar anisotropy at inclinations up to about 60° to the direction of maximum compression, but they will develop at significantly lower stress differences for these inclinations, as compared with that required at inclinations of 75° or 90°. The present study evaluates the effects on strength and deformational geometry in slate caused by abrupt changes in the orientation of the anisotropy (cleavage) with respect to the direction of maximum compression. All data are from triaxial compression tests run at 400 bars confining pressure--i.e., at the pressure equivalent of about one mile of dry rock overburden. The tests were run dry, at room temperature, and at a strain rate of 1.5×10^{-4} per second.

---

[1] Professor and Head, Dept. Geology, University of Illinois, Urbana, Ill. 61801

EXPERIMENTAL PROCEDURE

Compound specimens were prepared to simulate changes in attitude of anisotropy such as might be encountered in folded rock sequences, at intersections of cross-bedding and stratigraphic layering, at cleavage-bedding or fracture-bedding intersections, and in similar natural situations. Each compound specimen consisted of two right circular cylinders ½-inch diameter by ½-inch length, hereafter referred to as half-cylinders, held together by a small drop of Duco cement to form a single test cylinder one inch long. The half-cylinders were prepared from longer cores taken at 0°, 15°, 30°, 45°, 60°, 75°, and 90° to the plane of anisotropy.

Martinsburg slate (Ordovician; Bangor, Pennsylvania) was used in the study. This rock is characterized by a well developed mechanically effective rock cleavage that had been systematically evaluated in previous studies by the author (2,3). The drop of cement between the half-cylinders of slate was required to keep the cleavages properly aligned during handling of the specimens. The effect of the cement on the strength and deformational geometry was not evaluated; it was assumed to be negligible, an assumption that appears to be supported by the experimental results.

Illustrations of the method of obtaining oriented cores and of the test apparatus, and descriptions of other aspects of testing procedure and data reduction are provided elsewhere (3,6,8). The procedures followed in this study differ in two additional ways from those previously described: (1) marble spacers were substituted for the soft steel spacers normally placed between the test specimen and the piston and anvil; and (2) a Vidar 5206 Data Acquisition System was used to collect signals from the load cell, differential transformer, and other transducers during testing, for subsequent use in data reduction. Signals were collected from each of the several transducers at one-second intervals during each test, thus ensuring a virtually continuous record of differential stress versus longitudinal strain from the reduced data. Marble, rather than steel, spacers were utilized to permit faults initiated within the slate to propagate into the spacers. This is especially important for the 15-degree orientation, as faults developed in that orientation almost invariably intersect the spacers.

Because faulting normally occurs parallel to cleavage for inclinations of 15°, 30°, and 45° to the direction of maximum compression, these three orientations were used as standards--i.e., every specimen tested included a half-cylinder of one of these orientations. In addition, because it was not known whether differences would be observed if a particular half-cylinder were placed against the "active" piston instead of against the "passive" anvil, several of the tests were duplicated with the specimen configuration inverted, thus permitting evaluation of any possible active-passive effect.

The test series that were run in this study fall into two groups: specimens in which the intersections of the cleavage with the circular sections or ends of the specimen are parallel in each half-cylinder, the "strikes-parallel group"; and specimens in which the strikes in the upper half-cylinder are rotated with respect to those in the lower half-cylinder, the "strikes-rotated group". The first group included four test series, with the 15-degree orientation as the standard half-cylinder on both the bottom and the top, the 30-degree orientation as standard on the bottom, and the 45-degree orientation as standard on the bottom. The second group consisted of the following combinations: (1) a 15-degree half-cylinder above a 30-degree half-cylinder, with the strike of the cleavage in the upper half rotated 15°, 30°, 45°, 90°, and 180°, in different specimens, with respect to that in the bottom half-cylinder; (2) a 30-degree half-cylinder above a 45-degree half-cylinder, with the strike of the cleavage in the upper half rotated 30°, 60°, and 180°, respectively; and (3) two 45-degree half-cylinders rotated 60°, 90°, 135°, and 180° with respect to one another.

Individual specimen configurations will be referred to as 30/45 or 15/30-180, for example, in which the number preceding the slant indicates the cleavage inclination in the upper or top half-cylinder, the number following the slant indicates the cleavage inclination in the lower or bottom half-cylinder, and the number following the hyphen, when present, indicates the horizontal angular rotation of the dip vector in the upper half-cylinder with respect to that in the lower.

DEFORMATION IN STRIKES-PARALLEL GROUP

Deformational Modes.--Three modes of deformation were observed among the various combinations studied: (1) brittle faulting along a single shear fracture or along a few parallel shear fractures; (2) slip on numerous cleavage surfaces; and (3) kink banding. The results for tests on the strikes-parallel group of specimens are given in Table 1. The mode of deformation in each half-cylinder is designated by a letter; the letter preceding the slant indicates the mode in the top half-cylinder and that following the slant indicates the mode in the bottom half-cylinder.

TABLE 1.--MODES OF DEFORMATION IN STRIKES-PARALLEL SPECIMENS

Standard Half-Cylinder	Associated Half-Cylinder 0°	15°	30°	45°	60°	75°	90°
15° top	$\underline{K}$/N		B/$\underline{S}$	$\underline{S}$/$\underline{S}$	K/B,K/B	$\underline{K}$/B	K/N
15° bottom	N/K		$\underline{S}$/$\underline{K}$	$\underline{K}$/$\underline{K}$,$\underline{K}$/K	B/$\underline{K}$	B/$\underline{K}$	N/K
30° bottom	N/$\underline{S}$			$\underline{S}$/S	B/$\underline{K}$,$\underline{K}$/B*	B/$\underline{K}$,$\underline{K}$/B*	N/$\underline{S}$
45° bottom	N/$\underline{S}$	K/$\underline{K}$	$\underline{S}$/$\underline{S}$		B/$\underline{S}$	N/$\underline{S}$	N/$\underline{S}$

* 30-degree half-cylinder on top
B = brittle fault; S = slip; K = kink band; N = not deformed
Underscored S or K indicates that a fault is also present in the half-cylinder

As seen from Table 1, neither the 0-degree nor the 90-degree half-cylinders deformed permanently under the conditions and in the combinations tested (see, e.g., Fig. 1a). The reason for this is obvious when one considers the strength data. The yield stress for these two orientations is 2.4 to 5.9 times greater than that of any of the standard orientations (see Table 2), and the differential stress never reached the required levels for failure in the stronger orientations. The 75-degree half-cylinders fail by a well developed shear fracture when in contact with 15- and 30-degree half-cylinders (Fig. 1b), but did not fail in contact with the 45-degree half-cylinder. Failure in the 60-degree half-cylinders is by brittle faulting in each instance (Fig. 1c). Slip is the typical mode of deformation in the 30-degree and 45-degree half-cylinders (Fig. 1d), although poorly developed kink bands are present in four of the former and three of the latter orientation (Fig. 1f). The kink band is the preferred mode of deformation in the 15-degree half-cylinders (Fig. 1b,f). Kink bands are present

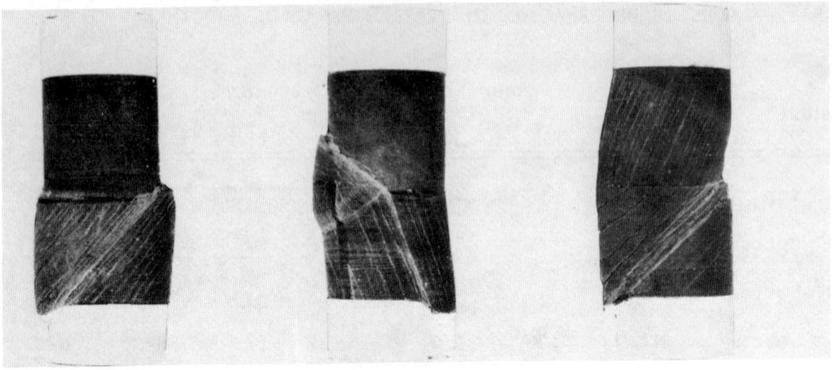

0/45 75/15 15/60

45/30 15/30 45/15

FIG. 1.--DEFORMATIONAL MODES IN STRIKES-PARALLEL SPECIMENS

Upper row, left to right: (a) 0-degree nondeformed half-cylinder above 45-degree half-cylinder deformed by slip and faulting; (b) brittle faulting in 75-degree half-cylinder above kink band and fault in 15-degree half-cylinder; (c) kink band in 15-degree half-cylinder above brittle fault in 60-degree half-cylinder.

Lower row, left to right: (d) slip and fault in 45-degree half-cylinder above 30-degree half-cylinder deformed by slip; (e) brittle faulting in 15-degree half-cylinder above 30-degree half cylinder showing slip and fault; (f) kink bands and faulting in 45-degree half-cylinder above 15-degree half-cylinder.

in all but two of the fifteen 15-degree orientations, although they are not uniformly well developed in the various combinations. A kink band normally does not develop without slip having occurred first; whether a kink band subsequently develops as a consequence of slip depends upon the rate and amount of displacement that occurs during the initial slip (4).

There appears not to be any active-passive effect on the mode of deformation in the 15/0, 15/60, 15/75, 15/90, 30/45, 30/60, or 30/75 configurations. In the 15/30 and 15/45 configurations, however, it appears that faulting or slip is favored in the 15-degree half-cylinder when it is next to the piston and kink bands are favored when it is next to the anvil. This result is consistent with the conclusion drawn from another study that the boundary stresses at the piston-specimen interface are influenced by the orientation of the anisotropy and most strongly affected for the 15-degree orientation (7).

Fault Attitudes.--The representative specimens shown in Figure 1 illustrate the fact that the geometry of deformation is strongly influenced by the inclination of the cleavage within each half-cylinder. This influence can be demonstrated graphically by plotting the inclination of faults against the inclination of the cleavage, as done previously (2,3). Of particular interest in the present study is the effect on fault geometry of abrupt changes in attitude of the anisotropy; this effect can be evaluated from the data presented in Figure 2.

Faulting in the 15-, 30-, and 45-degree standard half-cylinders is always parallel to the cleavage, and the symbols in Figure 2 are therefore used to indicate the relationship between faulting and

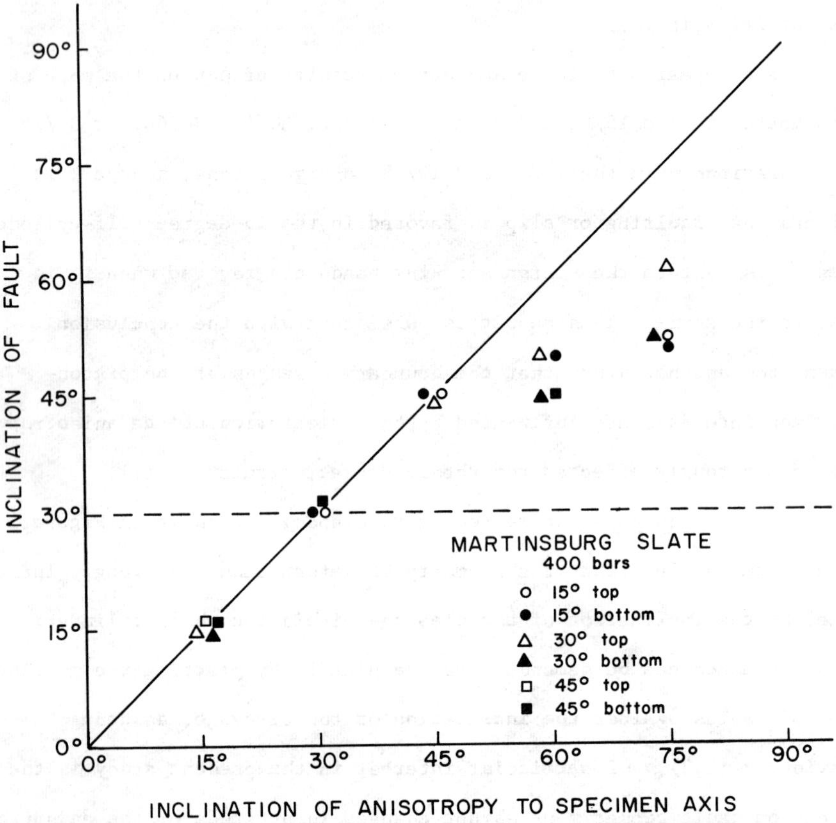

FIG. 2.--FAULT INCLINATION VERSUS ANISOTROPY INCLINATION

cleavage inclination in the associated half-cylinder. For example, the open triangle at the uppermost right of the diagram represents a 30° half-cylinder above a 75° half-cylinder. Faulting occurred parallel to the cleavage in the upper half-cylinder and, as indicated by the position of the symbol on the diagram, a fault developed at 61° to the direction of maximum compression in the 75-degree bottom half-cylinder. As shown in Figure 2, faults developed parallel to the cleavage in all 15-, 30-, and 45-degree orientations of the associated half-cylinders, and the inclinations of the faults were strongly affected by the inclination of the anisotropy in the 60- and 75-degree orientations. In homogeneous, isotropic rock one would expect the inclination of the faults to be approximately 30° to the loading axis--i.e., the symbols would fall along the horizontal dashed line. Figure 1e shows a typical specimen (15/30) in which the fault attitude changes abruptly where an abrupt change in the attitude of the anisotropy occurs.

Strength Variation.--Not only does the inclination of the anisotropy have a controlling effect on the inclination of the faults, but it also strongly influences the differential stress at which failure occurs. The variation in strength with different orientations of anisotropy for the Martinsburg slate is very pronounced, as demonstrated by previous work (2,3). Of interest here is the factor controlling the strength in instances in which an attitude discontinuity exists.

The maximum differential stress sustained by each of the strikes-parallel compound specimens tested in this study is plotted in Figure 3. It is clear from the data presented in Figure 3 that the strength of each compound specimen having either a 30-degree or a 45-degree standard half-cylinder is simply that of the respective standard half-

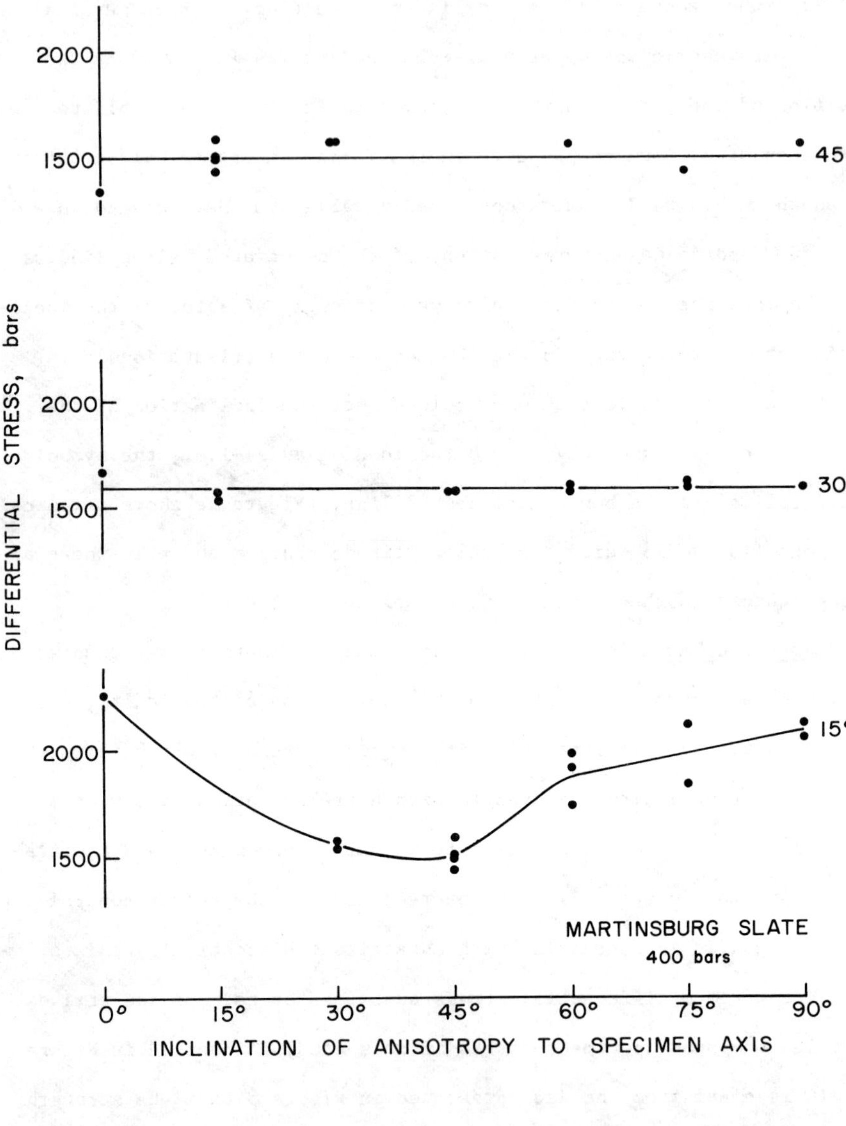

FIG. 3.--STRENGTH OF STRIKES-PARALLEL COMPOUND SPECIMENS

cylinder. Some variation in strength is present in compound specimens having a 15-degree standard half-cylinder. Slip or faulting parallel to the cleavage must intersect the spacers for this orientation, and the strength variation undoubtedly reflects boundary conditions. A comparison of strength variation in the compound specimens for the three standard orientations with that in whole cylinders of Martinsburg slate is given in Table 2. The data for the whole cylinders are from reference (2), and all data are expressed in terms of coefficients with the strength of the 30-degree orientation taken as 1.00.

TABLE 2.--ANISOTROPY STRENGTH COEFFICIENTS FOR MARTINSBURG SLATE

Specimen Type	Orientation 0°	15°	30°	45°	60°	75°	90°
whole cylinder*	4.65	1.94	1.00	1.41	2.24	3.53	5.88
15° half-cylinder	1.46		1.00	0.98	1.21	1.29	1.36
30° half-cylinder	1.05	0.98		1.00	1.01	1.03	1.01
45° half-cylinder	0.86	0.96	1.00		1.00	0.91	1.00

* based on 350-bar data

No definite conclusion can be drawn from comparison of the strengths of inverted configurations that were run to evaluate the active-passive effect. The only significant variation in strength among all configurations occurs in the 15/45, 15/60, and 15/75 specimens (Fig. 3). Although the compound specimen is stronger when the stronger half-cylinder is next to the piston in the 15/75 specimens, the reverse is true for the 15/60 configuration. Among the 15/45 specimens, one inverted specimen and its counterpart have virtually identical strengths; the strongest has the stronger half-cylinder (15-degree) next to the piston. As mentioned above, the observed variation undoubtedly reflects boundary conditions.

DEFORMATION IN STRIKES-ROTATED GROUP

Deformational Modes.--Effects caused by changes in the direction of the anisotropy, as well as in the amount of inclination, were evaluated by testing several compound specimen configurations (15/30, 30/45, and 45/45) in which the dip vector for the upper half-cylinder was rotated with respect to that in the lower half-cylinder. The modes of deformation that were observed for these special configurations are given in Table 3. For purposes of comparison, the modes observed for the strikes-parallel group (0° angular deviation) are included in the table.

TABLE 3.--MODES OF DEFORMATION IN STRIKES-ROTATED SPECIMENS

	Angular Deviation of Dip Vectors							
Specimen Type	0°	15°	30°	45°	60°	90°	135°	180°
15/30	B/S	B/S	B/K	B/K		N/S		K/S
30/45	S/S		S/S		S/S			K/K
45/45					S/S	S/S	K/K	K/K

The mode of deformation is clearly affected by changes in direction of the anisotropy, but the effect differs depending upon the magnitude of change in both inclination and direction. With increased rotation of the dip vector in the upper half-cylinder with respect to that in the lower for the 15/30 configuration, the 15-degree upper half-cylinder deforms by brittle faulting for angular deviations up to 45° (see Fig. 4a), it undergoes no permanent deformation when the cleavage in it dips at right angles to that in the bottom half-cylinder, and it develops a kink band when the cleavage in the two half-cylinders dips in opposite directions (Fig. 4b). Deformation in the 30-degree half-cylinder at the bottom is typically slip with faulting (Fig. 4b), except when the angular deviation of dip vectors is 30° and 45°, in which case kink bands develop (Fig. 4a). The mode of deformation in

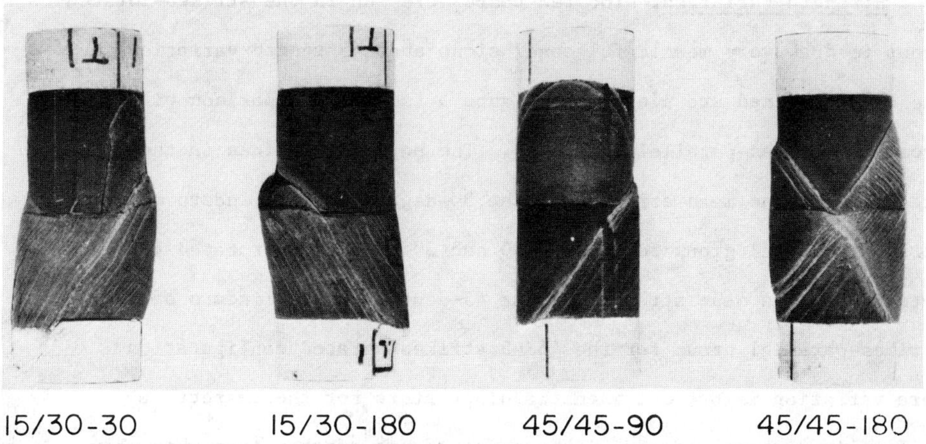

FIG. 4.--DEFORMATIONAL MODES IN STRIKES-ROTATED SPECIMENS

Left to right: (a) 15-degree upper half-cylinder showing brittle fault above and rotated 30° with respect to 30-degree lower half-cylinder showing fault and kink band; (b) 15-degree upper half-cylinder with kink band above and rotated 180° with respect to 30-degree lower half-cylinder deformed by slip and faulting; (c) 45/45-90 specimen deformed by slip and faulting in each half-cylinder; (d) 45/45-180 specimen showing kink bands and faults in both upper and lower half-cylinders.

the 30/45 specimens is observed to be slip in both half-cylinders for rotations up to 60°. Kink bands with faults developed in both half-cylinders, however, when the dip vectors were in opposite directions.

One might surmise from the results for the 30/45 and 45/45 specimens that slip with or without faulting in both half-cylinders would be characteristic for both configurations for angular deviations up to 90° (Fig. 4c), whereas kink bands would be favored when the dip vectors first begin to dip in opposite directions--i.e., when the angular deviation exceeds 90°. Kink bands, with faults, developed in both upper and lower half-cylinders of the 30/45-180, 45/45-135, and 45/45-180 configurations (Fig. 4d).

Strength Variation.--Too few tests were run in the strikes-rotated group to draw very meaningful conclusions about strength variation, but the data obtained are plotted in Figure 5 to allow comparison with data from the strikes-parallel specimens. The horizontal lines in the figure indicate the mean strength of the 30-degree half-cylinders of the strikes-parallel group for the 15/30 and 30/45 strikes-rotated configurations and the mean strength of the 45-degree half-cylinders of the strikes-parallel group for the 45/45 strikes-rotated configuration. More variation is present when the dip vectors for the respective half-cylinders are rotated with respect to one another than when they are in the same direction. However, in general, it appears that the strikes-rotated compound specimens which include a 30-degree half-cylinder tend to be slightly weaker than the 30-degree strikes-parallel counterparts, and the 45-degree strikes-rotated specimens are somewhat stronger than the 45-degree strikes-parallel counterparts. There is some suggestion of trends, and when additional data are available it may be worthwhile to determine the resolved shear stress on planes of failure and analyze the results in a more meaningful way. The strength variation caused by angular rotation of dip vectors that has been

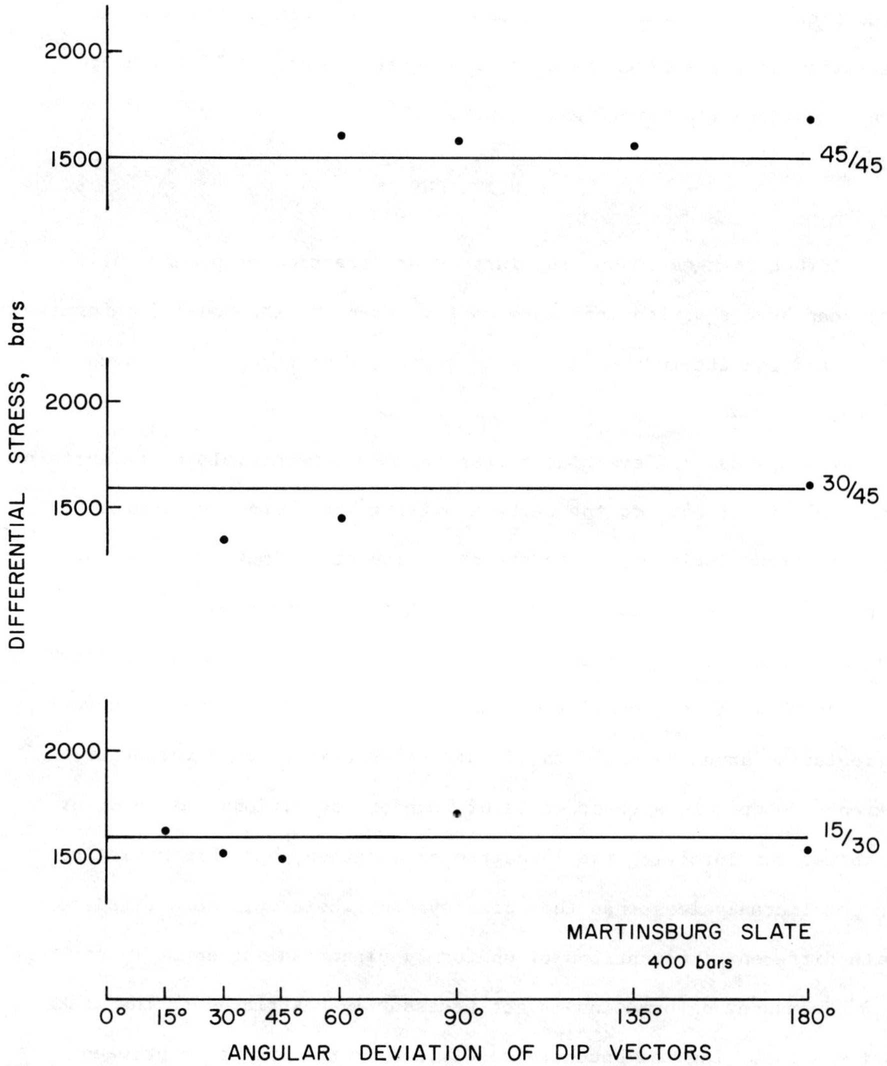

FIG. 5.--STRENGTH OF STRIKES-ROTATED COMPOUND SPECIMENS

observed thus far, although greater than in the strikes-parallel 30-degree and 45-degree standard half-cylinder configurations, is still significantly less than that caused by boundary conditions and dip variation in the strikes-parallel 15-degree standard half-cylinder configurations (compare Figs. 3 and 5).

CONCLUSIONS

Abrupt changes in the inclination or direction of planar anisotropy can have a marked influence on the strength, the mode of deformation, and the attitude of faults in strongly anisotropic rock such as slate.

The maximum differential stress that Martinsburg slate can sustain is equal, in effect, to the weakest orientation among any combination of anisotropy inclinations to the direction of maximum compression. Experimental results at 400 bars confining pressure show that all combinations which involve the 30-degree orientation have the strength of that orientation; similarly, combinations involving the 45-degree orientation have the strength of that orientation. Some variation in strength occurs as a consequence of boundary conditions, as shown by combinations involving the 15-degree orientation, but this variation is considerably less than that displayed by whole cylinders of slate with different inclinations of uniformly planar anisotropy.

Permanent deformation is not necessarily restricted to the weaker orientation. Induced boundary conditions at the interface between anisotropic half-cylinders of different inclinations, presumably related to step displacements and release of elastic strain energy caused by failure in the weaker orientation, can create the necessary conditions for failure in normally stronger associated orientations. For orientations associated with cleavage inclinations of 15°, 30°, or 45°, failure is restricted to these weaker orientations only for maximum compression perpendicular or parallel to the cleavage; brittle faulting occurs in slate half-cylinders in which the cleavage is inclined at 60° and 75°. The faults that develop in these half-cylinders are inclined 45° to 60° to the loading axis, appreciably more than the 30° to be expected in isotropic rock.

Slip or faulting occurs parallel to the cleavage in all 15-degree, 30-degree, and 45-degree half-cylinders. Kink-bands typically develop in the 15-degree half-cylinders, and are present also in several of the 30- and 45-degree orientations when associated with 15-degree, 60-degree, and 75-degree orientations. If the cleavage dip vectors of associated half-cylinders are rotated as well as varied in inclination, kink bands are favored over faulting and slip in the 30-degree half-cylinder of the 15/30 configuration with angular rotations of the dip vectors of 30° and 45°. Similarly, kink bands are favored over slip and faulting in both half-cylinders of the 30/45 and 45/45 configurations when the angular rotations are 135° and 180°.

The lowest confining pressure at which kink bands had formed in previous studies of this rock was 800 bars (5). Their development in this study at 400 bars serves to underscore an earlier statement (ref. 4, p. 469 and 491) that the effects of cohesion and other complicating factors (e.g., boundary conditions) will have to be evaluated if the phenomenon is to be of use as a pressure indicator. The slate used in this study is appreciably stronger than that used earlier, owing to its higher cohesive resistance (7). The strength of the 30-degree orientation at 500 bars confining pressure reported in reference (3) was about 770 bars differential stress. At 400 bars, the slate used for this study has a strength of about 1590 bars differential stress. As pointed out in reference (4), kink bands are favored in slate specimens having higher cohesive resistance and, hence, higher stored elastic strain energy at failure. In slate having lower cohesive resistance, kink bands would not develop at pressures as low as 400 bars nor would they form in the 30- and 45-degree orientations.

Results of this study indicate that the attitude of a fault propagating through anisotropic rock with an attitude discontinuity will be strongly affected by the anisotropy inclinations. Faults can be expected to parallel the anisotropy for all combinations of inclinations of 15°, 30°, and 45° to the direction of maximum compression. For inclinations of 60° and 75°, faults would be inclined at 45° to 60° to that direction. For small angular rotations of the dip vectors (e.g., 15° and 30°) with small changes in inclination (e.g., $15^{o}/30^{o}$), faults propagating across an attitude discontinuity tend to parallel the anisotropy across the discontinuity. For larger angular rotations, faulting, slip, or kink bands develop on opposite sides of the discontinuity, depending upon the amount of angular rotation and the respective inclinations of the anisotropy with respect to the deforming stresses.

ACKNOWLEDGMENTS

The experimental work was carried out in the Structural Geology High Pressure Laboratory at the University of Illinois, Urbana. The assistance in the laboratory of Jeffery H. Randall is gratefully acknowledged; Mr. Randall was supported during the summer of 1970 by Shell Aids funds granted to the Department of Geology by the Shell Companies Foundation, Inc.

APPENDIX.--REFERENCES

---

1. Chenevert, M. E., and Gatlin, C., "Mechanical Anisotropies of Laminated Sedimentary Rocks", Journal, Society of Petroleum Engineers, Vol. 5, Mar., 1965, pp. 67-77.
2. Donath, Fred A., "Experimental Study of Shear Failure in Anisotropic Rocks", Bulletin, Geological Society of America, Vol. 72, June, 1961, pp. 985-990.
3. Donath, Fred A., "Strength Variation and Deformational Behavior in Anisotropic Rock", State of Stress in the Earth's Crust, W. R. Judd, ed., American Elsevier, New York, 1964, pp. 281-297.
4. Donath, Fred A., "The Development of Kink Bands in Brittle Anisotropic Rock", Igneous and Metamorphic Geology, Geological Society of America Memoir 115, 1968, pp. 453-493.
5. Donath, Fred A., "Experimental Study of Kink Band Development in Martinsburg Slate", Proceedings, Conference on Research in Tectonics, Geological Survey of Canada Paper 68-52, 1969, pp. 255-288.
6. Donath, Fred A., "Some Information Squeezed Out of Rock", American Scientist, Vol. 58, No. 1, Jan.-Feb., 1970, pp. 54-72.

7. Donath, Fred A., "Effects of Cohesion and Granularity on Deformational Behavior of Anisotropic Rock", Studies in Precambrian Geology and Mineralogy, Geological Society of America Memoir 135, in press.

8. Donath, Fred A., and Fruth, Lester S., Jr., "Dependence of Strain-Rate Effects on Deformation Mechanism and Rock Type", Journal of Geology, Vol. 79, No. 3, May, 1971, pp. 347-371.

9. McGill, George E., and Raney, Jay A., "Experimental Study of Faulting in an Anisotropic, Inhomogeneous Dolomitic Limestone", Bulletin, Geological Society of America, Vol. 81, Oct., 1970, pp. 2949-2958.

10. McLamore, R., and Gray, K. E., "The Mechanical Behavior of Anisotropic Sedimentary Rocks", Transactions, American Society of Mechanical Engineers, Series B, Feb., 1967, pp. 62-76.

SHAPE OF THE COMPLETE STRESS-STRAIN CURVE FOR ROCK

John A. Hudson[1], Edwin T. Brown[2] and Charles Fairhurst[3]

INTRODUCTION

In order to calculate the stresses and strains induced by a change in the geometry or loading conditions of a rock mass, the mechanical characteristics of the rock mass components must be known. The initial elastic modulus, compressive strength and failure behavior of intact and jointed rock units can be determined from the appropriate complete stress-strain curve. This curve is the scaled complete force-displacement curve that is obtained when a rock specimen is tested from initial loading to final zero load-bearing capability in a stiff or servo-controlled testing machine.

It is important to establish whether the complete stress-strain curve is a real material property or a function of both the rock and the experimental technique. The purpose of this paper is to present experimental data and discussion on variability in the shape of the complete stress-strain curve for rock with particular emphasis placed on the influence of specimen geometry.

[1,3]Post-Doctoral Research Fellow and Professor, Department of Civil and Mineral Engineering, University of Minnesota, Minneapolis, Minnesota 55455.

[2]Senior Lecturer, Department of Engineering, James Cook University of North Queensland, Townsville, Australia.

THE COMPLETE STRESS-STRAIN CURVE

When a rock specimen is loaded in uniaxial compression, the complete functional relationship between axial stress and axial strain is represented by a complete stress-strain curve. In order to obtain the complete curve experimentally, the rock failure process must be controlled; this depends on the mechanical characteristics of the machine-rock system and the way in which the test is conducted.

During a compression test on rock, an attempt is usually made to achieve a constant loading rate or a constant displacement rate. If a constant loading rate is maintained, then force must be regarded as

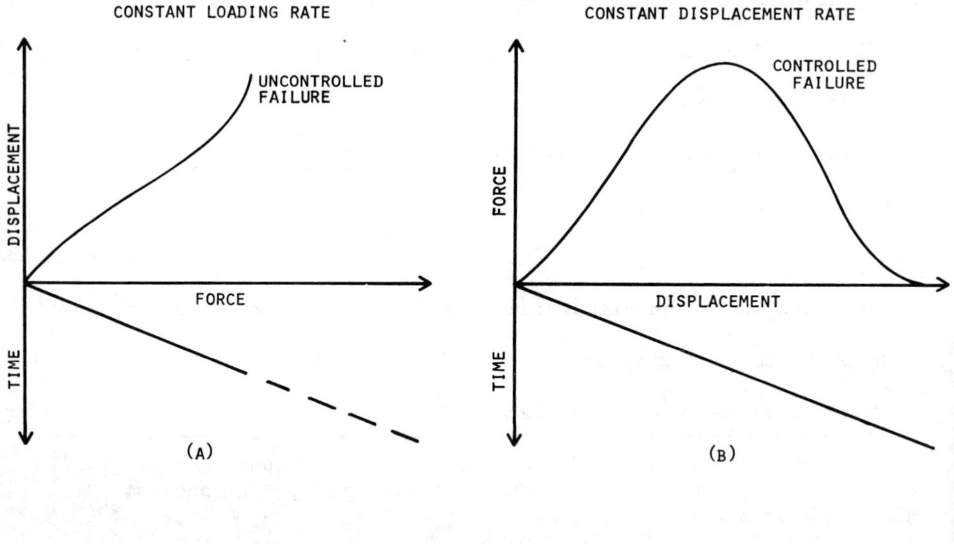

FIG. 1 - FORCE OR DISPLACEMENT CAN BE CHOSEN AS THE INDEPENDENT VARIABLE DURING A COMPRESSION TEST ON ROCK

the independent variable and displacement as the dependent variable (Figure 1A). A constant loading rate requirement dictates that the force must be increased, not only during the elastic portion of the force-displacement curve, but also in the failure region. Under these conditions, a specimen fails explosively when the compressive strength of the rock is reached.

Conversely, when displacement is regarded as the independent variable and a constant displacement rate is maintained, the dependent variable, force, rises to a maximum at the specimen's maximum load-bearing capability and then diminishes to zero (Figure 1B). The complete force-displacement curve that can be obtained under these conditions is composed of two portions: the loading path and the unloading path or failure locus. With a knowledge of the cross-sectional area and length of the specimen, the complete force-displacement curve can be scaled to a complete stress-strain curve.

The independent variable, displacement, can be linearly increased with time by two methods:

(1) By using a testing machine with a relatively high longitudinal stiffness. The strain rate is kept approximately constant by moving the testing machine platens together at a constant rate. The high stiffness is necessary in order that sudden displacement accelerations are avoided as the machine elastically unloads during rock failure.

(2) By using a closed-loop servo-controlled testing machine that can automatically reduce the load faster than the specimen breaks. A constant strain rate is achieved by the continuous operation of the closed-loop as the actual experimental

condition is adjusted to coincide with a programmed constant strain rate condition.

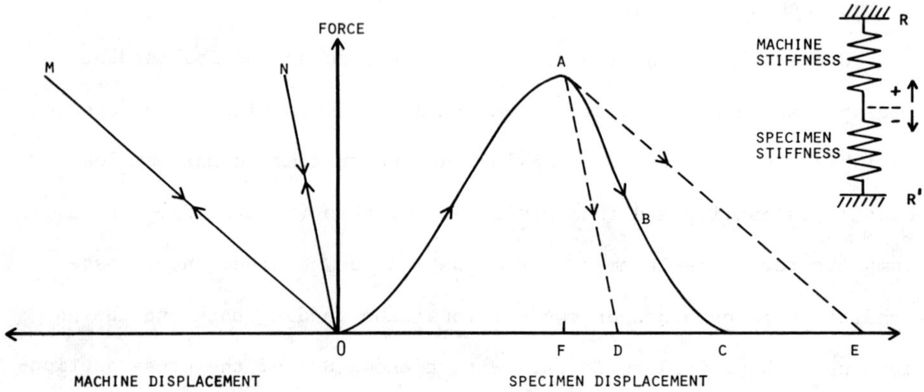

FIG. 2 - INFLUENCE OF TESTING MACHINE STIFFNESS ON THE FAILURE OF ROCK

The diagram in Figure 2 illustrates the influence of the machine stiffness on rock failure. The testing machine and rock specimen stiffnesses are represented by two springs fixed between the rigid, but moveable, supports R and R'. As the supports are moved, the testing machine is loaded to the peak force along the line OM (a soft machine) or ON (a stiff machine) and the specimen is loaded along OA*.

* The displacement sign convention is indicated in the diagram. Relative to the junction of the springs, R moves in the negative direction and R' moves in the positive direction.

Assume that the specimen force-displacement co-ordinates during the failure process form the locus ABC.

From the maximum load, a soft machine elastically unloads along the line AE (corresponding to MO) and supplies more energy than is required to fail the rock. At all points beyond F on the displacement axis, the load applied by the elastically unloading testing machine exceeds the load-bearing capability of the rock. This causes the typical abrupt and violent failure that is observed when a rock specimen is tested in a conventional testing machine. Conversely, the energy that would be released by the stiff testing machine in an increment of displacement is insufficient to break the rock; the failure process is stable and can be slowly advanced by moving the rigid supports R and R'.

Although complete force-displacement curves that are characterized by a monotonically increasing strain can be obtained in an appropriately stiff testing machine, a constant displacement rate cannot be achieved precisely. The specimen displacement monotonically increases during the generation of the curve OABC but the machine displacement acts in one direction during loading (ON in Figure 2) and in the opposite direction during failure of the rock specimen (NO or AD in Figure 2). The result is a variable strain rate unless special precautions are taken to adjust the drive mechanism.

A feedback control system forms the basis of a closed-loop servo-controlled testing machine. The main units in such a system are shown in Figure 3. The loading frame can be soft or stiff: it is the choice of feedback signal and the fast response time of the system that are the important control factors.

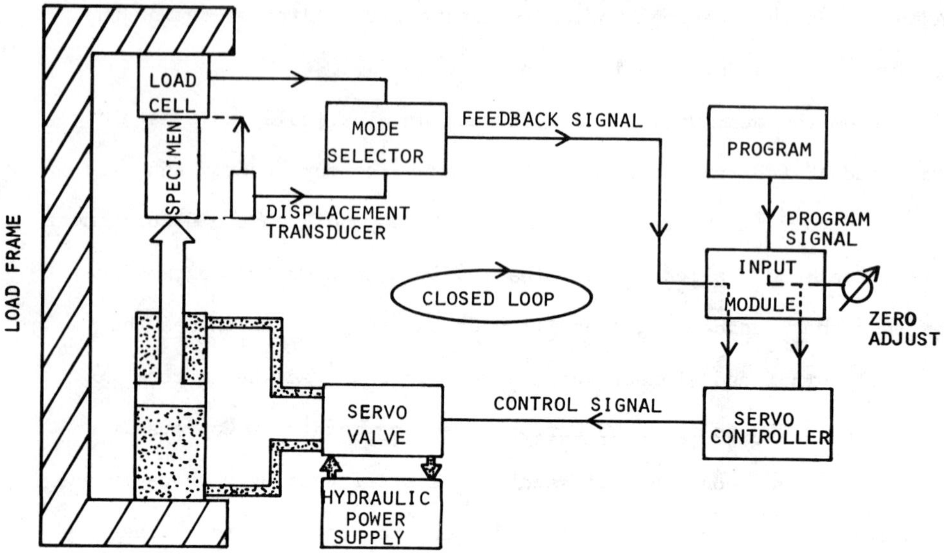

FIG. 3 - SCHEMATIC OF FAST-RESPONSE, CLOSED-LOOP, SERVO-CONTROLLED TESTING MACHINE. ADAPTED FROM MTS MANUAL (1)

A signal generated by a transducer and representing the value of any experimental parameter is electronically compared with a signal representing a pre-determined program value. If a difference exists, an error signal activates the servo-valve and the pressure on the specimen is adjusted until the actual and desired conditions coincide. In this way it is possible to control an experimental variable precisely continuously and automatically.

To achieve a constant displacement rate, and hence obtain a complete force-displacement curve, the feedback signal corresponds to the specimen displacement and the program signal is linearly increased with time. The continuous operation of the closed-loop then maintains a constant displacement rate regardless of any changes in the mechanical state of the specimen. In general, any function of one or more transducer signals can be used as feedback in the control loop. This provides a unique capability and makes servo-controlled systems extremely versatile compared to stiff machines which are inherently cumbersome and functionally inflexible.

Thus, control of rock failure can be achieved with a stiff or servo-controlled testing machine. It is emphasized, however, that the scaled complete force-displacement curve associated with structural collapse of the test specimen is not necessarily a true strain - strain curve for the rock. During elastic loading, the point values of stress and strain vary statistically, but homogeneously, from the gross stress and strain fields. The scaled force and displacement values are then averages for the stress-strain state of the rock. In the post-peak region, the stress and strain point variations are often severely biased by localized failure mechanisms; the scaled force and displacement values are still average values but not necessarily meaningful in terms of the true failure characteristics of the rock.

For example, if uneven loading causes premature shear failure, the scaled force and displacement values are not representative of the rock alone, but of the rock plus the particular shear failure mechanism. On the other hand, in an evenly loaded test, the effective cross-sectional area of the specimen can be reduced to practically zero towards the end of the failure process. In this case, the stress that is obtained by dividing the force by the original area is absurdly in error. For comparing the force-displacement curves that are obtained from specimens with different shapes and sizes, however, it is convenient to scale the curves according to the original specimen dimensions and use the term 'complete stress-strain curve'.

EXPERIMENTAL METHOD AND RESULTS

In order to determine the influence of specimen geometry on the complete stress-strain curve of intact rock, specimens of Georgia Cherokee marble were tested in a 1,000,000-lb. MTS closed-loop servo-controlled testing machine. Specimens with different shapes and sizes (Table I) were tested in uniaxial compression and three specimens were tested for each geometrical configuration.

TABLE I - SPECIMEN DIMENSIONS

Diameter 3/4"					
Length	1/4"	3/8"	3/4"	1 1/2"	2 1/4"
L:D ratio	1/3:1	1/2:1	1:1	2:1	3:1
Diameter 2"					
Length	2/3"	1"	2"	4"	6"
L:D ratio	1/3:1	1/2:1	1:1	2:1	3:1
Diameter 4"					
Length	1 1/3"	2"	4"	8"	12"
L:D ratio	1/3:1	1/2:1	1:1	2:1	3:1

The influence of specimen size can be studied by comparing the curves obtained from specimens with the same shape but different sizes (the same L:D ratio but different diameters). Note that the 4-inch diameter specimens were about 152 times the volume of the 3/4-inch diameter specimens. The shape effect can be studied by comparing the curves obtained from specimens with the same size but different shapes (the same diameter but diferent L:D ratios). In the latter case, the specimen size was not kept constant because all the specimens had the same diameter but the maximum variation was a factor of 9 as compared to 152 in the size effect.

In order to minimize the effects of different end conditions and strain rates caused by testing specimens with different sizes and shapes, the following precautions were taken:

(1) Each specimen was loaded through steel platens with an L:D ratio of 1:1 and the same diameter as the specimen.

(2) A constant strain rate was achieved throughout the complete stress-strain curve. A digital ramp generator was used to program a constant axial displacement rate and different rates were chosen for specimens with different lengths so that all specimens were tested at the same strain rate (1.5×10^{-6}/second).

The force was measured by a load cell that was in direct contact with the specimen and the displacement between the ends of the specimen was measured by a double cantilever arrangement.

The data is summarized in Figure 4 which shows the influence of specimen geometry on the complete stress-strain curve for marble loaded in uniaxial compression. The individual and average compressive strengths are given in Table II and photographs of four specimens are shown in Figures 5 and 6.

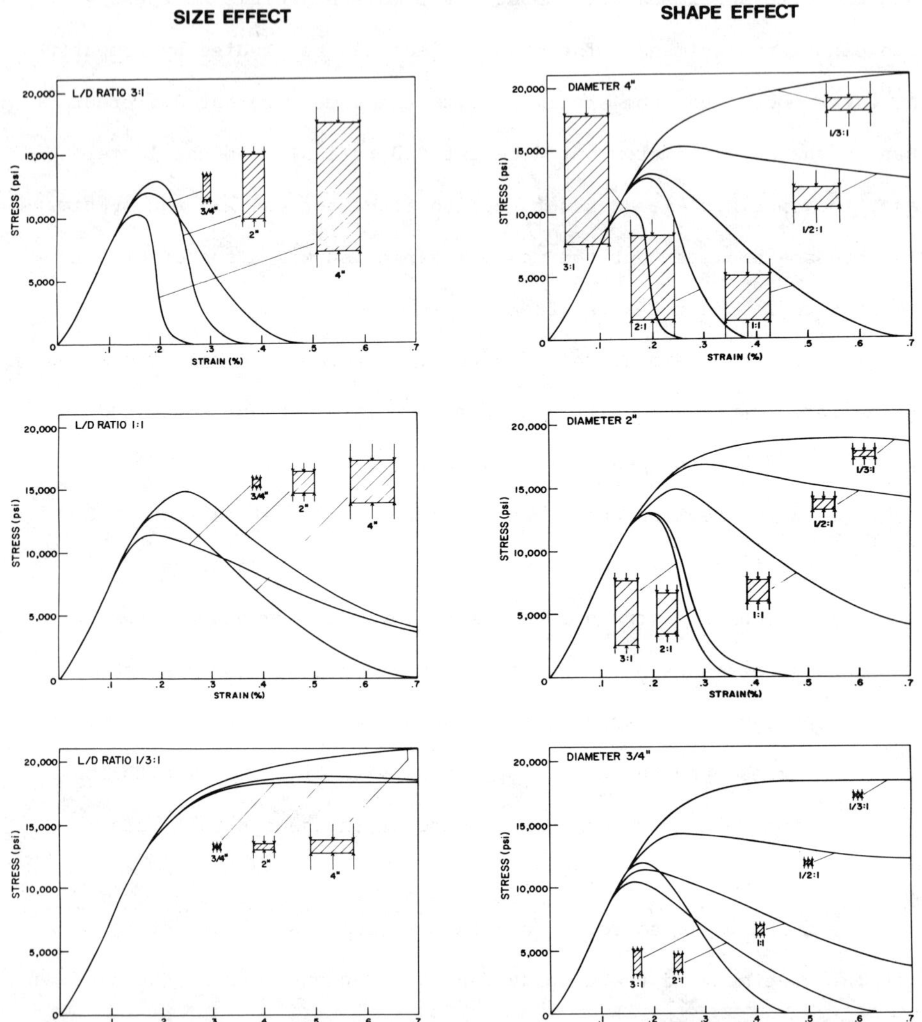

FIG. 4 - INFLUENCE OF SPECIMEN SIZE AND SHAPE ON THE COMPLETE STRESS-STRAIN CURVE FOR MARBLE LOADED IN UNIAXIAL COMPRESSION

TABLE II - INDIVIDUAL AND AVERAGE COMPRESSIVE STRENGTHS

(GEORGIA CHEROKEE MARBLE)

L:D Ratio \ Diameter	3/4-inch		2-inch		4-inch	
3:1	11,510 11,800 12,560	11,960	11,920 13,240 13,350	12,840	10,110 10,310 10,340	10,250
2:1	10,240 10,410 10,540	10,400	12,910 13,020 13,350	13,090	12,750 12,890 12,910	12,850
1:1	9,760 11,380 12,990	11,380	14,330 15,210 15,320	14,950	12,140 13,250 14,090	13,160
1/2:1	13,790 14,330 15,100	14,410	15,530 16,850 18,050	16,810	15,000 15,060 15,620	15,230
1/3:1	17,620 18,120 18,970	18,240	17,290 18,710 20,020	18,670	20,670 20,890 21,330	20,960

2-inch diameter

3/4-inch diameter

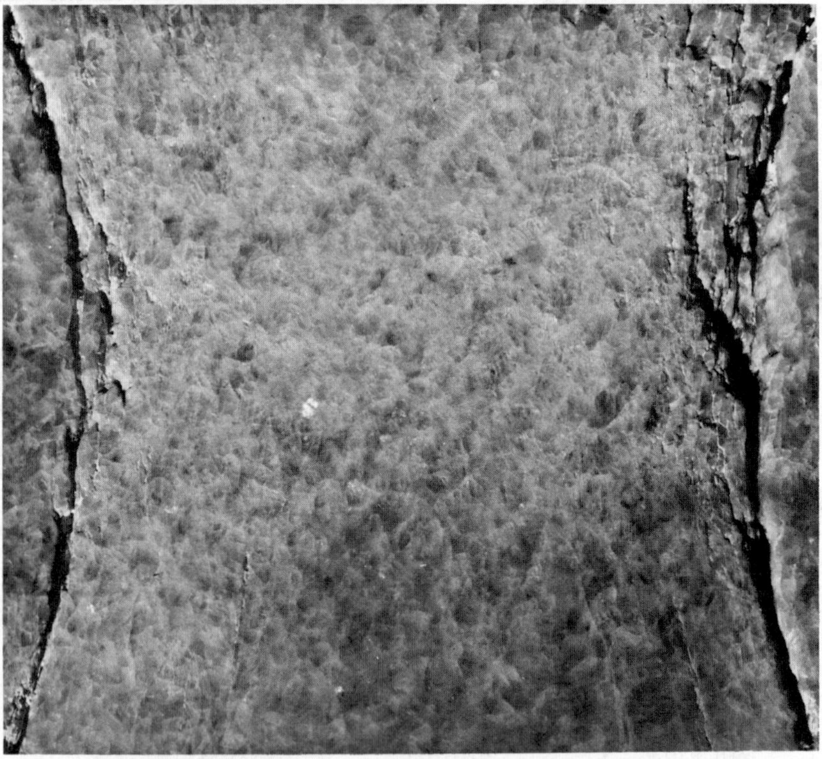

4-inch diameter

FIG. 5 - CROSS-SECTION OF 1:1 SPECIMENS AT AN ADVANCED STATE OF FAILURE. (GEORGIA CHEROKEE MARBLE)

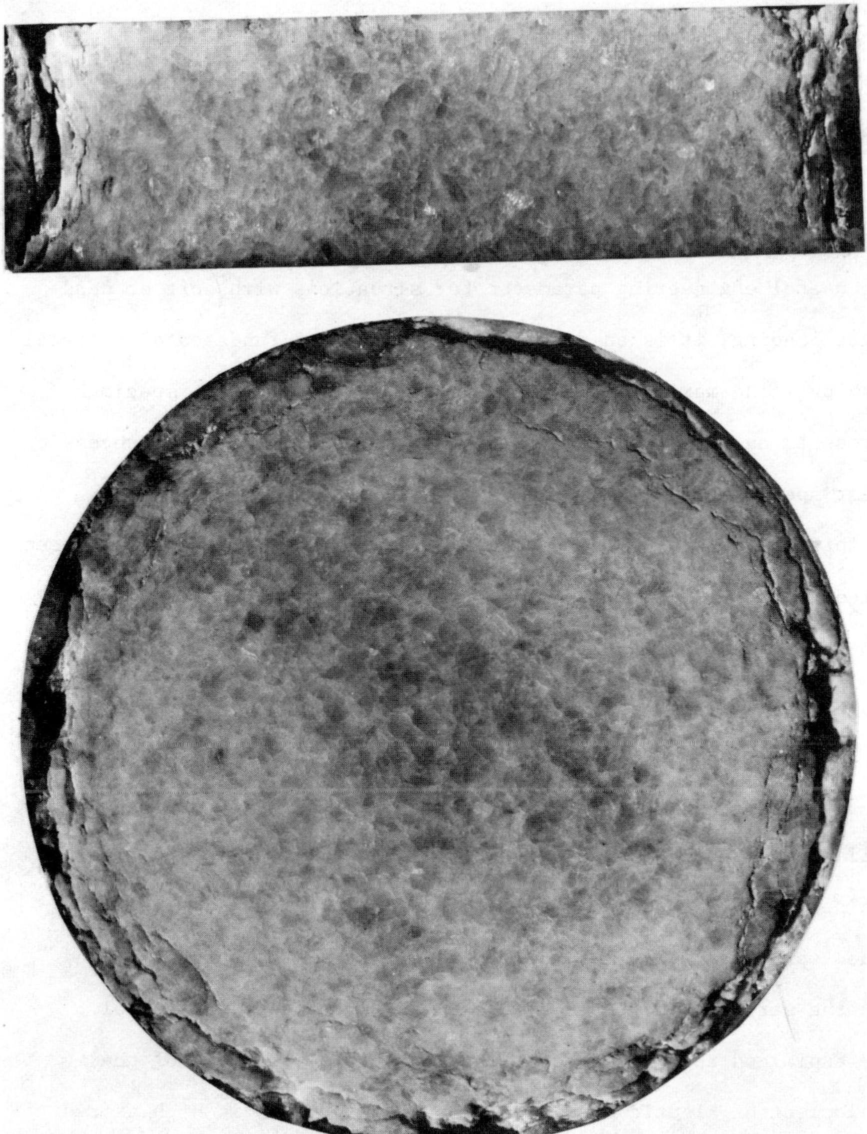

FIG. 6 - CROSS-SECTION AND END VIEW OF 1/3:1 SPECIMEN (4-INCH DIAMETER) AT ADVANCED STATE OF FAILURE. (GEORGIA CHEROKEE MARBLE)

DISCUSSION

Previous work concerning the influence of specimen geometry on rock properties has been concentrated on the compressive strength size effect; larger specimens are generally associated with lower compressive strengths. It should be noted that, although the strenth is a useful engineering parameter for situations with soft or dead-weight loading, it is an unreliable indicator of fundamental material behavior. The maximum load-bearing capability of a test specimen represents an arbitrary stage in the structural breakdown process of a test specimen.

Interpretations of the compressive strength size effect have been varied:

1. The quantity of stored strain energy influences failure: at a given stress level, large specimens contain more energy than small specimens.

If failure occurs when a certain amount of strain energy is available, the compressive strength is a function of the testing machine stiffness and the specimen size. Glucklich and Cohen (4) loaded springs in parallel with gypsum plaster specimens and found that the strength indicated by these tests was lower than usual. They explained this observed strength reduction in terms of the development of kinetic energy of propagating cracks in high stored strain energy systems. The kinetic energy, that is not present to the same extent in stiffer systems (or with small specimens), is available to do work against uncracked material, bringing about premature fracture.

2. The stress gradient influences failure: in a standard uniaxial compression test, the stresses applied to the specimen are neither uniform nor uniaxial. Larger specimens are associated with lower stress gradients.

As the size of the specimen is decreased, the stress gradient increases provided that geometrical similarity is maintained and, as the L:D ratio is decreased, the end effects influence a greater proportion of the specimen. Thus, if the brittle fracture of rock is stress gradient dependent, a change in specimen size or shape will change the nature of the test. Hodgson and Cook (5) found that the strengths of s shale and quartzite were virtually independent of size under the conditions of uniform stress but that the strengths were markedly size dependent in the punch bearing test where stress gradients prevail.

3. The failure of the specimen is dictated by the strength of the weakest link: larger specimens are more likely to contain weak elements.

This interpretation is not valid because multiple local failure precedes total collapse in a compression test (11) and the strength of the specimen is not dictated by the strength of its weakest link. The most comprehensive weakest link statistical theory was developed by Weibull (12, 13, 14). Unfortunately, the theory has been badly maligned in the literature having been accepted for an invalid reason [the ubiquitous log-log plot proved the theory, (7, 8)] and rejected for an invalid reason [a degenerate case of the general theory was too limited, (2, 6)].

4. There is no size effect: different strength values are the

result of inappropriate testing procedures.

Obert, Widnes and Duvall (9) tested cylinders of marble, limestone, granite, sandstone, slate and greenstone with diameters in the range of 7/8-inch to 2 1/8-inch and found no size effect. Hodgson and Cook (5) similarly found no evidence of size effect in compression tests on quartzite and shale cylinders of up to 6-inch diameter.

The compressive strength data presented in this paper indicates that there is no compressive strength size effect. Under uniform stress fields, hard rock that is relatively free of pre-existing microfracturing is unlikely to show a strength-size effect other than a possible one arising from the influence of strain energy stored within the failing rock. The data indicate that there is a significant compressive strength shape effect that is caused by severely non-uniform stress fields in the specimens with small L:D ratios.

The compressive strength is, however, a characteristic of only one point on the complete stress strain curve. It is more appropriate to study the effect of specimen geometry on the complete curve rather than isolate the variation of the stress associated with a zero tangent modulus.

The three main factors that determine the shape of the complete stress-strain curve are:

a) The mechanism of progressive structural breakdown of the test specimen.

b) The different structural breakdown mechanisms associated with specimens of different sizes and shapes.

c) Scaling the force by the original cross-sectional area of the specimen.

During compressive testing, a specimen suffers detectable permanent damage to the microstructure when the displacement exceeds about 50% of the displacement corresponding to the maximum load-bearing capability (3). The inhomogeneous and anisotropic nature of the microstructure induces local tensile stresses which produce small cracks aligned along the loading axis (11). As the displacement is increased, the density of cracks increases until gross slabbing and shearing occur on the descending side of the complete force-displacement curve (11).

The shape of the complete stress-strain curve will depend on the variability of the microstructure. If the rock is composed of a uniform structure of elements that have the same elastic-brittle characteristic, the test specimen will behave elastically until all the elements fail simultaneously and the load-bearing capability of the specimen drops instantly to zero. Conversely, if the rock is composed of different elements in a random structure and with extreme strength variations, the test specimen will fail very gradually as the displacement is progressively increased. It is this principle of progressive destruction that produces the basic shape of the complete stress strain curve.

The small axial cracks that develop during compressive failure are caused by the inhomogeneity of the microstructure and their size is independent of the specimen size. This is illustrated by the photographs in Figure 5. The effect of these cracks appears to be different for specimens with different sizes. In all cases, smaller specimens were associated with a more gentle slope of the descending section of the complete stress-strain curve (Figure 4).

In general the axial cracks are concentrated in the central portion of the specimen's length because lateral restraint at the specimen ends inhibits their growth near the specimen-platen interface. In the longer specimens, the central region is failed whilst the ends are elastically loaded and then unloaded. For an L:D ratio of 1/3:1, however, the cracks appear to be uniformly distributed throughout the length of the specimen. This causes a variation in the shape of the complete stress-strain curve as shown by the curves in Figure 7. The curves were derived from a simple, idealized model comprised of three elements loaded in series of which two (of equal length) remain intact

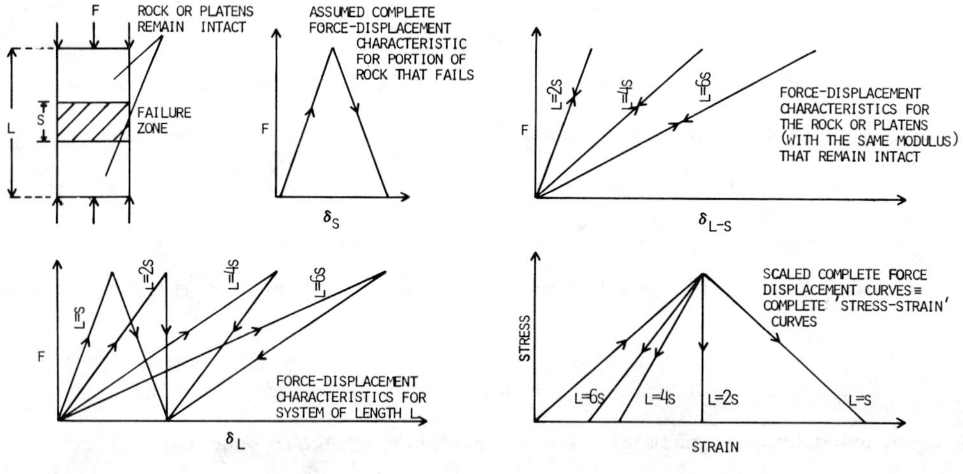

FIG. 7 - INFLUENCE OF SPECIMEN LENGTH ON THE SHAPE OF THE COMPLETE STRESS-STRAIN CURVE AS INDICATED BY AN IDEALIZED MODEL

and one fails. Different 'stress-strain' curves are indicated from tests in which the specimen length, (L), is varied whilst the extent of the failure zone, (s), remains constant.

The family of stress-strain curves in Figure 7 shows that more brittle behavior is exhibited when part of the specimen does not fail but unloads elastically after the peak force. If the intact zones represent two metal spacers rather than intact rock, the curves illustrate the importance of measuring the displacement between the specimen ends rather than between two arbitrary, but experimentally convenient, points. Also, in servo-controlled tests in which axial displacement is used to generate the feedback signal, failure control is optimized when the displacement is measured between the specimen ends. If the displacement of metal spacers is included in the feedback signal, the total force-displacement curve might not monotonically increase in displacement and failure will be uncontrolled.

In discussing the shape of the complete stress-strain curve as a natural consequence of the inhomogeneity of the microstructure, it was noted that the test specimen collapses as individual elements fail. Initially, there is a distribution of elemental strengths and this distribution gradually becomes truncated on the low strength side as the elements are successively eliminated in order of their strengths. Note, however, that if the element strengths are expressed in terms of stress, the last element to break will require the greatest stress and hence the end of the complete stress-strain curve should be associated with the highest value of stress and not zero!

The complete stress-strain curves in Figure 4 were obtained by scaling the force by the original cross-sectional area of the

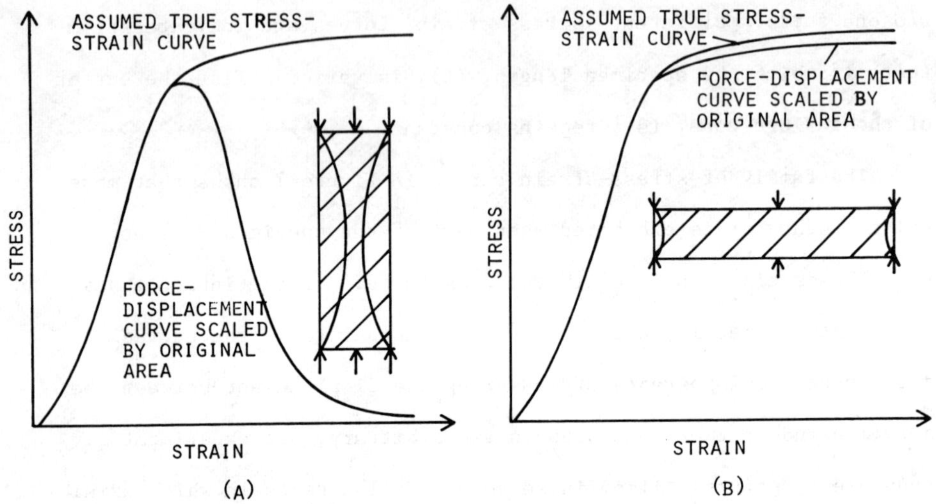

FIG. 8 - VARIATION IN SHAPE OF THE APPARENT COMPLETE STRESS-STRAIN CURVE CAUSED BY SCALING THE FORCE BY THE ORIGINAL CROSS-SECTIONAL AREA

specimen rather than the continuously decreasing cross-sectional area (10). As shown in Figure 8, it is likely that the difference between the apparent stress-strain curve and the true stress-strain curve depends on the L:D ratio of the specimen. For a large L:D ratio, the cross-sectional area of the specimen can be reduced to practically zero towards the end of the test (Figure 8A). The force also drops to zero but it is probable that the stress continues to rise. For a very small L:D ratio, (Figure 8B), the slabbing does not reduce the area by such a significant amount and the apparent curve is closer to the real curve. Note, however, that it is extremely difficult, if not impossible, to experimentally estimate the effective area at each stage in the failure process and hence obtain the true complete stress-strain curve.

CONCLUSION

The following conclusions were reached directly from the experimental data:

1. The initial elastic modulus did not vary with specimen geometry.
2. There was no significant variation in the compressive strength with specimen size.
3. A definite variation in compressive strength with specimen shape was observed. Specimens with low L:D ratios indicated high strengths.
4. For all L:D ratios, larger specimens were associated with a steeper slope in the post-peak portion of the complete stress-strain curve.
5. For all diameters, longer specimens were associated with a steeper slope in the post-peak portion of the complete stress-strain curve.
6. The size of the small axial cracks that developed during compressive failure was related to the microstructure and not the specimen size.

It is suggested that the shape of the scaled complete force-displacement curve (i.e. the apparent complete stress-strain curve) for rock is governed by three factors:

a) The mechanism of progressive structural breakdown of the test specimen

b) The different structural breakdown mechanisms that are associated with specimens of different sizes and shapes

c) Scaling the force by the original cross-sectional area of the specimen rather than the actual cross-sectional area that progressively diminishes as failure occurs.

ACKNOWLEDGEMENT

The authors thank Dr. Steven L. Crouch for his contribution to many discussions. The research was carried out in the Department of Civil and Mineral Engineering, University of Minnesota. It was supported by the Advanced Research Projects Agency of the Department of Defense and was monitored by the United States Bureau of Mines, Twin Cities Research Center, under contract number H0101610 (Principal Investigator, Professor Charles Fairhurst).

APPENDIX I - REFERENCES

---

1. An Introduction to MTS Closed-Loop Testing Systems, MTS Systems Corporation, Minneapolis, Minnesota, 1970.

2. Bieniawski, Z. T., "The Effect of Specimen Size on the Compressive Strength of Coal." Int. J. Rock Mech. Min. Sci., Vol. 5, July, 1968, pp. 325-335.

3. Crouch, S. L., "Experimental Determination of Volumetric Strains in Failed Rock." Int. J. Rock Mech. Min. Sci., Vol. 7, November, 1970, pp. 589-603.

4. Glucklich, J. and Cohen, L. J., "Size as a Factor in the Brittle-Ductile Transition and the Strength of Some Materials." Int. J. Fract. Mech., Vol. 3, No. 4, 1967, pp. 278-289.

5. Hodgson, K. and Cook, N. G. W., "The Effects of Size and Stress Gradient on the Strength of Rock". Proceedings, Second Cong. Int. Soc. Rock Mech., Belgrade, Vol. 2, 1970, Paper 3-5.

6. Lo, K. Y., "The Operational Strength of Fissured Clays." Geotechnique, Vol. 20, March, 1970, pp. 57-74.

7. Lundborg, N., "The Strength-Size Relation of Granite", Int. J. Rock Mech. Min. Sci., Vol. 4, July, 1967, pp. 269-272.

8. Lundborg, N., "Strength of Rock-Like Materials." Int. J. Rock Mech. Min. Sci., Vol. 5, September, 1968, pp. 427-454.

9. Obert, L., Windes, S. L. and Duvall, W. I., "Standardized Tests for Determining the Physical Properties of Mine Rock." U. S. Bur. Min. Rept. Investn. 3891, 1946.

10. Peng, S., Personal Communication.

11. Wawersik, W. R., Detailed Analysis of Rock Failure in Laboratory Compression Tests. PhD Thesis, University of Minnesota, 1968.

12. Weibull, W., "A Statistical Theory of the Strength of Materials." Ing. Vetenskaps. Akad. Handl., Vol. 151, 1939, pp. 5-44.

13. Weibull, W., "The Phenomenon of Rupture in Solids." Ing. Vetenskaps. Akad. Handl., Vol. 153, 1939, pp. 5-55.

14. Weibull, W., "A Statistical Distribution Funtion of Wide Applicability." J. Appl. Mech., Vol. 18, 1952, pp. 293-297.

DYNAMIC BEHAVIOR OF ROCKS TO SPHERICAL STRESS WAVES[1]

R. P. Swift[2] and M. W. McKay[2]

ABSTRACT

Two distinct types of experiments are employed to examine the behavior of Westerly granite below 25 kbar to explosively created spherical impulses. One type examines radial stress-time histories of the rock with manganin piezoresistive stress transducers; the other examines particle velocity-time histories using the electromagnetic velocity gauge technique. The observed response indicates that yielding and dilatant effects are occurring, although they are less prevalent than previously observed at stress levels above 25 kbar. An indication that the material response could be rate-dependent as well as dilatant is given by the good agreement with measurements predicted by a linear visco-plastic model. Also, a model based on triaxial static data with dilatancy and work-hardening provides fair agreement with the observed response.

INTRODUCTION

The behavior of rocks under the influence of dynamic loading, where only one non-zero strain component exists, has been exten-

---

[1] This work was supported by the Defense Nuclear Agency.

[2] Applied Mechanics Department, Physics International Company, San Leandro, California 94577.

sively studied using plane waves generated from explosives and plate impact methods. Although constitutive models can often be determined from experiments of this type, their validity when applied to other geometric configurations can only be inferred.

Dynamic loading of rocks with spherical stress waves provides a means of investigating their behavior under states of stress that are unattainable in conventional plane wave experiments. Two distinct components of stress and strain exist; the radial components and the circumferential components. Any mathematical model describing the material response associated with spherical waves must be employed in terms of its three-dimensional formulation. This implies that constitutive models constructed from accurate spherical data tend to be more suitable to predict the response for other complex states of stress than models developed solely from uniaxial data. Spherical stress wave data can also serve to check the ability of existing models to predict correctly stress wave propagation in a nonplanar geometry, where a three-dimensional formulation of the model is important.

The analysis and experiments presented here were carried out to examine the ability of various models to describe the behavior of Westerly granite to spherical stress waves. Experiments were conducted to observe the stress wave attenuation by measuring radial stress-time histories and particle velocity-time histories at various radii. The spherical loading was generated by detonating a high-explosive sphere in an enclosed cavity of a rock sample. This provided stress levels ranging from about

300 kbar at the cavity wall down to a few kilobars out at several cavity radii. Earlier studies (References 1 and 2) were concentrated mainly on the high stress level response of Westerly granite and several other rocks. In this paper, we direct our concern to the behavior below 25 kbar. The measurements in this stress range indicate that some anelastic effects are still occurring. Constitutive models with and without dilatancy and with and without rate-dependency are examined.

EXPERIMENTAL METHOD

The basic experimental configuration is described in detail in Reference 1. Briefly, it consists of placing a two-inch- or four-inch-diameter sphere of high explosive (LX 04-01) in a block of sample material about 1 ft^3. The explosive sphere is detonated with the aid of a 1/2-inch-diameter tetryl booster contained at its center, which was detonated by a mild detonating fuse at its center. The resulting spherical stress wave induces one-dimensional radial motion in the medium surrounding the explosive. The two principal field variables measured in these experiments were radial stress-time history and particle velocity-time history. Both types of measurements were carried out in separate shots, and the arrangement of the gauges was different for each case.

RADIAL STRESS MEASUREMENT

The configuration shown in Figure 1 was used to measure the radial stress-time histories. The explosive sphere was placed

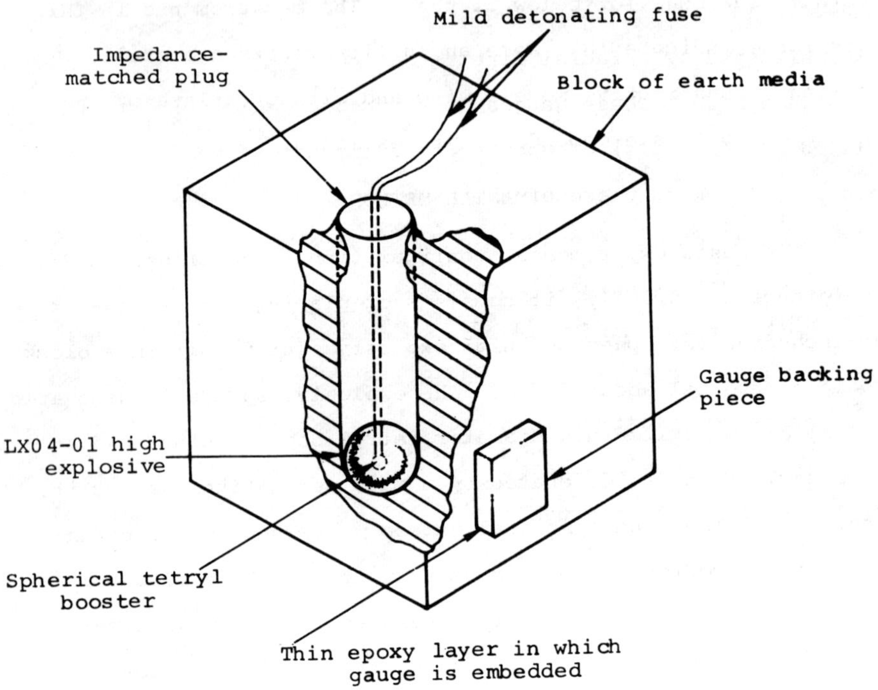

Figure 1 Experimental configuration for spherical stress wave studies.

off-center in the block so that measurements of stress could be obtained at five different radii, corresponding to the four side faces and bottom face of the block. Piezoresistive stress transducers, 50-ohm photo-etched manganin foil gauges, were mounted on these surfaces with their active elements perpendicular to and bisected by a radial line emanating from the center of the explosive sphere. These gauges were embedded in a thin layer (between 2 and 3 mils) of epoxy and sandwiched between the sample block and a backing piece of the same material. The dimensions of the backing piece were such as to allow at least 6 μsec of unperturbed one-dimensional response.

The electronic circuitry used with the manganin gauges is shown in Figure 2. The power supply used in the shot was a multi-channel constant-current capacitive discharge type. Each channel employed was discharged across a balanced bridge circuit with a 50-ohm manganin gauge as one leg in the circuit. Initially, the bridge is in a balanced position so that V = 0. When the stress wave arrives at the gauge, its resistance changes and the bridge becomes unbalanced. Voltage measurements were made at three locations in the bridge: (a) the total voltage across the gauge was measured at G, namely the turn-on voltage and the change caused by the change in resistance of the gauge; (b) the voltage across the adjustable leg was measured at I which gave a measure of the current through the part of the bridge containing the gauge; (c) the potential, V, between the two legs of the bridge was monitored at D. The manganin gauge exhibits a fractional change in resistance that is linearly related to the stress acting normal to the gauge. This stress is related to the measured value of V, (Reference 3), by

$$\sigma = \frac{1}{K} \frac{R_g}{R_{g_o}} \left(\frac{4 \frac{V}{V_o}}{1-2 \frac{V}{V_o}} \right) \tag{1}$$

where V_o is the voltage of the power supply, R_{g_o} is the zero-pressure resistance of the gauge without leads, R_g is the initial combined resistance of the gauge and leads, and K is the piezo-resistive coefficient of the gauge. Sample traces for 50 Ω gauges in granite at scaled radii of $R/R_o = 3.99$ and 5.89, where $R_o =$ 2.54 centimeters, are shown in Figure 3.

PARTICLE VELOCITY MEASUREMENT

Measurements of particle velocity were made with electro-magnetic velocity gauges. The electromagnetic gauge technique is well known (References 4, 5, and 6), and consists of embedding a fine wire or foil in a nonconducting sample and placing the assembly in a magnetic field. The active element of the gauge, the particle motion of the sample, and the magnetic field are oriented mutually orthogonal to one another. When the stress wave propagating through the sample arrives at the gauge, the active element equilibrates to the material velocity. The EMF, ε, in volts, induced by the movement of the gauge element through the magnetic field, is related to the material particle velocity u, in cm/sec, by Faraday's law of induction in the form

$$\varepsilon = 10^{-2}\, u\, \ell\, B \tag{2}$$

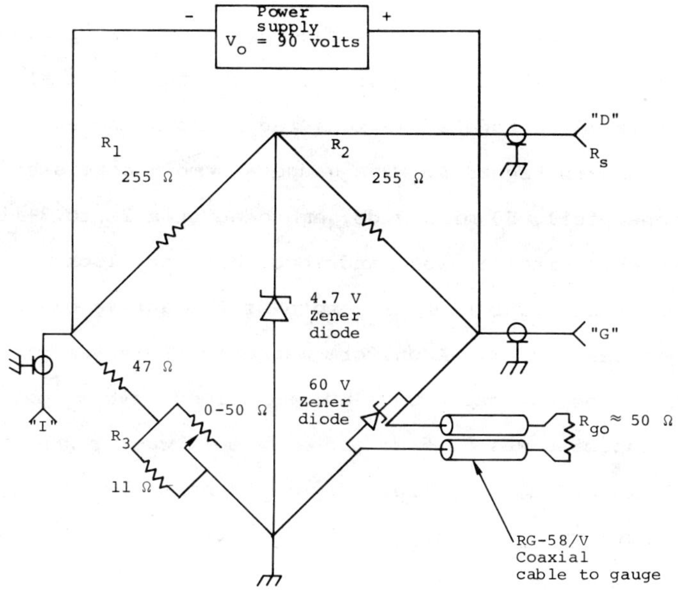

Figure 2 Bridge circuit used for manganin-gauge measurements of stress.

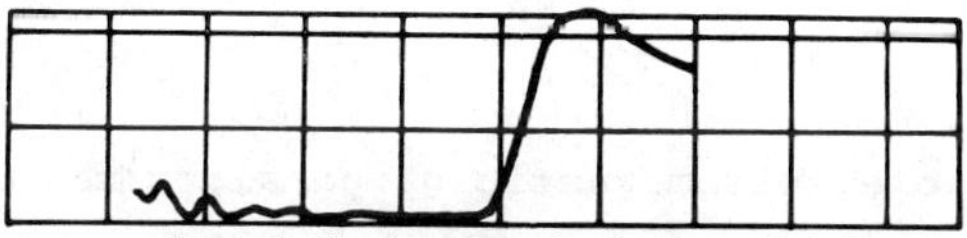

a. Sensitivity: 0.5 volts/cm
Timing: 2.0 μsec/cm
R/R_o = 3.99, Peak stress ≈ 24.2 kbar

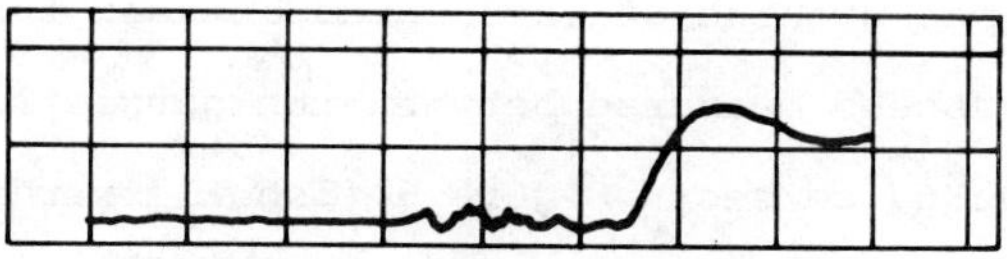

b. Sensitivity: 0.5 volts/cm
Timing: 2.0 μsec/cm
R/R_o = 5.89, Peak stress ≈ 13.8

Figure 3 Manganin gauge records of radial stress for spherical waves in Westerly granite.

where ℓ is the length of the gauge element in centimeters and B is the magnitude of the applied field in gauss.

The application of this technique to spherical geometry is described in detail in Reference 3. A configuration allowing measurements to be made at various radii along the same radial line is shown in Figure 4. The gauge assembly consisted of a 1-mil copper foil, 50 mils wide, embedded in a 2- to 3-mil layer of epoxy, which in turn was sandwiched between blocks of the sample material. The nominal length of the active element for the gauges was 1.2 cm. A uniform magnetic field was created in the vicinity of the gauge by a pulsed Helmholtz-coil pair. The coil pair spacing was 31.5 cm and each coil was wrapped with 169 turns of #18 AWG copper wire. Figure 5 shows the coil system and pulsing circuitry. Discharging 10 kV through six 14.5 μF capacitors provided a peak field of 1145 gauss = ± 20 gauss with a half-period of 8.3 msec. This provided an essentially constant peak field intensity for the 20 μsec window where measurements were made.

Traces of two electromagnetic gauge signals in granite are shown in Figure 6 for gauges located at R/R_o = 5.29 and R/R_o = 6.39, where R_o = 2.54 cm. The traces started at the same time and gave a good measurement of wave speed between the two gauges. The average wave speed measured between the gauges in this experiment was 0.567 cm/sec. Figure 6 also illustrates clearly the loading, unloading, and peak attenuation of the particle velocity associated with the diverging stress wave. The high-

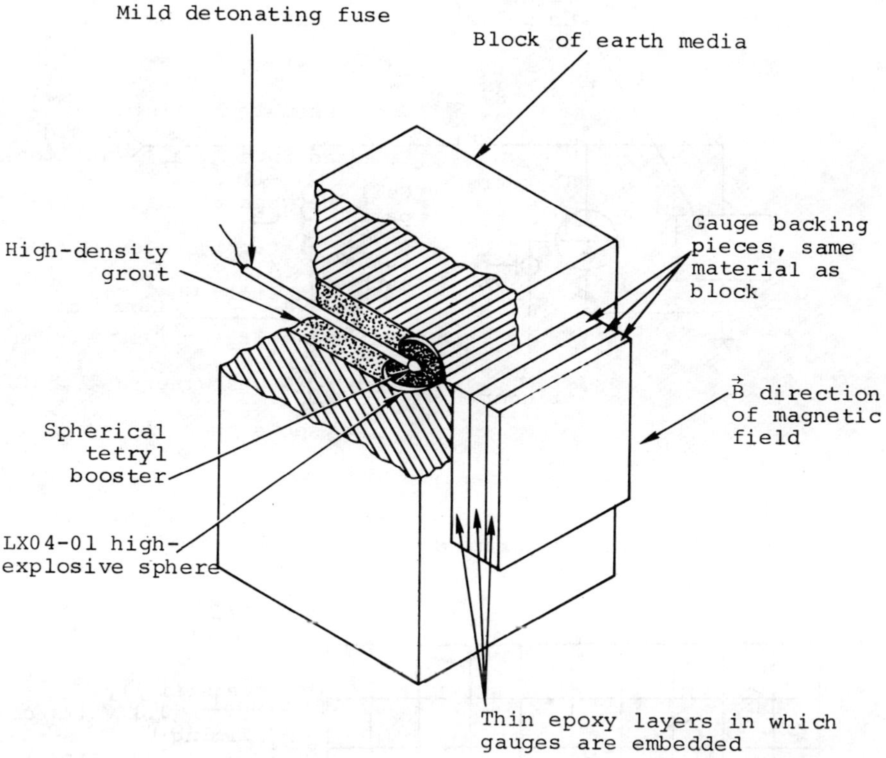

Figure 4 Experimental configuration for measuring spherical stress-wave response along a single radial line.

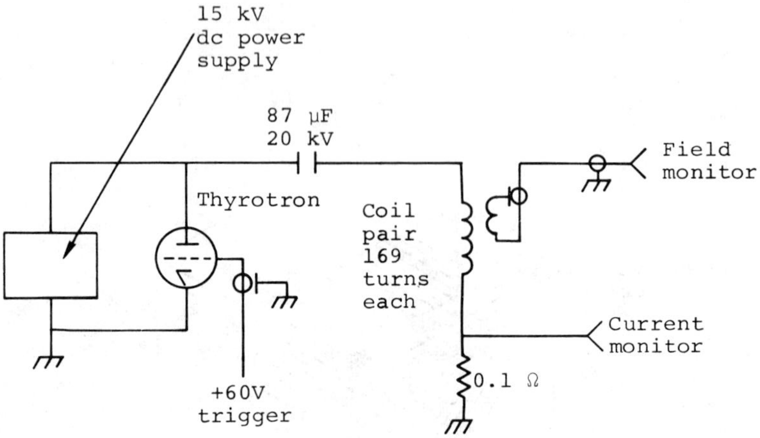

Figure 5 Schematic diagram of coil pulsing circuitry.

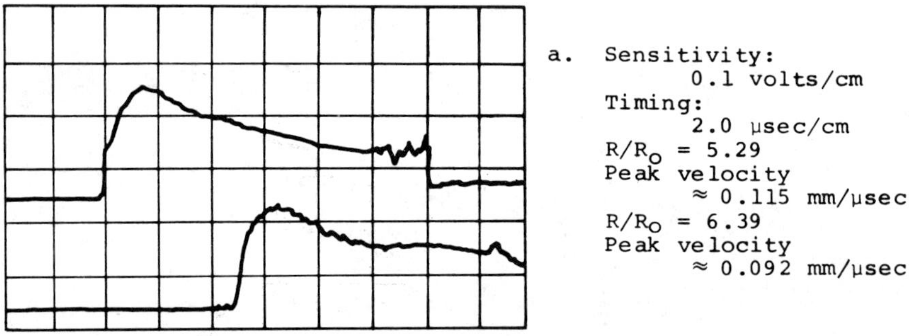

Figure 6 Electromagnetic gauge records of particle velocity for spherical waves in Westerly granite.

frequency noise superimposed on the signal is due to the piezoelectric disturbance of the quartz by the shock wave in the quartz-bearing granite.

TEST RESULTS

Stress-time profiles deduced from the manganin gauge bridge circuit using Equation 1 are shown in Figure 7. The value of the piezoresistive coefficient K used to obtain these profiles was determined in Reference 3 to be equal to 0.00245 Ω/Ω/kbar. Time is referenced to the initiation of the mild detonating fuse. The time-to-peak stress is essentially constant, about 1.6 μsec, as the wave propagates outward. There is no evidence of a two-wave structure as was observed in the higher stress level measurements in Reference 1. However, the unloading portions of the stress profiles show some broadening as the wave proceeds radially outward. This decrease in the rate of unloading is not characteristic of an elastic response.

The electromagnetic measurements of particle velocity-time profiles are shown in Figure 8. These correspond to stress levels below 25 kbar (i.e., $R/R_o \gtrsim 4$ for $R_o = 2.54$). Similarly, as observed for the stress profiles, the time-to-peak particle velocity is consistently about 1.6 μsec and the unloading portions of the velocity profiles exhibit a broadening with increasing radius. The difference in two measured velocity profiles in separate shots at nearly equivalent scaled radii of 5.19 and 5.29 is on the order of eight percent. Some of this difference can be attributed to the various inhomogeneities of the material. Also,

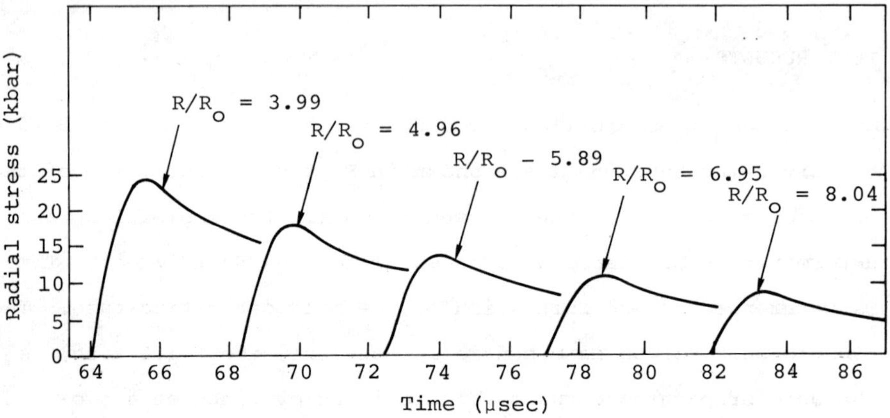

Figure 7 Radial stress profiles from spherical wave experiment using 50 ohm manganin gauges (R_o = 2.54).

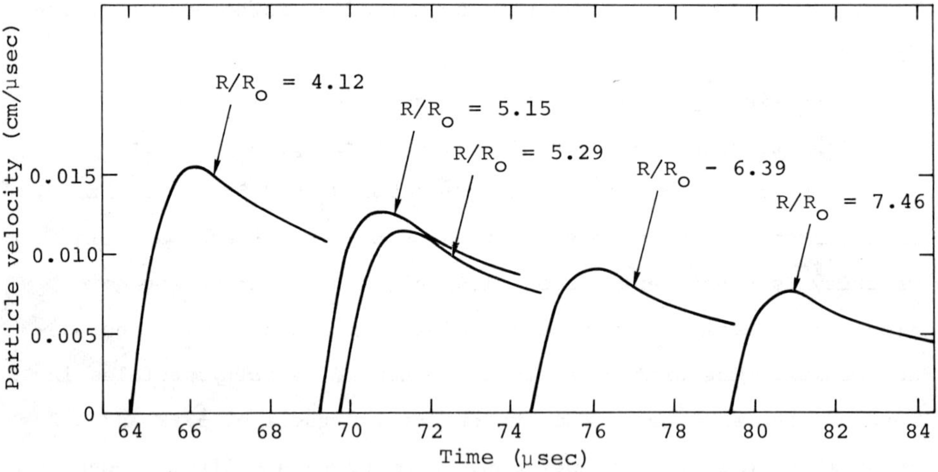

Figure 8 Particle velocity profiles from spherical wave experiments in Westerly granite (R_o = 2.54 cm).

to a small but nevertheless observable extent, some error is introduced by the piezoelectric noise from the quartz-bearing granite being superimposed on the signal.

In Figure 9, the decay of the peak radial stress for the low stress levels is shown along with earlier data from Reference 1. The decay of the peak particle velocity is also shown. The peak stress attenuation with increasing radius beyond $R/R_o = 4$ is still greater than that nominally associated with spherical geometry decay, i.e., $(R/R_o)^{-1}$. The peak particle velocity also attenuates at a rate slightly larger than $(R/R_o)^{-1}$. This behavior, along with the broadening of the stress and velocity profiles, implies that some anelastic mechanism, such as dilatancy, is still controlling the response of the material below 25 kbar. However, that this decrease in the rate of unloading is due solely to dilatancy of the material cannot be completely substantiated. As will be seen later, time-dependent effects can also exhibit a similar broadening of the wave.

ANALYSIS OF MATERIAL BEHAVIOR

The behavior of earth media subjected to dynamic loading, such as explosive-generated shock loading, is generally treated from one of three basic constitutive viewpoints. For the region close to the explosive source, stresses are sufficiently large that a hydrodynamic description of the material response is usually adequate. At the other extreme, far away from the source, the stresses are small and an elastic description is frequently employed. In the intermediate stress region, between

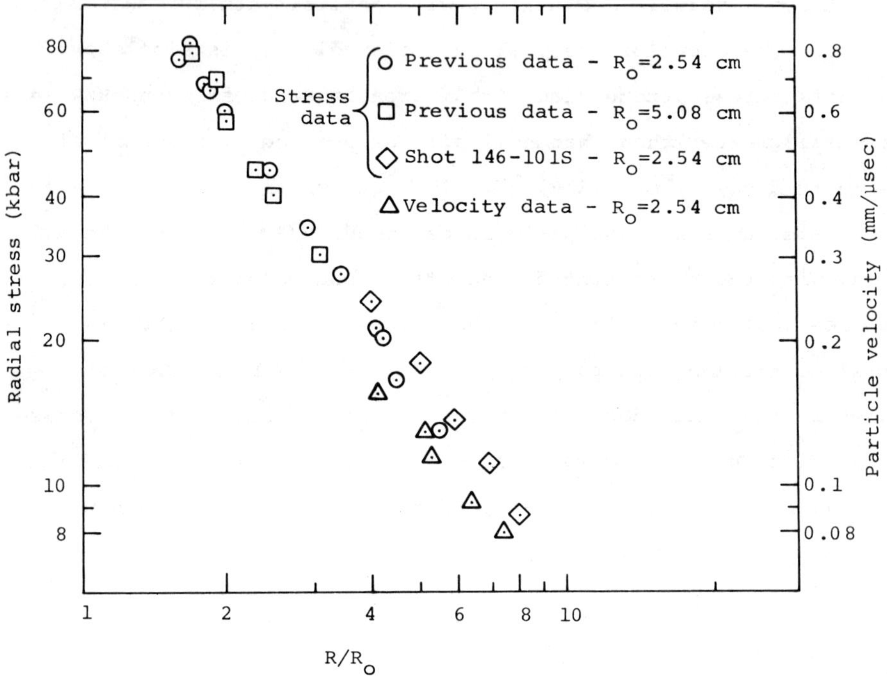

Figure 9 Peak stress and peak particle velocity attenuation in Westerly granite.

the high and low extremes, the rigidity of the material is significant and the response is of an anelastic nature. The three regions are not clearly separated, and the stress levels in the intermediate region may vary from less than one kilobar to a few hundred kilobars. Usually, more than one constitutive model is required to bridge this region. As previously noted, emphasis is directed here to the low intermediate region. However, for completeness, a brief summary is given of the response of Westerly granite for the high intermediate stress range.

Measurements of stress-time histories for the high intermediate stress range, obtained in Reference 1, showed three important features of the granite response: (a) a marked change in shock velocity occurred at a radius corresponding to a radial stress near 50 kbar indicating the onset of some physical process, most likely a phase transition; (b) a possible occurrence of some dissipative processes, such as phase transition and ductile flow, was indicated by the more rapid decay of peak radial stress with radial distance than could be predicted by a pure hydrodynamic response; and (c) the observed rate of decay exhibited by the unloading portion of the stress-time profiles was slower than normally predicted by standard constitutive models, thus indicating possible dilatancy.

MODEL CONSIDERATION FOR LOW STRESS LEVEL RESPONSE

The measurements below 25 kbar indicate that some yielding effects and possibly dilatancy are still occurring. It is probable that these effects, are, to some extent, rate dependent.

Although the present data are insufficient to ascertain to what degree rate effects are important, some of the observed behavior of the response can be predicted well with plastic rate-dependent models. One model considered, hereby denoted as the dilatant rock model, describes dilatancy due to microcracking in a rate-independent manner. Other models examined are plastic rate-dependent descriptions with and without dilatancy. Also an elastic prediction of the response is compared to the experimental data.

The dilatant rock model is applied here to Westerly granite using the triaxial data from Reference 7. For competent rock, a yield surface and a failure surface, both of which are pressure-dependent, are assumed. The yield surface is pushed out by work-hardening toward the failure surface. As the failure surface is approached, increasing numbers of microcracks are formed, i.e., the rock exhibits dilatancy. The number of cracks that form at failure is linearly dependent on the current pressure. After some maximum pressure is exceeded, plastic flow continues without any increase in dilatancy. The maximum dilatancy possible for the material, permanent volume change at failure, is assumed to be a linear function of pressure. A more detailed description of this model is given in Reference 8.

The rate-dependent models considered postulate the material to be rate-dependent only after some yield surface has been exceeded. Below this yield surface the material is assumed to respond in an elastic manner. Such models are commonly referred to as elastic-viscoplastic models. The particular form, proposed by Perzyna (Reference 9), is one in which the plastic strain rate, $\dot{\varepsilon}^{p}_{ij}$, is an arbitrary function of the excess stress above the static yield state. A special case of this general form introduces the additional assumption of isotropic plastic behavior and includes the yield criteria of von Mises and Mohr-Coulomb.

The constitutive equations describing the foregoing behavior are

$$\dot{\varepsilon}_{ij} = \frac{1}{2\mu}\dot{S}_{ij} + \frac{1}{3K}\dot{S}\,\delta_{ij} + \gamma < \Phi(F) > \left(\alpha\delta_{ij} + \frac{S_{ij}}{2\sqrt{J_2}}\right) \quad (3)$$

where the superposed dot denotes the partial derivative with respect to time, S_{ij}, S, and ε_{ij} are the components of the deviatoric stress, mean stress, and strain, μ is the shear modulus, K is the bulk modulus, γ is a constant associated with the relaxation time of the material, α is a constant corresponding to the coefficient of friction of the material, and J_2 is the second invariant of the deviatoric stress tensor. The symbol $<\Phi(F)>$ is defined as follows:

$$\Phi(F) = \begin{cases} 0 & F \leqslant 0 \\ \Phi(F) & F > 0 \end{cases} \quad (4)$$

The function F represents the static yield condition of inviscid plasticity in the form

$$F = \frac{\alpha J_1 + \sqrt{J_2}}{k} - 1 \quad (5)$$

where $J_1 = 3S$ is the first invariant of the stress tensor and k is the initial yield in simple shear. The above relation denotes the Mohr-Coulomb yield law and when $\alpha = 0$ it becomes the von Mises condition.

The last term in Equation 3 represents the plastic strain rate. The explicit form of the function $\Phi(F)$ in this term is dictated by the desire to represent the characteristics associated with a material under dynamic loading. There are numerous forms, some that are in a sense physically motivated. The simplest form--the one examined in this work--is the linear form $\Phi(F) = F$. This particular form can be traced back to the work of Bingham (Reference 10); however, its generalization and application to three-dimensional stress states in a solid is contained in work of Hohenemser and Prager (Reference 11). The linear form of the plastic strain term in Equation 3 is given by

$$\dot{\varepsilon}^{P}_{ij} = \gamma \left(\frac{\alpha J_1 + \sqrt{J_2}}{k} - 1 \right) \left(\alpha \, \delta_{ij} + \frac{S_{ij}}{2\sqrt{J_2}} \right) \tag{6}$$

If the dilatancy is to occur, $\alpha \neq 0$ and the plastic volumetric rate of dilatation is given by

$$\dot{\varepsilon}^{P}_{ii} = 3\alpha\gamma \left(\frac{\alpha J_1 + \sqrt{J_2}}{k} - 1 \right) \tag{7}$$

For the limiting case where $\gamma = 0$, Equation 3 describes elastic behavior. For the opposite extreme, as $\gamma \to \infty$, $\alpha J_1 + \sqrt{J_2} \to k$, and we obtain the perfectly plastic Mohr-Coulomb dilatant model formulated by Drucker and Prager (Reference 12),

$$\dot{\varepsilon}^{P}_{ij} = \lambda \left(\alpha \delta_{ij} + \frac{S_{ij}}{2\sqrt{J_2}} \right) \tag{8}$$

where λ is a constant of proportionality independent of any relaxation time.

COMPARISON WITH EXPERIMENTAL DATA

Calculations using the dilatant rock model in spherical geometry were made with POD, a one-dimensional, finite-difference code. The calculations for the elastic and rate-dependent models were made with a one-dimensional characteristic code which is similar to the one used in Reference 13. The value of density used for Westerly granite was 2.63 gram/cm^3. A linear equation of state was used in all the calculations to facilitate the semilinear requirement of the rate-dependent constitutive models imposed by the characteristic code. For these low stress levels, the hydrostat of Westerly granite can be approximated reasonably in a linear manner. The bulk modulus was equal to 0.51 Mbar and the value of the shear modulus was 0.25 Mbar.

The boundary condition used in the calculations was the measured stress-time profile obtained from a manganin gauge located at a scaled radius of $R/R_o = 3.99$. This profile was selected because it gave a more authentic description of the material response for the study of stress-time histories at larger radii than could be obtained using the cavity boundary condition together with some assumed model for predicting the stress wave attenuation down to levels below 25 kbar.

A comparison of the stress-time profiles predicted by the elastic and dilatant rock models with experimental data at scaled radii of 4.96, 5.89, and 6.95 is shown in Figure 10. Time is referenced to the loading boundary $R/R_o = 3.99$. Both models predict higher peak stresses than were measured. The dilatant rock model predicts slightly higher peak stresses than the elastic model, indicating that the reduction in volume due to dilatancy is effected by the loading part of the wave and retards the decay of the peak stress. The unloading portion of the stress profiles predicted by the dilatant model, although higher in amplitude, show the same attenuation behavior that the experimental profiles do, while the elastic model predicts a much more rapid attenuation. This suggests that dilatancy is still occurring at the lower stress levels.

Figure 11 compares experimental results with the stress-time profiles obtained from calculations using the elastic-viscoplastic models. The model using a von Mises yield, where $k = 0.5$ kbar and $\gamma = 100/sec$, provides very good agreement with the experimental profiles. The same model, where $\gamma = 1000/sec$, shows considerably more decay in the peak stress and a much slower attenuation behind the peak. This indicates the relaxation behavior of the model to be strongly dependent on the rate of loading. The model using the Mohr-Coulomb yield condition and associated rate-dependent flow rule, where $k = 0.5$ kbar, $\gamma = 100/sec$, and $\alpha = 0.15$, predicts peak stresses higher than measured and gives an attenuation higher in amplitude but similar in shape to that observed. The difference in predicted amplitude with the data increases as the wave propagates outward.

The above results show that the apparent dilatant behavior of the measured stress-time profiles can be described quite well by a rate-dependent model in which dilatancy is excluded. It is possible that adjusting the parameters in the rate-dependent Mohr-Coulomb model can result in an equally good description of

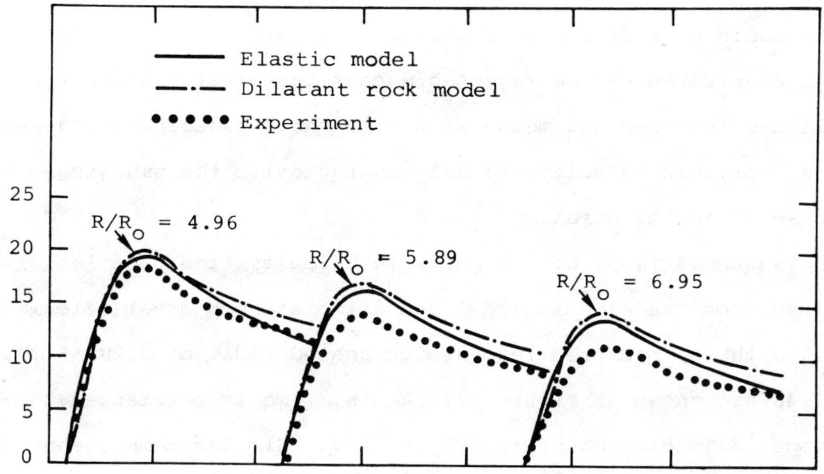

Figure 10 Experimental stress-time histories compared with results obtained using the dilatant rock model and elastic theory.

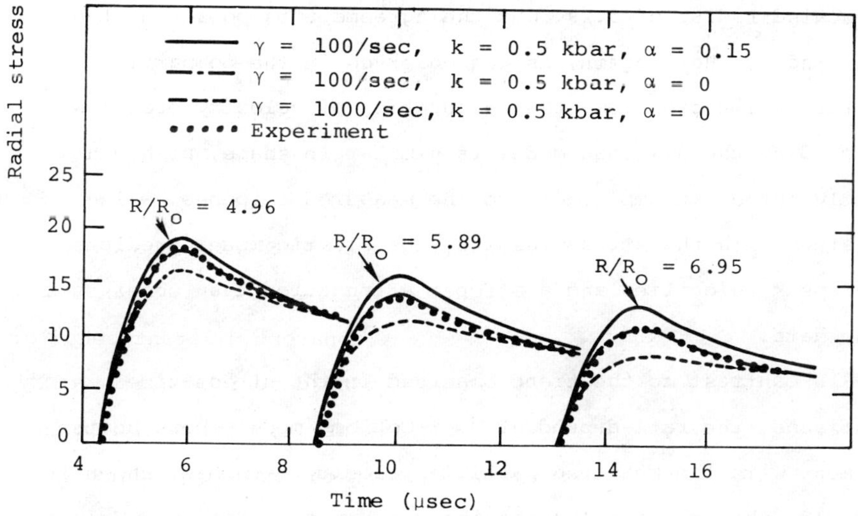

Figure 11 Experimental stress-time histories compared with results obtained from elastic-viscoplastic theory.

the response. Likewise, the parameters in the dilatant rock model, to within the static data scatter, could probably be adjusted to give a better fit to the observed response. Therefore, on the basis of radial stress measurements alone, none of the models considered can be rejected except the elastic model. This underlines the need for measurements of another dependent variable, such as particle velocity, to help characterize the spherical response of the material.

The comparisons of the particle velocity-time profiles, obtained from the elastic model and dilatant rock model calculations, with the measured response at scaled radii of 5.29, 6.39, and 7.46 are shown in Figure 12. Again, time is referenced to the imposed loading boundary $R/R_o = 3.99$. Similar comparisons using the elastic-viscoplastic models are given in Figure 13. In Figure 12, good agreement is given by the dilatant rock model at a scaled radius of 5.29, but the agreement is poorer at the larger radii. Here again, as was observed in the comparison of the stress-time profiles, the attenuation of velocity profiles predicted by the dilatant model is similar in shape, although slightly larger in amplitude, to the measured response. Also consistent with the stress results, the elastic model predicts higher peak velocities and a steeper attenuation than observed in the data.

In contrast to the trend observed in the stress-time history comparisons, the rate-dependent Mohr-Coulomb model shows better agreement with the measured particle velocity response, shown in Figure 13, than the model using the von Mises condition. Values

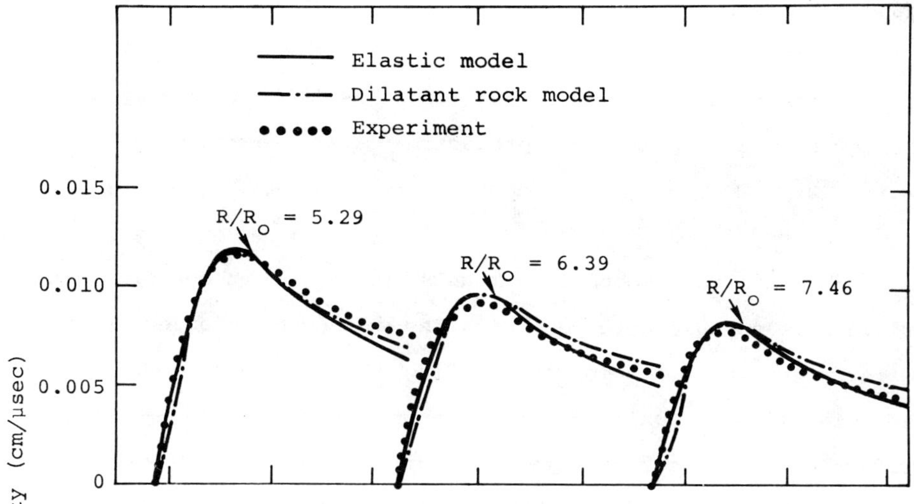

Figure 12 Experimental particle velocity-time histories compared with results using the dilatant rock model and elastic theory.

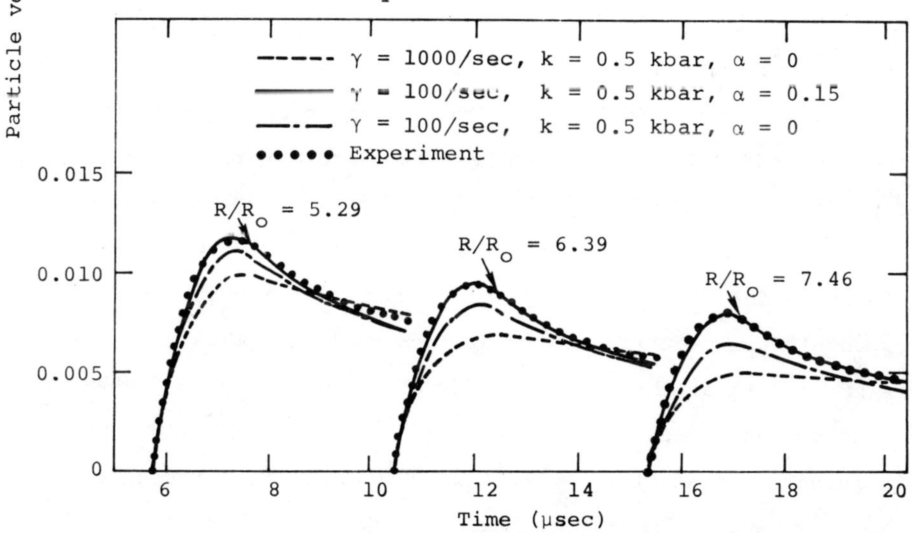

Figure 13 Experimental particle velocity-time histories compared with results obtained from elastic-viscoplastic theory.

for k, γ, and α were the same as those used for the stress predictions. The difference in the velocity response given by these two models is consistent with their difference in stress predictions. The dilatancy associated with the Mohr-Coulomb model tends to retard the viscous effect as is indicated by the smaller amount of relaxation predicted for the response.

CONCLUSIONS

The elastic-viscoplastic models, with and without dilatancy, provides reasonable agreement with the measured response of Westerly granite. The model including dilatancy best describes the velocity data, while the nondilatant model provides a better prediction of the stress data. The difference in the results given by the two models is slight and appears to affect only the amplitude and not the shape of the time histories. Using a more complex form of the function $\Phi(F)$ could undoubtedly result in a better overall fit to both sets of data. Many variations of $\Phi(F)$ are given in Reference 9. However, physical motivation is lacking for many of these forms and the strain rate data necessary to confirm them is unavailable.

Significant in the above results is the ability of simple rate-dependent models (i.e., linear function $\Phi(F)$) to describe a response usually attributed to dilatancy alone. This enhances the prospect that the effects of rate on the governing mechanisms are important. Such an inference can best be substantiated by examining the character of the stress-wave material response under various loading and rate of loading conditions, particularly in a nonplanar geometry where dilatancy is more markedly exhibited.

The dilatant rock model, being more firmly based on physical arguments and static data, predicts higher responses for both stress and particle velocity but shows good agreement with the shape of the measured responses. Taking into account the experimental scatter of the static data used to evaluate the parameters in the rock model, combined with the uncertainty associated with the measured dynamic responses, makes this model appear favorable.

REFERENCES

1. M. W. McKay, C. S. Godfrey, "Study of Spherically Diverging Shock Waves in Earth Media," DASA-2223, Physics International Company, San Leandro, California, (March 1969).

2. M. W. McKay, R. R. Shadel, R. P. Swift, C. Young, "Experimental Studies of Stress-Wave Propagation in Earth Media," DASA-2525, Physics International Company, San Leandro, California (September 1970).

3. Robert P. Swift, "The Dynamic Response of Westerly Granite to Spherical Stress Waves," DASA-2717, Physics International Company, San Leandro, California (August 1971).

4. C. F. Petersen, W. J. Murri, R. W. Gates, "Dynamic Properties of Rocks," DASA-2298, Stanford Research Institute, Menlo Park, California (June 1969).

5. S. J. Jacobs, D. J. Edwards, "Experimental Study of the Electromagnetic Velocity Gage Technique," presented at the Fifth Symposium on Detonation, Pasadena, California (August 1970).

6. C. Young, R. Fowles, R. P. Swift, "An Electromagnetic Stress Gauge," in Shock Waves and the Mechanical Properties of Solids, Proc. 17th Sagamore Conference, September 1970.

7. W. S. Brown, S. R. Swanson, "Constitutive Equations for Westerly Granite and Cedar City Tonalite for a Variety of Loading Conditions," DASA-2473, University of Utah (March 1970).

8. J. E. Reaugh, "Yield Model MINERAL ROCK/MINE UNDER Test," PIPR-242-3 and PIPR-252-3, Appendix, Physics International Company, San Leandro, California (January 1971).

9. P. Perzyna, "Fundamental Problems in Viscoplasticity," Advances in Applied Mechanics, 9, Academic Press, New York (1966).

10. E. C. Bingham, Fluidity and Plasticity, McGraw-Hill, New York (1932).

11. K. Hohenemser, W. Prager, "Uber die Ansatze der Mechanik Isotroper Kontinua," ZAMM, 12, 216-226, 1932.

12. D. C. Drucker, W. Prager, "Soil Mechanics and Plastic Analysis or Limit Design," Quart. Appl. Math., 10, 157-165, 1952.

13. R. P. Swift, I. M. Fyfe, "Elastic/Viscoplastic Theory Examined Using Radial Cylindrical Stress Waves," J. Appl. Mech., 34, 1134-1140, December 1970.

PORE FLUID AND THE MECHANICAL BEHAVIOR OF ROCK

by

F. H. Cornet*

Abstract

Pore fluid influences the behavior of rock subjected to a triaxial state of stress. The fluid has a direct action on the matrix of the rock: the unit surface energy value and coefficient of friction are altered and the matrix can be chemically dissolved. In addition, if the rock is saturated, pore pressure modifies the state of stress and decreases the principal stress values.

Theoretical and experimental aspects of the pore fluid effect are discussed and it is suggested that the presence of pore fluid will accentuate the brittleness of the rock. The pore pressure could be used as a parameter to represent, qualitatively or quantitatively, the state of failure in a specimen tested in undrained conditions.

1. INTRODUCTION

In design considerations of structures supported by or contained within a rock mass, the behavior of the intact rock must be established. Rocks containing internal fluids are frequently encountered and this fluid can significantly affect the mechanical properties of the rock. The principal aspects of this phenomenon are described here and it will be shown that the fluid can act in two ways:

1. A physical and chemical action on the rock

---

* Graduate Student, Department of Civil and Mineral Engineering, University of Minnesota, Minneapolis, Minnesota, 55455.

2. An indirect action due to a change in the stress state caused by the pore fluid.

These effects can be studied by three types of laboratory tests:

1. Tests on air-dried specimens
2. Drained tests in which the pore pressure is kept constant
3. Undrained tests in which the pore fluid is sealed and the pore pressure changes during loading.

2. THEORETICAL ASPECTS

2.1 Direct Action

The direct action of pore fluid on rock properties has been described by BARON et al. (1963), BOOZER et al. (1963), and ALDRICH and MERRITT (1969). It is generally agreed that the compressive strength of rock is significantly influenced by the nature of the pore fluid. For example, ALDRICH et al. report that oil saturated specimens of Berea sandstone were 15% stronger than water saturated specimens. Similar results were obtained by BARON et al. (1963) for Vosgian sandstone and Hassi-Messaoud sandstone. The authors attributed this phenomenon to the fact that water could weaken the matrix of these sandstones which were partly composed of clay minerals. They also suggested that the two fluids had a different effect on the free surface energy of the rock.

A change in the sliding coefficient of friction could also contribute to the difference in compressive strengths. MCCLINTOCK and WALSH (1962) introduced the coefficient of friction into a criterion for rock failure in compression and, thus, if the pore fluid has a lubricating effect, the magnitude of the compressive strength will depend on the nature of the fluid. This idea is strengthened by the fact that the shear strength of saturated specimens is lower than the shear strength of unsaturated or dry specimens, even though the shear stress is independent of the pore pressure.

Laboratory studies of the pore fluid effect should utilize the appropriate in-situ fluid, whether it is water, oil or gas. If another fluid is used for convenience, however, it is important to remember that the direct action of the fluid may differ from that of the in-situ fluid.

2.2 Indirect Action

The indirect action of the pore fluid is a modification of the stress state in the rock caused by the pore pressure. This change in stress state has been described and calculated (e.g. HANDIN et al., 1963) via the effective stress concept. If σ is the stress acting across a surface dS and u is the fluid pressure acting across the surface dS, the effective stress acting on dS is $\bar{\sigma} = \sigma - u$. Throughout the following development, compressive stresses will be considered positive.

In the case of a saturated specimen subjected to a hydrostatic pressure, P_c, an additional axial pressure, P_v, and a pore pressure P_p, the effective stress concept indicates that the effective octahedral normal stress acting on the matrix is

$$\sigma = (P_v/3) + P_c - P_p$$

However, HOLLAND and ROBINSON (1969) found that the formula is incorrect for limestone and shale and proposed to use the boundary porosity concept to adjust the formula. TERZAGHI (1945) defined the boundary porosity as

... the ratio of that part of the area

of the potential surface of failure which is in contact with the internal liquid and the total area of this surface.

The influence of boundary porosity is described in the next few paragraphs. The treatment follows that proposed by HOLLAND et al. (1969).

If σ and τ are the octahedral normal and shearing stresses in the rock matrix then:

$$\sigma = (\sigma_1 + \sigma_2 + \sigma_3)/3 \quad \ldots\ldots (1)$$

and

$$\tau = 1/3\{(\sigma_1 - \sigma_2)^2 + (\sigma_2 - \sigma_3)^2 + (\sigma_3 - \sigma_1)^2\} \quad \ldots\ldots (2)$$

where σ_1, σ_2 and σ_3 are the principal stresses in the specimen and functions of the pore pressure.

If a body is in equilibrium, the summation of all the vertical forces acting across a lateral section $\overline{AA}$ of the specimen must be equal to zero:

$$\sum_n P_{Gn}A_{Gn} + \sum_m P_{Pm}A_{Pm} - (P_V + P_C)\, A_{\overline{AA}} = 0 \quad \ldots\ldots (3)$$

where

$\sum_n P_{Gn}A_{Gn}$ is the sum of the vertical forces transmitted across the grains (n grains in the cross section $\overline{AA}$)

$\sum_m P_{Pm}A_{Pm}$ is the sum of the vertical forces transmitted across the pore spaces by the pore fluid (m pores in the cross section $\overline{AA}$)

P_{Gn} is the stress in the nth grain

P_{Pm} is the pore pressure in the mth pore space

P_V is the uniform axial stress applied to the specimen

P_C is the confining pressure

A_{Gn} is the area of the nth grain in the plane $\overline{AA}$

A_{Pm} is the area of the mth pore space in the plane $\overline{AA}$

$A_{\overline{AA}}$ is the total cross-sectional area created by plane $\overline{AA}$.

The same consideration applies to an axial cross-section $\overline{BB}$:

$$\sum_{n'} P_{Gn'} A_{Gn'} + \sum_{m'} P_{Pm'} A_{Pm'} - P_C A_{\overline{BB}} = 0 \quad \ldots\ldots\ldots\ldots\ldots(4)$$

Where the notation is identical except for summation over n' and m' instead of n and m.

Equation (3) can be written as

$$\sigma_1 = (\sum_n P_{Gn} A_{Gn})/A_{\overline{AA}} = P_V + P_C - (\sum_m P_{Pm} A_{Pm})/A_{\overline{AA}}) \quad \ldots\ldots(5)$$

HOLLAND and ROBINSON (1969) assume that the pore pressure is uniform in the specimen and hence deduce the relationship

$$(\sum_m P_{Pm} A_{Pm})/A_{\overline{AA}} = P_P(\sum_m A_{Pm})/A_{\overline{AA}} = P_P(A_{\overline{AA}} - A_G)/A_{\overline{AA}} = P_P(1 - A_G/A_{\overline{AA}}) \ldots(6)$$

Where A_G is the total area of the rock matrix in the plane $\overline{AA}$ and A_S is the cross-sectional area of the octahedral plane.

From equations (5) and (6)

$$\sigma_1 = P_V + P_C - P_P\,(1 - A''_G/A_S) \quad \ldots\ldots\ldots\ldots\ldots\ldots\ldots(7)$$

If the rock grains and pore spaces are distributed randomly and isotropically,

$$A_G/A_{\overline{AA}} = A'_G/A_{\overline{BB}} = A''_G/A_S \quad \ldots\ldots\ldots\ldots\ldots\ldots\ldots\ldots(8)$$

where A'_G is the total area of the rock matrix in the plane $\overline{BB}$ and A''_G the corresponding area in the octahedral plane.

From equations (4) and (8), we obtain

$$\sigma_2 = \sigma_3 = (\sum_{n'} P_{Gn'} A_{Gn'})/A_{\overline{BB}} = P_C - P_P(1 - A''_G/A_S) \quad \ldots\ldots(9)$$

The octahedral normal and shear stresses can be found from equations (1), (2), (7) and (9) and are given by:

$$\sigma = P_V/3 + P_C - P_P(1 - A''_G/A_S) \quad \text{(10)}$$

and $$\tau = \sqrt{2}\, P_P/3 \quad \text{(11)}$$

Since liquids cannot support a shear stress, τ is independent of the pore pressure.

If A_{SF} is the area of the plane of fracture, SF, then the definition of boundary porosity (TERZAGHI op. cit., 1945) yields

$$n_B = 1 - A'''_G/A_{SF} \quad \text{(12)}$$

where n_B is the boundary porosity.

Since it was assumed that the rock grains and pore spaces are distributed randomly and isotropically

$$A'''_G/A_{SF} = A''_G/A_S \quad \text{(13)}$$

and hence from equations (10), (12) and (13)

$$\sigma = P_V/3 + P_C - n_B P_P \quad \text{(14)}$$

The boundary porosity, n_B, can be measured if the maximum octahedral shear stress failure criterion is assumed. Failure occurs for a certain relationship between τ and σ, say $\tau = f(\sigma)$ and thus a specimen can fail at a certain value of the normal stress (and hence shear stress) but at different values of the confining pressure and pore pressure. Assume that the specimen fails in one state σ, τ, P_C, P_P and in another state σ', τ', P_C', P_P', but with the same P_V. Then

$$\sigma = P_V/3 + P_C - n_B P_P \quad \text{(15)}$$

$$\sigma' = P_V/3 + P_C' - n_B P_P' \quad \text{(16)}$$

If $\tau = \tau'$, and $\sigma = \sigma'$, we have from equations (15) and (16)

$$n_B = (P_C - P_C')/(P_P - P_P') \quad \text{(17)}$$

If the rate of loading is low, then the pressure gradient in the specimen can be neglected and P_p can be considered as a constant; the higher the permeability, the better the assumption. It was also assumed that the pore distribution was isotropic but, when failure initiates, micro-cracks are developed in the material. Although their distribution may be random in the major part of the specimen but their orientation is not. This invalidates equation (8) but it may be considered correct to a first approximation.

3. EXPERIMENTAL ASPECTS

The influence of pore fluid on the mechanical behavior of rock can be experimentally investigated in the triaxial test. An axial load is applied to a cylindrical specimen whilst it is subjected to a lateral confining pressure: $\sigma_1 > \sigma_2 = \sigma_3$.

Initially, air-dried specimens are tested so that the influence of the internal fluid can be evaluated with reference to the behavior of a specimen containing no fluid. Subsequent tests with saturated specimens can then be carried out under constant pore pressure conditions (the drained test) or under variable pore pressure conditions (the undrained test). The pore pressure changes due to the sealing of the internal fluid circuit in the latter test.

3.1 Tests on Air-Dried Specimens

In order to compare the results of tests on air-dried specimens with those of tests on saturated specimens, it is essential to keep the same loading rate, temperature and specimen geometry for all the tests.

WAWERSIK (1968) has conducted unconfined compression tests in a high rigidity testing machine and has demonstrated that it is possible to prevent rapid disintegration of the specimen by limiting the energy input to the specimen once the peak load is reached. In some testing situations with some rocks it was necessary to reverse the direction of platen deformation to achieve slow disintegration. Apparently, under such loading conditions the energy stored in the specimen at the peak load is sufficient to cause rapid collapse if this energy is released into the collapsing parts of the specimen. Reversal of platen direction allows the stored energy to be released into the platen. A typical stress (i.e. mean force) - strain (i.e. mean displacement) curve for such a specimen and test is shown in Figure 2. WAWERSIK referred to such behaviors as Class II. Obviously, since the strain must be reversed, it is not possible to observe Class II behavior in constant strain rate tests. Methods for controlled deformation during "Class II" tests are discussed by HUDSON et al. (1971). For situations where the specimen could be deformed monotonically to complete disintegration were given the description Class I (Figure 2). WAWERSIK observed that an increase in the confining pressure was always associated with a flattening of the post peak portion of the curve. This can be described as a decrease in brittleness.

3.2 Drained Triaxial Compression Tests

In a drained test, a saturated specimen is axially loaded while the confining pressure and pore pressure are kept constant. It is relatively easy to maintain the confining pressure at a constant value but it is much more difficult to keep the pore pressure constant, particularly if the rock has a low permeability. BARON et al. (1963) found that the time required to maintain a uniform pressure varied from a few minutes to several hours. To ensure that the pore pressure is uniform in the specimen, the pressure at the entry and exit of the pressure vessel must be compared. The pressure is uniform when gauges 1 and 2 in Figure 1 indicate the same pressure.

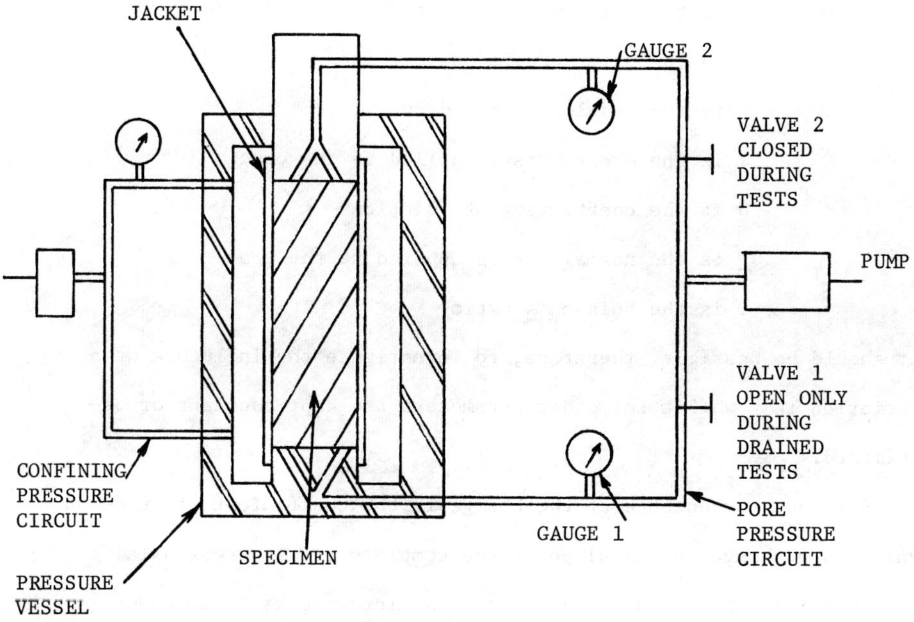

Figure 1. Experimental arrangement for drained triaxial compression test.

The presence of the pore fluid alters the value of the coefficient of friction and hence the influence of the fluid on the complete stress-strain curve should be studied. Assuming that the physical and chemical action of the fluid (partial dissolution of the matrix and change of the unit surface energy value) upon the residual strength is negligible when sliding occurs, a comparison of the residual strengths of an air-dried specimen and saturated specimen should provide a measurement of the lubrication effect.

Using this information it is possible to evaluate the change in the compressive strength due to the physical and chemical action of the fluid. An expression for the critical crack length for the initiation of failure in simple compression is given by COOK (1965).

$$c = 8\gamma G/\pi\{(1- \ )(\tau-\mu\sigma n)^2\}$$

where c is the critical crack length

γ is the unit surface energy

G is the modulus of rigidity

τ is the shear stress applied to the crack

μ is the coefficient of friction

σ_n is the normal stress applied to the crack

ν is the poisson's ratio

It should be possible, therefore, to investigate the influence of a variation in γ whilst the other parameters are kept constant or determined.

A second consequence of the change in the coefficient of friction value is a change in the slope of the complete axial stress-axial strain curve. Since the friction across micro-cracks is decreased, the amount of energy necessary to deform the specimen is decreased. Therefore, the slope of the stress-strain curve should be decreased. A flattening of the curve in the inelastic pre-peak range should be observed and the slope of the post-peak region should be steeper. It might even be possible for a rock to exhibit Class I behavior when it is air-dried and Class II behavior when it is saturated.

These effects have been experimentally observed and attributed to an increase in "brittleness" caused by a decrease in σ_2. If, however, the slope of the descending portion of the stress-strain curve varies with air-dried and saturated specimens <u>tested at the same value of σ_2</u>, the effect is caused by a change in the coefficient of friction.

In the previous section it was noted that, to a first approximation, the principal stresses in the specimen are:

$$\sigma_1 = P_V + P_C - n_B P_P$$

$$\sigma_1 = \quad _3 = P_C - n_B P_P$$

Now, if σ_1 and σ_2 are kept constant as P_C and P_P are varied, the boundary porosity, n_B, can be measured.

ROBINSON (1959), BARON et al. (1963) and ALDRICH and MERRITT (1969) found that the classical effective stress concept was applicable to sandstone. It seems, therefore, that the boundary porosity is equal to unity for sandstone, but for other rocks it may be less. This is confirmed by the results of HOLLAND and ROBINSON (1969) who found that for limestone and shale the boundary porosity was not unity. They also observed that an increase of confining pressure was followed by an increase of n_B so that, for relatively high confining pressures, n_B can always be considered as unity. Nevertheless, further investigations in this direction are required.

3.3 Undrained Triaxial Compression Tests

In these tests the internal fluid circuit is kept closed so that the pore pressure changes as the experiment is performed.

In Figure 2, the value of $\sigma_1 - \sigma_2$ is plotted on the y-axis, and the value of the vertical strain (curve A), lateral component of the volumetric strain (curve B), volumetric strain (Curve C) are plotted on the x axis following BIENIAWSKI (1969), LANE (1969) and CROUCH (1970). In addition, the curve D is obtained by plotting in abscissa the value of the volumetric strain and in ordinate the value of the pore pressure.

As the value of $\sigma_1 - \sigma_2$ increases, the volumetric strain of the specimen decreases because of the elastic compressibility of the

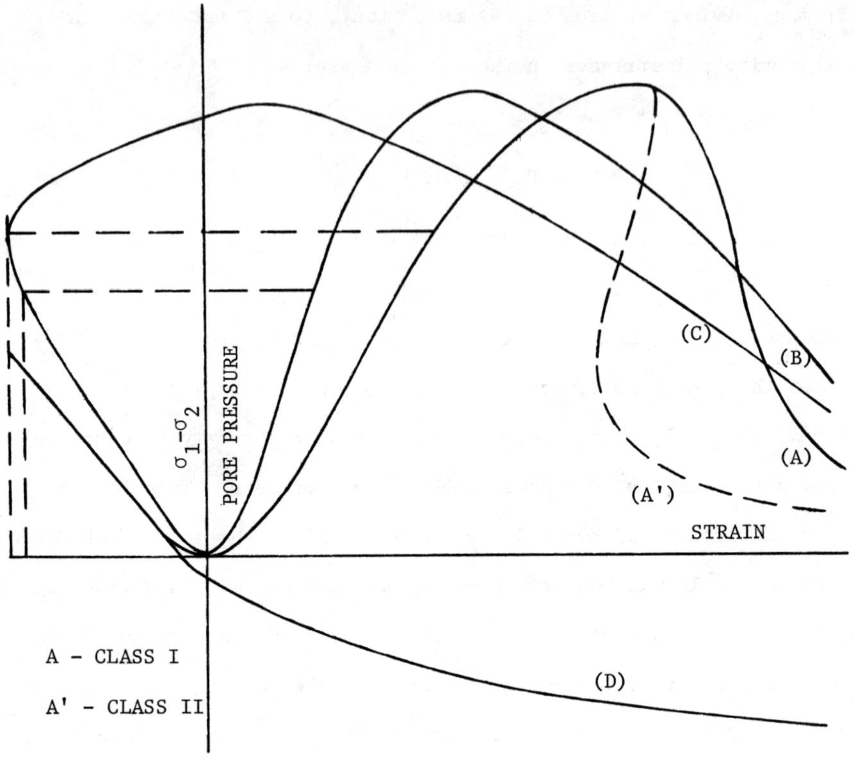

Figure 2. The complete stress-strain curve and associated curves.

specimen. This problem was treated with reference to porous rocks by LUBINSKI (1954), BIOT (1955) and GEERTSMA (1957). They showed that, if V_P is the pore volume and V_B is the bulk volume, then, during elastic deformation

$$dV_B/V_B = C_r dp + C_B(d\sigma - dp)$$

where C_r is the compressibility of the rock matrix

C_B is the bulk compressibility of the porous structure

$d\sigma$ is the variation of the external stress

dp is the pore pressure variation.

If the porosity, $f = V_P/V_B$, then

$$dV_P/V_P = -(1/f)(C_B-(1+f)C_r)dp+(1/f)(C_B-C_r)d\sigma$$

They have also shown that the total change in pore and bulk volumes resulting from a deviatoric stress system are zero. This indicates that the measurement of the change in pore pressure allows a determination of the change in pore volume. The increase in pore pressure will be initially linear and then non-linear since it follows the change of pore volume during deformation of the specimen. The volumetric strain increases as the axial strain is increased until macrofracture occurs.(1) This is due to the fact that micro-cracks are being developed in the specimen. The pore pressure could be used as a parameter to quantitatively represent the behavior of the specimen: an increase in pore pressure corresponds to elastic behavior; the maximum value of the pore pressure corresponds to the ultimate elastic deformation; a decrease in pore pressure is associated with micro-cracks in the matrix. From the point that the pore pressure decreases, the volumetric strain continuously increases and hence the pore pressure will continuously decrease. When macro-fracture occurs (by sliding along one or more planes of macro-failure), the pore pressure reaches a constant value because no change in volume is associated with macro-failure.

Experimental undrained tests have been performed by ALDRICH and MERRITT (1969) on Berea sandstone and the theoretical ideas and experimental data are in good agreement to the limit of their data at the peak strength.

As was noted earlier, the presence of a fluid in a specimen

---

(1) This applies when the behavior of the specimen is no longer elastic.

changes the strength and the post-failure behavior of the specimen. The effect due to the fluid itself is the same in drained and undrained tests. Therefore, comparison of the complete stress-strain curves obtained in drained and undrained conditions should provide information on the contribution of pore pressure alone to the pore fluid effect.

In the post-peak failure region, the increase in pore volume is caused by two phenomena: a partial decrease in axial strain caused by the decrease in the value of the axial stress and an increase in the number and length of micro-cracks. This is followed by a decrease of the pore pressure and therefore an increase in the magnitude of σ_2 and σ_3. Since an increase in σ_2 is associated with an increase in the strength of the specimen and less brittle behavior, the complete stress-strain curve obtained in undrained conditions should lie above the one obtained in drained conditions as shown in Figure 3, at least during its descending part.

Field conditions are more likely to approximate the ideal laboratory undrained conditions and therefore the measurement of strength in drained conditions should give an underestimate of the strength of porous rocks.

If the axial strain of the specimen is known, the pore pressure can be used to determine the state of failure of the core. This can be achieved by reducing the actual value of the axial stress σ_1 to the fixed value of $\sigma_2 = \sigma_3$ so that the specimen is subjected to the same hydrostatic state of stress as at the beginning of the experiment. The difference between the value of the pore pressure at this stage and the initial pore pressure at the beginning of the experiment allows the increase in pore volume to be computed. The increase in

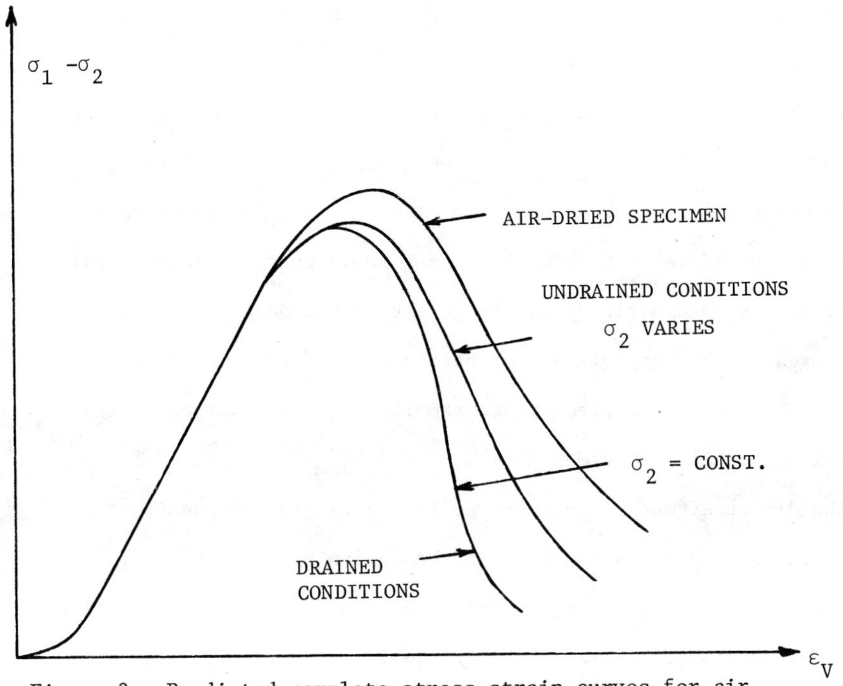

Figure 3. Predicted complete stress-strain curves for air dried specimen and saturated specimens tested in drained and undrained conditions.

pore volume is associated with the state of failure in the specimen. In addition, a measurement of the time that is necessary for the pore pressure to become constant after the axial stress has been reduced should provide a method of measuring the variation in permeability.

For all types of tests, the complete axial stress-axial strain curve must be obtained and special methods of controlling and monitoring the failure process are necessary. Servo-controlled testing systems have been used to control failure in uniaxial conditions (RUMMEL and FAIRHURST, 1970; HUDSON et al., 1971) and an extension of this method is proposed for triaxial testing.

It was shown by WAWERSIK (1968) that the complete stress-strain curve can either monotonically increase in strain (Class I behavior) or be a multiple valued function of strain (Class II behavior). In the latter case it is impossible to obtain the complete stress-strain curve with a program that dictates a monotonic increase in axial strain. However CROUCH (1970) has shown that the lateral component of the volumetric strain increases continuously throughout all stages of the complete stress-strain curve. Failure of all specimens can be controlled by programming a linear increase of this lateral component with time.

The direct effect of an increase in the lateral component of the volumetric strain is an increase in the confining pressure in the pressure vessel. Therefore, in order to maintain a constant confining pressure, an amount of fluid exactly equal to the incremental increase in the lateral component of the volumetric strain must be extracted for each incremental increase in axial strain. The feedback control system in the servo-controlled testing system operates on a closed-loop principle. A primary closed-loop activating the main hydraulic actuator that applies the axial load will be programmed to maintain a constant rate of increase in the lateral component of the volumetric strain. A secondary closed-loop, with a slower response time than the primary loop will ensure that the confining pressure is maintained at a constant value. This double closed-loop control system will operate until gross shear failure occurs and the experiment is terminated.

4. PORE FLUID EFFECTS

The presence of a pore fluid has many interesting ramifications.

For example, if the pore pressure is high enough, it is possible to have negative values for the effective confining stresss, $\sigma_2 = \sigma_3$. Under these conditions, the specimen can be subjected to a vertical compression and a lateral tension. This phenomenon has been studied by HAIMSON and FAIRHURST (1967) and has interesting applications in studies of hydraulic fracturing in porous rocks.

Since different fluids can influence the value of the coefficient of friction, comparison of the results obtained with different liquids should provide an experimental determination of the influence of this coefficient of internal friction on the failure process of a rock specimen. Also since pore pressure and pore volume are directly related, the relationship between a variation in the pore pressure and the state of failure in the reservoir rock could be established, for laboratory and field experiments.

In the case of jointed rocks, the presence of a pore fluid should appreciably affect their behavior, particularly since most jointed rocks are characterized by a relatively low shear strength. If the rock system has a low permeability, undrained conditions prevail. If the rock system is permeable, drained conditions can be assumed and there should be no pressure gradient.

The phenomenon of stick-slip friction should be more noticeable with rocks of low boundary porosity. If n_B is equal to one, the fracture plane should pass between the constitutive grains of the matrix leaving them unbroken. In this case, the fracture plane is relatively rough. As the value of n_B decreases, however, the percentage of broken grains increases and therefore the macro-fracture planes should be smoother. "Stick-slip" friction occurs when sliding takes place along smooth surfaces rather than rough ones and hence the phenomenon will be more pronounced with low boundary porosity rocks.

5. CONCLUSION

It has been shown that there is a difference in the behavior of air-dried and saturated specimens in the triaxial test. One effect of the pore fluid is due to the direct physical and chemical action of the fluid: the matrix is partially dissolved, the unit surface energy of the rock is changed and the value of the coefficient of friction is also changed. The latter change significantly influences the post-failure stress-strain characteristics of a saturated specimen. There is an increase in the brittleness and a decrease in the shearing strength. These will be manifested to a high degree in jointed materials whose behavior is critically dependent on the shearing strength and sliding properties.

The development of a pressure in the pore fluid changes the state of stress in a specimen of porous rock. It was indicated that a parameter known as boundary porosity plays a very important role in the amount by which the stress state is changed. It seems that for sandstone n_B can be assumed to be unity so that the stress state is changed significantly. For limestone and shale n_B is less than unity and the effect is correspondingly less. A more complete study is required for a better knowledge of the possible variations of the boundary porosity.

The two types of triaxial compression tests on saturated specimens were described. The drained test can be used for measuring the boundary porosity and a comparison of results from both types of test and tests on air-dried specimens allows the influence of a variation in the coefficient of friction to be evaluated.

In addition, it was suggested that variations in the pore pressure in undrained conditions are a good indication of variation in the pore volume as failure develops. This in turn allows a qualitative or quantitative estimate of the state of failure and change in permeability to be made.

6. ACKNOWLEDGEMENT

The author wishes to express his sincere thanks to Professor C. Fairhurst and Doctor J.A. Hudson for their suggestions and corrections.

This review is the result of research work supported by the American Petroleum Institute (Project 67A; Principal Investigator, Professor Charles Fairhurst).

7. REFERENCES

1. ALDRICH, G. R. and MERRITT, J., (1969), "Pore pressure effect on Berea sandstone subjected to experimental deformation". Bull. Geol. Soc. Am. 80, 8.

2. BARON, G., CASTEL, P. and HABIB, P., (1963). "Influence de la pression interstitielle sur les caracteristiques mecaniques des roches en condition de fond". Revue de l'I.F.P., 18.

3. BIENIAWSKI, Z. T., (1967). "Mechanism of brittle fracture of rock". Int. J. Rock Mech. Min. Sci., 4, 4.

4. BIOT, M. A., (1955). "Theory of elasticity and consolidation for a porous and isotropic solid". J. Applied Physics, 26, 182-5.

5. BOWDEN, F. P. and TABOR, D., (1964). "The friction and lubrication of solids, Vol. II". Clarendon Press, Oxford.

6. BOOZER, G. D., HILLER, K. H. and SERDENGECTI, S., (1963). "Effects of pore fluids on the deformation behavior of rocks subjected to triaxial compression". Proc. 5th Symp. on Rock Mechanics, Univ. of Minn.

7. COOK, N. G. W., (1965). "The Failure of Rock". Int. J. Rock Mech. Min. Sci. 2, 384-403.

8. CROUCH, S. L., (1970). "Experimental determination of volumetric strains in failed rocks". Int. J. Rock Mech. Min. Sci., 7.

9. GEERTSMA, J., (1957). "The effect of fluid pressure decline on volumetric changes of porous rocks". Petr. Trans. Am. Inst. Min. Eng., 210, 12.

10. HAIMSON, B. and FAIRHURST, C., (1967). "A criterion for hydraulic fracturing in porous rock". 4th Conf. on Drilling and Rock Mech., Austin, Univ. of Texas.

11. HANDIN, J., HAGER, R.V., FRIEDMAN, M. and FEATHER, J. M., (1963). "Experimental deformations of sedimentary rocks under confining pressure: pore pressure tests". Bull. Am. Assoc. Petr. Geol., 47.

12. HOLLAND, W. E. and ROBINSON, R. H., (1969). "Some interpretations of pore fluid effects in rock failure". 11th Symp. on Rock Mechanics, Univ. of Calif., Berkeley.

13. HUDSON, J. A., BROWN, E. T. and FAIRHURST, C., (1971). "A method of optimizing the control of rock failure in servo-controlled laboratory tests". Rock Mechanics, in press.

14. LANE, K. S., (1969). "Discussion of the article of Aldrich and Merritt". Bull. Geol. Soc. Am., 80, 8.

15. LUBINSKI, H., (1954). "The theory of elasticity for porous bodies displaying a strong pore structure". Proc. 2nd Nat. Cong. Appl. Mech.

16. MCCLINTOCK, F. A. and WALSH, J. B., (1962). "Friction on Griffith cracks under pressure". 4th U.S. Nat. Cong. Appl. Mech., Berkeley, 1015-21.

17. ROBINSON, R. H., (1959). "The effect of pore and confining pressure on the failure process in sedimentary rocks". Colorado School of Mines Quarterly, 54, 3.

18. RUMMEL, F. and FAIRHURST, C., (1970). "Determination of the post-failure behavior of brittle rock using a servo-controlled testing machine". Rock Mechanics, 2.

19. TERZAGHI, K., (1945). "Stress conditions for the failure of saturated concrete of rocks". Proc. Am. Soc. Test. Mat., 45.

20. WAWERSIK, W., (1968). "Detailed analysis of rock failure in laboratory compression tests". PhD thesis, University of Minnesota.

MECHANICAL BEHAVIOR OF ROCK UNDER CYCLIC FATIGUE

By Bezalel C. Haimson[1] and Chin Man Kim[2]

INTRODUCTION

Rock and rock structures such as open-pit benches, mine escavations, bridge abutments, dam and road foundations undergo cyclic loading caused by earthquakes, traffic, drilling, blasting, etc. This type of loading often causes a material to fail at a stress lower than its determined strength, a phenomenon called fatigue. Fatigue characteristics are usually presented in the form of an S-N curve (stress level versus the number of cycles required to bring about failure). The top stress value that appears to be unaffected by cyclic loading is sometimes called the fatigue limit. Cyclic fatigue in structural materials such as metals, concrete and soil has been thoroughly investigated in the last decades. Rock fatigue, however, has received only little attention. An early attempt to establish fatigue characteristics of a limestone was inconclusive mainly due to a very limited number of tests undertaken (3). Burdine (1), however, ran an extensive series of compressive cyclic loading tests in Berea Sandstone, discovered that the rock was definitely weakened by repetitive loading, and determined S-N curves for different testing conditions. Hardy and Chugh (4) used modern equipment to improve testing

[1] Assistant Professor, Department of Metallurgical and Mineral Engineering, University of Wisconsin, Madison.

[2] Graduate Student, Department of Metallurgical and Mineral Engineering, University of Wisconsin, Madison.

procedures and concluded that three additional rock types were fatigue prone even under a low maximum number of cycles per test (10,000). The present paper reports the results of the first phase in a comprehensive experimental study of rock cyclic fatigue. The main objectives of the investigation are to provide engineers quantitative results in the form of S-N curves obtained for different rock types under various stress conditions, to determine the fatigue strength of failed rock for the use of mine designers and earthquake researchers, and to provide a better understanding of the fatigue mechanism by strain measurements and careful observations of fabric changes.

LABORATORY EQUIPMENT AND TEST PROCEDURES

Rock Specimens

Two rock types have been used in the reported study. White Tennessee Marble was chosen because of its uniformity, isotropy and fine grain size. It has been extensively tested and is known to have very consistent mechanical behavior (6). Georgia Marble was selected for the ease with which it can be controlled in its post failure mode (the descending part of the complete stress-strain curve).

Cylindrical specimens, 1.0 inch in diameter and 2.5 inch long, were cored out of a large rock block by diamond drilling in one direction only. The automatically-fed coring yielded straight smooth rock cylinders that did not require further machining. Specimen ends were surface ground until flat and parallel faces were obtained to within 0.001 inch. The specimens were then oven dried at $120^{\circ}F$ for a week prior to testing.

Apparatus

Specimens were loaded in an electro-hydraulic servo-controlled loading machine of 100,000 lbs. capacity. A general view of the entire apparatus is shown in Fig. 1. The machine can be programmed

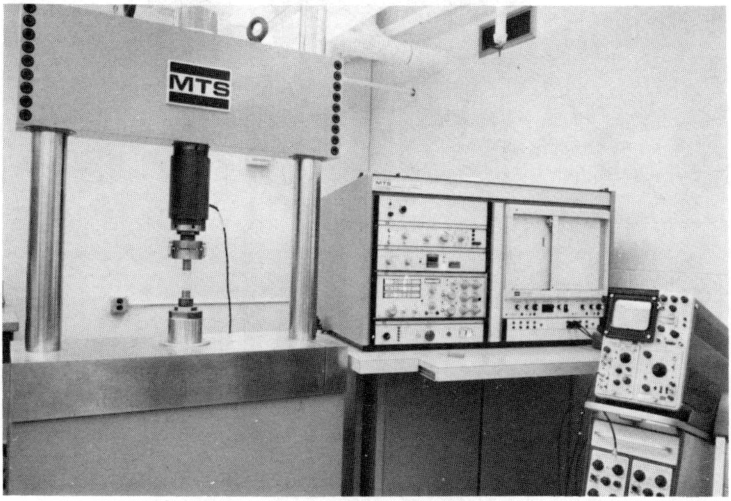

FIG. 1 LOADING SYSTEM, ELECTRONIC CONTROLS AND DATA ACQUISITION INSTRUMENTS.

through a function generator to apply cyclic loading to a specimen by controlling either stress or strain rates. In the experiments described a constant rate of compressive load was applied to specimens using a triangular wave shape. A digital counter gave the number of cycles per test, and an X-Y recorder was used to plot stress-strain or strain-time curves. The load applied to specimens was sensed by a dynamic load cell mounted on the bottom part of the machine crosshead. The longitudinal strain was measured through strain gages mounted on a system of double cantilevers (Fig. 2). The rings holding the cantilever device were threaded on to the platens in contact with the rock specimens and held in place by locking nuts. This method of attachment proved superior to the previously used set-screws which tended to slip in long duration tests.

FIG. 2 LOADING JIG WITH SPECIMEN, CANTILEVER SET AND SWIVEL HEAD.

The platens had a diameter only slightly larger than that of the specimens and were part of a loading jig especially built fot these tests. The lower platen was rigidly attached to the hydraulic ram; the upper platen contained a swivel head mechanism to ensure that complete contact was made with the specimen. To prevent slippage of the swivel head, six bolts were used to lock it to the part of the loading jig rigidly attached to the crosshead. The locking of the bolts was a tedious job, carefully performed to ensure good alignment between specimen and platens. Overtightening of one of the bolts could cause misalignment and hence premature failure in cyclic testing. Utmost care was, however, taken in specimen installation and the consistency of the results testify to the success of the method.

Experimental Program

The portion of the fatigue testing program that has been carried out so far, and is described in this paper, was limited to uniaxial compression. The intent was to test rock cylinders under compressive loading increased steadily from zero load to a certain upper peak, decreased at the same rate to the initial point, and cycled until failure occured. To prevent loss of contact between sample and platens at zero load, the lower peak of the compressive cycle was moved up to around 200 psi and was kept approximately constant throughout the testing program. The upper peak, however, was set anew for each test.

The cyclic rate used was intended to simulate the frequencies of the major pulses in earthquakes (1-2 cps) and blasting (10 cps). In the tests described the frequencies were kept at 1-4 cps. The higher frequencies were preferred because of shorter test duration, but the lower frequencies (1-2 cps) were necessary when stress-strain recordings were made, due to the X-Y plotter response limitations. The maximum number of cycles per test was kept at 10^6 with a few exceptions when the figure was exceeded. To obtain a representative S-N curve the fatigue life of specimens (number of cycles to failure) was determined for different upper-peak stress values. The general procedure was to reduce the maximum applied load from test to test until the fatigue life of the specimens reached 10^6 cycles.

Cyclic tests were also run on specimens that had been first loaded up to their peak compressive stress carrying capacity. In Georgia Marble, specimens were actually brought to 70% of the peak stress capacity on the descending side of the complete stress-strain curve, and only then cyclicly loaded. In order to test the fatigue characteristics of these failed specimens quasi-static strain-controlled loading was used to bring the rock to the initial condition for cyclic stress-controlled loading.

EXPERIMENTAL RESULTS

The most extensive testing in the program has been carried out in White Tennessee Marble. The first stage of the experimental work was to determine the compressive strength of the rock, a value to be used as the upper limit to the maximum applied stress in cyclic loading. Specimens were loaded stress-controlled, at four different rates, two in the quasi-static range, and two in the dynamic range used in the cyclic loading. The results (Table 1) show that the

TABLE 1.--COMPRESSIVE STRENGTHS OF WHITE TENNESSEE MARBLE AT DIFFERENT LOADING RATES

Loading rate (psi/sec)	No. of Specimens	Mean Compressive strength (psi)	Standard Deviation	
			psi	%
40	9	19,740	355	1.8
100	11	21,150	900	4.2
50,000	12	23,285	840	3.6
200,000	7	24,890	425	1.7

compressive strength varies considerably with the rate of loading. Hence, whenever this parameter is employed it should be accompanied by the conditions under which it was determined. Because of the high response required, a storage type oscilloscope was employed to directly record the stress strain curve in each of the tests. By photographing the oscilloscope trace a permanent record was obtained.

The first important conclusion drawn from cyclic testing of White Tennessee Marble is that it is definitely weakened by repetitive loading. The fatigue effect can best be verified from the S-N curve shown in Fig. 3. The stress is given in percentage of the compressive strength

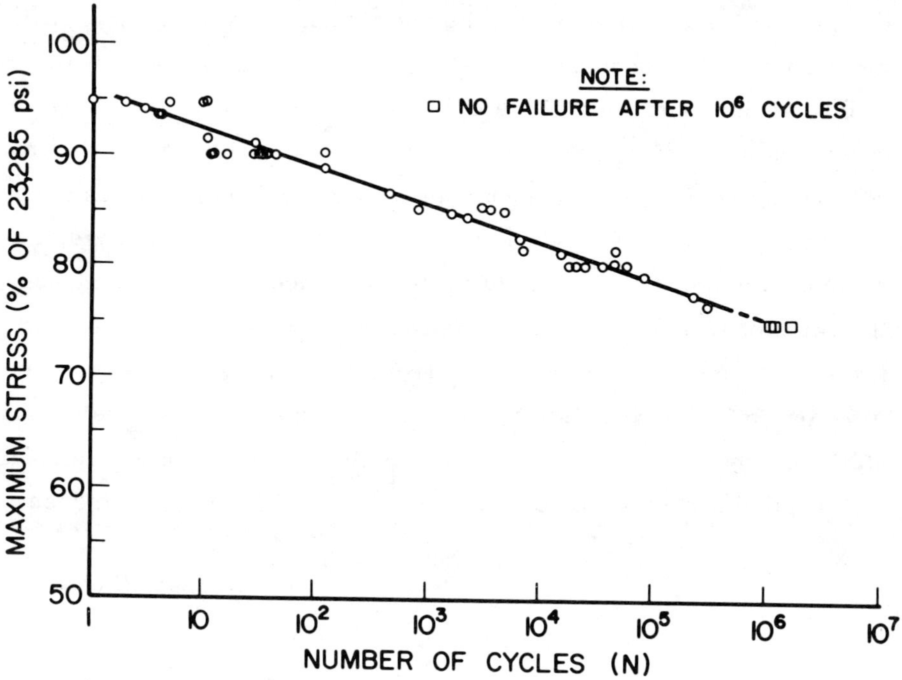

FIG. 3 S-N CURVE FOR WHITE TENNESSEE MARBLE

at 50,000 psi/sec. Most of the tests were run at 4 cps, while some were run at 1 or 2 cps for the purpose of recording stress-strain characteristics. No apparent difference was found between the frequencies used as far as specimen fatigue life was concerned. The S-N curve clearly shows that as the maximum compression decreases the life expectancy of a specimen increases. Fig. 3 also indicates that the spread of experimental points was surprisingly limited, due to the uniformity of the rock as well as the great care taken in the preparation and operation of the tests. In a semi-logarithmic plot, as the one shown in Fig. 3, the average relationship between the maximum applied stress (S) and the number of cycles needed to cause failure (N) is given by a straight line. At 75% of the dynamic compressive strength used specimens did not fail within the experimental limit of 10^6 cycles. Moreover, when loaded monotonically to failure at the conclusion of the cyclic test, they did not show any sign of weakening or strengthening effects. It is hardly expected that rock will be subjected to a larger number of similar loading cycles during the expected life of an engineering structure. In an earthquake, for example, no more than 100-200 cycles are encountered (2). Hence, the value of 17,450 psi (=0.75 x 23,285 psi) can be used as the fatigue limit or the critical compressive strength of White Tennessee Marble. A design in intact rock using this value, not only is protected against static and dynamic stresses but also against all kinds of cyclic loading.

Typical stress strain curves are shown in Fig. 4. Permanent

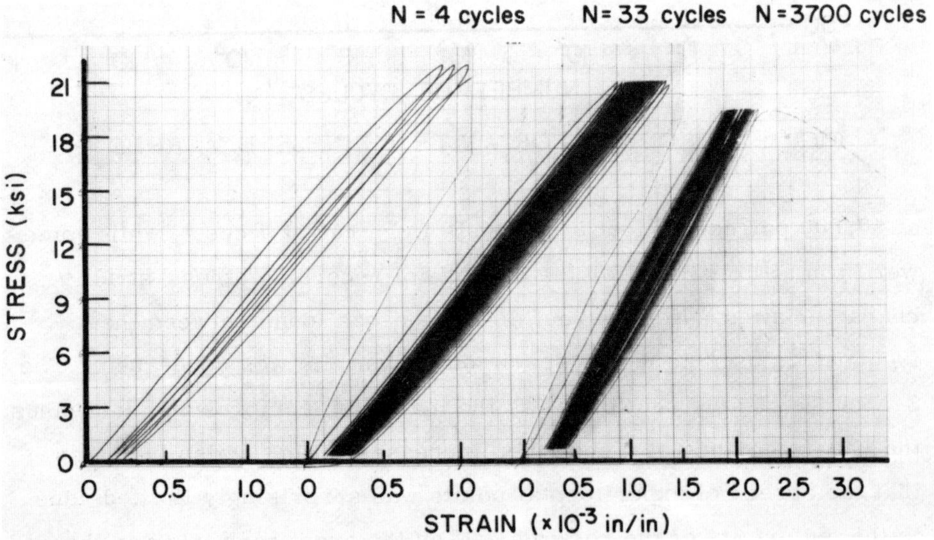

FIG. 4 STRESS-STRAIN CURVES FOR WHITE TENNESSEE MARBLE CYCLICLY LOADED TO DIFFERENT MAXIMUM STRESS VALUES

strain appears to accumulate in the rock with the dissipation of a relatively high amount of energy. Common to all tests is the phenomenon of rather large hysteresis in the first few cycles, an almost complete closure in the next group of cycles, and a reopening in the last several cycles prior to failure. The hysteresis in the first cycles is probably due to the very high peak loads which take the rock to beyond its linear elastic limit and into the zone of irreversible structure changes. In the very short life tests, the damage done by the high stresses is irreparable and the specimens break before reaching the second stage. For lower applied maximum stresses the phenomenon of hysteresis closure occurs seemingly due to micro-crack propagation being blocked and the linear elastic limit being lifted. As cyclic loading continues, the actual fatigue phenomenon eventually takes place, internal cracking reinitiates, the hysteresis grows and the specimen finally fails.

The amount of strain difference between the upper peaks of the last and first stress-strain cycles has been closely followed. No hard conclusions can be drawn as yet, but it appears that it is definitely limited by the strain difference between the ascending and the descending parts of the complete stress-strain curve for the same value of stress. This implies that the complete stress-strain curve actually defines the limit of strain that can be applied to a rock at a certain stress level, without producing failure. Quantitatively, the amount of strain difference varied between $25 - 35 \times 10^{-5}$ inch/inch, which is within the limit set by the complete stress-strain curve.

The behavior of strain versus time was often recorded and a typical result is shown in Fig. 5. The curve shaped by the upper peak points clearly resembles that of creep behavior, and can be divided into three stages. There is a primary stage in which the upper peak strain increases at a decelerating rate. It is followed by a steady state stage which appears in Fig. 5 as an ascending straight line. In medium and long life tests this stage is invariably the longest. The third and final stage is that of accelerating upper peak strain culminating in specimen failure. The stages observed in the strain-time behavior match those in the stress-strain characteristics.

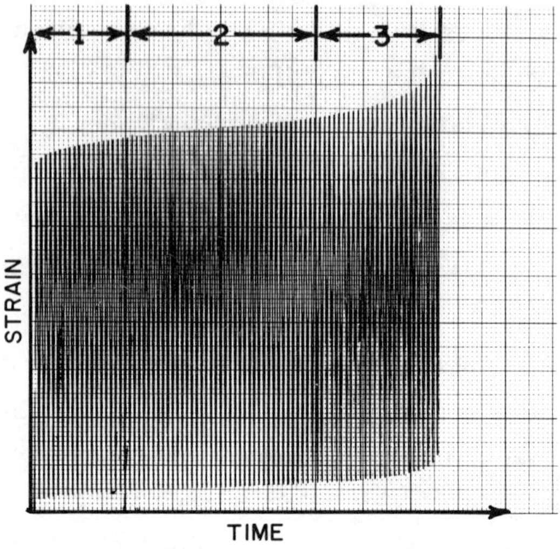

FIG. 5 TYPICAL STRAIN-TIME CURVE

It can be easily verified from Fig. 5 that the lower peak strain points undergo a much more reduced amount of increase. This implies that the value of the tangent modulus per cycle generally decreases during the test. Indeed, the average modulus of the ascending portion of the first cycle was 11.1×10^6 psi, while that of the last cycle was 9.9×10^6 psi. Similar results had been found in concrete (5).

Typical fatigue failed specimens are shown in Fig. 6. No apparent external difference was found between this type of failure and that encountered in quasi-static loading. Several samples were removed from the loading machine during different stages of the cyclic loading. The samples were vertically sectioned, polished and photographed. No damage was observed in specimens that had not reached the accelerated strain increase stage. Those that had reached the last stage prior to failure showed extensive structural damage dominated by vertical cracking. A photomicrograph of the central part of a vertical section in a specimen removed during thelast stage is presented in Fig. 7.

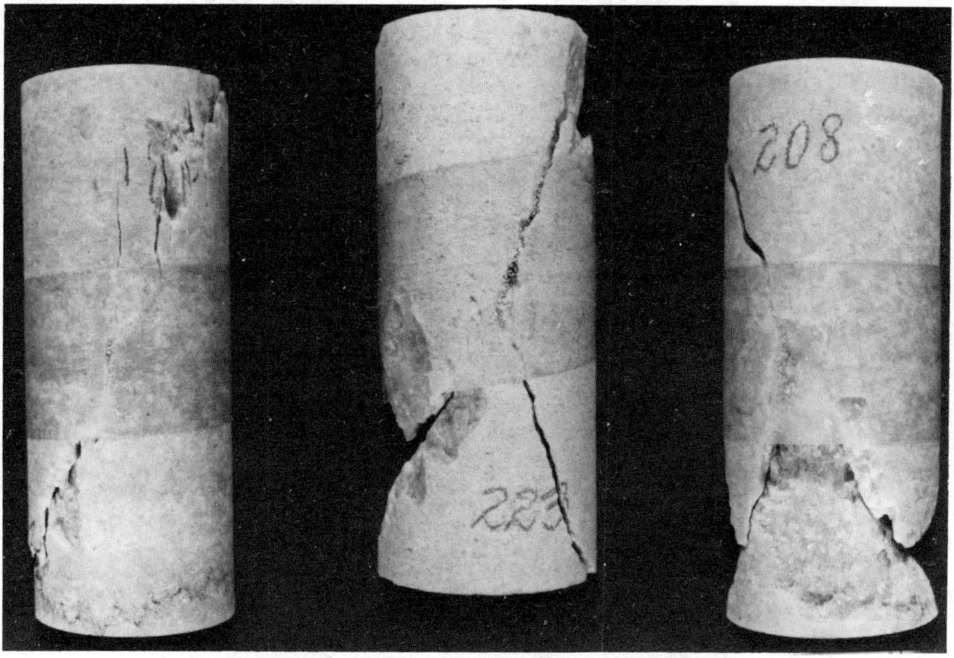

FIG. 6 FATIGUE FAILED WHITE TENNESSEE MARBLE SPECIMENS

A series of cyclic loading tests was run in White Tennessee Marble that had first been loaded in strain control to its compressive stress carrying capacity limit. This is the value commonly referred to as the compressive strength. In stress-controlled loading rock specimens will fail violently at this point as they cannot take any additional compressive force. However, if the loading is performed by controlling strain rate the rock will not collapse as it can usually continue to shrink in length while its load supporting capacity is gradually lowered. In this fashion a complete stress-strain curve can be obtained. Fig. 8 shows such a curve in addition to four other plots. In each of the plots a specimen was loaded up to its compressive strength, unloaded, and then cycled

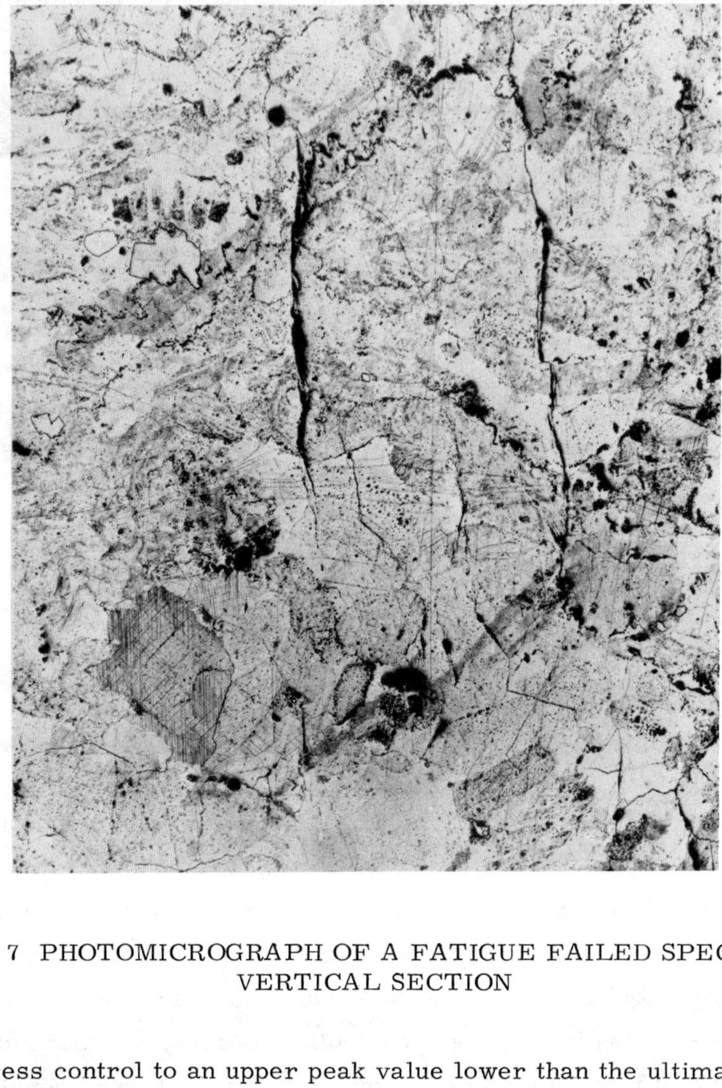

FIG. 7 PHOTOMICROGRAPH OF A FATIGUE FAILED SPECIMEN--VERTICAL SECTION

in stress control to an upper peak value lower than the ultimate strength. The purpose was to verify whether such failed rock could support fatigue type loading. Surprisingly the rock appeared rather strong, and could probably still perform useful work, although a comparison with the S-N curve in Fig. 3 will show that it had considerably weakened. As expected, the number of cycles increased as

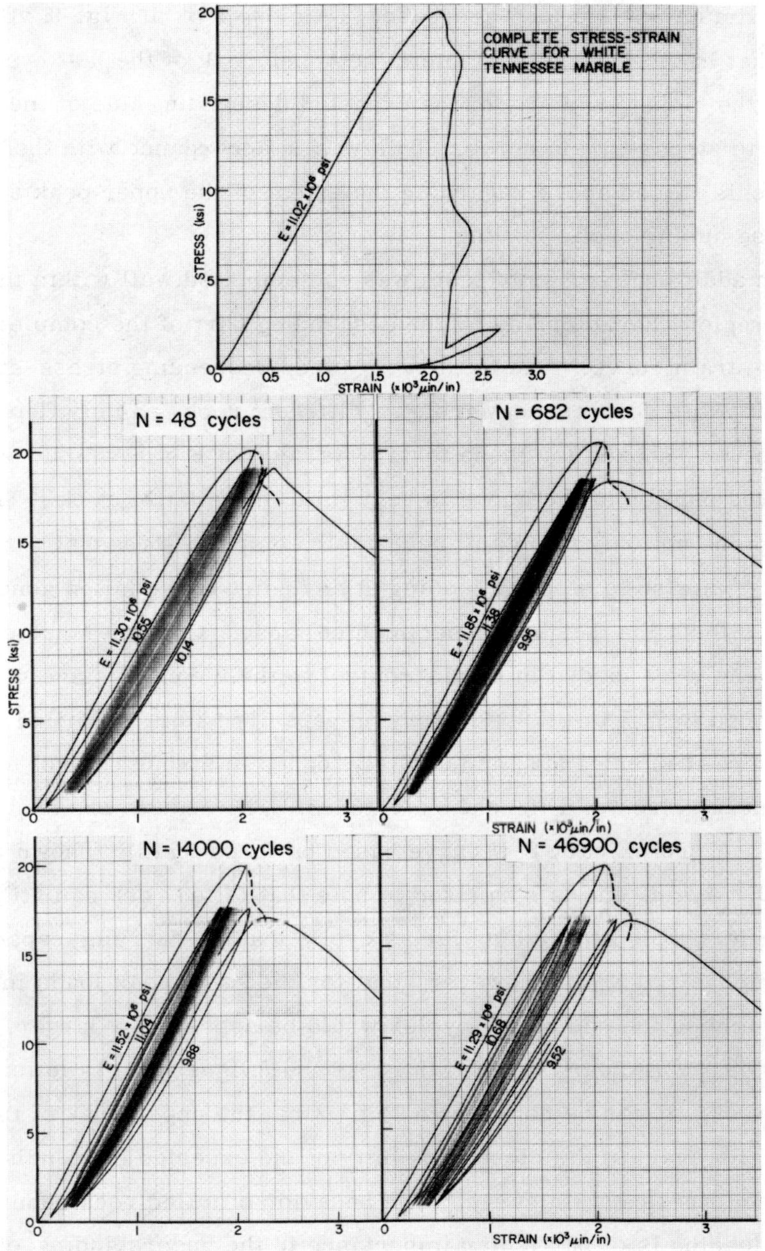

FIG. 8 STRESS-STRAIN CURVES FOR FAILED WHITE TENNESSEE MARBLE (DASHED LINES ARE EXPECTED COMPLETE STRESS-STRAIN CURVES)

the maximum applied load decreased. A close look at Fig. 8 will show that in three of the four plots the upper peak of the last cycle was in the close proximity of the expected descending line of the complete stress-strain curve. This was in accordance with the hypothesis raised above regarding the extent of the upper peak strain increase during cyclic loading.

An additional series of tests was contemplated well within the failed region characterized by the descending part of the complete stress-strain curve. Due to the very steep descending stress-strain curve in White Tennessee Marble, a different rock, namely Georgia Marble, was chosen for these tests. As seen in Fig. 9 the slope of the failed portion of the curve is very mild in the latter rock. This enabled one not only to obtain repeatable complete stress-strain curves, but also to stop the loading at any point in the failed zone and unload without the danger of specimen collapse. Georgia Marble specimens were loaded in strain control to their compressive strength and beyond, to 70% of the load carrying capacity within the failed zone. They were then unloaded, and cyclicly loaded in stress control to upper peak values of different magnitudes. The results are shown in Fig. 9. A preliminary S-N curve based on these tests is shown in Fig. 10. Again, one is surprised to note that failed rock can still show remarkable strength as far as cyclic loading carrying capacity. It is emphasized again, however, that for the same peak loads in the unfailed mode the number of cycles would be appreciably larger. Moreover, it is clear that in both rocks the fatigue limit is lowered in failed rock. It can be noted from Fig. 9 that the upper peak of the last cycle falls again in the close proximity of the expected descending complete stress-strain curve. The behavior of failed rock under cyclic loading is of particular importance to the understanding of failed underground pillars and walls subjected to fatigue type stresses. Because failed rock is internally fractured, the study may also be indicative of the behavior of jointed benches or slopes and faulted formations subjected to earthquake, blasting, or traffic loadings.

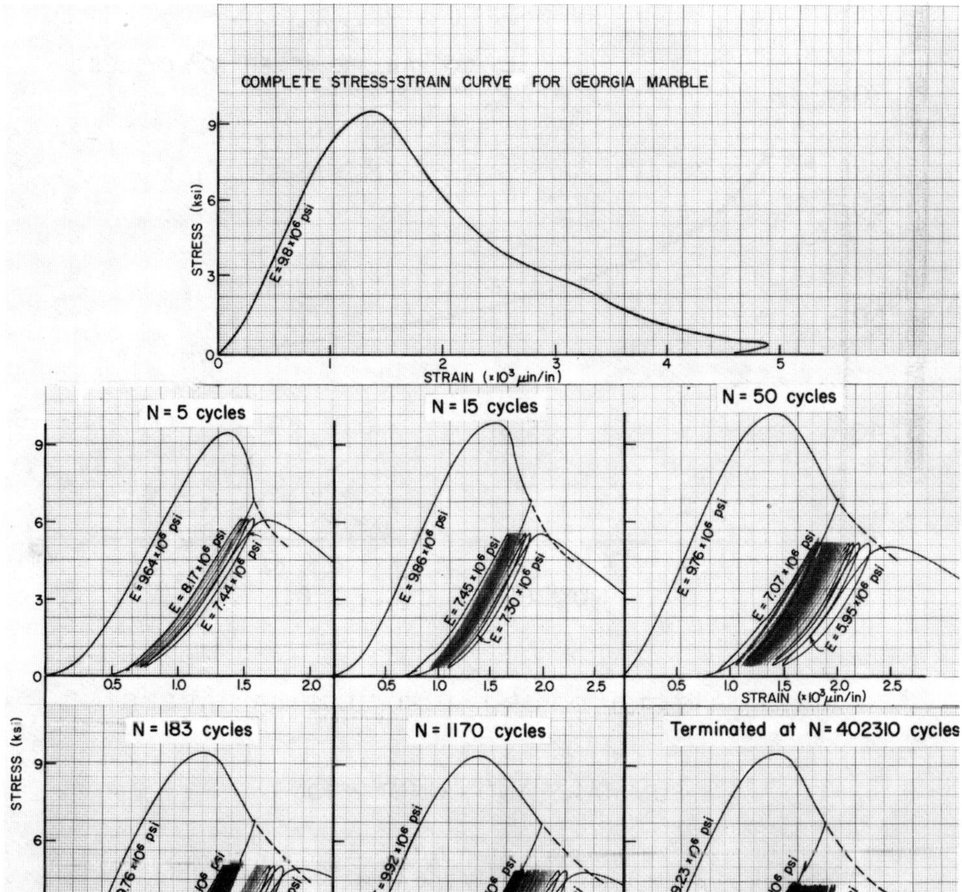

FIG. 9 STRESS-STRAIN CURVES FOR FAILED GEORGIA MARBLE (DASHED LINES ARE EXPECTED COMPLETE STRESS-STRAIN CURVES)

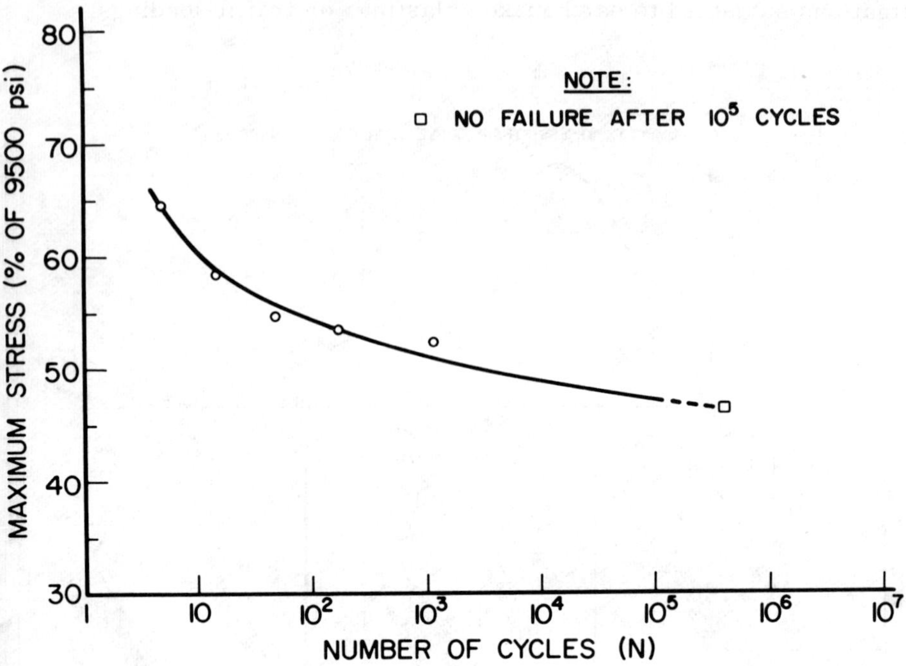

FIG. 10 S-N CURVE FOR FAILED GEORGIA MARBLE

SUMMARY AND CONCLUSIONS

The work reported here is only the first phase of an extensive investigation into the behavior of rock under cyclic loading. The main results obtained thus far in cyclic uniaxial compression of White Tennessee Marble are:

1) Cyclic loading has a definite weakening (fatigue) effect on the rock if the maximum applied stress is in the 75-100% range of the compressive strength.

2) Stress-strain records reveal three major stages occuring during cyclic loading: decreasing hysteresis, no hysteresis, increasing

hysteresis. The last stage is associated with fatigue affected internal fracturing.

3) The increase in upper peak strain between the first and the last cycles is always within the limit set by the complete stress-strain curve for the same maximum applied load.

4) The variation of the cyclic upper peak strain with time closely resembles that of creep, with primary, steady state, and tertiary or accelerated stages.

5) No visible structural damage can be observed in specimens removed prior to the accelerated strain-increase stage. However, polished sections of specimens that had reached the tertiary stage show an abundance of internal fracturing dominated by vertical cracking along and across grain boundaries.

6) Failed rock, although weakened, can sustain a certain amount of fatigue loading depending on the level of applied stresses. This conclusion is supported by testing of both Tennessee and Georgia Marbles.

The results obtained thus far show that rock cyclic fatigue is a phenomenon that cannot be ignored by engineers and rock mechanics scientists. A direct recommendation emerging from the study is that in surface and underground design the apparent fatigue limit, as determined by tests similar to those described above, be used instead of the commonly employed compressive strength value. In the design of structures that eventually reach their peak load carrying capacity, use could be made of the fatigue characteristics of failed rock. These characteristics are believed to also be indicative of jointed and faulted rock behavior under cyclic loading. The mechanism of fatigue is not yet clearly understood but the findings are promising enough to justify continuation of the research. Studies on the behavior of rock in uniaxial compression, tension and triaxial compression subjected to cyclic fatigue are currently underway.

ACKNOWLEDGEMENTS

The authors wish to acknowledge the financial assistance of the Wisconsin Alumni Research Foundation in the early stages of the investigation. The research was supported by the Advanced Research Projects Agency of the Department of Defense and was monitored by the United States Bureau of Mines, Twin Cities Re search Center, under contract no. H0210004. J. A. Hudson and F. Rummel, University of Minnesota, provided invaluable suggestions and assistance.

APPENDIX--REFERENCES

1. Burdine, N. T., "Rock Failure Under Dynamic Loading Conditions", Society of Petroleum Engineers Journal, March 1963, pp. 1-8.
2. Ellis, Willard and Hartman, V. B., "Dynamic Soil Strength and Slope Stability", Journal of the Soil Mechanics and Foundations Division, A. S. C. E., vol. 93, No. SM4, July 1967, pp. 355-375.
3. Grover, H. J., Dehlinger, P., and McClure, G. M., "Investigation of Fatigue Characteristics of Rocks", Report by Battelle Memorial Institute to Drilling Research, Incorporated, Nov. 30, 1950.
4. Hardy, H. R., Jr. and Chugh, Y. P., "Failure of Geological Materials Under Low-Cycle Fatigue", Proceedings of the Sixth Canadian Symposium on Rock Mechanics, Montreal, Canada, May 1970.
5. Nordby, Gene M. "Fatigue of Concrete--A Review of Research", Journal of the American Concrete Institute, Aug. 1958, pp. 191-219.
6. Wawersik, Wolfgang R., "Detailed Analysis of Rock Failure in Laboratory Compression Tests", thesis presented to the University of Minnesota, Minneapolis, in 1968, in partial fulfillment of the requirements for the degree of Doctor of Philosophy.

Stress Measurement by Retractable Cable Method

by

Rodolfo V. de la Cruz*

I. Objective.

This paper proposes a method of determining the initial state of stress and the mechanical properties of rocks at great depth (1000 to 5000 feet). Accurate determination of the initial state of stress and mechanical properties of rock at depth provides a key to the solution of many problems in rock mechanics and has wide application in such areas as the design of defense installations, underground waste disposal, gas or liquid storage, earthquake research, as well as in the fields of mining, civil and petroleum engineering.

II. Introduction.

Numerous investigations have been carried out over the last three decades on the determination of the initial state of stress in rocks, and these are summarized in Table 1 (from reference [1]). Current methods either have limited depth applicability (less than 40 feet) or are ambiguous and do not provide adequate data to determine the complete stress tensor at the points of measurement. An extensive review of stress-measuring methods, together with their theoretical and field restrictions and limitations, are given in references [2] and [3].

*Civil Engineer
U.S. Geological Survey
National Center for Earthquake Research
Menlo Park, California 94025

Table 1. Methods of Measuring Stresses in Rock [1]

Method	Key Reference
COMPLETE STRESS RELIEF BY OVERCORING	
Measure diametrical displacements	Merrill (1967), Panek (1966), Leeman (1967), Merrill et al (1964)
Measure stresses in stiff inclusion	Wilson (1961), Hult (1963), Hast (1958), Leeman (1964), Coutinho (1949)
Measure strains on bottom or walls	Leeman (1964), Leeman and Hayes (1966) Hoskins (1967)
PARTIAL STRESS RELIEF BY EXCAVATION	
Measure displacement on cutting a slot and pressure in slot to cancel displacements (flat jack method)	Tincelin (1958), Alexander (1958), Hoskins (1967)
Measure displacement of points inside a hole on drilling a contiguous hole	Kawamoto (in preparation)
Measure diametrical displacement of points inside a hole on deepening it	de la Cruz and Goodman (1969)
CALCULATIONS FROM STRUCTURAL BEHAVIOR	
Measure displacement of tunnel walls, or roof, or invert on advance of face	Obert, Merrill and Duval (1958, 1967)
Measure strain or stress in tunnel resulting from new mining	Morgan and Panek (1963)
ROCK FRACTURE STRESS	
Measure pressure to fracture rock with borehole jacks	Jaeger and Cook (1963)
Measure water pressure to initiate and propagate fracture in hole	Fairhurst (1965)
Core discing	Obert and Stephenson (1956)
INDIRECT METHODS	
Seismic velocity	Marsh and Turston (1951)
Seismic frequency	None
Electrical resistivity	Sen (1962)
Gamma ray	Borecki and Kidybinske (1966)

The approach to measuring the initial state of stress in rocks suggested in this paper consists of the improvement of an existing method to extend its application well beyond its present capability. This is accomplished by the utilization of techniques completely unrelated to the original stress measuring method, but which when used in combination with the original makes the measurement feasible.

The original stress measuring technique, usually considered the standard method for near-surface measurement, is the borehole deformation method. In this method, a smaller instrumented pilot hole is overcored with a larger concentric hole. The near-surface limitation of this technique is due to the practical difficulty of transmitting the deformation measurements from the pilot hole through cables running inside the drill string at distances greater than about 40 feet; the cables would get ripped and entangled in the rotation of the drill rods during the overcoring operation. The introduction of a specially designed cable hookup unit to be used in conjunction with a greatly modified borehole deformation transducer allows the measurement of deformation at depth and the transmission of this data to the surface recorder.

The measurement of the mechanical properties of the rock, specifically, the Young's modulus and the Poisson's ratio, would be accomplished by use of a pressure-deformation transducer together with the same or similar cable hookup unit mentioned above.

III. Technical Discussion.

A. Borehole Deformation with Retractable Cable Method of Measuring Initial Stresses in Rocks.

The primary problem to be solved by the proposed method is the determination of the initial state of stress in rocks at great distances from either the surface or the free face of an underground opening. No major difficulties in the research, development or application of the technique are anticipated except the usual problems encountered in drilling and coring at great depths.

This method of measuring the initial state of stress in rocks consists of a borehole-deformation gauge unit and a cable hookup unit. The borehole-deformation gauge unit is essentially the same as those currently used for near-surface measurements with some modifications, such as retractable sensors and clamps as well as rigid clamping. This gauge measures the diametrical deformation of the pilot hole when overcoring is carried out [4]. The cable hookup unit is similar to the overshot and wire-line sheave and stuffing box unit of the underground wire-line core drilling system [5] wherein cores are obtained at the end of a drill string without bringing the entire drill string to the surface. The difference is in using shielded electronic instrumentation cables instead of ordinary wire.

The technique of measuring stresses at depth by the proposed method is as follows (For instrumentation, see Figure 1):

A pilot hole slightly larger in diameter than the deformation gauge is drilled. A concentric overcoring hole is drilled close to the point of measurement. With the gauge clamped and nested inside a modified swivel-type core barrel with a central hole, the overcoring bit and associated drill string is brought down to the depth of the desired stress measurement. Details of the improved borehole deformation gauge are shown in Figure 2. The cable hookup unit is then pumped through the drill string until the overshot engages the borehole gauge spearhead and simultaneously completes the electronic connection between gauge sensors in the hole to the surface readout units (Figure 3). The gauge unit is now unclamped from the barrel, either electrically, hydraulically, or by simple tension on the wire-line,

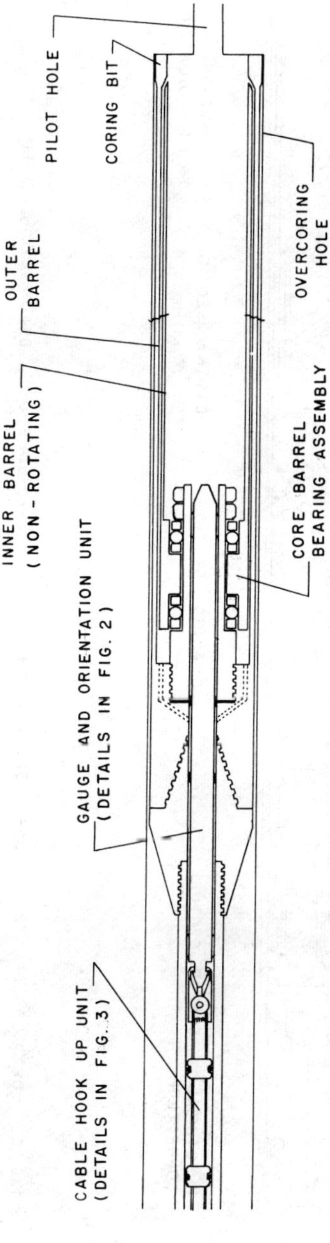

FIG. 1 SETUP OF INSTRUMENTATION FOR INITIAL STRESS MEASUREMENT BY BOREHOLE DEFORMATION WITH RETRACTABLE CABLE METHOD.

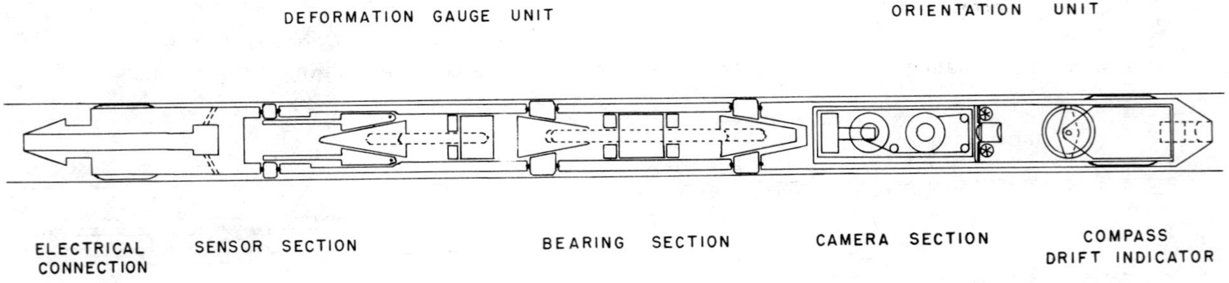

FIG. 2 DETAILS OF THE GAUGE – ORIENTATION UNIT OF THE BOREHOLE DEFORMATION WITH RETRACTABLE CABLE METHOD.

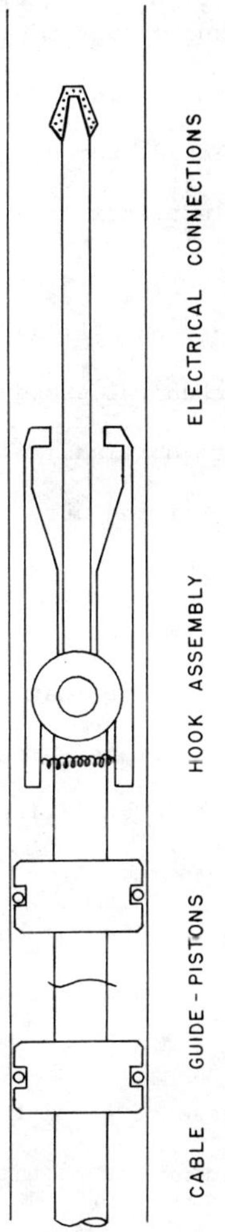

FIG. 3 DETAILS OF THE CABLE HOOK UP UNIT

and can now be installed inside the pilot hole by further pumping or by simply lowering the gauge if the hole is downward--inclined or vertical. At the desired measurement point the gauge is clamped in place and the sensors extended to contact the sides of the hole. The initial gauge readings are obtained and the hookup system detached and removed from the drill string. Usual overcoring can now be carried out; when the desired depth of overcoring has been completed, the cable hookup is again pumped into the drill string and, upon electrical connection, the final gauge readings are obtained.

From these two sets of readings, as well as the gauge orientation in the hole and the predetermined mechanical properties of the rock, some components of the stress tensor are obtained. If two other measurements in differently oriented holes are carried out, the complete state of stress can be determined. If axial stress is assumed to be a principal value and its magnitude isknown or can be estimated, then the stress tensor can be obtained from a single measurement. Note that repeated measurements at close intervals to the initial point of measurement could be easily carried out. The number of measurements would be limited only by the length of the core barrel and the interval of measurement.

The foregoing technique of retracting the cable solves the problem of having the cables running through the drill string during drilling as is currently done.

To determine the state of stress at a point in the rock mass completely, it is necessary not only to determine the magnitude of the components of the stress tensor, but also to define their orientations. To determine the stress orientation, the orientation of the measured deformations must be obtained. It is possible to calculate the magnitude of the stresses without doing this, but their orientation in the material would not be known.

The orientation of measurement at depth cannot be determined as easily as in surface measurement. A sensor orientation unit will be required and can be designed and incorporated with the stress measuring system. Such a unit would consist of a magnetic compass-drift indicator and a camera. By prealigning the sensors with the compass-drift indicator (e.g., sensor no. 1 coinciding with North markings in the compass, etc.), a picture taken of the compass-drift indicator once the gauge unit is installed will allow the determination of the gauge orientation in the hole. Since electronic connection to the surface can be achieved by the cable hookup unit of the borehole deformation gauge, the necessity of predetermined multi-shot photography of the orientation mechanism used in oriented-core drilling could be eliminated completely, making the mechanism and operation of this unit much simpler and more economical. The sensor orientation unit can be attached to the gauge unit, preferably ahead of the sensor for the compass to be further removed from the influence of the magnetic coring bits, core barrel and drill string (see Figure 2).

The borehole deformation with retractable cable method of measuring stresses in rocks at great depths is believed to be a feasible method of obtaining results as accurate as those that are currently obtained in near-surface measurements. The feasibility of this method is indicated by the wide acceptance and use of the component units individually. The operation is not appreciably complicated by combining these units. Some of the advantages and novel features of this technique are listed below:

1. Established techniques cannot determine the initial state of stress at depth without ambiguity or without making very restrictive assumptions and requirements. This method allows the determination

of the initial state of stress at any depth where diamond drilling is feasible using the same acceptable assumptions made for surface stress measurement.

2. The cable hookup unit can be attached to the gauge for reading output and then detached from the gauge and removed from the drilling system during the overcoring operation.
3. The cable hookup unit can be used for installing and retrieving the gauge.
4. The gauge is designed with remote-control positive extension of sensors to insure good contact with the sides of the hole and positive retraction of the sensors to avoid dragging these sensitive instruments against the sides of the hole during installation or removal of the gauge.
5. The gauge is designed with remote-control extension and retraction of the bearing unit to hold the gauge more firmly against vibration and circulating water in the pilot hole--replacing the unstable spring wire guide currently used. The addition of the bearing unit and its location further down the pilot hole than the sensor section insures continuous solid support of the gauge even after overcoring the region of influence of the sensors. These improvements will result in more accurate measurement.
6. A simple sensor orientation unit has been incorporated into the borehole deformation gauge unit which consists of a camera, magnetic compass and drift indicator.
7. A unique feature is the use of detachable and wire-line unit for plucking data from remotely located transducers as well as for installing or retrieving them. This cable could consist of

electronic, hydraulic, or ordinary wire lines and the gauge could be any type of transducer.

8. Numerous measurements can be readily taken, a feature of utmost importance in dealing with rock and rock masses due to their variable nature.

B. Determination of the In-Situ Elastic Constants of Rocks at Great Depth.

The determination of *in-situ* mechanical properties of rocks is by itself significant and its rationale will not be discussed here. In the measurement of the initial state of stress, knowledge of the elastic constants of the rock is required so that the measured deformation can be reduced to the required stress values. Mechanical properties of rock materials are currently obtained from laboratory measurement of core samples. Mechanical property determination of core samples is much less desirable than *in-situ* determination because the two values are seldom the same and the *in-situ* values are more realistic.

To obtain a more accurate transformation of the measured deformation into corresponding stresses, the *in-situ* elastic constants must be determined and this is the secondary problem to be solved by this investigation.

There are no major difficulties anticipated in the design, fabrication, testing or calibration of the device for measuring the *in-situ* elastic constants of rocks at depth.

The method proposed for the measurement of *in-situ* elastic constants of rock at depth is by the use of borehole jacks, specifically the Goodman jack (See reference [6] for its principles of operation and theoretical basis). Briefly, this device consists of two curved steel plates forced unidirectionally apart by race-track shaped pistons. Two LVDT's give the diametrical deformation at either end of the plates corresponding to a given pressure at the pistons.

The Goodman jack, or a modified version, together with the cable hookup unit discussed above, can be used to measure the mechanical properties of the rock at great depths. Its use could be simplified by incorporating the jack into the borehole deformation gauge unit of the stress-measuring device and could be made even more convenient by replacing the clamping mechanism of the stress measuring device with the borehole jack. Thus, the jacking system would serve a dual role: 1) it would fulfill the clamping requirements for the stress measuring system and 2) it would apply the necessary pressure to the rock if and when elastic property determination is undertaken. A sketch of the working components of the clamping-jacking system together with the deformation meters is shown in Figure 4.

A check of the in-situ mechanical property measurement could be readily carried out on the recovered cores by laboratory methods. Note specifically, that the technique allows for the recovery of a substantial size of core, 4" core for a 1-1/2" pilot hole. This core could be useful primarily for geologic study and secondarily for compressive, tensile and shear strength measurements, porosity and permeability tests, creep tests and residual stress measurements.

IV. Discussion and Conclusion

The successful application of Parts A and B combined will make it possible to obtain, in a single package and installation, all of the required measurements for determining the state of stress in a rock mass in-situ--borehole deformation, sensor orientation, and mechanical property determination. This procedure would save considerable time and money in the measurements project. More importantly, it would reduce the accumula-

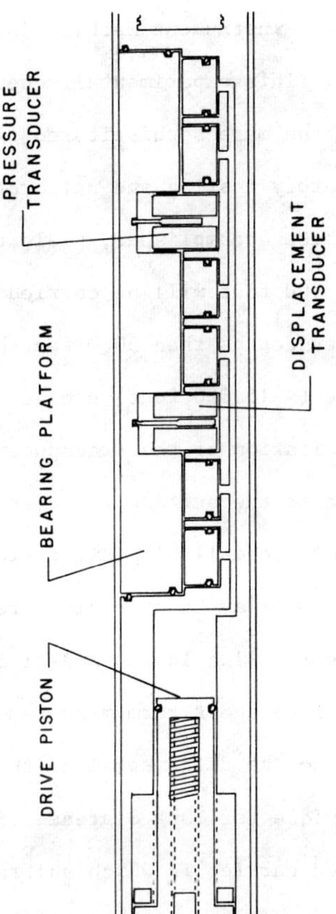

FIG. 4 DETAIL OF THE CLAMPING – JACKING UNIT

tive errors likely by performing separate measurements.

A prototype, incorporating a 3-component borehole deformation gauge unit, an orientation unit, and a clamping-jacking deformability unit, is presently being made in the Model Instrument Machine Shop of the National Center for Earthquake Research. This experimental model will be used to test several design details of the more sophisticated field instrument.

Design, fabrication, laboratory testing and calibration of the field prototype will follow. With these accomplished, field testing at successively deeper intervals up to 1000 feet will be carried out.

Measurements at much greater depths than 5000 feet but where diamond core drilling is still feasible ($\leq$ 15,000 feet) can be made possible if and when very long distance excitation of the transducer and eventual transmission of the output data to the surface recorders can be carried out. There are a few methods of accomplishing excitation and transmission of signals for long distances. One way is to apply more power than is normally required to compensate for line losses and also generate more signal at the transducer.[7] Since short term measurements only are involved, this procedure would not be detrimental to the electronic systems. The simplest conventional technique for long distance transmission is by the use of a frequency-modulated carrier in which shifts in the DC output of the transducer are transmitted as shifts in frequency.

The foregoing procedures suggest still another possible method of deep stress measurement. By miniaturization and fully automatic operation using servo mechanisms and the like it is possible to completely encapsulate within the device the power supply and recording systems. This would thus completely eliminate even the need for the retractable cable hookup unit envisioned in the current device. For example, with the stress measuring instrumentation located at the core barrel section of the drilling system, we would initially require a mechanical system of installing and retrieving the device in the pilot hole. Provisions for relocating the device at successively deeper intervals along the pilot hole for multiple measurement would also be desirable. Either telemetry or especially wiring the drill string would accomplish electrical connection to the mechanical installer. Once the device is installed in the pilot hole, strain and orientation readings can be obtained either continuously or at some pre-determined interval. The strain readings may be recorded in a magnetic tape while orientation data are recorded on film at the camera section. When mechanical property measurements are desired, other recorders for the applied pressures and corresponding wall deformations would be required. Admittedly, there are numerous other problems associated with deep hole measurements. For instance, very high temperatures will undoubtedly be encountered in deep holes. The delicate electronic instrumentations such as strain transducers and tape recorders would require heat shields for their proper operation in this hostile environment. Encapsulation of the device in an insulated vacuum sealed container would probably solve this problem. The foregoing innovations, though technologically beyond the present state-of-the-art are conceptually straightforward.

The necessity of stress and mechanical property determinations at very great depths (> 5000 feet) is not as apparent to civil and mining engineers but the significance of this information to the petroleum industry is evident. The knowledge of the initial state of stress at very great depths is, however, probably much more important to the geologist, geophysicist, and seismologist in connection with studies of orogenetic processes such as mountain building and slow deformation of the earth's crust. More directly related would be earthquake analysis and its possible prediction.

Finally, it is concluded that the problem of determining the initial state of stress and in-situ mechanical properties of rocks at depth can be considered as solved. The expense involved in the process of measurement does not add appreciably to the basic cost of drilling and coring a borehole. Since this coring operation is anyway mandatory in an extensive geologic/geophysical investigation, the initial state of stress and mechanical property determination can be carried out concurrently and thereby obtain not only geologic information from the hole but also initial conditions of loading and the rocks' in-situ mechanical properties.

V. Acknowledgments

The author gratefully acknowledges the valuable suggestions and encouragement of Dr. C. B. Raleigh of the OER/CS, USGS. He is also indebted to A. E. Augustson, Mercury Christiansen Diamond Products Co. and C. Holmstrom, OER/CS, USGS for their advice in the preliminary design of the instruments.

Disclosure of the method and device presented in this paper has been made on behalf of the government.

VI. References

(1) Goodman, R. E., 1968, Mimeograph notes for a University of California extension course, U. C. Berkeley, pp. 11-12.

(2) Leeman, E. R., 1964, The measurement of stress in rock: Parts I, II, and III, J. So. African Inst. Mining Metall., Sept. 1964, pp. 45-114, Nov. 1964, pp. 254-284.

(3) Fairhurst, C., 1968, Methods of determining in-situ rock stresses at great depths, Technical Report No. 1-68, Mo. River Div., Corps. of Engrs, Omaha, Neb., February, 1968.

(4) Merrill, R. H., 1967, Three-component borehole deformation gauge for determining the stresses in rock, U.S. Bureau of Mines, R. I. 7015.

(5) E. J. Longyear Company, undated, A revolutionary development in diamond core drilling: Underground wire-line system, Bull. 3290, Minneapolis, Minnesota.

(6) Goodman, R. E., Tran Van, Heuze, F., 1968, The measurement of rock deformability in boreholes, Proc. 10th Symp. Rock Mechanics, AIME, 1968.

(7) Hall, J., May, 1971, PREMMCO/PH, Long Beach, California, Personal Communication.

13th SYMPOSIUM ON ROCK MECHANICS

August 30 - September 1, 1971

ROSTER

Kakuichiro Adachi	U. of I., graduate student, 2074-B Orchard Street Urbana, Illinois 61801
Edjardo Alonso	Northwestern University, graduate student, 1915 Maple Ave., Apt. 522, Evanston, Illinois 60201
Robert Alperstein	6 Tulip Drive, Fords, New Jersey 08863 Joseph S. Ward and Associates
Fred A. Anderson	P.O. Box 809, Juneau, Alaska 99801 U.S. Army Corps of Engineers, Alaska
Thomas Anderson	377 Elliot Street, Newton Upper Falls, Massachusetts Goldberg-Zoino and Associates, Inc.
Thomas C. Atchison	P.O. Box 1660, Twin Cities, Minnesota 55111 U.S. Bureau of Mines
Asher Atkinson	525 S. Decatur Blvd., Las Vegas, Nevada 89107 Birdwell, Division of Seismograph Service Corporation
Raymond E. Aufmuth	1103 Holiday Drive, Champaign, Illinois 61820 Construction Engr. Research Laboratory
Nolan B. Aughenbaugh	P.O. Box 479, Rolla, Missouri 65401 University of Missouri
R.W. Baltosser	6200 East 41st Street, Tucxa, Oklahoma 74102 Birdwell Division Seismograph Service Corporation
Don C. Banks	502 Longview, Vicksburg, Mississippi 39180 USAE Waterways Experiment Station
Michal Bardecki	Department of Geography, University of Guelph, Guelph, Ontario, Canada
William Baron	Clemson University, Clemson, South Carolina 29631
Charles L. Bartholomew	U. of I., graduate student, 1001 Rainbow, Urbana, Illinois 61801
Manfred Baudendistel	12, Kaiserstrasse, University, Institut fur Bodenmechanik und Felsmechanik, D 75 Karlsruhe, Germany
William Bawden	U. of I., student, 1002 S. 2nd St., Champaign, Illinois 61820
Newcomb B. Bennett	1714 S. Wright Street, Denver, Colorado 80228 U.S. Bureau of Reclamation

Gordon Benson	Box 154, R.R. #6, Springfield, Illinois 62707 Illinois Division of Highways
Alan D. Black	815 East 4th South, Salt Lake City, Utah 84102 Terra Tek, Inc.
Albert N. Bove	2101 Constitution Avenue, NW, Washington, D.C. 20418 National Academy of Sciences - National Academy of Engineering
Richard C.E. Bray	21 Sydenham Street, Apt. 2, Kingston, Ontario, Canada Smith and Bray Engineering Ltd.
Gary Brierley	238 Main Street, Cambridge, Massachusetts 02142 Haley and Aldrich, Inc.
Carl Broadbent	5750 Glenbrook Street, Salt Lake City, Utah 84121 Kennecott Copper Corporation
Alan D. Brown	133 East Broadway, Butte, Montana 59701 Anaconda Company
Edwin T. Brown	528 20th Avenue South, Minneapolis, Minnesota 55404 James Cook University of North Queensland
Robert L. Brownfield	1841 W. Iles Avenue, Springfield, Illinois 62704 Illinois Division of Highways
William Brumund	School of Civil Engineering, Georgia Institute of Technology, Atlanta, Georgia 30332
Charles Glenn Bufe	5072 Elrod Drive, Castro Valley, California 94546 U.S. NOAA Earthquake Mechanism Lab, San Francisco
Thomas G. Bumala	3668 Albion Pl. North, Seattle, Washington 98103 Slope Indicator Company
James D. Byerlee	4570 Alpine Road, Portola Valley California 94025 U.S.A.S.
Richard D. Call	Department of Mining and Geological Engineering, Geological Engineering, College of Mines, University of Arizona, Tucson, Arizona 85721
Manuel M. Casteleiro	Northwestern University, graduate student, 1671 N. Prospect Ave., Milwaukee, Wisconsin 53202
Enrique Castillo	Northwestern University, graduate student, 1915 Maple Ave. Evanston, Illinois 60201
Roland Christensen	824 Univ Village, Salt Lake City, Utah, Univ. of Utah
Riley M. Chung	Northwestern Univ., graduate student, 1596 Oak, Evanston, Ill.
J.D. Campbell	1425 Broad Street, Clifton, New Jersey 07012 Woodward-Moorhouse & Associates, Inc.

Giuseppe Coletti	1000 Liberty Ave., Pittsburgh, Pennsylvania 15222 Corps of Engineers - Pittsburgh Dist.
Wilfrid Comeau	112 Ridge Ave., State College, Pennsylvania 16801 Penn State University, University Park, Pa.
Robert J. Conlon	3516 Harvard Ave., Niagara Falls, Ontario, Canada H. G. Acres Limited
Edward J. Cording	2221 C.E. Building, University of Illinois, Urbana, Ill. 61801
Francois Cormet	501 15th Ave., S.E., Minneapolis, Minnesota 55414 University of Minnesota
Charles Corns	5713 Guy Place, Springfield, Virginia 22151 U.S. Army Corps of Engineers
Fabiano A. Cossich	(CEMIG, Belo Horizonte, Brazil) c/o Lynch, Wilde & Co, Inc. 1346 Connecticut Ave., N.W., Washington, D.C. 20036 Centrais Electricas de Minas Gerais (CEMIG)
James Coulson	7025 Gainesborough Dr., Knoxville, Tennessee 37919 Tennessee Valley Authority
Bernard F. Cronin	1218 West 23rd, Casper, Wyoming 82601 U.S. Bureau of Reclamation - North Platte River Proj.
Everett B. Crowell	1840 Morning Canyon Road, Diamond Bar, California 91765 Los Angeles County Flood Control District
Rodolfo De La Cruz	212 Lexington Drive, Menlo Park, California 94025 U.S. Geological Survey
Don Deere	2210 Fletcher Street, Urbana, Illinois 61801 University of Illinois
Albert J. Depman	124 St. James Ave., Merchantville, New Jersey 08109 U. S. Army Corps of Engineers
Robert Dimock	#12 Pine Street (P.O. Box 484), Ruth, Nevada 89319 Kennecott Copper Corp.-Nevada Mines Division
Norman A. Dixon	8725 McNair Drive, Alexandria, Virginia 22309 Office Chief of Engineers
Samuel Doak	Clocktower Building, Rock Island, Illinois 61201 Rock Island Dist. - Corps of Engineers
Jerry S. Dodd	12780 West 75th Ave., Arvada, Colorado 80002 U. S. Bureau of Reclamation
R. Kenneth Dodds	520 S.W. Sixth Avenue, Portland, Oregon 97204 Foundation Sciences, Inc.

Gabrielle Douglas	1761 Clifford, Montreal, Quebec, Canada Dept. of Geology, Loyola College
Robert Dowdell	403 Steele Street, Kimberley, British Columbia Cominco Ltd.
John R. Ege	U. S. Geological Survey, Federal Center, Denver, Colorado 80225
Paul Engeling	2306 Carlisle Dr., Champaign, Illinois 61820 University of Illinois
Robert Fark	Box 10003, Pittsburgh, Pennsylvania 15233 Richardson, Gordon & Associates
Harry Ferguson	1000 Liberty Avenue, Pittsburgh, Pennsylvania 15222 Corps of Engineers - Pittsburgh District
Wayne E. Ferree	1136 Little Oak Circle, San Jose, California Terratech, Inc.
Richard L. Fischer	P. O. Box 1660, Twin Cities Airport, Minnesota 55111 U. S. Bureau of Mines
Charles Fondelier	8175 Ashgrove Drive, Cincinnati, Ohio 45244 U. S. Army Corps of Engineers
Fred Fox	52 Sycamore Avenue, Berkeley Heights, New Jersey 07922 GEONICS
A. G. Franklin	Dept. of Civil Engr., Northwestern University, Evanston, Illinois 60201
John A. Franklin	999 N. Main Street, Glen Ellyn, Illinois 60137 Seelec Division - Labquip Corporation
Louis R. Frei	U. S. B. R., P. O. Box 1309, Auburn, California 95603 Bureau of Reclamation
Ronald L. Friestad	U.S. Army Engineer District, Omaha, 7410 U.S. P. O. & Court House, Omaha, Nebraska 68102
John W. Gadsby	1632 McGuire Ave., North Vancouver, British Columbia, Canada Piteau, Gadsby, MacLeod LTD.
James C. Gamble	931 Harvard Road, Monroeville, Pennsylvania 15146 General Analytics, Inc.
Harold Ganow	305 Kelly Court, Champaign, Illinois 61820 University of Illinois
Richard Gates	745 Hanover Street, Livermore, California 94550 U. S. Army Engineer Explosive Excavation Research Office

Max W. Giger	Northwestern University, research assistant 1915 Maple Avenue, Evanston, Illinois 60201
Pablo Girault	Av. Goldsmith 37-301, Mexico 5, D.F., Mexico Girault and Company
Mark Goetzinger	2028 E. Newberry Blvd., Milwaukee, Wisconsin 53211 University of Wisconsin
Richard Goodman	715 Arlington, Berkeley, California 94707 University of California
Alan G Goodfield	1109 Interlacken Drive, Springfield, Illinois 62704 Illinois Division of Highways
Richard E. Gray	508 Lenox Avenue, Pittsburgh, Pennsylvania 15221 General Analytics, Inc.
Vernon Greenwood	Clocktower Bldg., Rock Island, Illinois 61201 Rock Island District Corps of Engineers
David Gross	Illinois Geological Survey, Urbana, Illinois 61801
Benjamin Grote	Defense Nuclear Agency, Test Command, Kirtland Air Force Base, New Mexico 87115 Dept. of Defense - U.S.A.
Giuseppe Guzzetta	Instituto di Geologia - Universita, L.S. Marcellino, 10 Naples, Italy 80138 C.N.R.-I.R.P.I.
Charles Haas	University of Missouri, Rolla, Missouri 65401
Karim Habibagahi	P.O. Box 1963, Harrisburg, Pennsylvania 17105 Gannett Fleming Corddry and Carpenter
Joseph Hagerty	3912 Hillsboro Road, Louisville, Kentucky 40207 University of Louisville
Bezalel Haimson	1509 University Avenue, Madison, Wisconsin 53706 University of Wisconsin
D.J. Hammel	1356 Kennecott Bldg., Salt Lake City, Utah 84111 Kennecott Copper Corp., SECC
James Hamel	Civil Engineering Department, South Dakota School of Mines & Technology, Rapid City, South Dakota 57701
John Handin	Texas A & M University, College Station, Texas 77843 Center for Tectonophysics
Milton Harr	2201 Indian Trail Drive, W. Lafayette, Indiana 47906 Purdue University

Michael Haverland	12080 W. Mexico Ave., Lakewood, Colorado 80228 U.S. Bureau of Reclamation
Roy J. Helfinstine	368 Natural Resources, Urbana, Illinois 61801 University of Illinois
Frank A. Henderson	1910 Oakdale Ave., Apt. 104, W. St. Paul, Minnesota 55118 U. S. Bureau of Mines
Francois Heuze'	472 Davis Hall, University of California, Berkeley, California 94704
Frank Horino	Fed. Ctr. Building 20, Denver, Colorado 80225 U. S. Bureau of Mines
Harry M. Horn	1425 Broad Street, Clifton, New Jersey 07012 Woodward-Moorhouse & Associates, Inc.
Earl R. Hoskins	2115 Jane Drive, Rapid City, South Dakota 57701 South Dakota School of Mines
John Hudson	528 20th Ave., S., Minneapolis, Minnesota 55404 University of Minnesota
William Hustrulid	Dept. of Mining, Colorado School of Mines, Golden, Colorado 80401, Colorado School of Mines
Brima Ibrahim	2202 #102 Orchard South, Urbana, Illinois 61801 University of Illinois
Herbert O. Ireland	Civil Engineering Building, Urbana, Illinois 61801 University of Illinois
Alan Jacobs	264 NR, Illinois Geological Survey, University of Illinois
Jeremiah E. Jennings	24 Seymour Avenue, Parktown West, Johannesburg, S. Africa University of the Witwatersrand
Klaus W. John	Stresemannstr. 12, 75 Karlsruhe 21, Federal Republic, Germany Ruhr University Bochum
Ronald Jones	714 South Race Street, Urbana, Illinois 61801 University of Illinois
William Judd	200 Quincy St., W. Lafayette, Indiana 47906 Purdue University
William Kelly	Civil Engineering Dept., Grad Student, Univ. of Notre Dame Notre Dame, Indiana 46556
Bruce A. Kennedy	1849 West North Temple, Salt Lake City, Utah 84116 The ANACONDA Co.

Abdul Khair	M.I. Bldg, Pennsylvania State University, University Park, Pennsylvania 16801
George A. Kiersch	119 Warwick Place, Ithaca, New York 14850 Cornell University
Ran Kim	1006 Golfview Ave., State College, Pennsylvania 16801 The Pennsylvania State University
Michael S. King	Dept. of Geological Sciences, Univ. of Saskatchewan, Saskatoon, Saskatchewan, Canada
David E. Kleiner	2120 Hickory Ln., Palatine, Illinois 60067 Harza Engineering Co.
Kyung Ko	1515 Mineral Square, Salt Lake City, Utah 84111 Kennecott Copper Corporation
Stephen R. Korbay	Box 3030, San Rafael, California 94902 Harding-Miller-Lawson & Assoc.
William D. Kovacs	629 Waldron St., W. Lafayette, Indiana 47906 Purdue University
George Kruse	2624 Carmichael Way, Carmichael, California 95608 California Dept. of Water Resources
Frank W. Kuehnel	620 Rose Court, Lisle, Illinois 60532 U. S. Army Corps of Engineers - Chicago Dist.
Alan Kuhn	1001 Fairlawn Drive., Urbana, Illinois 61801 University of Illinois
Kenneth Lane	415 S. Howes, Ft. Collins, Colorado 80521
Thomas A. Lang	120 Montgomery Street, San Francisco, California 94104 Leeds, Hill and Jewett, Inc.
Hendrik Lauga	570 Dunsmuir St., Vancouver, British Columbia, Canada International Power & Engineering Consultants, Ltd.
Fitzhugh T. Lee	1710 Alkire Court, Golden, Colorado 80401 U. S. Geological Survey
Gerald A. Leonards	School of Civil Engineering, Purdue University, Lafayette, Indiana 47907
Jerome J. Leszkiewicz	1550 Northwest Highway, Park Ridge, Illinois 60068 Dames & Moore
Thomas S. Lewis	3012 S. Park, Springfield, Illinois 62704 State of Illinois - Div. of Highways

David A. Lienhart	7229 Longfield Dr., Cincinnati, Ohio 45243 U. S. Army Engineer Division, Ohio River
John G. Livingston	U.S.B.R., P. O. Box 1309, Auburn, California 95603 Bureau of Reclamation
K. Y. Lo	Faculty of Engineering Science, University of Western Ontario, Canada
Robert Chung-Yi Lo	119 Glenwood Ave., #3B, Yonkers, New York 10701 TAMS Engineers and Architects
Charles Logie	3097 A Colonial Way, Chamblee, Georgia 30341 Dames & Moore
William Lounsberry	316 Valentine Street, Highland Park, New Jersey 08904 N. J. Dept. of Transportation
Cole R. McClure, Jr.	1729 Newcastle Drive, Los Altos, California 94022 Bechtel, Incorporated
Dougal R. McCreath	1259 Dorchester Road, Niagara Falls, Ontario, Canada H. G. Acres Ltd.
Wilson McCurry	c/o The Anaconda Co., P. O. Box 127, Sahuarita, Arizona 85629 The Anaconda Co.
David McKittrick	24 Torrey Road, S. Braintree, Massachusetts 02185 American Drilling & Boring Co., Inc.
Richard Mabry	2100 Locust Street, Philadelphia, Pennsylvania 19103 Woodward, Gardner & Assoc.
Dennis K. Marker	506 Sunrise, Urbana, Illinois 61801 University of Illinois
Gleny S. Marshall	203 N. Live Oak Drive, Moncks Corner, South Carolina 29461
Robert H. Merrill	9120 W. 3rd Pl., Lakewood, Colorado 80226 U. S. Bureau of Mines
Andrew Merritt	c/o H. G. Acres Ltd., Apartado Aereo 2366, Cali Colombia, South America
Peter Messian	2205 Lee St., Evanston, Illinois 60202 Soiltest, Inc.
Donald Miller	4560 Old Pineville Road, Charlotte, North Carolina 28209 Law Engineering Testing Company
Eugene J. Miller	1807 Liffy Lane, Jeffersonville, Indiana 47130 Corps of Engineers, Louisville District

R. P. Miller	1105 North 38th St., Seattle, Washington 98103 Shannon & Wilson, Inc.
Richard W. Miller	1227 South 6th St., Springfield, Illinois 62703 Walter E. Hanson Company
Robert Miller	1027 San Mateo S.E., Albuquerque, New Mexico 98108 Ken O'Brien and Associates
Richard A. Miller	1425 Broad Street, Clifton, New Jersey 07012 Woodward-Moorhouse & Associates, Inc.
Nestor T. Mirafuente	2730 Adeline Street, Oakland, California 94607 Woodward-Lundgren & Associates
Dewayne L. Misterek	630 S. Taft St., Denver, Colorado 80228 Bureau of Reclamation
Bijan Mohraz	2213 Civil Engineering Building, Urbana, Illinois 61801 University of Illinois
Jose Monte	1663 N. Prospect, Milwaukee, Wisconsin 53202 Northwestern University
Norbert R. Morgenstern	Dept. of Civil Engineering, University of Alberta, Edmonton, Alberta, Canada
Donald J. Murphy	91 Roseland Avenue, Caldwell, New Jersey 07006 Joseph S. Ward & Associates
Alberto Nieto-Pescetto	Grad Student, University of Illinois, Urbana, Illinois 61801
Ray Nitzsche	900 South Bishop, Rolla, Missouri 65401 Rock Mechanics & Explosives Research Center
Leonard Obert	Bldg. 20 Denver Federal Center, Denver, Colorado 80225 U. S. Bureau of Mines
L. L. Oriard	2730 Adeline Street, Oakland, California 94607 Woodward-Lundgren & Associates
Rual V. Orozco	Apartado Postal #41, Jardines De San Mateo, Estado De Mexico, Mexico, Secretaria De Recursos Hidraulicos
Lars Osen	2420 Miss, Joplin, Missouri 64801 Atlas Chemical Ind., Inc.
Louis Panek	Bldg. 20, Federal Center, Denver, Colorado 80225 U. S. Bureau of Mines
Constantin Papantonopoulos	3263B Van Horne, Montreal, Quebec, Canada Ecole Polytechnique, Montreal

William Pariseau	514 West Mercury St., Butte, Montana 59701 Montana Tech.
Harvey W. Parker	203 Flora Drive, Champaign, Illinois 61820 University of Illinois
James Passage	461 Yorkshire Square, Bolingbrook, Illinois 60439 Harza Engineering Company
Franklin D. Patton	3160 Benbow Road, West Vancouver, British Columbia, Canada
Ralph Peck	1911 G.H. Baker Drive, Urbana, Illinois 61801 University of Illinois
David Pentz	224 West 8th Ave., Vancouver, British Columbia, Canada Golder Brawner & Assoc.
Ronald Perisho	401 South Sixth Street, Champaign, Illinois 61820 University of Illinois - graduate student
Howard J. Pincus	Dept. of Geological Sciences, Sabin Hall, University of Wisconsin, Milwaukee, Wisconsin 53201
Douglas R. Piteau	1632 McGuire Avenue, N. Vancouver, British Columbia, Canada Piteau Gadsby MacLeod Limited
Darrell Porter	3790 Melody Lane, Hagerstown, Maryland 21740 DuPont
Howard R. Pratt	815 East 4th South, Salt Lake City, Utah 84102 Terra Tek, Inc.
Donal M. Ragan	Department of Geology, Arizona State University, Tempe, Arizona 85381
Clyde Renz	Clocktower Building, Rock Island, Illinois 61201 Rock Island District - Corps of Engineers
Brian J. Reynolds	602 Lang's Drive, Preston, Ontario, Canada University of Gruelph
Charles Henry Rich, Jr.	13451 W. Ohio Drive, Lakewood, Colorado 80228 U. S. Bureau of Mines
Frank E. Richart, Jr.	2210 Hill Street, Ann Arbor, Michigan 48104 University of Michigan
Robert E. Riecker	Farrar Road, S. Lincoln, Massachusetts 01773 Air Force Cambridge Research Laboratories
Kenneth Rodda	1901 S. Navajo Street, Denver, Colorado 80223 ECI International, Inc.

J. Lyndon Rosenblad	23 Surrey Road, Norwood, Massachusetts 02062 Stone & Webster Engr. Corp.
Dennis Rubin	266 Bayview Drive, Oakdale, Long Island, New York 11769 American Electric Power Service Corporation
James E. Russell	916 Clark Street, Rapid City, South Dakota 57701 South Dakota School of Mines & Technology
Antonio Santos	1663 N. Prospect Ave., Milwaukee, Wisconsin 53202 Northwestern University
Jucks Schwellnus	Dept. of Mining Engineering, McGill University, P. O. Box 6070, Montreal 101, Quebec, Canada
Ben L. Seegmiller	1849 West North Temple, Salt Lake City, Utah 84116 The Anaconda Co.
Robert Semple	705 W. High, Urbana, Illinois 61801 University of Illinois
Shosei Serata	1229 8th, Berkley, California 94710 Serata Geomechanics
Thomas Sexton	Box 306, Weston, Massachusetts Weston Geophysical
John Sharp	224 West 8th Avenue, Vancouver, British Columbia, Canada Golder Brawner & Assoc.
Walter C. Sherman	1829 Edna Drive, Vicksburg, Mississippi 39180 USAE Waterways Experiment Station
George Sinnamon	1106 Civil Engr. Bldg, Urbana, Illinois 61801 University of Illinois
Edward Slebir	2200 Jackson Street, Golden, Colorado 80401 U. S. Bureau of Reclamation
Jas. E. Smiley	R. R. #3, Box 37, Pana, Illinois 62557 Tunnel & Mining Supplies, Inc.
Terrence J. Smith	13 Chestnut Ave., Clarendon Hills, Illinois 60514 U. S. Army Corps of Engineers
Enrique Socias	1915 Maple Ave., Evanston, Illinois 60201 Northwestern Univ., - graduate student
Arthur Soderberg	4601 Plaza Hills, Hixson, Tennessee 37343 Tennessee Valley Authority
Antonio Soriano	1915 Maple, Evanston, Illinois 60201 Intecsa - Espana
John J. Speaker	6311 Apex Drive, Louisville, Kentucky 40219 Corps of Engineers, Louisville District

Gregory Spencer	5624 Jackson, Omaha, Nebraska 68106 U.S. Army Corps of Engineers
Stephen R. Swanson	2887 Evergreen Ave., Salt Lake City, Utah 84109 University of Utah
Robert Swift	5227 Crestline Way, Pleasanton, California 94566 Physics International Company
James R. Talbot	2902 S.E. 160th Avenue, Portland, Oregon 97236 USDA, Soil Conservation Service
Peter J. Tarkoy	U. of I., graduate student, R.R. 1, Urbana, Illinois 61801
Kuppusamy Thirumalai	P.O. Box 1660, Twin City Airport, Minnesota 55111 U.S. Bureau of Mines
Harry E. Thomas	9610 Tiverton Way, Louisville, Kentucky 40222 Corps of Engineers, Louisville District
Thomas H. Thornburn	U. of I., 2230 Civil Engineering Bldg., Urbana, Illinois 61801
Robert E. Tipton	790 Vista Lane, Lakewood, Colorado 80215 U.S. Bureau of Mines
Rogelio Torres	Harza Engineering Company, Bolingbrook, Illinois 60439
Robert Van Nostrand	Box 334, Alexandria, Virginia 22313 Teledyne Geotech
Roy J. Van Ryswyk	U. of I., student, 309 S. New St., Champaign, Illinois 61820
George B. Wallace	11200 Linda Vista Drive, Lakewood, Colorado 80215 Bureau of Reclamation
Wolfgang R. Wawersik	1239 Gilmer Drive, Salt Lake City, Utah 84105 University of Utah
Larry West	1414 Dexter Avenue N., Seattle, Washington 98109 Dames & Moore
Terry R. West	824 Essex St., W. Lafayette, Indiana 47906 Purdue University
Karl D. Willig	12405 Brooklyn, Kansas City, Missouri 64146 U.S. Army Corps of Engineers
Eiju Yatsu	Department of Geography, University of Guelph, Guelph, Ontario, Canada
George A. Young	8939 S. Sepulveda, Los Angeles, California 90045 Agbabian-Jacobsen Associates

Mehmet Semih Yucemen	U. of I., research assistant, 502 W. Griggs #205, Urbana, Illinois 61801
Timothy Zeigler	103 Colleen Dr., Vicksburg, Mississippi 39180 USAE Waterways Exp. Station

LATE ARRIVALS

Win Carpenter	P.O. Box 24012, Minneapolis, Minnesota 55424 MTS Systems Corporation
Douglas C. Moorhouse	1425 Broad Street, Clifton, New Jersey 07012 Woodward-Moorhouse & Associates, Inc.
Jerry Parola	**Lake States Engineering Corp. 1410 Higgins Road Park Ridge, Illinois 60066**
Paul Schlenker	**Lake States Engineering Corp. 1410 Higgins Road Park Ridge, Illinois 60066**

PRE-REGISTERED BUT DID NOT ATTEND

Hans M. Ewoldsen	758 North Batavia, Orange, California 92668 Woodward-McNeill & Associates
Allan Falconer	653 College Avenue, West, Guelph, Ontario, Canada University of Guelph
Richard T. Frankian	2301 West Alameda Avenue, Burbank, California 19506 R. T. Frankian & Associates
Axel M. Fritz	3401 48th Avenue North, Minneapolis, Minnesota 55429 Bison Instruments, Inc.
Ken Gray	3930 Sierra Drive, Austin, Texas 78731 University of Texas at Austin
Pierre Habib	Laboratoire de Mecanique des Solides, Ecole Polytechnique 17 - rue Descartes, Paris, Paris V, France
Earl Hall	P. O. Box 959, San Rafael, California 94902 Geo-Testing, Inc.
Boise Macon	R. R. #1, Box 304B, Tell City, Indiana 47586 J. A. Jones Construction Co.
Efrahim Murati	G.P.O. Box 2948, San Juan, Puerto Rico 00936

Leonard Pearson	36 Queen Anne's Gate, London, England, S.W.1. H 9AR Foseco - Minsep Limited
Donelson A. Robertson	66 Long Clove Road, Congers, New York 10920 Appalachian Stone
J. Ronald Salley	Box 744, El Dorado, Kansas 67042 Shannon & Wilson, Inc.
Piero Sembenelli	8 Via Chiabrera, 8, Milano 20151 Italy ELC-electronconsult spa
Theodosios Tassios	43 Acocdimias, Athens, Greece Geoereuna S.A.

SUBJECT INDEX

AUTHOR INDEX

Subject Index

Author Index